THE HUMAN GENOME

A USER'S GUIDE

SECOND EDITION

THE HUMAN GENOME

A USER'S GUIDE

SECOND EDITION

Julia E. Richards
University of Michigan
Ann Arbor, Michigan

R. Scott Hawley
Stowers Institute for Medical Research
Kansas City, Missouri

ELSEVIER
ACADEMIC
PRESS

AMSTERDAM • BOSTON • HEIDELBERG • LONDON
NEW YORK • OXFORD • PARIS • SAN DIEGO
SAN FRANCISCO • SINGAPORE • SYDNEY • TOKYO

Acquisition Editor: Jeremy Hayhurst
Project Manager: Paul Gottehrer
Editorial Assistant: Desiree Marr
Marketing Manager: Linda Beattie
Cover Design: Cate Barr
Composition: SNP Best-Set Typesetter Ltd., Hong Kong
Printer: China Translation & Printing Services Ltd.

Elsevier Academic Press
200 Wheeler Road, Burlington, MA 01803, USA
525 B Street, Suite 1900, San Diego, California 92101-4495, USA
84 Theobald's Road, London WC1X 8RR, UK

This book is printed on acid-free paper. ∞

Library of Congress Cataloging-in-Publication Data
Application submitted

British Library Cataloguing in Publication Data
A catalogue record for this book is available from the British Library

ISBN: 0-12-333462-4

For all information on all Academic Press publications
visit our Web site at www.academicpressbooks.com

Printed in China
04 05 06 07 08 09 9 8 7 6 5 4 3 2 1

To Jesse

and the many unsung heroes
who have helped create modern medicine
through their participation in research

Table of Contents

Acknowledgments

Many people have contributed to the existence of this book, and each of them has our profound gratitude. We thank our families for their love, patience, and support throughout the process of writing this book. They are the foundations in our lives that make such endeavors possible.

We especially thank Jeremy Hayhurst, Desiree Marr and Elsevier for giving us the opportunity to share this book with you. We also thank Catherine Mori, for her efforts both in the creation of the concept of *The Human Genome: A User's Guide* and for her work on the first edition. That work was a substantial stepping-stone as we embarked on creating the current version of this book.

Thanks for reading chapters and for many excellent suggestions go to Beverly Yashar, Paula Sussi, Paul Gelsinger, James Knowles, Jill Robe-Gaus, Randy Wallach, Rick Guidotti at Positive Exposure, Jerrilyn Ankenman of NOAH, Julie Porter of the Hereditary Disease Foundation, Linda Selwa, Marcy MacDonald, Sayoko Moroi, Christina Boulton, Leeann Weidemann and Alice Domurat Dreger. Thanks go to Carl Marrs and the students in Epidemiology 511, including Miatta Buxton and Gail Agacinski, for reading the whole book and providing feedback. We also thank the artists who contributed to this book including Kathy Bayer, Ed Trager, Sophia Tapio and Sean Will and the artists at Dartmouth Publishing, Inc.

We thank the members of our research groups who have helped in numerous ways over the years, and we want to express special appreciation for the efforts of our administrative assistants, Linda Hosman, Nina Kolich, and Diana Hiebert. Scott Hawley also thanks both the Stowers Institute and its president, Dr. Bill Neaves, for support and encouragement during the writing process, and Julia Richards wishes to thank Dr. Paul Lichter for creating an environment in which genetics can bridge the gap between basic research and clinical practice.

We often use the first person in this book, but when speaking of scientific findings ("We now know that ..."), we do not mean to lay claim to this vast body of work we discuss. Many researchers have expended great amounts of time and energy for more than a century to arrive at the frankly amazing body of knowledge presented here. Although we are both active researchers in the field of genetics, in this book we speak as users of the human genome, teachers of genetics, and continual students of this fascinating topic.

We owe thanks to the individuals and families who contributed the stories in the book, each of which was included not only because it makes some scientific or educational point but also because these are stories that have touched our hearts. We want to offer special thanks to Jim Knowles, for letting us share Brenda's tale with you, and to Paula Sussi and Paul Gelsinger, who each continue working in education and policy areas to try to ensure that what happened to their children Marlaina and Jesse will not happen again. Others who shared their stories anonymously are just as much deserving of our thanks, even if we must leave them unnamed here. For some of those stories, we have simplified the tale to keep it focused on the lesson to be learned from the tale, and in some cases we have changed minor details where necessary to preserve confidentiality, such as through avoiding use of real names. In general, where we use no names or only first names, these are still true stories unless we have indicated otherwise. In rare cases in which we present a hypothetical situation derived from many similar stories, we try to indicate that we are doing this by stating that the tale is hypothetical or saying, "What if we looked at a family with these characteristics?" With many of the families we encountered the hope that helping other people understand what has happened will help them cope with the genetic situations in their lives. We also encountered the hope that the sharing of their tales would keep someone else from going through the same thing that had happened to their families. If this book accomplishes that goal for even one family, the writing will have been well worth all of the effort.

Preface

Huge changes in health care and in our understanding of ourselves will emerge during the first half of the twenty-first century, and those who take the time to understand the issues will be in the best position to take advantage of what is to come. If you have picked up this book expecting to find the biological cousin of the books used in organic chemistry and calculus classes, you may be surprised by the material that follows. We interweave personal anecdotes, discussions of ethical issues, historical remarks, and our own opinions right alongside an eclectic mix of scientific facts, molecular models, and cartoon figures. If you are not a student of the sciences but would really like to know more about genetics, this book was written with you especially in mind. It offers all of the fundamental concepts without requiring that you know anything about hydrogen bonding, hybridization kinetics, or differential equations. Keep in mind as you read that we are all astonishingly complex organisms and that there are exceptions to almost everything we will tell you since it is difficult, if not impossible, to arrive at generalizations that can truly encompass that complexity.

It is not our intention to turn any of you into geneticists, although that wouldn't be such a bad thing. Our real hope is to impart enough knowledge that you will be able to bring this subject into your own lives. It is hoped that by the end of this book you will know when and why to seek the council of a medical geneticist or genetic counselor, should you ever need one. It is also hoped that you will have become sophisticated enough to sort out some of the myths and misconceptions about human heredity that pass for simple truths in folklore and in the press. To the extent that we achieve even a small measure of success with either of these goals, we will consider this book a success.

Science is often presented as a dry recitation of objective facts so devoid of opinions and feelings that it is hard to derive a mental image of the author of the work. In many cases, this objectivity is a good thing. After all, there are powerful reasons for identifying solid facts and distinguishing them from opinions. To us, genetics is highly personal and not some abstraction removed from ourselves, so we have made a point of interjecting ourselves into this book about the genome that we share with each other and with all of you. We, authors and readers alike, are the end users of the information in our own genomes. So join us on a journey through this user's guide to the human genome.

Section 1

THE BASICS OF HEREDITY

This section provides a description of how traits are inherited and introduces the concept of the gene. We talk about how some of the basic genetic concepts apply to human inheritance and about how patterns of inheritance can look very different depending on the trait you are studying.

1 SLAYING MOLECULAR DRAGONS: BRENDA'S TALE

"To dream . . . The impossible dream . . ."
—Don Quixote in Man of La Mancha

Healthy young people aren't supposed to die. Even amidst the many dangers that arise from the exuberance and hazards of youth, the death of someone young is always a shock. And when the blow is delivered from some direction we never expected, were not waiting for, had never considered, when someone young is felled by an illness such as leukemia, we are left feeling stunned. It seems impossible to understand such an outcome, and we find ourselves asking, "How could this have happened?" And the next question that comes to mind is, "What can be done so that this does not happen again?"

Brenda Knowles was a graduate student in Scott's lab back in the late 1980s (Figure 1.1). She was bright and funny and totally unimpressed by Scott's supposed seniority. She was trained as a chemist and had begun graduate school doing biochemistry. However, Brenda had a strong connection to biology and the organisms that embody so much more complexity than simple biochemistry. Soon she found her way into a lab where there were organisms to work on, maybe just fruit flies, but organisms nonetheless.

She shared her time in Scott's lab with the usual array of characters that populate a "working lab". Science is a business that cherishes eccentricity, even

FIGURE 1.1 Brenda Brodeur Knowles (1962–1996). (Photo courtesy of James Knowles.)

encourages it. A healthy, growing lab will have its share of unusual characters. The basic foundation on which any new lab is started is unusual and novel ideas. Such ideas often come from and attract unusual and novel people.

In some ways, Brenda resembled the classic image of a young scholar. Her radio played classical music or National Public Radio, drowning out the competing styles of rock music from other desks or that much-ridiculed country music emanating from Scott's office. Her desk was neat and her ideas were equally well organized. She was rigorous in her critical thinking and tenacious in her pursuit of answers to scientific questions. She wrote (on her own) two papers from Scott's lab and went on to continue her scientific training by taking on a postdoctoral fellowship at Yale.

On her way to that fellowship, she married a handsome young doctor and they bought a beautiful little house in rural Connecticut. If you sense a fairy tale being told here, there's a reason: Brenda's life always seemed a bit of a fairy tale to Scott. This fairy tale was unusual only in the sense that Brenda was enough of a feminist to slay most of her own dragons.

That is, until Brenda got sick. Sometime in the early 1990s, Brenda acquired acute myelogenous leukemia (AML). We'll talk more about leukemia later in this book. The disease results from a rather nasty genetic alteration that occurs in one of the stem cells that produce the circulating cells in our blood. The result is an instruction for the altered stem cell to divide repeatedly. Leukemia was the ultimate dragon in Brenda's life, and she committed all of her resources to slaying it. She tried everything that was available, or even close to available. She suffered more than our words can convey. In the end, she lost the battle.

The battle she lost was just one battle in what the press often refers to as the "war on cancer". In 1969 a full-page ad in the *New York Times* urged President Nixon to begin a war on cancer, saying ". . . We are so close to a cure for cancer. We lack only the will and the kind of money and comprehensive planning that went into putting a man on the moon." The war on cancer was proposed in 1969. Brenda lost her battle with cancer in 1996.

There have been too many such battles. For most of history the idea of a cure for cancer has seemed like an impossible dream. We daresay that there will not be a single reader of this book who does not know someone touched by cancer. After all, one in four of us will get cancer in our lifetimes. But not all the battles are lost. There are some cures, many remissions, and many cases in which the cancer is simply held in check for years at a time. Still, Brenda died.

With impatient excitement, we watch advances in cancer treatment begin building on the results coming out of genetic studies of cancer. Breakthroughs in understanding of the molecular mechanisms of various forms of leukemia have led to breakthroughs in the development of new treatment approaches. Scientists have begun creating molecular "lances" aimed at slaying the monsters that are the various kinds of leukemia. Their molecular lances are drugs designed based on an understanding of what has gone wrong at the molecular level in the leukemia cells. How wonderful that these weapons against leukemia are emerging; how terrible that they will come too late for Brenda. Increasingly, we are seeing "magic bullets" emerge based on breakthroughs in our understanding of the underlying mechanisms of diseases caused by defects in genes. Some of these new cures use gene therapy, but we are going to see a lot of other pharmaceutical treatments emerge that will not use gene therapy even though they will be based on the information gained from the study of genes.

In a very real sense the scientists who are developing these new genetically based anti-cancer drugs are having to decipher a "lock" smaller than a thousandth of a pinpoint. That lock had been created by a change in the genes of a human cell. That lock committed that cell to a future of unrelenting cell division. The cure comes from building a "key" that releases that lock. If you understand that metaphor, that's wonderful. It would be even better if you understood the "magic bullet" and the "dragon-slaying" metaphors that we used before.

But we hope, we really hope, that you find such trite metaphors to be entirely unsatisfying. We hope you want to know what we mean by cells, and genes, in order to understand what all of these metaphors really mean. Because the scientist who builds this "magic bullet" isn't a wizard or a magician, he or she is a biologist. And as much magic as we biologists do see in the living world, we need to describe living systems, and manipulate them, in terms of molecules that interact with and within structures called cells.

That need to describe the chemistry of molecules and the structures of cells has been interpreted by others as a need to use terminology that requires a bachelor's degree in biology (and, better yet, chemistry!) to comprehend. However, we think that we can keep the chemistry in hand, by focusing on the processes that go on in a cell and on the functions that certain types of molecules play in the cell. We don't need to understand polymer chemistry to play with Legos made of plastic polymers. Similarly, to understand molecular genetic processes, we need only to know what overall structure the cell is trying to build, what pieces we have in our toy box, and how to snap them together. This does not mean that the chemical terms and structures are unimportant. Such details are in fact critical to anyone who is going to carry out studies of these systems. However, a lot of the concepts unveiled by such studies can then be understood without needing the expertise that was required to make the discovery in the first place.

Using that kind of framework, we will build you the verbal equivalent of Lego models of cells and, more importantly, of genes. We'll try to show you how genes work and how they control the activity of the cell. In time, we'll build a model of an "engine" that controls when cells divide and describe the "lock" that forces that engine to be locked "on." And we'll tell you how scientists are finding keys that disarm some of the locks that commit cells to relentless division and growth.

Treatments for leukemia are just one such example of the kinds of "genetic" medicines that will emerge with increasing frequency in the future. There will be ever so many more. The sad news is that the "cure" will have come too late for Brenda; the good news is that it will come at all! There will be more Brendas, but now we can dream the impossible dream, that there will be cures and the outcome will be better. Much better.

THE ANSWER IN A NUT SHELL: GENES, PROTEINS, AND THE BASIS OF LIFE

2

There are always those who ask, what is it all about? For those who need to ask, for those who need points sharply made, who need to know "where it's at," this: —*Harlan Ellison*[1]

Our genes provide a blueprint for our bodies. In doing so they set some upper and lower limits on our potential. Our interaction with the world and others defines the rest. —*R. Scott Hawley*

Marlaina Susi was a beautiful little eight-year-old girl who was active and friendly. She was an energetic child who was filled with a love of life and embraced everyone she encountered. She earned above average grades, participated in a variety of sports and other activities, and had not missed a single day of school due to illness during the previous school year. She has also been described as a picky eater, but no one realized at the time that her aversion to dietary protein might have been protecting her by helping her avoid high levels of protein that could be harmful to people with some types of metabolic defects. In 1999 her happy and seemingly healthy life was interrupted one day by a brief illness and fever from which she should have recovered, as young children normally recover from the usual array of "bugs" that get passed around an elementary school. Instead of recovering and rejoining her friends at school, she developed elevated levels of ammonia in her blood, was hospitalized, and died thirty three days later. After her death, her grief-stricken family continued their search for an answer to what had caused her death. They were told that she had a defect in the ornithine transcarbamylase (OTC) gene, one of several genes responsible for helping our bodies cope with the ammonia (NH_3) that forms as a normal part of metabolizing protein that we consume. If her OTC defect had been diagnosed during her hospital stay, there were medical remedies that would have been available to help her. But getting a correct diagnosis on time was complicated by several things: OTC defects are rare, they usually manifest in infants, they are usually seen in boys rather than girls, and Marlaina's defect was partial rather than complete. So what is an OTC defect, how can it have such a devastating effect, and why did the problem not show up until Marlaina was eight years old? To understand what happened to Marlaina, and to eventually find ways to protect other children with similar gene defects, we need to understand how a defect in a gene can lead to such devastating consequences.[2]

Marlaina Susi (1991–1999)
(Photo courtesy of the Susi family)

[1] From "Repent Harlequin!" Said the Ticktockman by Harlen Ellison.

[2] On the web site for the National Urea Cycle Disorders Foundation (NUCDF), there is a page that talks about Marlaina and the two memorial marches that have been held in her name to raise money for the Foundation. Information on OTC and other urea cycle disorders can be obtained from NUCDF, the Canadian Society for Metabolic Disease, or the National Organization for Rare Disorders.

THE BLUEPRINT INSIDE EACH CELL

Our bodies contain billions of cells, intricate little factories that carry out their own internal functions, as well as carrying on complex interactions with surrounding cells and the rest of the body. Almost all of those cells have a *nucleus* that contains most of the information required to make a complete human being (Figure 2.2). We refer to this set of information contained in the nucleus as our genome. It is composed of a chemical called deoxyribonucleic acid (DNA). Our genome doesn't function as a single entity but rather is comprised of tens of thousands of subunits of information called *genes.*

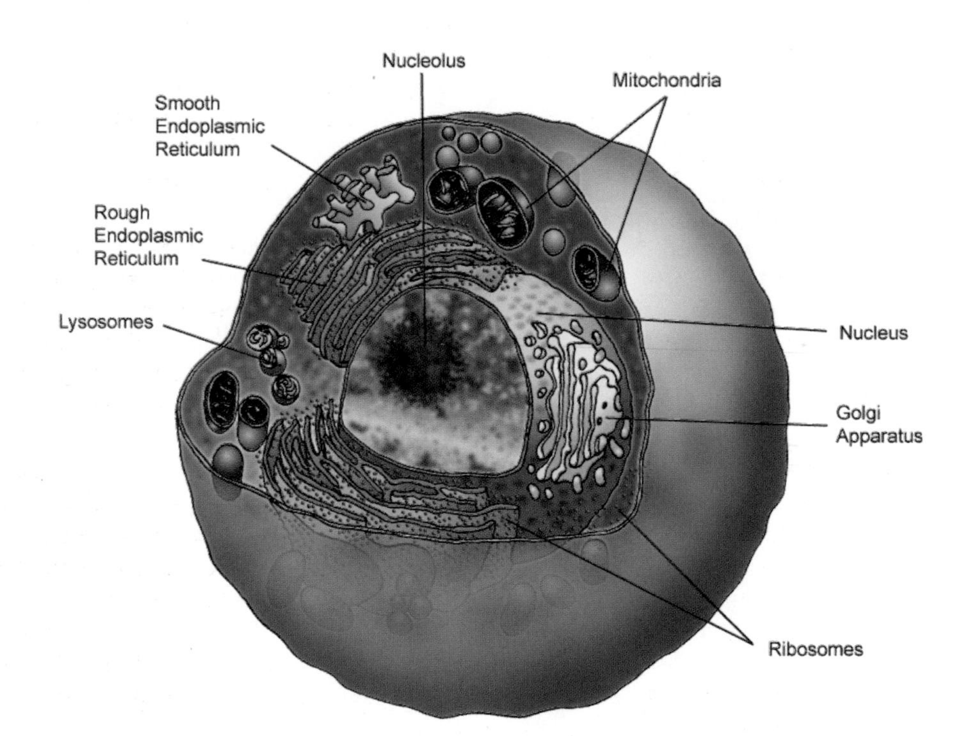

FIGURE 2.1 A microscopic view shows that a cell is a sack-like structure made of a membrane filled with *cytoplasm* in which structures called *organelles* are suspended. The largest organelle, the nucleus, contains the information used to run the cell and produce its structures. Actively growing cells contain a large inclusion within the nucleus called the *nucleolus,* the source of information used to construct ribosomes. Outside of the nucleus in the cytoplasm, millions of ribosomes use genetic information received from the nucleus to produce proteins. The endoplasmic reticulum and the Golgi apparatus are folded membrane structures where proteins may get additional chemical modifications and where key steps direct proteins to their final destinations. Thousands of mitochondria produce energy to run the cell. Membrane-bound containers called *lysosomes* hold molecules whose specialized functions need to be kept separated from the cytoplasm, like proteins that digest other kinds of molecules. This picture does not show all of the organelles in the cell or even all of the types of organelles in a cell, but it does show samples of organelles of importance to things we talk about in this book. How many of which organelles are present can vary for different cell types and different situations such as very active cell growth. *The key concept here is that the genetic information is located inside of the nucleus and the ribosomes that will "read" that information are located outside of the nucleus.* (Courtesy of Edward H. Trager.)

Virtually all of the cells in our bodies contain exactly the same full set of genes. Genes themselves are little more than repositories of information that tell the cell how to produce a gene product that carries out an essential function. Most often gene products are large, complex chemicals called *proteins* that actually do the work for and provide the structure of our cells. Proteins are the business end of cellular processes. The cell uses some proteins, such as tubulin, keratin, and collagen, as structural pieces of scaffolds and skeletons that are both inside and outside of cells. Other proteins called *enzymes* carry out a host of essential biochemical reactions, such as digestion and energy production.

You see colors and detect smells because of receptor proteins such as color opsins and odor receptors. Your heart or skeletal muscles move because of proteins called actin and myosin. Your body fights off infection with the help of proteins called immunoglobulins. Thus, our cells differ in size and shape. They carry out different functions such as transmitting pain signals or producing stomach acids because of the differences in the proteins they produce. In fact, one type of cell may even make different proteins at different points in the life span of a human being.

Many, if not most, of the differences that exist between us reflect the fact that the information in a gene can be permanently altered by a process called mutation, and changing the information in a gene by mutation changes the protein product that it creates. Although many think of mutation as a term for something negative or harmful that can cause birth defects and genetic disease, mutations can also be neutral (having no detectable effect) or even beneficial. They can cause differences in many of the characteristics by which we recognize each other: height and build; hair color and texture; and shapes of face, nose, ears, eyes, and eyebrows. Mutations can affect things that are harder to define, such as behavior. Mutations are responsible for differences that are very important even if they are invisible to us on a daily basis, such as blood type. Without mutations, we would all have exactly the same set of genetic information and billions of us would all resemble each other in much the same way that identical twins resemble each other. The vision of billions of identical humans is a chilling thought that leaves us quite pleased with the amount of diversity we see around us.

The term *mutation* refers to a startlingly large array of different types of processes that can permanently change the structure, and thus the information content, of genes. Although mutation occurs rarely, there are an awful lot of us, we breed well, and we have been breeding for a very long time. Thus there has been ample opportunity for mutations in each of our genes to occur and in many cases to be spread widely throughout our population. These altered genes may produce an altered protein or produce no protein at all. Although missing proteins often turn out to cause severe or even lethal phenotypes, altered proteins may cause a broad range of phenotypes, in some cases severe and in other cases almost undetectable. Mutations that result in altered proteins are responsible for much of the diversity we see around us.

Accordingly, genes affect our form, appearance, physical abilities and limitations, talents, and many aspects of our behavior as well. Each of us received one complete copy of the "human genome" from our mom and one copy from our dad. Thus each of us carries two copies of each gene. When we make gametes (sperm or eggs), we place only one of our two copies of each gene in each gamete. This trick sees to it that each generation will always have two copies of each gene, and it introduces an amusing bit of randomness to the process. Each sperm or egg that we produce consists of a different combination of genes derived from our own mothers and fathers. However, when we pass genes along to the next generation, some of the genes we pass along are the copy we got from mom, and for other genes we pass along the copy that we got from dad. Thus each new baby is the result of implementing a set of genetic instructions created by two rolls of the genetic dice, one that took place in the father and one that took place in the mother.

Genetic diseases, or inborn errors, result from cases in which the DNA blueprint is incorrect or incomplete, usually because a specific gene is damaged or missing. In such a case, the cells of an individual bearing such a genetic defect will make a damaged version of that protein or perhaps not make the protein at all. For example, people like Scott who lack functional copies of a gene that makes one of the color opsins will be unable to distinguish colors. So genetic disorders are not always lethal, and may not even make you sick. Many differences between copies of the genome present in different people cause no harm at all. In some cases they may cause simple cosmetic differences. In other very rare cases, they may even give someone a desirable characteristic not shared by their neighbors, such as resistance to an infectious disease. All too often, though, differences in the genetic blueprint are not just neutral changes; they are considered defects because they cause a problem.

A DEFECT IN THE OTC GENE CAUSES ALTERED PROTEIN METABOLISM

To look at how defects in the genetic blueprint borne, by a developing zygote result in loss of an essential function in the body, lets look at a serious gene defect that is sometimes found in the human genetic blueprint. Many harmless biochemicals that make up our bodies can become harmful if we have too little or too much of them. Examples include blood sugar, cholesterol and nitrogen.

Normally, nitrogen levels in our bodies are regulated by a set of biochemical reactions called the urea cycle, the process by which our bodies convert excess nitrogen from food into a compound that can be excreted from the body (Box 2.1). A protein called ornithine transcarbamylase, or OTC, carries out one of the critical steps in the urea cycle.

Babies who are born with a defect in their genetic blueprint at the point that contains the information needed to make the OTC protein cannot properly control ammonia levels in their blood because they don't correctly metabolize proteins from their food (Figure 2.2). If there is no OTC protein, excess nitrogen does not get carried through the urea cycle the way it should. One consequence is that excess ammonia accumulates, and the ammonia is toxic. When a baby is born who is completely lacking in functional OTC protein, symptoms within the first three days of life may start with problems with breathing and eating. If these babies are not treated, ammonia levels build up in their blood and their brain, they go into a coma, and they die. Other children like Marlaina, with a partial defect in which OTC levels are reduced but not gone, may live healthy lives for years because the small amount of

BOX 2.1 DEFECTS IN THE UREA CYCLE

When we eat protein, nitrogen enters the body. The body uses some of the nitrogen but some of it needs to be eliminated. The protein ornithine transcarbamylase carries out one of several critical steps in the urea cycle. In babies with a normal copy of the gene that makes the OTC protein, the urea cycle uses dietary nitrogen to produce urea and the extra nitrogen from the diet is thus excreted. A baby who has only damaged information for making OTC protein cannot use the urea cycle to turn nitrogen into urea to be excreted. These babies have problems that include accumulation of nitrogen-containing ammonia, which can be toxic. Excess ammonia can lead to problems such as brain damage, liver damage, coma and even death. How severe the problems are depends on whether the OTC protein is completely missing or whether the protein is damaged but still able to carry out its job at a low level. Defects in genes controlling the other steps in the urea cycle can cause similarly terrible consequences. Children with genetic defects affecting other steps in the urea cycle may not all have identical health problems, but one of the common problems for urea cycle disorders is the build up of ammonia. Diets and treatments exist that can help limit build-up of ammonia, but there is no cure and the treatments themselves are difficult and limited in how much they can help. According to the National Urea Cycle Disorders Foundation, 1 in 10,000 children are born with a urea cycle disorder, and some cases of Sudden Infant Death Syndrome may actually be undiagnosed urea cycle disorder cases. Many children with urea cycle disorders are seriously harmed within days of birth, and many more die before their fifth birthdays. More information about urea cycle disorders and prenatal screening can be obtained from the National Urea Cycle Disorders Foundation.

OTC activity in their bodies is enough to handle the very small amounts of nitrogen coming in from their low-protein diet. Thus, they might live a long time without being diagnosed until an illness or consumption of too much protein causes a crisis that requires prompt diagnosis and treatment to survive. All too often, in these later onset cases, the need for treatment during a crisis is urgent and great harm can occur during any delay while doctors carry out tests and struggle to sort out a diagnosis that can be difficult to make.

One fundamental point must be made here: the altered information in the damaged OTC gene does not directly do any harm or cause the disease. Rather the disease results because the child lacks intact, functional OTC protein needed to carry out an essential function. Although the primary event in the disease may be the damaged gene, the direct cause of harm is in the failure of the gene product produced by that gene. A damaged gene, like the blueprint for a cruise missile, is in and of itself pretty harmless. It is the product of that blueprint (either the cruise missile or the defective or absent protein) that poses a problem.

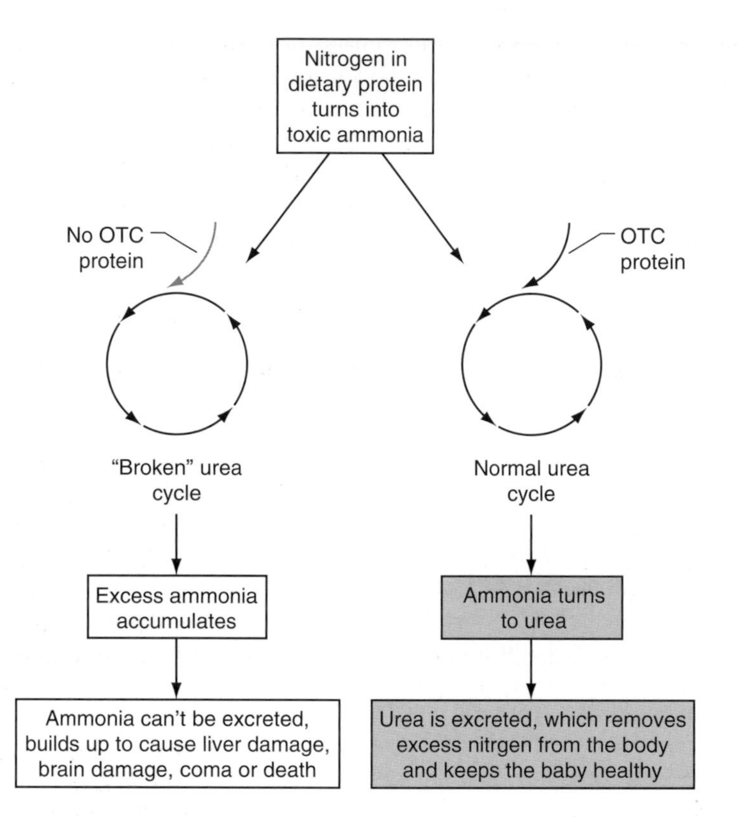

FIGURE 2.2 If there is a defect in the part of the genetic blueprint responsible for the ornithine transcarbamylase protein (OTC), the consequences can be a serious illness that can lead to death if toxic levels of ammonia are not controlled. The different diseases of the urea cycle are complex and accumulation of ammonia is only part of the problem, but we show it here because it is a central key to the problem.

THE ANSWER IN A NUTSHELL

So this is our answer in a nutshell: *Usually, no one dies because of a defect in his or her genes; they die because that genetic defect alters the gene product so that it no longer performs its function correctly.* This is the foundation for everything else we will discuss—that information, in the form of DNA located in the nucleus, directs the production of gene products (which are mostly proteins) that actually carry out the cell's functions. And many differences we find between different human beings trace back to a change in how some function was carried out (or not carried out at all) by a damaged (or missing) gene product. There are in fact exceptions to this generalization, as there are exceptions to almost everything we will tell you about in this book, but keeping this core concept will give you a framework for everything else we will say.

In the case of Marlaina, we see that her death was the result of a defect in her genetic blue print at the point that contained the information needed to make a functional OTC protein. We also see that in her case the defect ended up with her having reduced levels of OTC activity rather than a complete absence of OTC activity. This explains why she managed to remain healthy through the eating habits that kept her protein intake low. Much remains to be learned about why some kinds of illness can throw off this tenuous equilibrium maintained by someone with a partial OTC deficiency and result in an increase of ammonia in the blood. Eventually, newborn genetic screening may allow for the identification of children like Marlaina who seem quite healthy but actually suffer from a deficit that threatens their lives. Knowing that these children are at risk of losing control of their ammonia levels under certain conditions can make a big difference in the kinds of preventive measures that can be taken and can also be important in allowing doctors to make a rapid, knowledgeable response to the kind of crisis that Marlaina experienced. This is just one example of ways in which genetics can make a powerful, even life-saving, difference for the many people who have a typographic error in their blueprint.

In this book, we hope to share the fascination with genetics that has led so many to spend their lives investigating genetic blueprints and the way they work. We will explore how information in the DNA blueprint translates into proteins and functions. We will look at how changes in the DNA blueprint come about and the consequences of different kinds of changes. We will talk about how we go about studying DNA and telling whether or not a particular change in the DNA blueprint can account for a disease or some other characteristic that differs between individuals. Once we have explored some of the fundamentals of how the DNA blueprint does its job, we will talk about how particular genes affect fundamental processes—such as what makes someone male or female- and the different ways in which particular genes can lead to disease. We will raise questions about what constitutes "normal" and examine the broad array of human characteristics that are affected by the DNA blueprint. However, this book will take you far beyond the simple facts of cell biology to explore how genetic testing, gene therapy, and other advances in genetics affect our lives and the lives of those around us. The emerging technologies we will discuss have tremendous power to accomplish good, relieve pain, and improve peoples' lives, but only if used with an eye to ensuring that no harm is done. In time we will return to the various ethical, legal, and social issues that complicate modern genetics. Before we can discuss them, we really need to have a more detailed knowledge of genes themselves.

Perhaps surprisingly, our story starts not in a modern lab but in a nineteenth-century Austrian garden where a monk cultivating pea plants started a quiet scientific revolution . . .

MENDEL AND THE CONCEPT OF THE GENE

3

"In the beginning . . ."
—Genesis 1:1

We suspect that people have been curious about how heredity works ever since they figured out where babies came from. It is important to note that our current sophistication in these matters is of fairly recent origin. There is an old saying that "like begets like," but this seemingly obvious knowledge that children will be like their parents would have been heresy to some ancient Greeks who wrote about the progeny that resulted from mating members of different species, such as swans and sheep.

We also know that children share similarities with both of their parents. For long periods in our history, people imagined that children were the offspring of only one parent (either the mother or the father). There were schools of thought in which children were preformed only in their mothers; the father was thought to provide only a "vital spark" (much like jump-starting a dead battery). However, the early microscopists, most of whom were men, imagined that babies were preformed in the father and sailed in sperm down the vaginal canal into awaiting uterine incubators (Figure 3.1). Indeed, there are existing drawings dating back to the seventeenth century that show these tiny preformed individuals (now known as homunculi) inside the sperm.

These myths persisted despite the realization by farmers that animal offspring often appeared to be a mixture of both of their parents. This philosophy of heredity, known as blending, took a long time for humans to incorporate into our views of our own heredity. Although it seems to have taken a long time for such ideas to catch on in a world in which there were many examples of apparent blending, by the mid-nineteenth century, most people were willing to accept the concept that the traits observed in children were some mix of those observed in both parents and in both sets of grandparents.

As silly as it seems today, blending really was not an unreasonable model to propose. If you mix red paint and white paint, you get pink paint. If you mix hot water and cold water, you get warm water. People imagined that there was some kind of substance, such as blood, that blended in the offspring to produce a mixture of traits in the child. (Note the term "blood relative," which implies a shared ancestry, not relationship by marriage.)

Still, there were some surprises that blending did not explain: blue-eyed kids born to brown-eyed parents, blond children of raven-haired moms and

FIGURE 3.1 Artist's conception of the tiny preformed individuals envisioned by early microscopists viewing magnified images of sperm.

dads, kids who are taller than either parent, and so on. Blending, although a useful way to understand some traits such as height and weight, did not explain everything.

It was into this rather curious intellectual environment that Gregor Mendel was born in 1822. During his lifetime, this man's intellect would boldly go where no person's mind had gone before. Like Galileo, Newton, Freud, and Einstein, Mendel's vision would change the course of human understanding. That vision results from one simple set of experiments. We will describe one of those magical moments in human cognition when a new set of concepts became beautifully obvious and clear.

WHAT MENDEL DID

Mendel, a monk with a garden plot, began with a specimen, the pea plant, which was simple to cultivate. He chose to study the inheritance (the passing of a characteristic from one generation to the next) of seven simple and obvious traits that could clearly be distinguished between different pea strains:

- Seed shape—round vs. wrinkled
- Seed color—yellow vs. green
- Flower color—white vs. colored
- Seedpod shape—inflated vs. constricted
- Color of the unripe seedpod—green vs. yellow
- Flower position—along the stem vs. at the ends
- Stem length—short vs. tall

These were simple "yes-or-no" traits and not quantitative traits such as weight that can vary over a wide range of different values. Some traits, such as stem length, can vary under different conditions such as rich vs. poor soil, so Mendel selected traits that were so severely different in the different strains that even substantial environmental differences could not make one strain appear to have the characteristics of the other. For instance, for the stem-length experiments, one strain was selected that is consistently 6 to 7 feet tall and the other strain was one that is always less then $1\frac{1}{2}$ feet tall. So tall plants were not always the same height, but they were so much taller than the short plants that the two categories were never mistaken for each other.

Mendel also began with pure-breeding populations of plants. For example, he had a bunch of plants with yellow seedpods that produced only plants with yellow seedpods when bred to each other. Similarly, he had a bunch of plants with green seedpods that produced only plants with green seedpods when bred to each other (Figure 3.2).

Please note that in the first generation, when Mendel crossed plants with green seedpods to plants with yellow seedpods, all he saw in the progeny were plants with seedpods identical in color to those of the green seedpod parent. *None of the plants had seedpods of an intermediate color* (Figure 3.3).

This experiment helped rule out several of the old ideas about inheritance. A real adherent to blending would have expected the progeny of the first generation to have yellowish-green, not true green, seedpods. Mendel's observations were simply incompatible with a blending hypothesis. An

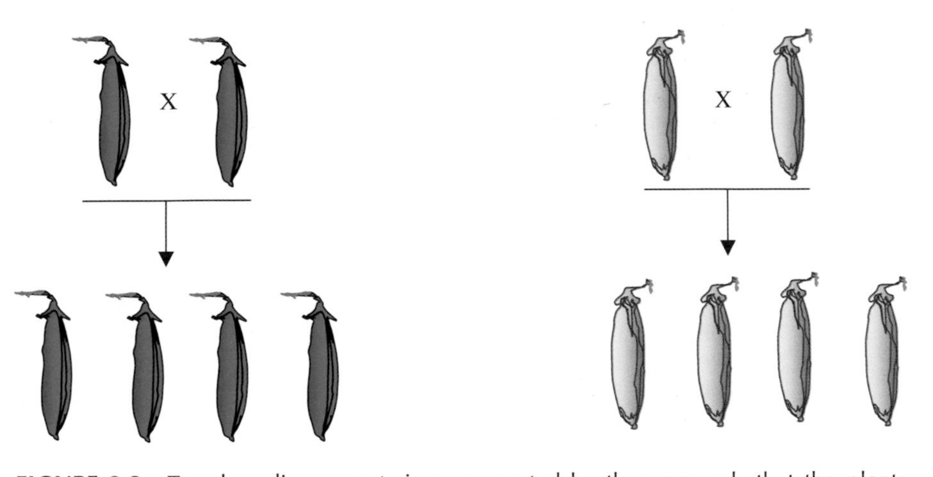

FIGURE 3.2 True-breeding pea strains represented by the pea pods that the plants produce. Although peas often show a lot of variation in characteristics such as colors of flowers, peas, peapods, and size or shape of peas, peapods, and stems, Mendel said that this was because of pollen from one type of plant getting to a plant with a different character to produce hybrid offspring that contained characteristics inherited from each separate parent. By protecting plants from random pollination by other plants and then carrying out artificial pollination, he was able to control which plants were being crossed to each other. He identified some strains that bred true for a particular characteristic, with each succeeding generation of the true-breeding strain producing progeny that were exactly like all of the preceding generations for the particular characteristic being looked at. The color of the immature peapod is one example of a characteristic for which he found true-breeding strains.

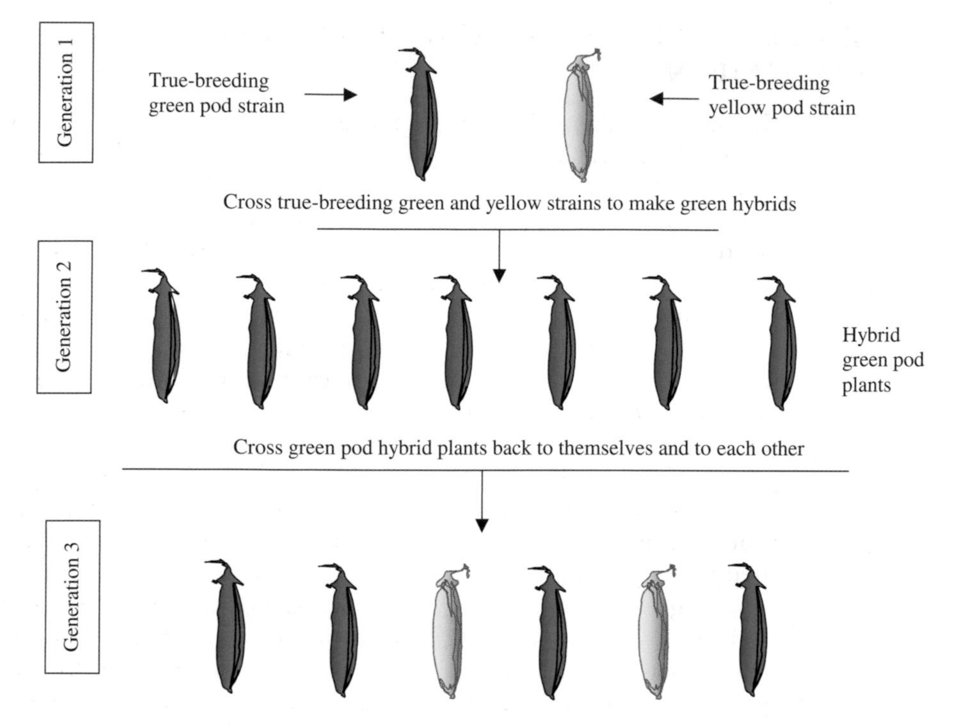

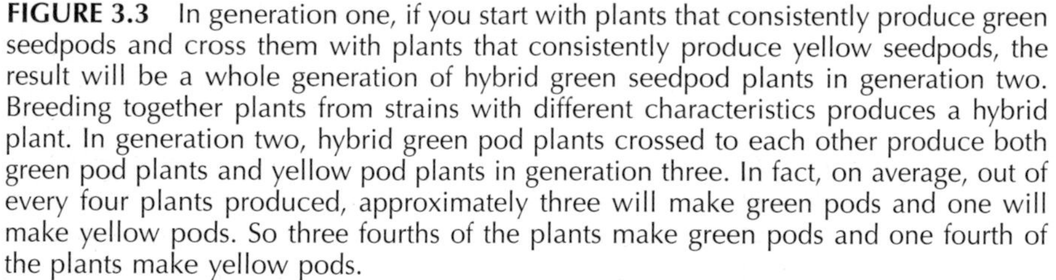

FIGURE 3.3 In generation one, if you start with plants that consistently produce green seedpods and cross them with plants that consistently produce yellow seedpods, the result will be a whole generation of hybrid green seedpod plants in generation two. Breeding together plants from strains with different characteristics produces a hybrid plant. In generation two, hybrid green pod plants crossed to each other produce both green pod plants and yellow pod plants in generation three. In fact, on average, out of every four plants produced, approximately three will make green pods and one will make yellow pods. So three fourths of the plants make green pods and one fourth of the plants make yellow pods.

adherent of the vital spark or homunculus theories might have expected the offspring to always resemble just the maternal or just the paternal parent. However, it turns out that it didn't matter which way the cross was made (i.e., green males crossed to yellow females or vice versa); all of the offspring had green seedpods. So much for the theories that traits come from only the male or only the female parent.

When green seedpod plants from generation two were crossed to themselves or each other, they produced both yellow- and green-seedpod plants (see Figure 3.3). The blending hypothesis doesn't work to explain two green pod parents making a yellow pod offspring.

Notice that the yellow pod characteristic from generation one disappeared in generation two and reappeared in generation three. One of the things this tells us is that the yellow pod trait from generation one got passed along to generation three without being evident in generation two.

How do we explain all of this? Mendel's explanation made use of several concepts, and no one of those concepts alone was enough to explain what he was seeing.

WHAT PASSES FROM ONE GENERATION TO THE NEXT IS INFORMATION

In the first of Mendel's three conceptual breakthroughs, he separated the information that produced a given trait (which we will call the genotype) from the physical manifestation of the trait itself (which we will call the phenotype). In the case of the pea plant, the yellow-pod recipe (genotype) produces a seedpod that appears to our eyes to be yellow (phenotype). If we were cooking, the words of the cake recipe on the page of the cookbook would be the genotype, but the lemon flavor of the cake would be its phenotype (Figure 3.4). We can carry this analogy further and point out that some phenotypes can be rather more complex, as in a cake with a chocolate genotype having several different characteristics (brown color, chocolate flavor) that are part of its phenotype.

Mendel argued that there were discrete units of heredity (now called genes) that were immutable pieces of information and were passed down unchanged from generation to generation. He argued that these genes specified the appearance of specific traits but were not the traits themselves. This insight gave rise to Mendel's concept of the purity and constancy of the gene as it passes from one generation to the next. In simple terms, Mendel said that genes received from the parents are passed along to the offspring in a precise and faithful fashion. So what gets passed from one generation to the next is the recipe, not the cake.

In order to explain differences in traits, Mendel supposed that genes could take different forms (now called alleles) that specified different expressions of the trait. For example, Mendel claimed that there was a gene that gave seedpod color, and that there were two different forms or *alleles* of that gene: one specifying green color and one specifying yellow color. We will refer to those alleles that specify green color as G alleles and those that specify yellow color as g alleles. All individual plants that breed true for production of green seedpods must have only the G allele that causes green seedpods. Further, they must have gotten these G alleles from their parents and will pass them along to their offspring. Similarly, plants with yellow seedpods must have only the g allele that causes yellow pod color. They must have gotten the g alleles from their parents and will pass them along to the offspring. So let's reexamine what we saw in Figure 3.2, where we just looked at the phenotype

FIGURE 3.4 Genotype vs. phenotype. The distinction between information and what can be produced using that information is one of the most important concepts in genetics. So the recipe (genotype) is distinct from the cake (phenotype). Another key concept is that changes in the information (change the word "lemon" to "chocolate") can give you change in the phenotype (flavor of the cake).

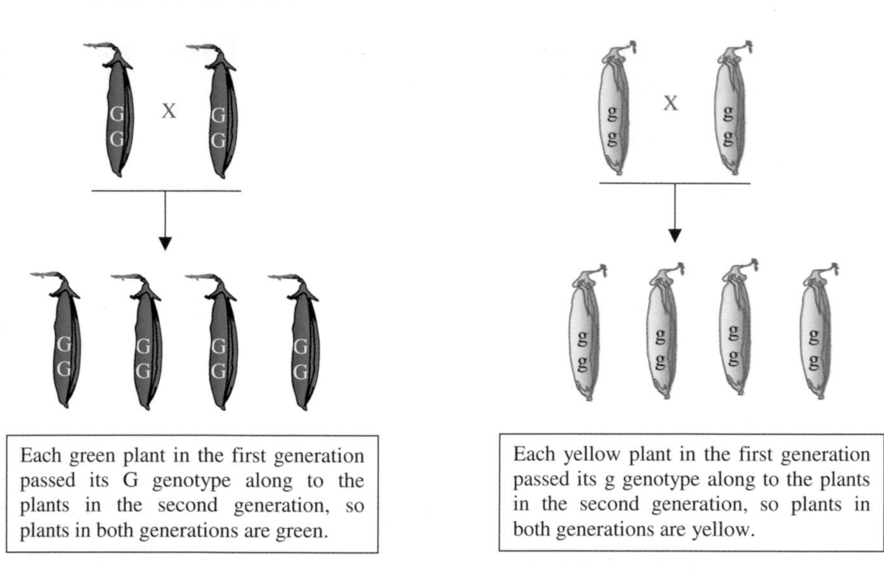

FIGURE 3.5 Genotypes that go with the phenotypes when true-breeding characteristics are passed from one generation to the next. Notice that every true-breeding green plant has two copies of the G allele, and every true-breeding yellow plant has two copies of the g allele.

of true breeding green and yellow plants that produced only plants like themselves. This time, let's add in information about the genotype of those plants, which will be expressed as a listing of the alleles of the pod-color gene that are present in the plants (Figure 3.5).

DOMINANT TRAITS MASK THE DETECTION OF RECESSIVE TRAITS

The idea of a genotype (information) that matches the phenotype (the trait produced by using that information) is easy to see and understand when the same strain of plant is bred to itself over and over, always producing plants just like the parents. Plants that only have G alleles produce offspring that only have G alleles, and they are all green. Plants that have only g alleles produce offspring that have only g alleles, and they are all yellow. However, this idea was not enough to explain what happened in Figure 3.3 when he crossed the hybrid plants in generation two to each other and got back both yellow and green offspring. *Mendel explained this by saying that an individual must also be able to carry genetic information for a trait it does not express.*

This was Mendel's second big insight: that this pattern of inheritance can be explained if some traits can mask our ability to detect other traits. Thus he hypothesized that green (G) alleles could mask the expression of the yellow (g) alleles, such that individuals getting a green allele from one parent and a yellow allele from the other would be just as green as those that got only green alleles from both parents.

Mendel introduced the terms *dominant* and *recessive* to identify traits that predominate or recede into an undetectable state. In this case, the G gene allele is said to be dominant (its green phenotype predominates) and the g allele recessive (its phenotype recedes into an undetectable state) because

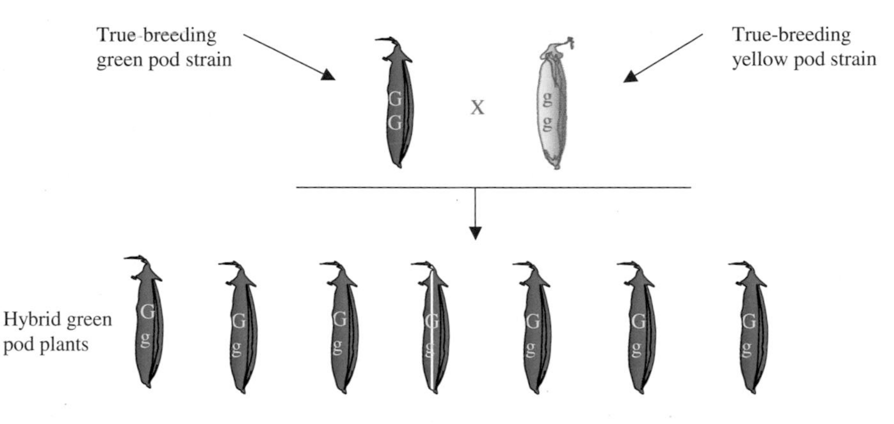

FIGURE 3.6 Masking of traits. Notice that when true-breeding green plants with only G alleles are bred to true-breeding yellow plants with only g alleles, the offspring all have one G allele and one g allele. As can be seen from the heterozygous Gg genotype of the green hybrids in the second generation, the green G allele masks our ability to tell that the yellow g allele is present.

Gg plants had green seedpods that looked just like the seed pods in true-breeding plants that only have the G allele. Individuals who carry two different forms of a gene, or two different alleles, are called *heterozygotes* because they carry two different alleles (Gg), whereas GG and gg individuals, who carry pairs of identical alleles, are called *homozygotes.*

To see an example of this kind of masking of information, let's take a look again at the creation of the green hybrid plants in the second generation. As we discussed previously, when the true-breeding green and yellow parents were bred, only green plants resulted, and the color was the true green of the green parent and not an intermediate color midway between the colors of the two parents. Let's look at the genotypes that went with the phenotypes in this cross as diagrammed in Figure 3.6. A cross of the true-breeding green strain to the true-breeding yellow strain produces heterozygous green hybrids that all have genotype Gg.

Is it obvious yet how the hybrids ended up being heterozygous and having one copy of each allele? Let's look at Mendel's next insight, which explains how this happens.

ONE ALLELE COMES FROM EACH PARENT AND ONE ALLELE PASSES TO EACH CHILD

Mendel's third insight was to assume that, although there is but one gene for each trait, the offspring gets one copy of any given gene from each parent, and when that offspring reproduces, it transmits one and only one copy of each gene to each gamete. On a very simple level, this explains how the hybrids in Figure 3.6 ended up having one G allele and one g allele.

So far, so good. But how do we explain the result in Figure 3.3, in which the offspring of the heterozygous hybrids produced so many more green offspring than yellow? Although the numbers in Figure 3.3 are small, we can tell you that if you do this experiment a lot of times, once the numbers are quite large, you will continue to see an excess of green offspring.

It turns out that each individual has two copies of a gene, but they put only one copy into each gamete that gets used in the formation of offspring. So an individual of genotype Gg does not make gametes with both a G allele and a g allele. Rather, that Gg individual makes some gametes that contain only the G allele and some that contain only the g allele. Furthermore, the G gametes and the g gametes get made with about equal frequency. It turns out that when two Gg individuals mate, the odds of producing a gg offspring are only 1 in 4, or 25%. There are four different combinations of genotypes that can be produced, and each new offspring has an equal chance (1 in 4) of getting one of the four genotypes:

Mom's G with Dad's G gives a GG genotype and a green phenotype.

Mom's G with Dad's g gives a Gg genotype and a green phenotype.

Mom's g with Dad's G gives a gG genotype and a green phenotype.

Mom's g with Dad's g gives a gg genotype and a yellow phenotype.

We now have the concepts needed to explain what happened when we crossed a true-breeding green strain to a true-breeding yellow strain to create a green hybrid that was capable of producing progeny of both colors. A G allele from one parent plus a g allele from the other parent created the heterozygous Gg hybrid that was green because the G allele is dominant and masks the recessive g allele. When a Gg hybrid was crossed to a Gg hybrid, each gamete had an equally likely chance of getting a G allele or a g allele. Thus, GG, Gg, gG, and gg genotypes could all be created and result in homozygous green plants, heterozygous green plants, and homozygous yellow plants. Sometimes this is easier to follow if you look at a diagram like the one in Figure 3.7, which shows the genotypes that go along with the phenotypes.

How did Mendel figure all of this out? In fact, he did experiments that took years and involved more than 10,000 plants. If you would like to see more about the details of his experiments and some of the thinking that arose from his results, you could check out his published work (Box 3.1).

BOX 3.1 MENDEL'S ORIGINAL WORKS

A lot of information is available about Mendel, his life, his education, the world he lived in, and what happened to his discoveries before they were brought to light again many years after his death. Although we describe enough about Mendel's experiments to help you understand the ideas he arrived at, the whole set of many experiments he did were complex and involved many different characteristics of the plants he was studying. If you want to know more about him, or if you want to read his original scientific writings (in English or in German), check out Mendelweb at www.mendelweb.org. This site does a very nice job of annotating his works and providing links to helpful items such as glossaries, reference materials, and related sites. This page does a lot to make Mendel's work easier to understand. Even if you don't want to read Mendel's writings in detail, it is worth checking out this excellent site.

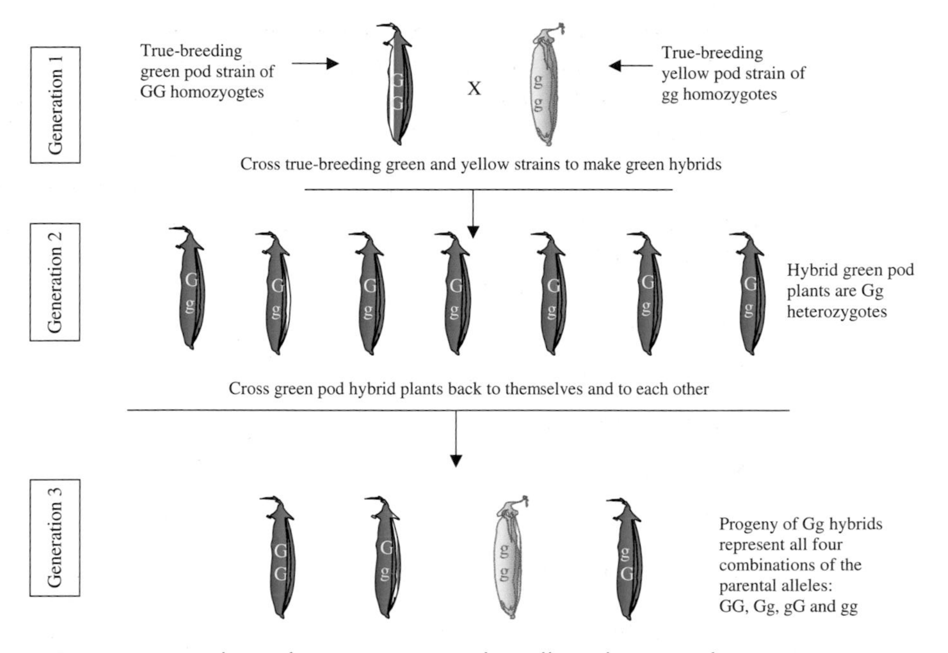

FIGURE 3.7 Masking of a recessive trait. The yellow phenotype from generation one disappears in generation two and reappears in generation three because the dominant G allele masks the recessive g allele. When a female parent from generation two contributes her invisible "g" and a male parent from generation two contributes his invisible "g" allele, an unmasked gg homozygote is created and manifests the yellow trait that could not be seen when a G allele was present. Notice that the Gg genotype (G from mom and g from dad) and the gG genotype (g from mom and G from dad) are equivalent and result in the same phenotype. Thus we expect the cross of two Gg heterozygotes to produce one green GG homozygote, two green Gg heterozygotes, and one yellow gg homozygote.

SUMMARY

One of the punch lines we want you to take away from this chapter is that the genetic rules in humans, and in many other complex organisms, operate by the same three rules laid out by Mendel to explain pea genetics:

1. Genotype and phenotype are distinct, with different alleles of individual genes corresponding to different phenotypes.
2. Some traits can mask other traits, leading to the concepts of dominance and recessiveness.
3. An organism has two copies of each gene, receiving one copy from each parent and passing one copy along to each offspring.

Mendel's ideas indicate that we will be able to predict the phenotype from the genotype, but when we look at human beings, we discover that the real-life situation is actually rather complex. Just as with alleles that produce seed pod color, some of the variant information in our two copies of the genome are bound to have certain outcomes—blue eyes vs. brown, for instance—whether our parents raised us on burgers or health food, in rich times or poor, on a farm or in a big city. However, other characteristics—how tall each of us ended up, for example—may vary depending not only on a difference in our genetic information but also on the environment in which we live.

Genetics is the study of characteristics that differ from one individual to the next and the transmission of those variable characteristics from one generation to the next. There are limits to how tall or short each of us could have become under the greatest environmental extremes of feast or famine. Those limits are set by genotype. Any feature that differs between individuals can be a valid point of genetic study, whether the variable feature is something visible, behavioral, or assayed by a biochemical test. In Chapter 4, we look at inheritance of human characteristics from the perspective of Mendel's laws. . . .

HUMAN MENDELIAN GENETICS

4

When Scott's daughter Tara was born, Scott and his wife were immediately surrounded by the expected group of close relatives, as well as many other relatives Scott had never met. Some of them brought to his daughter's crib the most extraordinary bits of genetic folklore. He can remember one of them staring at his daughter's eyes and saying, "She has her grandfather's eyes, but then girls always get their grandfather's eyes." Wait a minute; Scott's a geneticist, and this was big news to him! Such wisdom kept coming for the next few days, a continuous stream of different bits of genetic folk wisdom. Some of it was just folklore, but some of it had good basis in fact. The point is that people have long known that traits move through families in patterns, patterns that we now call modes of inheritance. One of the tricks we face in modern genetics is figuring out which pieces of genetic folk wisdom are actually true and then understanding why they happen.

Diversity is one of the greatest gifts granted to humanity. Some of that diversity is obvious when we look around us and find ourselves surrounded by unique individuals rather than carbon copies. Much of the diversity that gets noticed the most takes the form of physical characteristics—skin color, hair texture, height, weight, strength, speed, or facial features. The first thing everyone asks when a new baby is born is, "Is it a boy or a girl?" However, much important diversity has nothing to do with the kinds of characteristics that determine whether or not we get offered a modeling contract. A lot of important diversity takes place at the molecular and cellular level. Some of it seems so complex that we have to wonder if what we see in human diversity bears any resemblance to what happened with Mendel's peas. The kinds of characteristics we want to understand in humans may be as diverse as:

- Color-blindness
- Differences in the rate of aging
- Being male or female
- Risk-taking behavior
- Life-saving resistance to malaria
- Hairy ears
- Control of blood sugar levels

- Sneezing in response to sunlight
- Schizophrenia
- Susceptibility to a particular kind of cancer
- Perfect pitch in a talented musician
- Rejection of a transplanted kidney

Which of those characteristics do you think are genetic, and which ones do you think result from things that happen to you that originate from outside sources? Evidence exists to show that genetics plays a part in everything on this list, even in the behavioral traits. This chapter will introduce human inheritance and show how many (but not all) human traits show Mendelian inheritance, that is, transmission from one generation to the next in patterns that resemble what Mendel saw when he studied peas.

RECESSIVE TRAITS IN HUMANS

As we mentioned earlier, many of the early ideas about how inheritance in humans works turned out to do a poor job of explaining what actually happens. But when we take Mendel's ideas about inheritance in peas and apply them to humans, a lot of confusing things start to make sense. By proposing that information is different from what gets made using that information, by proposing the existence of the particles that we now call genes and by proposing that some traits are dominant over other traits, Mendel provided ideas that help us understand many things we see in human patterns of inheritance.

Let's look at the human trait albinism, specifically a form known as oculocutaneous albinism, which is manifested in people who make little or no melanin pigment (Figure 4.1). The common perception of albinism, people with stark white skin and hair and red eyes is over-simplified and incorrect. In fact, there is some variation in how pale people with albinism are, and the stereotype of red eyes is wrong. Some may even have yellowish hair or other signs of coloration, such as freckles. Most commonly, they have blue or gray eyes. Sometimes, their eyes may take on a purplish or reddish tint if the light is just right, resulting from the red tints from the retina showing through the pale coloring of the iris. This does not normally happen in most individuals with blue or gray eyes because the pigment in a pigment epithelium layer behind the iris normally blocks the red tints in the retina from being seen.

In most ways, people who lack melanin are just like the rest of us. But even though they are highly diverse in terms of a variety of traits, they do have some features in common, such as their unusual coloring and vision problems. The lack of melanin during development of the eyes causes abnormal routing of the optic nerves into the brain and results in inadequate development of the retina. They often use glasses, but their vision often cannot be corrected to 20/20 acuity with either glasses or surgery. They are unusually sensitive to bright light. Some are legally blind, but others see well enough to drive a car when using special lenses. Some are not blind but have vision problems that can't be helped by corrective lenses. In some cases, skin cancer can

FIGURE 4.1 Rick Guidotti's elegant photo spread in Life magazine featured individuals with albinism as unique and beautiful, emphasizing the need to avoid stuffing them into some prejudicial box with a label on it just because they happen to share a genetic trait. In contrast to the attitudes during the nineteenth century when people with albinism were featured in circus sideshows, this twentieth-century article succeeded in communicating a great sense of the unique and positive value in each of those photographed, including the beautiful woman featured here.* (Courtesy of Rick Guidotti for Positive Exposure.)

be a problem, especially in equatorial regions, if they don't use sunscreen and take other steps to protect themselves well enough from sunlight.

Although most individuals with simple albinism are as healthy as the rest of us, two very rare forms of syndromic albinism are associated with serious medical problems. Hermansky Pudlak syndrome includes albinism and multiple other characteristics, including problems with bleeding. Chediak-Higashi syndrome characteristics include susceptibility to infection and development of malignant lymphoma.

According to the National Organization on Albinism and Hypopigmentation, one person in every 17,000 has some form of albinism. Based on those numbers, that would mean that more than one in every 100 individuals may be a carrier, some one with one normal copy of an albinism gene and one defective copy of that same gene.

Albinism is hereditary, which may not seem obvious if you look at the pigmented families into which most people with albinism are born. When we

* At www.rickguidotti.com, there is more information about Rick Guidotti's Positive Exposure campaign which is aimed at challenging stigmatization of genetically unusual individuals and celebrating the differences that are the result of genetic diversity.

look at how this trait is passed along in Figure 4.2, we see that Mendel's ideas are not unique to pea plants. The same rules apply here.

Thus there is a human trait, absence of pigmentation, that is recessive to the dominant trait, presence of pigmentation. If someone has a pigmented version of the albinism gene (the pigmented allele) along with the albinism allele, the albinism trait is not manifested and they have color in their skin and hair. If the individual is homozygous for albinism alleles, skin and hair color will be white. As with the pea plants, the heterozygous individual's coloring is determined by their dominant pigmented allele and is not a blended average of the two traits. As happens with the true-breeding yellow pod strain

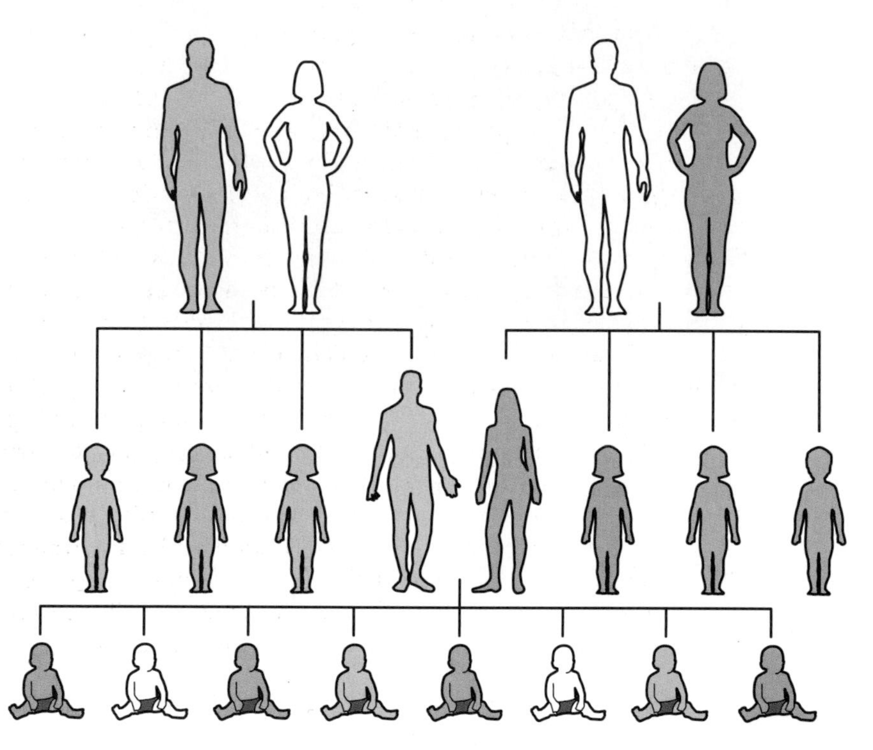

FIGURE 4.2 A family tree for a family with the albinism trait shows what happens when someone with the albinism trait marries someone pigmented. In the first generation, we see two different couples in which an individual with albinism marries someone who does not have the albinism trait. Notice that in the second generation, all of their children have pigment in hair and skin, but the marriage of two individuals who each have a parent with albinism can lead to about one quarter of their children having the albinism trait. The pigmented individuals in the second generation are considered carriers, individuals who lack the trait but carry the information, which can then be passed along to the next generation. In many families into which an child with albinism is born, there may be no known ancestors with albinism. In rare cases, the trait might be reported for a distant ancestor, but most often there will be no evident family history of albinism. Thus recessive information can pass through many generations before a carrier chances to marry another carrier to a produce a child with the trait. In this case, for the information to result in the trait, the child must receive an albinism allele from each parent so that they end up with two albinsim alleles. So what is wrong with this picture? The individuals with albinism in this picture are depicted as having skin as white as their hair, but the very pale skin of someone with albinism is not white like a sheet of paper.

of peas, if someone with albinism marries someone else with the same form of albinism, their children will also fail to make pigment.

There are individuals in the population with normal levels of skin and hair color who do not know that they carry an albinism allele. In fact, they won't ever know it unless they pick someone else with an albinism allele as their mate. Having one albinism allele does not dilute out the coloration brought about by the other skin color genes.

The term *carrier* is used for people who carry a recessive allele without showing any phenotypic differences to indicate that they have the recessive allele. Another way to think about it is that carriers are individuals who carry the information (genotype) without manifesting the trait (phenotype), just like the heterozygous green pea pods.

We talk about albinism as if this were some uniform condition, but in fact there are different forms of albinism, some of which do not look at all like our classic concept of someone with albinism. For instance, some individuals have what is called "yellow albinism," which may involve some coloration in both skin and hair. There is also a form of ocular albinism that only affects pigmentation in the eyes. Also, although syndromic forms of albinism can involve features other than coloration, simple albinism does not cause uniformity of anything outside of coloration. Thus individuals with simple albinism are as diverse as the rest of the human race in terms of intelligence, talents, temperament, agility, strength, and health status.

We talk about albinism as a recessive trait rather than talking about pigmentation as a dominant trait because, when we are talking about an unusual or rare characteristic, we are usually trying to identify what is going on with the mode of inheritance of the unusual phenotype within a family or population. We could actually say that pigmentation is dominant over albinism. The inheritance of skin pigmentation may look simple: if you have two copies of the albinism allele, you have the coloration of the albinism trait. However, if you have at least one copy of the normal dominant copy of the albinism gene, there are actually a lot of other genes that contribute to coloration and make inheritance of skin and hair color rather complex.

EPISTASIS

Homozygosity for the albinism allele results in presence or absence of pigment. If the pigment is present, it can cause pale beige skin with blond hair or it can cause dark brown skin and black hair. The genes that determine what the pigmentation will be, hair color genes and skin color genes, are different from the albinism gene that determines whether there will be pigmentation. Thus mutations in exactly the same albinism gene can cause a child with white skin and white hair be born into a family of pink-skinned, Scandinavian blonds, into a family of freckled, Irish redheads, or into a family of dark-skinned, black-haired Nigerians. If you have color in your skin and hair (pink, tan, dark brown skin; blond, red, brown, or black hair) you have the pigmented allele of the albinism gene, and your skin and hair color indicate things about which alleles you have at several other genes that determine coloration. If you have white skin and hair, you have two nonpigmented alleles of the albinism gene, and you cannot tell which alleles you carry at the genes that determine skin and hair color. However, if you were to look at your parents, siblings, and children, you might predict which alleles for those other color genes you are most likely carrying. For instance, an individual without pigmentation born into a family of African ancestors can look around him to tell that he would be more likely to pass his children genes for dark skin and black hair than genes for red hair and creamy skin, but he can only tell that by looking at his relatives and not by looking at himself. A gene that can block the manifestation of a trait caused by a different gene is said to have epistatic effects (Box 4.1).

Just to complicate this picture: there is more than one form of albinism, and the different forms of albinism are caused by different genes. Does this imply that Step 1 in Box 4.1 actually consists of more than one step that could be interrupted before you can get to Step 2, the determination of which color will be present?

In fact, there are actually at least three different genes that we know of so far that can cause oculocutaneous albinism. Think about what would

BOX 4.1 EPISTASIS

Sometimes something can have an overriding effect that keeps you from being able to detect or distinguish other characteristics. For instance, some flashing lights at intersections are red, and some are yellow. If the power for that section of the city goes out, the light does not turn on and you cannot tell what color it would be. Thus something that can block the light from obtaining electricity has an epistatic effect that blocks the manifestation of the different colors that the light could be. However, factors that affect the color of the light are separate from the factors that determine whether there is light at all. In the case of albinism, an albinism allele blocks manifestation of a completely separate set of coloration traits, so the albinism mutation is considered epistatic. Each organism has some genes that can have epistatic effects that mask our ability to tell what other genes would be doing.

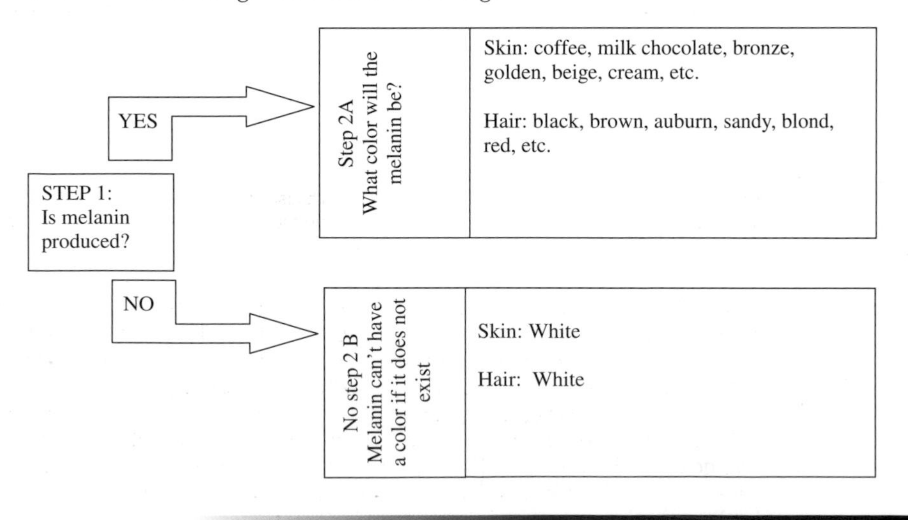

happen if a man with two defective copies of the first gene OCA1 married a woman with two defective copies of the second gene OCA2 (Figure 4.3, A). He would provide the children with a good copy of OCA2, and she would provide the children with a good copy of OCA1. The result is that the children would be carriers for both genes but would not have the albinism trait! If the man and the woman both have defects in the same gene, OCA1, they will only be able to pass along defective copies of OCA1 and all of their children will have the albinism trait (Figure 4.3, B). If two individuals with albinism have children who do not have the albinism trait, the parents have defects in different genes and the genes are said to be *complementary* (able to cover for each other) but not *allelic* (present in the same gene). Most individuals with albinism have two different mutant alleles of the gene that is causing their albinism. So the great amount of variability in pigmentation and visual acuity problems in albinism result from a combination of the following:

- There are multiple different albinism genes.
- There are many different mutations in those genes.
- Different combinations of those mutations result in differences in levels of pigmentation in different individuals.

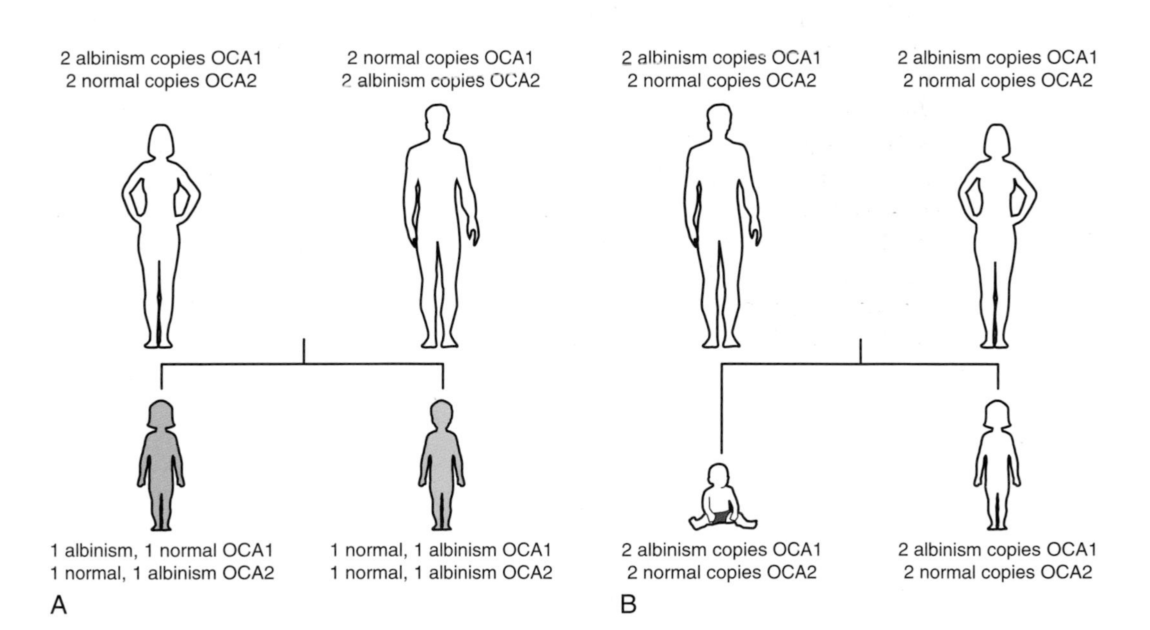

FIGURE 4.3 **A,** Affected parents with unaffected children. When mutations in different genes are responsible for the trait in each parent, the normal copy from one parent can complement the defective copy contributed by the other parent. Thus, even though both parents have the same trait, because the responsible genes are not allelic, the children end up heterozygous at each of the two genes. **B,** Affected parents with affected children. When both parents have the same trait due to mutations in the same recessive gene, their children will all inherit only defective copies of that gene, so they will have the trait. In this case, we would say that the gene causing albinism in the father is allelic to the gene causing albinism in the mother.

All of us have many recessive characteristics that never come to light because the presence of the dominant normal alleles mask the recessive alleles, and the recessive alleles only come to light if we produce children with two copies of the recessive allele. Albinism is an obvious example of recessive characteristics that can result from getting two recessive alleles together. Some recessive characteristics may cause miscarriages at the earliest stages of life or diseases that are fatal very late in life. Also, we will never know about most of the hidden recessive alleles we carry unless we are lucky enough or unfortunate enough (depending on the trait) to produce offspring with someone else who has the trait or is a carrier for the recessive allele that can cause the trait.

As with Mendel's peas, a pair of carrier parents can produce offspring with the recessive phenotype. So what happens in people? The pigmented father with genotype Aa and the pigmented mother with genotype Aa, can produce four different genotype combinations, with a child having an equal chance of getting any of the four combinations:

Mom's A with dad's A gives the AA genotype and a pigmented phenotype.

Mom's A with dad's a gives the Aa genotype and a pigmented phenotype.

Mom's a with dad's A gives the aA genotype and a pigmented phenotype.

Mom's a with dad's a gives the aa genotype and an albinism phenotype.

If we use a device called a Punnett square, we can separately identify what genotypes will be present in the gametes produced by the parents. We can then look at the different combinations of genotypes that can be produced by different combinations of sperm and egg. When we do this for two individuals who are carriers for mutations in the same albinism gene, we similarly find a prediction that about one fourth of their children would have the albinism trait (Figure 4.4).

Thus, for rare recessive traits, it is quite common to have a child with the trait born into a family in which no one else has the trait. If the child has a lot of brothers and sisters, another child might also have the trait, in which case the involvement of genetics might be easier to figure out. When children with albinism grow up and have children, they usually have no children with the trait except in the very rare cases in which they marry a carrier. Thus rare recessive traits often pop up unexpectedly in families with no history of the trait, or even families who have never heard of the trait in question, because the gene defect has been passed from one generation to the next without anyone being aware of it. The evidence is sometimes there if the family goes searching far enough back in their family history (Figure 4.5), but there is often no family history, especially for something rare. Such traits may look like they are not hereditary when in fact the genetic information is getting passed along because there often is no way to tell if someone is a carrier for a genetic defect by just looking at them.

If your newborn child has a trait that you have never seen in your family or maybe even in your community, how would you be able to tell that it is an inherited trait that is not the result of an infection, diet, or something else unknown that happened during the pregnancy? If the trait in question were

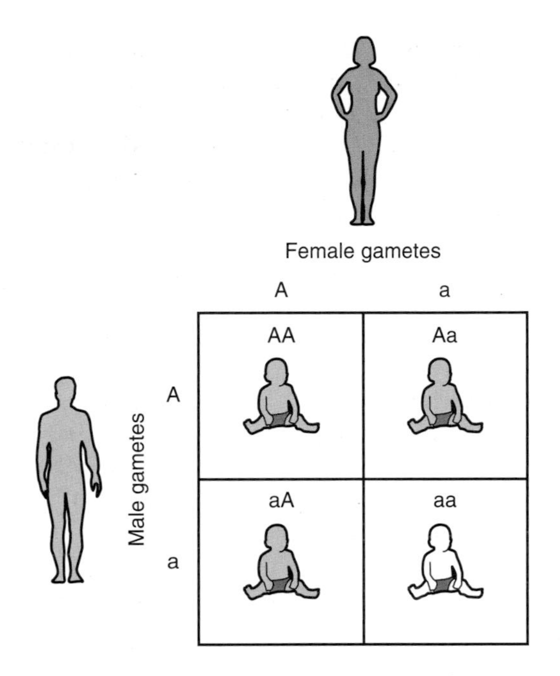

FIGURE 4.4 Punnet square to show inheritance of albinism in one family. Along the side, list the genotype of each type of gamete that one of the parents can make. Along the top, list the genotype of each type of gamete that the other parent can make. In each square, combine the alleles from the side with the alleles from the top to find out the genotype and phenotype of the offspring that would result from combining those gametes. When a man with normal skin coloration and a woman with normal skin coloration who each have one normal allele and one albinism allele have children, the Punnet square can be used to predict the chance that they will have a child with the albinism trait. Notice that we are not specifying what color of skin and hair the parents or the normal children have, since their coloration could range from very pale to very dark and have no effect on the outcome here. The Punnett square might seem like a lot of trouble for tracking the possible combinations of two alleles from each parent, but this system can be used to track more complex combinations of gametes representing different alleles at more than one gene, in which case this exercise can take a problem that is very complex and turn it into something simple and easy to visualize.

albinism, and you did not know of anyone with the kind of white hair and skin that your child has, how would you figure out whether they are inherited? If 1 in 17,000 individuals have this trait, you might look all around your community and not see a single case of someone else like your child. One clue can come from examining your own family history to see whether you have other relatives with the same trait, but there will often be no sign of the trait in earlier generations.

For many inherited traits, your doctor will be able to tell you whether it runs in families even if you can't find any relatives with the trait. Researchers working in medical genetics have learned a lot about many thousands of hereditary traits, often through studies of families or in some cases through studying populations of twins (Box 4.3). So when someone arrives in a doctor's office with no family history of the trait causing the problem, the doctor is not limited to the information about the patient or the patient's

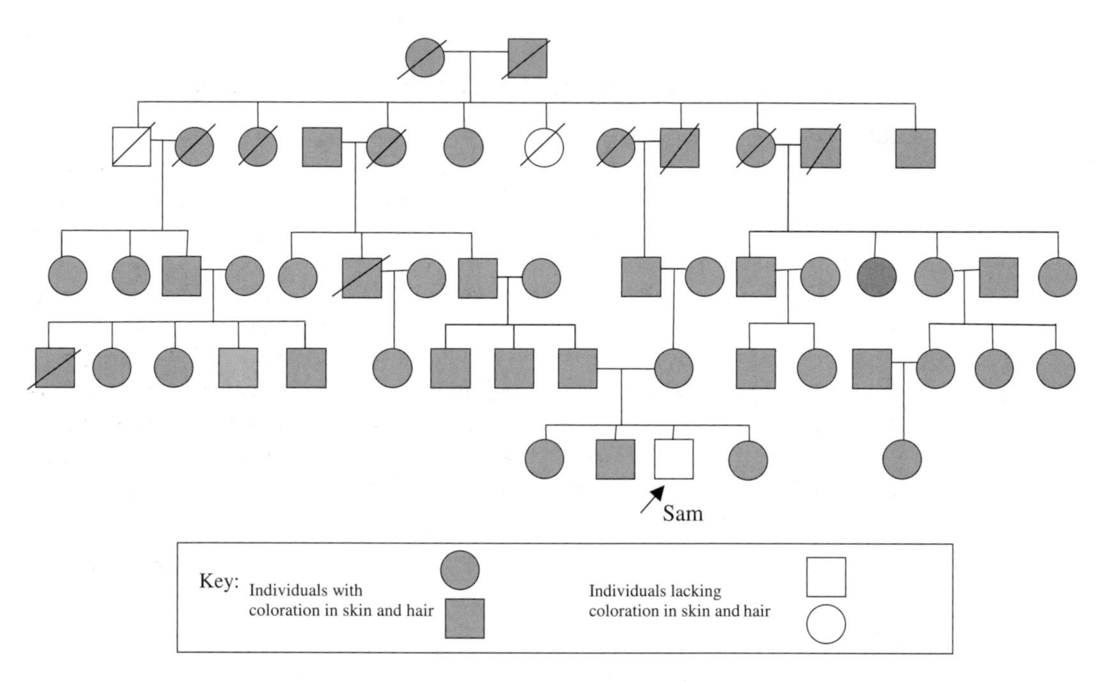

FIGURE 4.5 Hypothetical albinism pedigree shows what family history might look like for a family of five generations in which the youngest sibship includes a child with albinism. Compare with Figure 4.2 in which pictures took up a lot of space to appreciate this more efficient notation system that makes it possible to put a lot of information into a small space. Standard sets of pedigree symbols include squares for males, circles for females, and diagonal lines for deceased individuals. Pedigrees most often use filled symbols for people with the unusual trait or disease and open symbols for people who are healthy, normal, or have the common form of the trait, but in this case white was used for individuals with the trait because of its similarity to the characteristics of those individuals. The arrow points at the person called the proband (Box 4-2), the first person in the family who was identified. If both of the proband's grandfathers migrated from Europe to the United States, would you expect that the proband would have information about those earlier generations thousands of miles away? As you think about the answer to this, consider how much you do or don't know about the brothers and sisters of your great great grandparents. From this pedigree alone, you can tell the genotype of the couple at the top, individuals 1 and 2, who must have each been carriers for defects in the same albinism gene. Do you know whether you have any relatives with the albinism trait?

family but can draw on information gained from the study of many other rare individuals and families.

The study of rare families is not the only source of information that has helped doctors understand the hereditary factors contributing to many human diseases. Because we don't breed people like pea plants to answer questions about the mode of inheritance (fortunately!), it sometimes takes conceptual breakthroughs from experimental work with peas and flies and other nonhuman organisms to arrive at an understanding of recessive inheritance in humans. Mendel's work may make it easier to understand how something can be considered hereditary when it has not been seen over the course of many generations of a family. Similarly, mice with cancer, yeast with enzymatic defects, and infertile flies can all tell us things that translate quite

BOX 4.2 PROBAND: FIRST POINT OF CONTACT WITH A FAMILY

It turns out that it is actually important to keep track of who the first person is by which a family is identified in a genetic study. The term *proband* is one of several terms used to indicate the first member of a family who has contact with the doctor or researchers in a study. Sometimes the proband is someone who has the inherited characteristic that is being studied, as was the case for Sam in Figure 4.5. However, sometimes the proband may be a relative who first brought a family to the attention of researchers when they went to the doctor to say, "Other members of my family have this inherited disease, and I am worried about whether I might end up getting it, too." Sometimes, someone may show up in the doctor's office because they are worried about whether they can give their children a genetic disease present in their relatives but not themselves. Look for the small diagonal arrow to find the proband in families shown in this book.

BOX 4.3 IDENTICAL TWINS AND FRATERNAL TWINS

Why do researchers compare identical twins (formed when one sperm fertilizes one egg and then the embryo splits to form two embryos) to fraternal twins (formed when two different sperm fertilize two different eggs) instead of comparing identical twins to brothers and sisters born at separate times? Because whenever the siblings are twins, they have something in common besides their genetic makeup—they have also shared the same environment in the uterus for the whole nine months. If siblings were conceived and born at different times, the differences between them might include not only their genetic differences, but also differences in their pregnant mother's nutrition, exposure to smoke, encounter with physical trauma, or consumption of medications. Twin studies that compare identical to fraternal twins are often used to try to determine how much of a characteristic can be attributed to genetics and how much can be attributed to environmental causes.

directly into a better understanding of problems faced by human beings. We hope that the things we talk about in the rest of the book will clarify just what happens at the molecular and cellular levels to bring about the patterns of inheritance that we see in the characteristics of the people in a family. And we hope that an understanding of those underlying mechanisms will leave you thinking, "Of course it happens that way. It all makes sense." Arriving at a point at which it makes sense can be especially important for recessive traits, which often leave individuals and their families perplexed at something that seems to have just popped up out of nowhere.

ONE MAN'S TRAIT IS ANOTHER MAN'S DISEASE

5

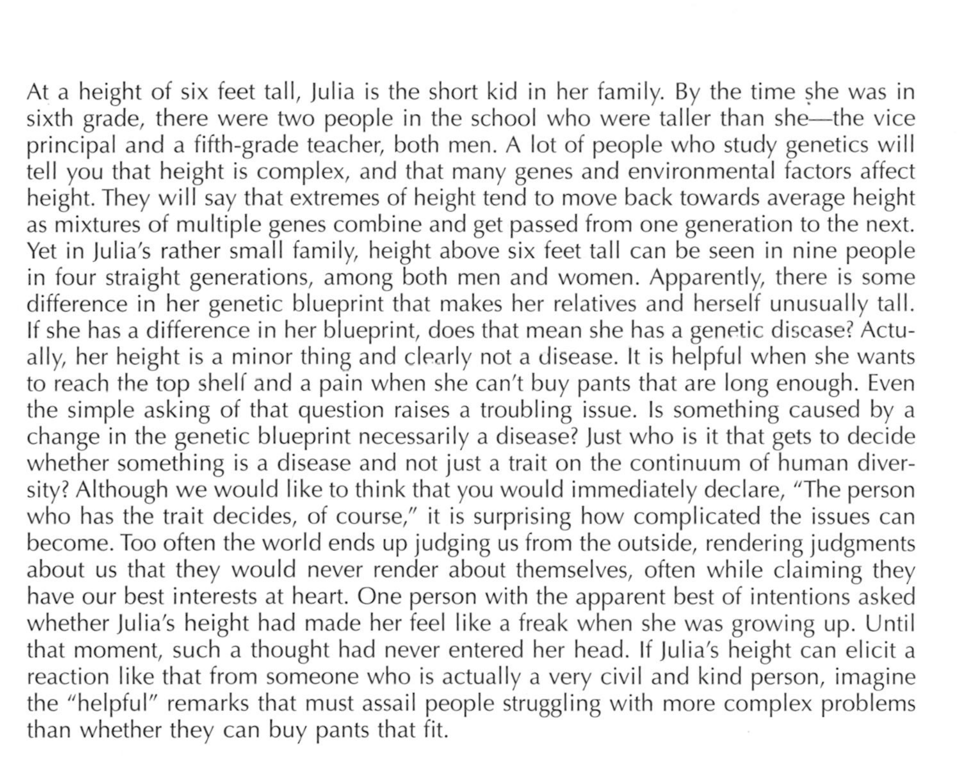

At a height of six feet tall, Julia is the short kid in her family. By the time she was in sixth grade, there were two people in the school who were taller than she—the vice principal and a fifth-grade teacher, both men. A lot of people who study genetics will tell you that height is complex, and that many genes and environmental factors affect height. They will say that extremes of height tend to move back towards average height as mixtures of multiple genes combine and get passed from one generation to the next. Yet in Julia's rather small family, height above six feet tall can be seen in nine people in four straight generations, among both men and women. Apparently, there is some difference in her genetic blueprint that makes her relatives and herself unusually tall. If she has a difference in her blueprint, does that mean she has a genetic disease? Actually, her height is a minor thing and clearly not a disease. It is helpful when she wants to reach the top shelf and a pain when she can't buy pants that are long enough. Even the simple asking of that question raises a troubling issue. Is something caused by a change in the genetic blueprint necessarily a disease? Just who is it that gets to decide whether something is a disease and not just a trait on the continuum of human diversity? Although we would like to think that you would immediately declare, "The person who has the trait decides, of course," it is surprising how complicated the issues can become. Too often the world ends up judging us from the outside, rendering judgments about us that they would never render about themselves, often while claiming they have our best interests at heart. One person with the apparent best of intentions asked whether Julia's height had made her feel like a freak when she was growing up. Until that moment, such a thought had never entered her head. If Julia's height can elicit a reaction like that from someone who is actually a very civil and kind person, imagine the "helpful" remarks that must assail people struggling with more complex problems than whether they can buy pants that fit.

For some traits in humans, it is easy to see that the trait is inherited. In the previous chapter, we compared the albinism trait to the yellow seed pods in Mendel's experiments and talked about albinism as a recessive trait relative to the genotype found in most people. What happens if someone has a trait that is dominant relative to what is present in most of the population? For example, what if most pea plants had yellow pods (the recessive form) and a dominant green-pod pea plant were introduced into the population? A recessive trait can effectively disappear for many generations in a population of individuals with the dominant characteristic. A dominant trait will be evident in generation after generation, wherever one copy of the dominant allele is present. The dominant trait will be very easy to detect and trace through later generations if that dominant allele exists in a population in which most

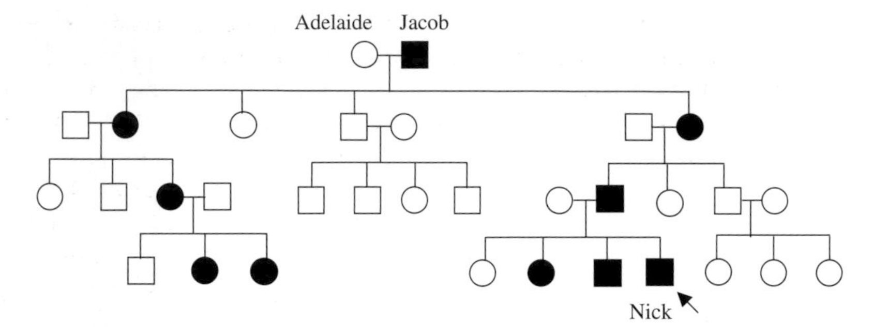

FIGURE 5.1 Autosomal dominant transmission of deafness in a hypothetical four-generation family would show that each person with the trait passes the trait along to about half of their children, both sexes can have the trait, and both sexes can pass the trait along. Individuals who do not have the trait do not pass deafness along to their descendants. Stop and consider this: every time someone deaf in this family has a child, there is a 50% chance that their child will be deaf. Does that mean that a deaf individual will have exactly 50% of their children turn out to be deaf? No. On average, about 50% of those who are at risk (have a deaf parent in this family) will end up deaf. This is like flipping coins. If you flip a coin many times, you will find that about half of the coin flips come out heads, but they will not exactly alternate heads, tails, heads, tails, etc. In fact, in some of the families we work with, we sometimes see a case in which someone with a trait will have as many as five children in a row who all have the trait, but when we look at many families, we find that they are balanced out by some other family who had five at-risk children who did not end up with the trait. In between, most of the families we work with have about 50% of the at-risk children turn up affected if it is a dominant trait. Before Nick was born, his parents might have thought that the next child would probably not be deaf since they already had two deaf children. Was this valid? No. *This is a key concept: with each new genetic flip of the coin, the chance is once again 50%, and that chance is not affected by whether some, none, or all of the previous children have the trait.*

individuals have the recessive allele and the accompanying recessive trait. Thus a recessive trait may suddenly pop up in a family that has never heard of that trait before, but a dominant trait is usually evident across a series of successive generations of a family.

Consider what a family would be like with deafness in all four generations for which information is available (Figure 5.1). Individuals who inherit the family's hearing loss trait are deaf from birth. If you look at descendants of Jacob, who is deaf, and his wife Adelaide, who is not, you find that about half of their children are deaf and half are not. Like the heterozygous pea plant with both green and yellow alleles, a deaf member of this family has one "deaf" allele and one "hearing" allele and has an equal chance of passing either along to their child. The difference is that individuals with one "deaf" copy and one "hearing" copy of the gene have the affected phenotype, so they do not have to marry someone else who is a carrier to have a deaf child. Thus, even though the trait is dominant rather than recessive, the transmission of information between generations is actually consistent with Mendel's model that half the progeny get one allele and half the progeny get the other allele. There are no carriers in this family because anyone who gets the information manifests the trait. Thus we also see consistency with one of Mendel's other predications: that some traits are dominant over others.

In fact, there are dozens of different genes that can cause hearing loss that is transmitted like this, in a dominant manner. However, there are dozens more that can cause recessive hearing loss. In theory, it is easy to make the kinds of predications we make here, that the mode of inheritance is dominant and that the risk to children of affected individuals is 50%. However, in real-life situations, things are often more complicated. Small families, adoption, divorce, early death of some family members, geographic distances between family branches, and other complications can sometimes limit the information that is available to help sort out the level of risk to a new child.

COMPLEX SYNDROMES

Deafness is a simple trait affecting one main characteristic, but many things that run in families are complex and are often complex enough that family members may not have even noticed that several separate traits are turning up together in several different family members. It might seem obvious that a genetic disease is an illness that can be inherited. However, there are some things that are genetic for which no one calls a doctor because there does not seem to be any medical problem. In some cases, a family may not have figured out that an obvious trait with strictly cosmetic implications may actually be an indicator of more serious problems. In especially complex situations involving a variety of seemingly unrelated traits affecting multiple different organ systems or cell types, we may end up using the term *syndrome*.

Nail-patella syndrome, which affects multiple different parts of the body including the thumbnails and the kneecaps, is caused by a defect in one copy of one gene. How could a defect in just one gene affect apparently unrelated things? In the case of the nail-patella syndrome gene, we know that the protein it makes carries out actions that affect a lot of other genes, so it is actually not surprising if the phenotype is complex. When one gene affects so many different phenotypic features we say that the gene has *pleiotropic* effects.

Consider the family in Figure 5.2, in which some members of the family are missing all or part of their thumbnails. As you can see, the proband Mary

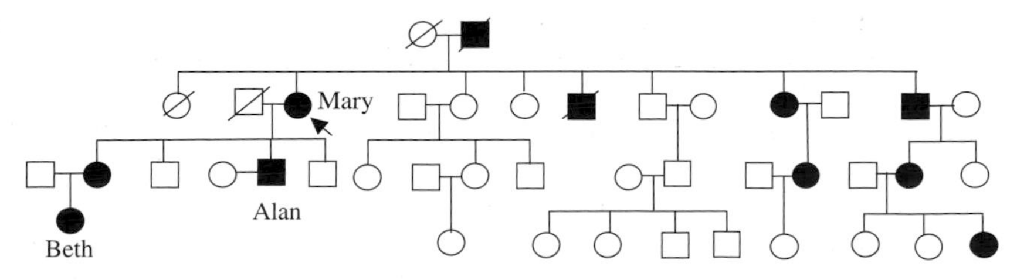

FIGURE 5.2 Transmission of the "no-nail" phenotype through four generations in a nail-patella syndrome family. As you can see, inheritance is autosomal dominant with a 50% risk of passing along an affected allele to the next generation. Both sexes are affected and both sexes can pass along the trait. Unaffected individuals do not appear to pass nail-patella syndrome along to their children.

who told us about this family knows of four successive generations in which the "no-nail" phenotype has appeared, including her father's generation in addition to her children and grandchildren.

For many generations, members of this family did not realize that what they referred to as the "family thumbs" (thumbs on which all or part of the thumbnail was missing) were just one sign of a phenomenon complex enough to be considered a syndrome. The formal name for the syndrome is nail-patella syndrome, and not everyone with this syndrome has the same symptoms. However, the thumbnails are a consistent and obvious feature that every affected person has, and family members knew that any time someone with the "family thumbs" became a parent, there was a moment of anticipation when they waited to see whether the new member of the family had thumbnails or not.

Nail-patella syndrome receives its name from the two most consistent and obvious characteristics found in affected individuals: unusual or missing thumbnails and unusual or missing kneecaps (patellae). Another consistent feature, the presence of hornlike extensions on the pelvic bones, is often hard to identify without a doctor's exam and perhaps an x-ray.

There is more to nail-patella syndrome than the nails and kneecaps. Problems with joints such as the elbows may cause restricted movement, in some cases severely restricted movement, and may sometimes call for surgery. In some cases, more severe orthopedic problems such as clubfoot may be found. Some individuals go on to develop kidney problems, and some develop an eye disease called glaucoma. Some individuals apparently don't develop problems with their kidneys or their eyes! The two consistent features, the thumbnails and the kneecaps, make it easy to study the mode of inheritance because they are features that can be identified without a doctor's visit and they are features that are obvious even in very young family members.

In fact, Mary watched for whether her children had the "family thumbs" as a point of curiosity but no concern. Her relaxed attitude about the situation changed when one of her sons, Alan, turned out to have some of the more severe medical problems that can happen to someone with nail-patella syndrome. She has watched him struggle through multiple surgeries while growing up. She has watched him decide that he will never have children because he would not be willing to pass his problems along to the next generation. Nail-patella syndrome has quit being a mere curiosity in Mary's life.

What if Mary's granddaughter Beth decides to have children? What are her chances of having a child with normal thumbnails and kneecaps? Beth has partially missing thumbnails and no kneecaps. Beth's fiancé Geoff has normal thumbnails, normal kneecaps, and no family history of nail-patella syndrome. So we know what Beth's and Geoff's phenotypes are, but what are their genotypes? Just as with the pea plants, we can use simple letter symbols to track the genotypes of different individuals. In this case we will use N to designate the "no-nail" phenotype and n to designate the common (normal) phenotype. Thus Beth has genotype Nn (n from her normal father and N from her affected mother). Geoff, with a normal phenotype and no family history of the disease, would have genotype nn. What do we expect to see among the children of Beth and Geoff? The possible combinations of alleles are:

Beth's allele n plus Geoff's allele n gives a child with nn genotype and normal phenotype.

Beth's allele N plus Geoff's allele n gives a child with Nn genotype and a "no-nail" phenotype.

Because Beth has genotype Nn, the n allele will be passed along to a child about half the time and result in a child with normal thumbnails, and the N allele will be passed along about half the time and result in a child with partially or completely missing thumbnails. Geoff, with two copies of allele n, will only pass along allele n. The child has an approximately equal chance of getting either of the two genotype combinations, Nn or nn, so the child has about a 50% chance of ending with her mom's "no-nail" phenotype. We have not yet heard of a child who has genotype NN, but that doesn't mean there aren't any.

ONE MAN'S TRAIT IS ANOTHER MAN'S DISEASE

So what have we been talking about so far, genetic diseases or genetic traits? Clearly, there appears to be a change in the genetic blueprint involved in all three traits we have discussed so far—albinism, dominant deafness, and nail-patella syndrome. This comes back to the question we raised about Julia's height: does it have to be a genetic disease just because it is caused by a change in the genetic blueprint? Frankly, the answer is no.

That answer throws us onto shaky ground as we face the question of what does constitute a genetic disease. If finding out that something is caused by a change in the genetic blueprint doesn't tell us that it is a genetic disease, how can we tell?

Here we come to an incredibly important concept—the distinction between a trait and a disease. The term *genetic disease* may be applied if the trait results in medical problems. If the effects are simply cosmetic, we may end up referring to it as a genetic trait instead. However, we find that this actually leaves us with many traits that occupy some middle ground, perceived differently by different people.

Whether nail-patella syndrome is a trait or a disease varies from one person to the next. For some people, the only effects are cosmetic; for others, the effects can be crippling or even lethal. We tend to consider nail-patella syndrome a disease because the potential for the medical complications is there for all of the affected individuals, and in some cases we won't know until late in life whether someone with an apparently cosmetic case of nail-patella syndrome has really missed the serious medical consequences that could have arisen. Really, it is a trait, a trait that can sometime be serious enough to cause illness.

In the case of albinism, coloration seems like something cosmetic that should be considered a trait. However, vision problems are always part of albinism, and a true lack of pigment actually has medical implications in terms of susceptibility to damage from sunlight that is serious enough to make it a genetic disease. If you have a form of albinism that isn't one of the rare forms of syndromic albinism that causes other major medical problems, if you manage your vision problems and if you don't develop skin cancer, do you have a disease?

One parent of a child with albinism recently offered the vehement argument that albinism is not a disease. In many ways, this argument is valid. Even though there are some inconveniences associated with the vision problems that can be part of albinism, these are people who live normal, healthy lives, and many of them might be surprised that anyone might think that they are ill. If people with the albinism trait do not consider themselves ill, the rest of us should accept this self-insight and notice how much like the rest of us they are.

In the case of deafness, which you might think would be classified as a disease because of the functional repercussions, there is an alternative perspective (Box 5.1). Some people consider deafness a genetic disease, but

BOX 5.1 DEAFNESS—AN ILLNESS, A TRAIT, OR AN ETHNIC GROUP?

According to the American Association of Pediatrics, 1 in every 300 infants is born with hearing loss. There are more than 50 distinct genetic causes of deafness known so far, and hearing loss can result from a variety of nongenetic causes, including as one of the consequences of some types of infectious disease. Even as doctors work to restore hearing and researchers work to develop new technologies for those doctors to use, there are those who don't think the need for those efforts is so obvious. The deaf community has a large and thriving culture that includes its own separate language and is quite distinct from the culture of the hearing society in which this culture is embedded. Mannerisms, patterns of communication and interaction, even art forms have all emerged in unique ways that make them not just copies of the cultural patterns in the hearing world. When a deaf child is born, there can be very different reactions depending on whether the child is born into a deaf family or a hearing family. Some within the deaf community want their children to have the choice of whether to hear or not, and hearing parents usually wish there were a way to give their children the gift of sound. For all of them, the continued development of new technologies and the availability of medical assistance are incredibly important. There are some within the deaf culture who regard medical efforts to eliminate deafness as a threat of cultural genocide—an effort to eliminate an entire separate culture and people by forcing their assimilation into a different mainstream culture for which they hold great distaste. Many others in the deaf community hold much more mild views in this era in which moderate help is available to those who seek it and not required for those who don't. With technological aids available, such as cochlear implants, it is interesting to find that some who have received these implants and gained the ability to carry on aural communication with those of us who don't speak sign language have then decided to return to the world of silence they were born into. Advances in the quality of the technological results have others who resisted cochlear implants in the past taking another look at them. If you ever find yourself saying, "Of course they should all just have their hearing restored," first spend some time reading about deaf culture and exploring the idea that in silence they may have found some things that the hearing world lacks. You may or may not end up agreeing with them, but many who hear what they have to say come away changed by it.

others consider it simply a trait. In fact, the news that a newborn child is deaf may be greeted with anguish by some families and calmness by others. The thing that determines whether the news is distressing is often the "hearing" status of the parents.

Some families believe that their deaf child has a disease and they want someone to come forward with a cure. Other families think that their child has a trait, and they have no interest in altering that trait. Individuals who wish they could hear view advances such as the cochlear implant as a gift that can restore a missing sense. However, there are those within the deaf community who see the cochlear implant as a tool for carrying out cultural genocide, a technology that causes a deaf child to grow up as a marginalized individual on the fringes of a cruel "hearing" society instead of growing up safe and esteemed as a full peer within the deaf community. What a complex ethical problem to weigh and measure the gain or loss of hearing against the gain or loss of esteem and acceptance. We find ourselves wondering if there is such a thing as a right answer.

This complex set of ethical issues mirrors so many situations that we encounter in human genetics, where the answers are often terribly complex but often become at least a bit simpler if we fall back on the principle of self-determination. Julia can't tell someone deaf whether their deafness is a trait or a disease, and that deaf person can't tell Julia whether her height is a trait or a disease. Each of us knows which call constitutes the truth for our own situations, and there are others who would make a different judgment call.

In this chapter, we have deliberately selected traits that are usually not seen as diseases by the people who have the trait but that sometimes are labeled as diseases by others or by rare individuals whose cases are especially severe. We have done this expressly for the purpose of making the point that one man's trait may be another man's disease. Is the trait actually causing a problem? Sometimes yes. Often not. The final judge has to be the person with the trait, not those looking in from the outside. It becomes a terribly important point to make, since there can be many consequences to an individual who is perfectly healthy if people around them start telling them that they have something wrong with them.

Decisions about whether something is broken or needs fixing must rest in the hands of the person with the trait or disease. Some of the gravest ethical errors in genetics in the past have been made when society or medicine removed the rights of individuals to judge this for themselves. Is it broken, and should we fix it? If we look beyond the tricky issue of whether or not we *can* fix it, we find that the real answer to whether we should even try to fix it lies in the heart of the individual with the trait. One man's trait is another man's disease. Only the individual with the trait can judge for himself or herself whether the trait is severe enough to be considered a problem, and different individuals with the same trait may arrive at very different perspectives on the question.

THE CENTRAL DOGMA OF MOLECULAR BIOLOGY

This section is about the genetic information stored inside of our cells and the processes by which the cell uses that information to create the unique characteristics of the cell by making the functional components of the cell.

DNA: THE GENETIC ALPHABET

6

Light is the left hand of darkness
And darkness is the right hand of light.
—Ursula K. LeGuin[1]

When children play word games involving opposites, knowing the first word is all the clue they need to find the second word—dark/light, up/down, cold/hot, far/near. When you walk along a railroad track looking at the left-hand rail, you don't have to see or touch the other rail to automatically know its shape—the tightness of the curve, or the pitch of the slope. Think about how one side of a yin/yang symbol automatically defines the other side.

What an interesting concept, that if you know part of something, you automatically know the rest of it! The unknown part of the information that you can automatically fill in is not identical to the first piece of information, but it is complementary to it. Clearly this does not apply to many complex situations in life, but in the case of DNA, this is the key to how genetic information is disbursed into each new cell that grows in the human body as it develops from a fertilized egg. As the genetic information in DNA is copied, it is done using this principle—that knowing half of the information automatically tells you the other half. So stick with us as we start back in molecular kindergarten with the letters of the genetic alphabet. We are going to build you a picture of what DNA is like and tell you why this principle of complementary information is one of the keys to understanding heredity.

THE GENETIC ALPHABET

What form does the genetic information take? We want you to start out with this picture in your mind: the information in the human genome is written out in a linear chain of chemical building blocks or genetic "letters" in much the same way this book is written as a linear chain of printed symbols. These chains of chemical building blocks are located within the cell inside of a cellular organelle called the nucleus.

First, let's talk about the actual chemical building blocks. The genetic letters are commonly referred to as A, C, G, or T, the first letters of their chemical names (Figure 6.1). These building blocks form a long continuous molecule we know as deoxyribonucleic acid (DNA). A, C, G, and T each have a different chemical structure that can be "read" and recognized by the machin-

[1] From *The Left Hand of Darkness* by Ursula K. LeGuin, Ace Books, 1969.

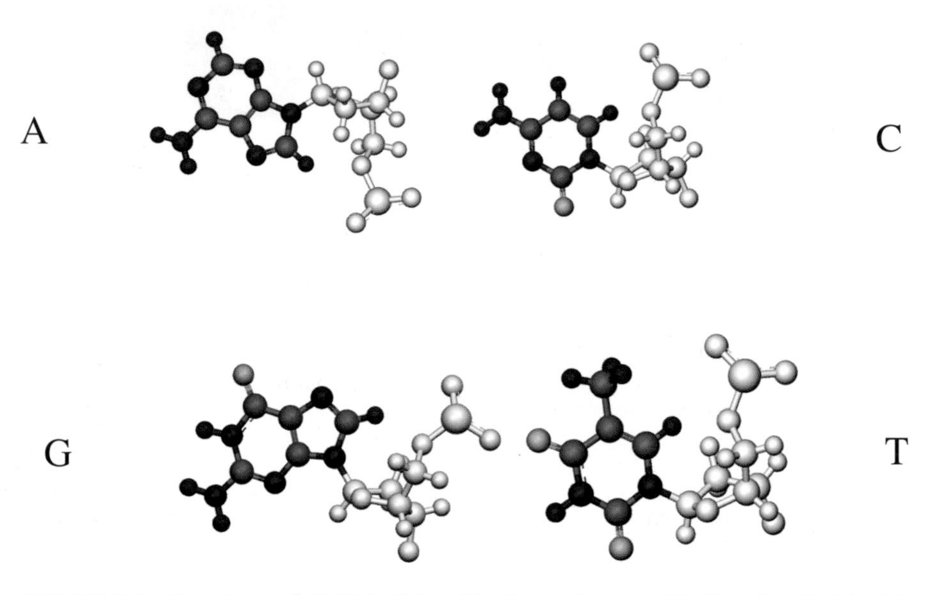

FIGURE 6.1 Structures of DNA building blocks A, C, G, and T. The chemical building blocks that act as genetic letters are called *bases*. The bases are adenine (A), thymine (T), cytosine (C), and guanine (G). Although we show you a cartoon of the chemical structures, you do not need to learn or remember these chemical structures to be able to talk about the important aspects of these building blocks, such as the fact that A, C, G, and T are distinct in ways that the cell machinery can tell apart, and that they function as building blocks that spell out the genetic information. The red spheres are the hydrogens, the blue spheres are the nitrogens, and the green spheres are the carbons that make up the genetic letters A, C, G, and T. The grey spheres are the atoms of the backbone that hook the DNA bases to each other. The backbone portion is the same for all four genetic letters but may look different here because they are being seen from different angles. (Courtesy of D. M. Reed.)

ery of the cell with an amazingly high level of accuracy. Although the cellular machinery reads these letters at the level of the hydrogens and nitrogens and carbons and sulfurs of which the letters are formed (as we display them in Figure 6.1), we actually need concern ourselves with very little about the real molecular structures of these letters to be able to understand the key features of the genetic alphabet. Instead, we can mostly consider these genetic letters as if they are simple building blocks, as in Figure 6.2.

The genetic letters are not printed on a page, so if the genetic letters are arranged in a linear order, what keeps them in that order? The bases A, C, G, and T are not hooked directly to each other. Instead, a long backbone structure runs the length of the DNA molecule, and each base is connected to the backbone. Thus the bases form a DNA chain in the same way that the crossties of a railroad track make a long row of ties, not by connecting directly to each other but rather all by connecting to the same rails that run along the edge. So each base is connected to a segment of the backbone, and the backbone pieces connect to each other, which brings the bases next to each other in a line without the actual bases touching each other. As you can see in Figure 6.2, this is an easy idea to diagram without using the chemical structures.

DNA is often referred to as double-stranded. The DNA inside of the nucleus of the cell is not normally a single strand of DNA such as that shown in Figure 6.2. Rather, it is consists of a pair of DNA strands, with two back-

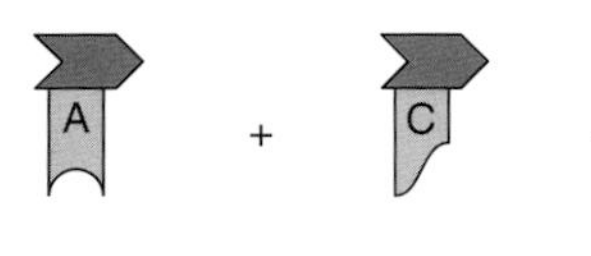

FIGURE 6.2 A backbone hooks the A, C, G, and T bases together into a long, continuous chain. Notice that the segments that make up the backbone are asymmetrical, something that causes the DNA chain to grow in only one direction as each new base gets added to the growing chain. That is, there is effectively a beginning and an end to the chain, and bases can only get added to the end but not to the beginning of the chain as new DNA gets made. This asymmetry or directionality to the DNA chain also gets referred to as the *polarity* of the chain. The backbones are made up of the gray parts of the molecules in Figure 6.1.

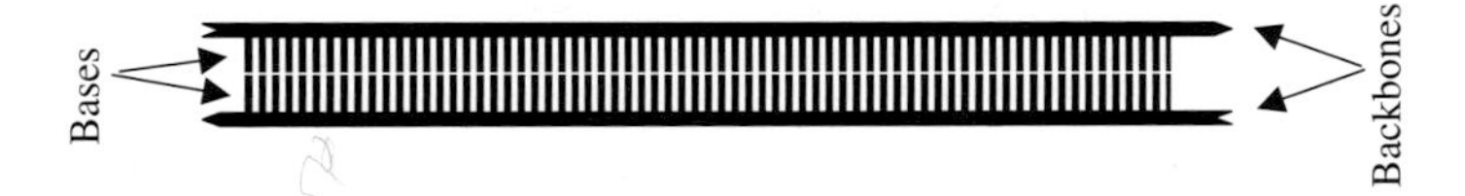

FIGURE 6.3 In an even more simplified format, we show that the double-stranded structure of the DNA has backbones that connect bases to each other. Bases meet in the middle to connect the two strands to each other. Because the backbones are an equal distance apart, you might think that they would be considered parallel to each other; however, because the backbones are directional, the backbones are considered antiparallel instead of parallel, since they run opposite directions.

bones paralleling each other like railroad tracks and the bases pointing inwards to contact the bases on the opposite strand to connect the backbones rather like ties connect parallel railroad tracks.

The completed structure of the DNA is referred to as the *double helix*. It is called double because of the property we already described: it consists of two strands arranged so that paired bases meet in the middle and backbones run up the outside edges, as in Figure 6.3. The double-stranded DNA does not lay out flat like a railroad track or the picture shown in Figure 6.3. To see what the real structure is like, imagine picking up the railroad track structure from Figure 6.3 and twisting it so that it resembles a spiral staircase in which the antiparallel handrails point in opposite directions (Figure 6.4).

REPLICATION OF DNA

The key to how DNA can be copied is the pairing of the bases that meet at the center of the double helix. The rule is simple to remember: A always pairs with T, and C always pairs with G (Figure 6.5), with a set of two bonds holding A and T together and a set of three bonds holding C and G together.

Because of the base-pairing rules, if you know the sequence or order of A's, C's, G's, and T's along one of the strands, you automatically know what

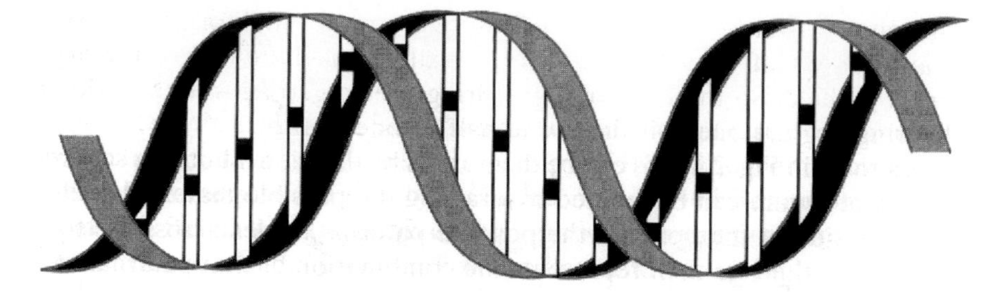

FIGURE 6.4 An artist's conception of the double helix, showing the pairing of the bases in the middle like steps in a staircase and the way in which the backbone structure spirals around the outside like the handrails of a spiral staircase.

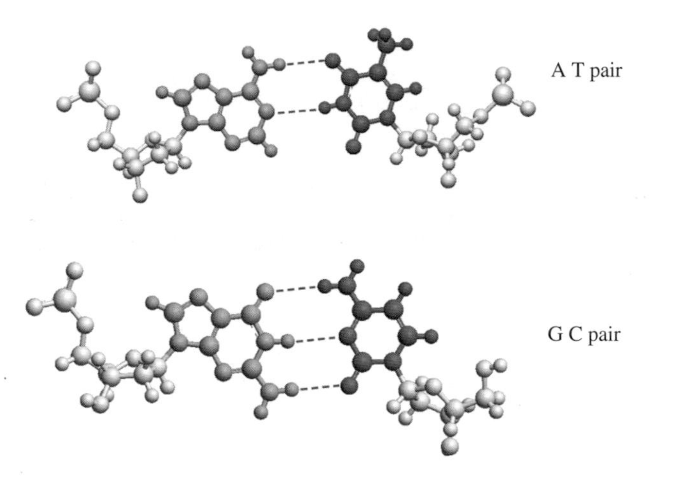

FIGURE 6.5 If we look at the molecular images of base pairs, we can see where the two bonds hold A to T, and three bonds hold C to G. For most purposes, we do not need to know where all of the different hydrogens and carbons and nitrogens and oxygens are located to be able to work with the basic concepts of AT and CG base pairing. (Image courtesy of D. M. Reed.)

A = T   T = A   C ≡ G   G ≡ C

FIGURE 6.6 A pairs with T and G pairs with C. Because there are only two chemical bonds holding A to T, but three holding G to C, G-C base pairs stick together more tightly than A-T base pairs.

the sequence of bases on the other strand must be. The term *complementary* is used to describe this relationship of the two strands that pair with each other, and you need recall little about the actual structure of an AT pair or a CG pair as long as you remember the basic base-pairing rule in Figure 6.6.

So when the two strands separate into single strands and then become double-stranded by filling in the complementary bases, the result is two new double helices whose sequence, or order of the bases, is identical to the original double helix (Figure 6.7). One new double helix passes into one of the new daughter cells, and the other new double helix passes into the other new

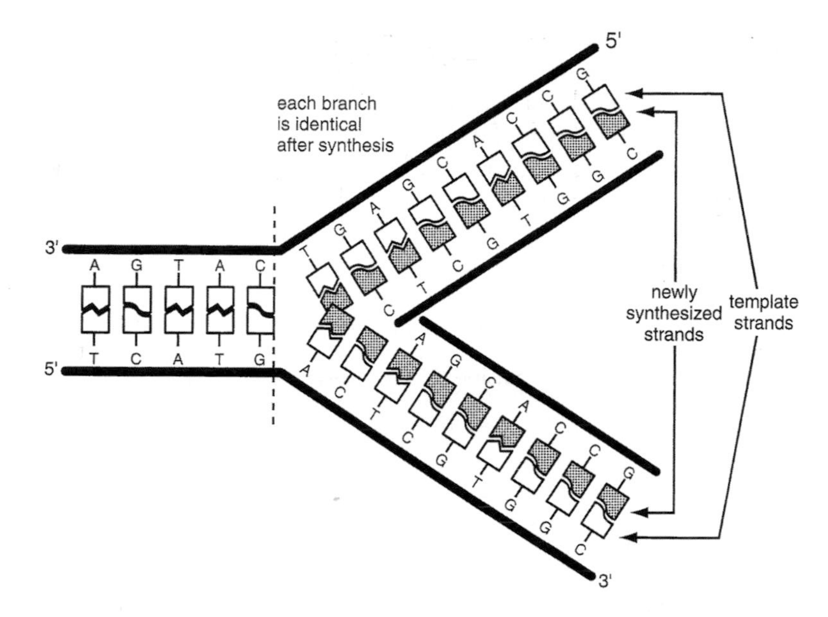

FIGURE 6.7 Copying the DNA. The original double-stranded material towards the left of the figure gets copied by separating the two strands from each other, and then creating a new strand that pairs with the prior single strand according to the rules of base pairing, A with T, G with C.

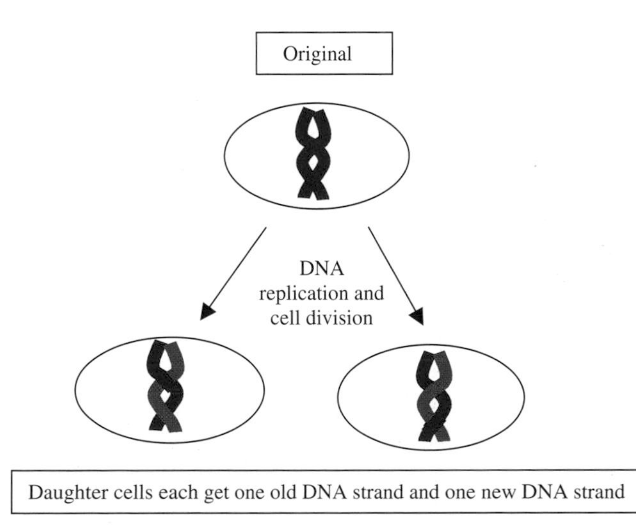

FIGURE 6.8 Replication of DNA goes hand in hand with cell division. DNA replication makes use of base-pairing rules to put "new" DNA with one old strand and one new strand into each new cell.

daughter cell (Figure 6.8). There is no "old" cell remaining because its pieces went into the construction of both of the two new cells. Also, there are no "old" DNA molecules left because each new helix contains one old strand of DNA and one new strand of DNA. This process is known as *semiconservative* DNA replication, meaning that each new double-stranded molecule has one old strand and one new strand (see Figures 6.7 and 6.8).

Copying of the cell's three billion base pairs of DNA (known as DNA replication) is carried out by proteins called DNA polymerases. The double-stranded structure of the DNA comes apart into two separate strands in the region that is being copied, and the polymerase "reads" the old strand of DNA and uses the base-pairing rules in Figure 6.6 to insert the bases of the new strand in the proper order to complement the sequence on the old strand. Meanwhile, the same thing happens to the other "old" strand of DNA.

DNA replication takes place during cell division, the process of making more of everything inside of the cell and then partitioning the increased amounts of cellular materials into two new cells. Before each cell division, these proteins must replicate the entire human genome *exactly* once.

DNA replication is so precise that mistakes in copying are made less than once in every 10,000,000,000 replicated bases. Each new base, complementary to the base on the original or template strand, is added sequentially to the growing end of the replicating strand. Copying of the DNA has to take place simultaneously at many places because the pieces of DNA inside the nucleus are very long. Otherwise, the cell would take a great deal of time to finish copying its genome and cell division would take far too long.

THE DOUBLE HELIX

This double-helix structure of DNA was deduced by James Watson and Francis Crick in 1953. It was an accomplishment filled with gossip, brilliance, and even a bit of intrigue. It was also a fundamental leap forward that deserved and won a Nobel prize (Box 6.1).

At the end of the paper announcing the structure of the DNA helix, Watson and Crick observed, "It has not escaped our notice that the specific pairing we have postulated immediately suggests a possible copying mechanism for the genetic material."[2] They were right, that the uncovering of the

BOX 6.1 DISCOVERY OF THE DOUBLE HELIX

Although Watson and Crick are famed as the authors of the paper that announced the double-helical structure of DNA, two other papers on the subject were published along with the Watson and Crick paper in that 1953 edition of the journal *Science*. One was by Maurice Wilkins, who later joined them in receiving the Nobel Prize in 1962. The third paper was by Rosalind Franklin, whose data helped Watson and Crick develop their model and in fact confirmed their model as something more than theoretical. We will never know whether history would have given Franklin recognition along with the others, since the Nobel Prize is not awarded posthumously and she was no longer alive by the time this particular prize was awarded. For a fuller version of the controversial story of this great discovery, we recommend three books: *The Double Helix* by James Watson, *The Eighth Day of Creation* by Horace Freeman Judson, and *Rosalind Franklin: The Dark Lady of DNA* by Brenda Maddox.

[2] From J. D. Watson and F. H. C. Crick, Molecular Structure of Nucleic Acids, *Nature*, 171, 737–738, 1953.

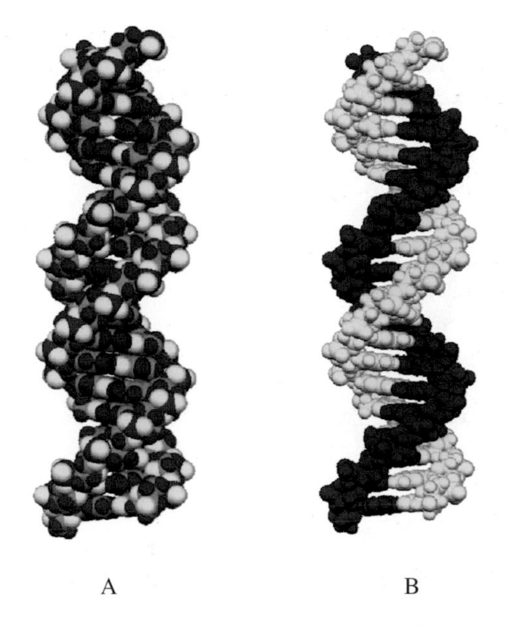

FIGURE 6.9 Ball-and-stick model of the DNA helix shows the base pairing that forms the "steps" of the spiral staircase and the backbones of the parallel coils of the two strands of the double helix. (This colorized image was generated by D. M. Reed with the use of Visual Molecular Dynamics (VMD) developed by the Theoretical and Computational Biophysics Group at the University of Illinois at Urbana-Champaign. (www.ks.uiuc.edu/research/vmd/) based on information from Schweitzer, B. I., Mikita, T., Kellogg, G. W., Gardner, K. H., Beardsley, G. P.: Solution structure of a DNA dodecamer containing the anti-neoplastic agent arabinosylcytosine: combined use of NMR, restrained molecular dynamics, and full relaxation matrix refinement. *Biochemistry* 33 (11460), 1994.)

structure of DNA (Figure 6.9), with its paired strands, pointed directly to the mechanism of DNA replication through base-pairing. Watson and Crick's model of DNA allows us to visualize the faithful replication of DNA as nothing more than simply separating the two complementary strands, the "Watson" strand and the "Crick" strand, and synthesizing two new complementary strands by using the rules of base pairing.

As you can see from the pictures we have shown, there are many different ways to diagram the DNA structures. Figure 6.9 shows two ball-and-stick models of DNA and Figure 6.10 shows a view that looks down through the center of the helix. To appreciate just how hard it was to figure out this structure that is too small to see with the most powerful microscope, consider this: in this spiral structure with ten stair-steps to every turn, the distance across the helix is 20 angstroms, with an angstrom covering a distance that is one hundred millionth of a centimeter!

SUMMARY

The simple beauty of using the base-pairing rule to copy DNA is this: by knowing the sequence of bases along one strand of DNA, we automatically know what the sequence of the other strand must be. Like in the children's game of opposites—left/right, yes/no, up/down—one piece of information

FIGURE 6.10 Model of DNA as seen looking down through the center of the helix. (Courtesy of D. M. Reed.)

automatically tells us the other. A tells us T, G tells us C, T tells us A, and C tells us G.

Each chain of DNA letters within the cell is tens or hundreds of millions of bases in length, with thousands of groups of letters scattered along that DNA chain spelling out the information content of genes. Each gene consists of many adjacent letters within the much longer chain of letters that may contain thousands of separate genes. Gene size can range from hundreds to millions of letters in length. If a cell wants to make use of one particular gene along that string of genes, it is faced with the problem that the long chain of DNA is located inside of the nucleus but the machinery for using that information is located outside of the nucleus. To solve this problem, the cell makes a transient copy of the information in the DNA, copying not the entire DNA molecule but rather just the region that contains the gene to be used. Interestingly, the process by which the genetic information gets copied into a message that can move outside of the nucleus operates on principles very similar to the mechanisms by which the DNA gets copied when the cell makes a new copy of the DNA, as we will describe in the next chapter.

THE CENTRAL DOGMA OF MOLECULAR BIOLOGY: HOW GENES ENCODE PROTEINS

7

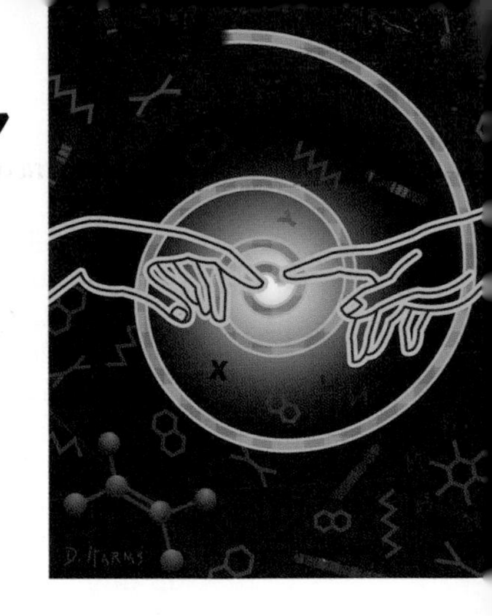

"What hath God wrought?"
—Samuel Morse

On May 24, 1844, using pulses of electricity traveling through a 41-mile-long telegraph line from Baltimore, Maryland, to Washington, D.C., Samuel Morse transmitted his first message using a code consisting of three characters: a short electrical pulse written out as . and called a "dot," a long pulse written out as – and called a "dash" equal to about three times the length of the dot, and spaces of various length depending on whether they denote a gap between characters, between letters, or between words. Using two symbols plus spaces, he could indicate all 26 letters of the English alphabet, many letters specific to other languages, the numbers zero through nine, and the common punctuation marks. The most familiar signal to people who don't know morse code is the distress code SOS, transmitted as ... – – – ... (three dots, three dashes, then three dots). Morse's original message over the first telegraph line would have taken the form of

.–– .... .– –   .... .– – ....   ––. ––– –..   .–– .–. ––– ..– ––. .... –   ..––..

which says "What hath God wrought?" when it is decoded. Morse revolutionized communication in the nineteenth century not only because of his engineering inventions that allowed for transmission of signals but also because of his realization that large amounts of complex information can be encoded and transferred from one place to another with even such a limited primary alphabet as the three Morse code characters—dot, dash, and space.

When we talk about DNA, we use the term *code*, in this case the genetic code. In Chapter 6, we told you that DNA has only four letters with which to keep track of the vast amounts of genetic information inside the cell. As we will discuss in this chapter, the cell uses a simple code not unlike Morse code: a code in which groups of the four DNA bases encode a single letter. Clearly the DNA code is not specifying letters of the English language and is not transmitted via electrical signals on a telegraph. So just how does the genetic code work—what does it "spell" and how does the code get translated?

PROTEINS: THE BUSINESS END OF THE CELL

The DNA code "spells out" proteins by designating the order of the amino acid subunits in a protein. If the DNA contains the cell's information, then proteins are the business end of cellular processes. A protein is a chemical chain made of building blocks called *amino acids*. The twenty different amino acids have very different properties—some are large, some are small, some are positively charged, some are negatively charged, and some are neutral and have no charge. A positive charge on one amino acid may interact with a negative charge on another amino acid elsewhere in the same protein or in a different protein. Some amino acids are larger, taking up more space within the protein. One amino acid is known for putting bends or kinks into the protein chain wherever it occurs, and another amino acid has properties that let it connect different parts of a protein together by forming a specialized chemical bond. Thus the order in which the amino acids are found along the protein chain can determine whether the protein has areas that are folded up in globular structures or have long, loose, linear sections. The order of amino acids can determine whether the protein has sections that can interact with other proteins. The order of amino acids can determine which parts of the molecule are chemically bonded to other parts of the molecule. Overall, the order of the amino acids ends up determining the folded-up three-dimensional shape of the protein (Figure 7.1) and the location of positive and negative charges within the three-dimensional structure. All of this determines the protein's function. Even the substitution of one amino acid for another can sometimes completely inhibit the ability of a protein to function. Thus, there are many different proteins and the differences between the functions of proteins are the direct result of the difference in the order of amino acids that make up the proteins.

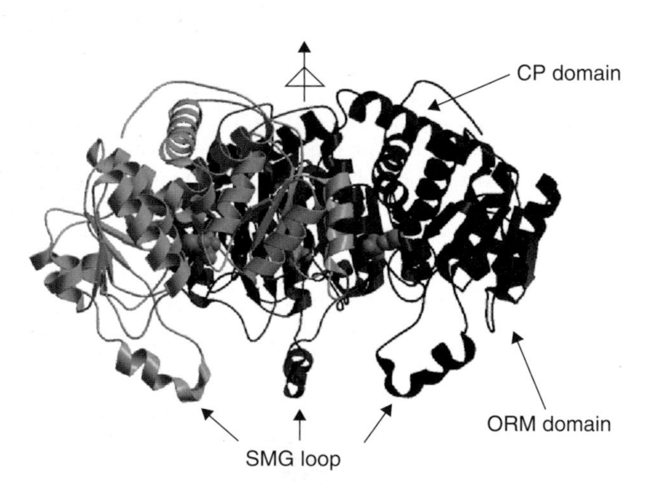

FIGURE 7.1 Information dictates form and function. The sequence of a protein determines the order of its amino acids. Each amino acid has different properties, such as shape, size, and charge, that affect how they fit together with the adjoining amino acids. The primary amino acid sequence determines the protein's shape and chemical properties. Those in turn determine what functions it can carry out. The ornithine transcarbamylase protein made by the OTC gene assembles three copies of itself (with the three copies appearing here in three different colors) to make the functional protein that carries out a step in the urea cycle. The straight and looped sections of the ribbon in this picture indicate how the protein is folded at that point in the protein. This particular image shows the conformation of the protein when it is complexed with its substrate, which is the molecule on which it acts. (Shi et al. *Biochemical Journal* 2001;354 (501–509).

TRANSIENT MESSENGER CARRIES INFORMATION FROM THE NUCLEUS

The question is then, how does information contained in the DNA get used to make a protein with a particular sequence and functional properties? The simple answer is that the sequence of base pairs in a given gene determines the protein product it can encode. The complicated part about this is that the genetic information is contained in the sequence of the DNA inside of the nucleus, but the machinery for translating the genetic information into a protein product resides outside of the nucleus, in cellular machinery called *ribosomes*. The DNA in the nucleus is separated from the millions of ribosomes that could translate their information by the nuclear membrane (Figure 7.2).

Why does the cell have the translation machinery for reading genetic information separated from the main repository of genetic information? The answer is: to protect that repository of genetic information, which has to last for the lifetime of the cell. Long-term information storage needs to be located in a very stable environment, which the nucleus is and the cytoplasm is not. The variable, sometimes volatile, processes of protein production need to be carried out in a dynamic environment, as in the cytoplasm. The first rule for running a healthy eukaryotic cell is, "Never let your DNA wander out into the

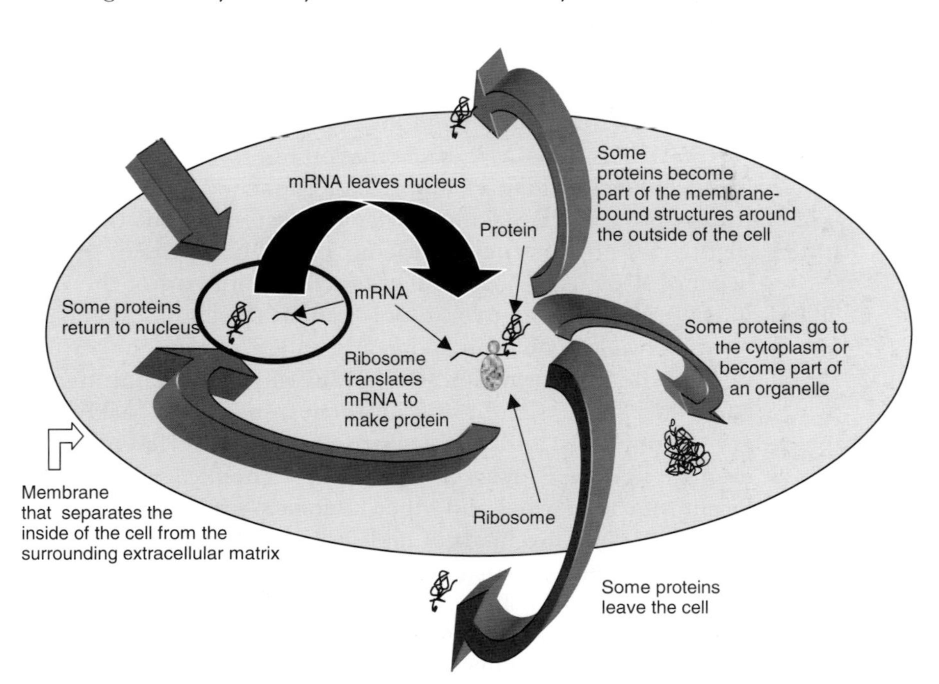

FIGURE 7.2 Structures and events in the cytoplasm of a cell. Information from a gene inside of the nucleus gets transferred into an mRNA molecule. The mRNA goes from the nucleus to the ribosome out in the cytoplasm, where the ribosome "reads" the information in the mRNA molecule and uses it to make a protein. The order of bases in the DNA strand determines the order of bases in the mRNA, and the order of bases of the mRNA determines the order of the chemical subunits that make up the protein. The resulting proteins can leave the cell, become bound to the outside membrane of the cell, become part of the cytoplasm, become part of a cellular organelle such as a ribosome, or return to the nucleus to help carry out functions such as creating mRNA or preparing the mRNA to be exported from the nucleus.

cytoplasm, where it can experience very bad things, such as being digested and eliminated from the cell!"

The cell solves the problem of how to connect the permanent repository of information to the dynamic processes for making new proteins in a simple, elegant manner. The cell creates a kind of messenger, a copy of the DNA sequence in the form of a short-lived nucleic acid called ribonucleic acid (RNA) that carries information from the nucleus to the ribosome (See Figure 7.2). Usually, this type of RNA is referred to as *messenger RNA (mRNA)* because of its information transfer function. As shown in Figure 7.2, this mRNA molecule then moves out of the nucleus into the cytoplasm, where it can direct the synthesis of a protein. Where the proteins go and what functions they carry out depend on the sequence of the protein—the order of the chemical subunits that make up the protein.

PRODUCTION OF RNA: TRANSCRIPTION

The function of the protein is dictated by the order of the amino acids. The order of those amino acids is dictated by the sequence of the messenger RNA. The sequence of the mRNA is dictated by the sequence of the DNA in the nucleus. Thus the order of information transfer goes:

DNA mRNA PROTEIN

This is sometimes considered the central dogma of molecular biology.

So DNA and RNA, the information reservoirs, are slightly different forms of nucleic acids, long chains of bases attached to backbone structures that connect the bases together. What is different about them? First, there is a slightly different structure to the backbones present in DNA and RNA. Second, RNA uses a base called uracil (U) instead of the thymine (T) used in DNA. We are not going to worry about the chemical and structural differences between a DNA backbone and an RNA backbone, or between T and U in the genetic information system. The important point here is that an RNA backbone "looks" significantly different from a DNA backbone to the machinery of your cells, and your cells recognize the difference between T and U even though they look very similar in a chemical diagram. Completely different sets of enzymes and other biological processes are associated with the handling of the two kinds of molecules, DNA and RNA.

Transcription (making an RNA copy of DNA) is like *replication* (making a DNA copy of DNA). It uses base-pairing rules to put a new RNA "letter" into position along a growing RNA chain by picking out the RNA "letter" that can pair with the DNA "letter" present on the DNA chain. In DNA A pairs with T, T pairs with A, G pairs with C, and C pairs with G. The pairs are the same between DNA and RNA, except that A in DNA pairs with U instead of T (Figure 7.3). So the DNA sequence ATGCTTCGA will end up as AUGCUUCGA in the RNA molecular that is made by "reading" off of the template strand shown in Figure 7.3.

The cellular machinery uses these rules to let it "read" the DNA into RNA. RNA molecules are synthesized from DNA molecules by a process called *transcription*. During transcription, only one strand of the DNA corresponding to that gene (known as the template strand) is copied into an RNA molecule

DNA strand	**ATGCTTCGA**	**AUGCUUCGA**	RNA strand
	: : : : : : : : :	: : : : : : : : :	
DNA strand	**TACGAAGCT**	**TACGAAGCT**	**DNA template**

FIGURE 7.3 Pairing of RNA bases with DNA bases works like pairing of DNA with DNA, except that RNA uses the base U in place of the base T, and the backbones that connect the bases to each other are different. In each case, if you know the order of bases along the strand of DNA that is being copied or read from, you know what the sequence of the new strand will be.

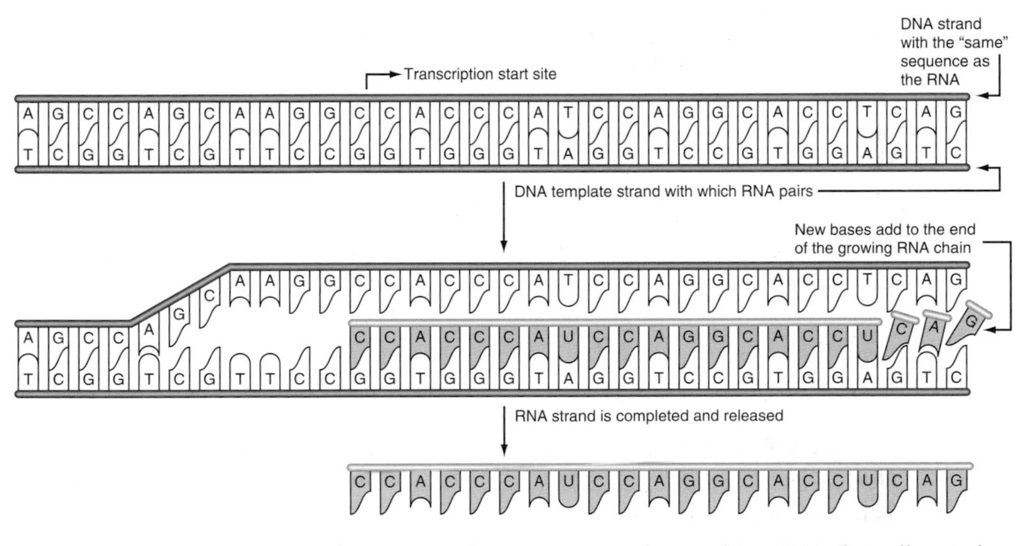

FIGURE 7.4 The process of transcription. When making RNA, the cell uses the same base-pairing system as it did when it made new DNA during replication.

(Figure 7.4). Each gene will produce a single-stranded RNA molecule that is complementary to the bases that compose the template strand of that gene, so RNA molecules "look" like the genes they are produced from but "look" different from each other.

Like DNA replication, the process of transcription is directional. The new RNA chain only grows in one direction from the start site. Once the first RNA base has been put in place by pairing with the DNA base, only one end is available to have another base added to it.

Unlike DNA replication, only one of the strands gets copied, and the RNA that results is single-stranded, containing the sequence present in only one of the two DNA strands. Moreover, genes and the RNA molecules produced by them are relatively short compared to the great length of the DNA molecules that encoded them. The amount of transcription at any given gene can be carefully controlled by the cell. This control is the molecular basis of gene regulation and is discussed in detail in Chapter 9.

mRNA MOLECULES DIRECT SYNTHESIS OF PROTEINS: TRANSLATION

Transcription, the process by which an RNA molecule is made, and translation, the process by which the information in the RNA is used to make a protein, are fundamentally different processes that take place at different points in the cell and make use of very different cellular machinery. Transcription happens in the nucleus. Translation is carried out by ribosomes outside of the nucleus. We present these fundamentally different processes together in one chapter because they are the critical linked elements of the central dogma of molecular biology:

DNA mRNA PROTEIN.

The term *genetic code* is commonly used these days in newspapers and magazines and on television. We have talked about the fact that the RNA "encodes" the information in the protein, but how does it actually do that? When scientists were first trying to decide which materials inside of the cell contained the genetic information, DNA was ruled "out" as the repository of genetic information because it only has four letters. Those studying the problem knew that there are thousands of different proteins in a human cell and twenty different amino acid building blocks used to make those proteins. They thought that four letters did not constitute a complex enough alphabet to produce that much different information.

One breakthrough came when scientists figured out that each "letter" in the genetic code is spelled out by three bases in a row. Just as the twenty-six letters of the English alphabet can be designated by Morse code combinations of dashes and dots, four genetic building blocks when combined in groups of three can produce sixty-four different letters. In fact, since there are only twenty different amino acids, there is actually some redundancy in the system and some amino acids get designated by more than one of the three-base letters, called codons. When the cell "reads" an RNA transcript with bases ACUAGA, it does not read A and then C and then U and then A and then G and A and so on. Instead, it reads ACU as one letter and then it reads AGA as a different letter.

Basically, the nucleotides of the mRNA molecule are read by the ribosome in such a way that each set of three nucleotides, called a *codon*, can specify a single amino acid. Thus the first three bases of the mRNA will encode the first amino acid of the protein, the second three bases the second amino acid, and so on. *Each codon specifies the incorporation of one and only one amino acid.*

Writing out all of the different ways in which a four-base alphabet can be used to write out letters that are three units long, produces sixty-four different codons. However, there are only twenty different types of amino acids. Thus, in some cases, different codons must code for the same amino acid (thus the code is said to be *degenerate*). So some amino acids such as methionine can be encoded by only one codon, AUG, but some amino acids such as leucine can be designated by as many as six different codons.

Note that no codon encodes more than one amino acid, even though in some cases one amino acid can come from more than one codon. Using Table 7.1, which contains a key for translating the genetic code, you can find out what amino acid will result from any of the sixty-four possible codons. You can also see that the amino acid encoded by UGU can be represented by the full-length name, cysteine, or by a three-letter symbol, Cys, or by the single-letter symbol, C. Some amino acids that are relatively rare are encoded by only one amino acid (such as tryptophan or methionine) but that other

TABLE 7.1 The genetic code uses a 3-base codon to specify each amino acid

Amino Acid	Three-Letter Symbol	Single-Letter Symbol	
Alanine	Ala	A	GCA, GCC, GCG, GCU
Arginine	Arg	R	AGA, AGG, CGA, CGC, CGG, CGU
Asparagine	Asn	N	AAC, AAU
Aspartic acid	Asp	D	GAC, GAU
Cysteine	Cys	C	UGC, UGU
Glutamic acid	Glu	E	GAA, GAG
Glutamine	Gln	Q	CAA, CAG
Glycine	Gly	G	GGA, GGC, GGG, GGU
Histidine	His	H	CAC, CAU
Isoleucine	Ile	I	AUA, AUC, AUU
Leucine	Leu	L	UUA, UUG, CUA, CUC, CUG, CUU
Lysine	Lys	K	AAA, AAG
Methionine	Met	M	AUG
Phenylalanine	Phe	F	UUC, UUU
Proline	Pro	P	CCA, CCC, CCG, CCU
Serine	Ser	S	AGC, AGU, UCA, UCC, UCG, UCU
Stop codon		*	UAA, UAG, UGA
Threonine	Thr	T	ACA, ACC, ACG, ACU
Tryptophan	Trp	W	UGG
Tyrosine	Tyr	Y	UAC, UAU
Valine	Val	V	GUA, GUC, GUG, GUU

amino acids such as arginine may be encoded by as many as six different combinations of three bases. Notice that Table 7.1 uses the version of the code that is spelled out with the bases used in RNA: A, C, G, and U.

It is cumbersome to try to look up any particular codon in Table 7.1. Several types of keys have been developed to try to make it easy to look up an amino acid to find out which codons can produce it, or to look up a codon to see which amino acid will result from its use. An especially nice format is shown in Figure 7.5, which uses a wheel-like structure to show the correspondence between the codons and the amino acids.

To use the translation key in Figure 7.5, start with the letter in the center and move outwards to assemble a codon, and then look at the amino acid on the outside ring that sits next to the third base of the codon. Thus we can see that TGG encodes tryptophan (Trp). If we start with G and then T, the third position of the codon can be occupied by any of the four bases to result in a codon that encodes valine (Val). Notice that this key uses T instead of U. Even though ribosomes read RNA that uses U, many technologies for determining the sequence of a gene "read" the sequence from DNA derived from the chromosome rather than "read" the sequence from the RNA copy.

Three codons do not specify the incorporation of any amino acid. Instead, they are placed at the end of the coding sequence contained on the mRNA to tell a ribosome to stop translating the message and release the assembled protein. These codons, referred to as UAA, UAG, and UGA in Table 7.1 and as TAA, TAG, and TGA in Figure 7.5, are appropriately called *stop codons* because they stop the translation process and cause the message and protein chain to be released from the ribosome.

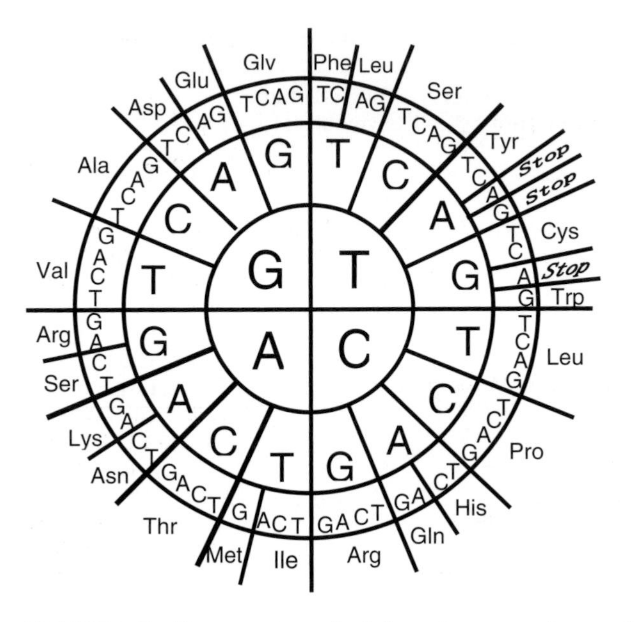

FIGURE 7.5 The genetic code. It is easier to translate codons via this table than when using Table 7.1. (Derived from *Chemical Biology. Selected Papers of H. Gobind Khorana (with introductions)* by H. Gobind Khorana in World Scientific Series in 20th Century Biology, Vol. 5.)

The mRNA is read, or *translated*, from one end to the other, beginning at the start and proceeding one codon at a time toward the end of the mRNA. For our purposes, we will consider that translation begins at a *start codon* (AUG) encountered by the ribosome as it reads along the message. At the point at which the start codon is encountered, the ribosome begins the chain with the insertion of the amino acid methionine that corresponds with AUG.

Right here we encounter an important concept: the translation from RNA to protein does not start at the first base on the RNA molecule, it starts at an AUG start sequence after reading past other bases present on the RNA strand before the AUG. After assembly of the growing protein chain begins, the ribosome keeps reading until it encounters a stop codon, which can be UAA, UAG, or UGA. Again, this is an important concept. The translation does not keep going until the end of the RNA molecule. The result is that the coding sequence, the part of the RNA used to direct synthesis of protein, is sandwiched in between two regions of non-coding sequence that precede and follow the coding sequence (Figure 7.6).

The AUG start codon directs the addition of the amino acid methionine (Met). As each successive codon is read, the ribosome incorporates the indicated amino acid into the growing protein. Translation stops when the ribosome encounters one of the three stop codons (UAA, UGA, or UAG) that do not specify the incorporation of an amino acid. Once the translation of the mRNA is completed, it is in some ways also a "translation" of the DNA sequence back in the nucleus even though the DNA and the ribosome never encountered each other (Table 7.2).

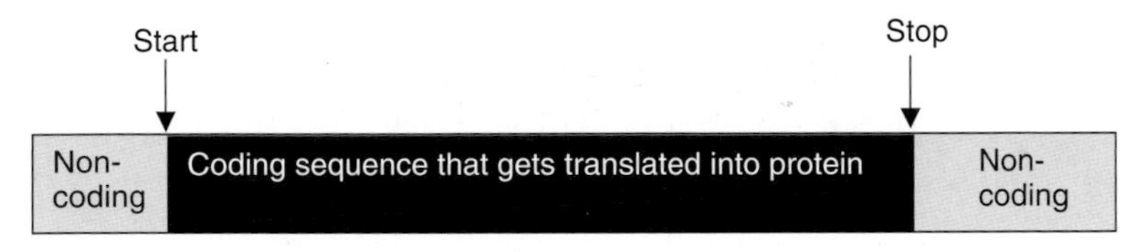

FIGURE 7.6 An mRNA starts out with non-coding sequence at the beginning, followed by a start codon, the coding sequence, a stop codon, and more non-coding sequence. This is the universal structure of an mRNA whether the gene is large or small.

TABLE 7.2 DNA ⇒ RNA ⇒ Protein

The DNA sequence:	ATTAGGTACGTATGTGAT TAATCCACGCATACACTA
Results in an mRNA:	AUUAGGUACGUAUGUGAU
Which gets read as:	AUU AGG UAC GUA UGU GAU
To produce the protein:	Ile-Arg-Tyr-Val-Cys-Asp

TRANSLATION REQUIRES AN ADAPTOR MOLECULE CALLED tRNA

The codons in a mRNA molecule cannot and do not directly recognize the amino acids whose incorporation they direct. This process instead depends on a class of small RNAs called *transfer RNAs* (*tRNAs*) that serve as adaptors. Basically, one end of this adaptor recognizes one of the codons on the mRNA and the other end of the adaptor has the amino acid that goes with that codon. The way the adaptor recognizes the codon is by having an anticodon, a set of three bases on the tRNA molecule that can base pair with the codon in the mRNA (Figure 7.7). It turns out that there is a specific tRNA molecule for each possible codon. The exceptions to this rule are the three stop codons; no tRNA molecules exist that can read these codons.

A ribosome ratchets along the mRNA reading it codon by codon (see Figure 7.7). At each codon, it searches for a tRNA molecule whose anticodon is complementary to that codon. So a tRNA that recognizes the AUG of a methionine codon is carrying a methionine. Each tRNA brings along the appropriate amino acid with it, which is then incorporated into the growing polypeptide chain. Once this is done, the spent tRNA molecule is discarded and the ribosome ratchets onto the next codon. This process continues until the ribosome reaches a stop codon, for which there is no corresponding tRNA, and the ribosome releases both the mRNA molecule (perhaps to be translated again) and the completed protein.

SUMMARY

The central dogma of molecular biology is that the direction of information transfer goes from DNA mRNA PROTEIN. The stable, long-term information located in the nuclear DNA gets transcribed to create a temporary messenger RNA that moves out of the nucleus to where ribosomes can use

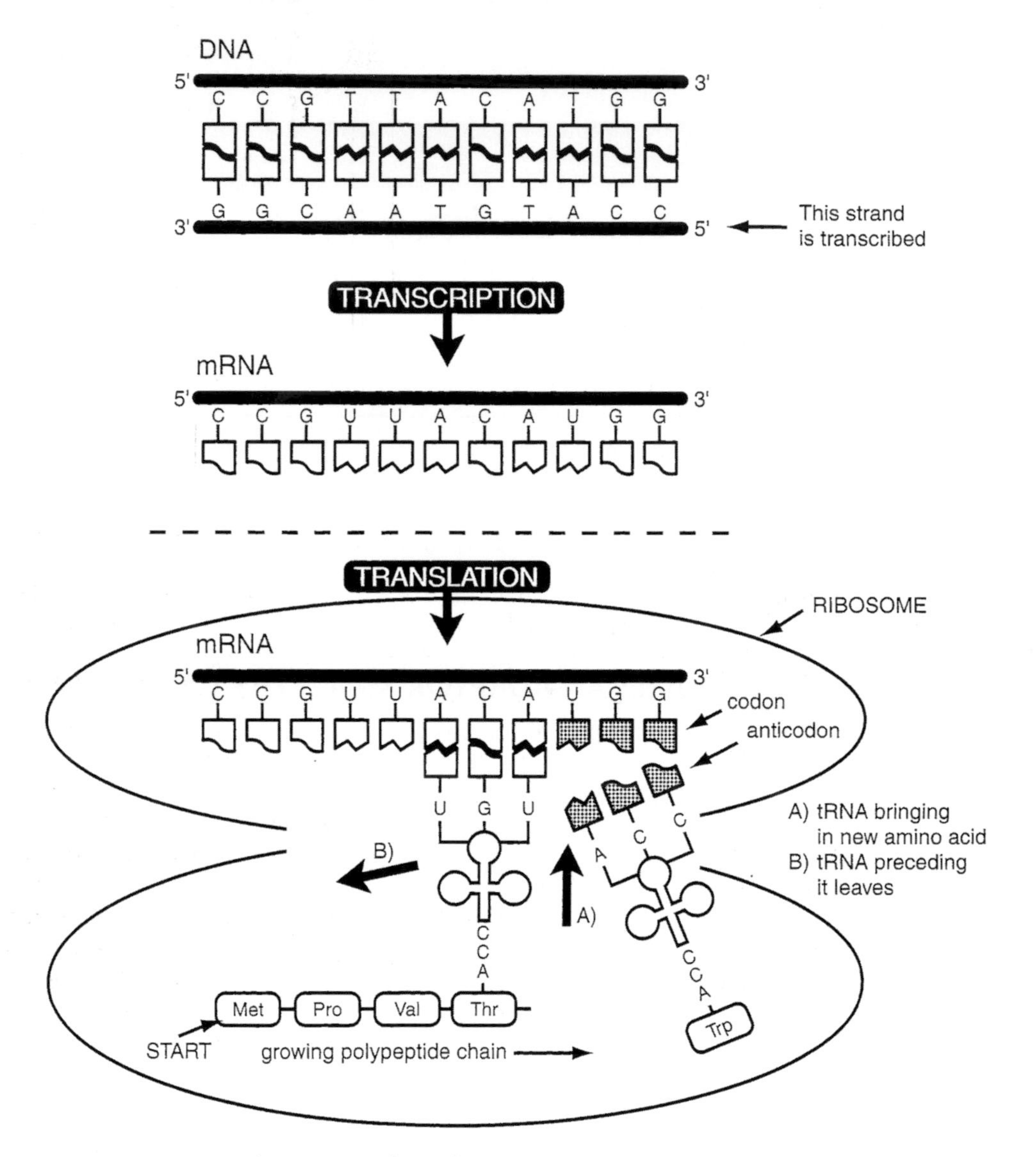

FIGURE 7.7 The process of translation.

the three-letter genetic code to "read" the message and synthesize a protein. The order of amino acid building blocks in the protein is dictated by the order of bases in the messenger RNA, which is dictated by the order of bases in the region of nuclear DNA that was transcribed to make the RNA, so knowing the DNA or RNA sequence tells us what the protein sequence will be. We have presented the gene and its transcript as being simple co-linear sequences, with the sequence in the DNA exactly corresponding to the sequence in the RNA. In fact, things are rather more complicated than that; Chapter 8 shows some of the complex ways in which the cell manages to make efficient use of some genes to accomplish more than one thing.

SPLICING THE MODULAR GENE

8

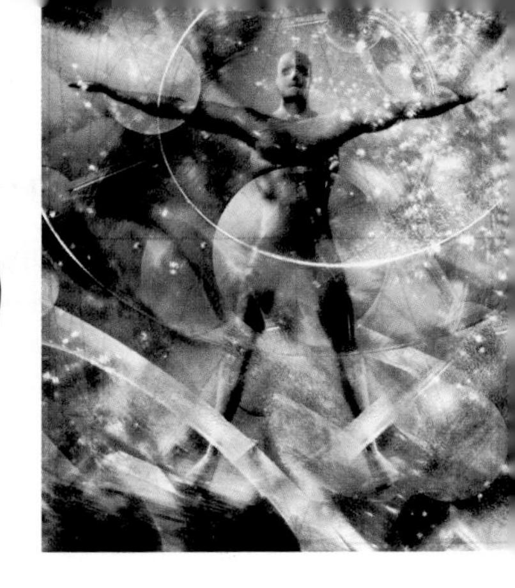

When Ed and Sophia bought their latest computers, they each found themselves faced with a variety of options: different speeds, different amounts of memory, and different peripherals such as speakers and scanners. By the time they were done making their selections, their computers had many things in common, but Ed the database manager had selected an automated tape-drive backup system that Sophia didn't think she needed, and Sophia the graphic artist had selected high-end graphics card options that Ed knew he didn't need. Because construction of computers is so modular, it is easy to optimize the features present on each computer without computer companies having to maintain separate lines of instruments with each possible combination of features.

The genome makes use of some similar efficiencies by designing some genes to be modular. Do we mean that the genome uses some genes together in a modular fashion? Well, yes, the genome does that, but that is not what we are talking about here. When we talk about the modular gene, we mean a gene that is made up of a set of separate pieces in such a way that not all of the pieces always get used, just as computers don't all end up with every option on the manufacturer's list. Sometimes genes do the same thing. Lets start by looking at how they do this through a process called *splicing*.

SPLICING

As we start talking about the process called splicing that makes *modular* genes possible, keep in mind that only some genes are spliced, and only a subset of those, called alternatively spliced genes, use modular combinations of information within the gene to produce different gene products that share some functions and not others.

The key to spliced genes is that, after an RNA transcript is initially made, a piece of the RNA may be removed before the transcript becomes mRNA that gets used to make protein. The piece that is cut out is called an *intron*. After the intron is removed, the cell splices the remaining pieces of RNA to each other so that what is left is still one continuous piece of RNA. The key to alternatively spliced, or modular, genes is that not always the same bit gets thrown out. So, under some circumstances, the cell uses some parts of a primary transcript to make the mRNA that will dictate the protein sequence, and under other circumstances, the cell uses different parts of the primary transcript. In these cases, the two different *splice variants* usually share most of the same sequence and differ in only some places.

The parts of the transcript that get left in the final mRNA and *ex*ported from the nucleus are called *exons* and the parts of the mRNA that get left *in*side of the nucleus and thrown away are called *introns* (Figure 8.1).

The example in Figure 8.2 shows a gene with just one intron flanked by two exons, but some genes have no introns so that no splicing step takes place. Other genes have many introns that have to be removed to make the final mRNA.

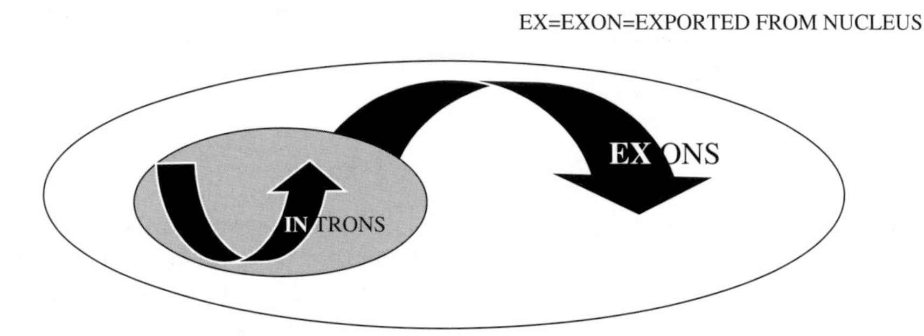

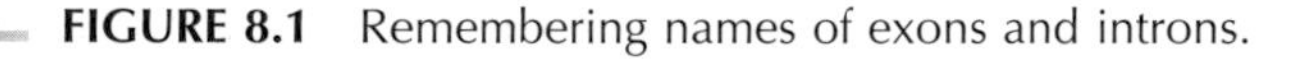

FIGURE 8.1 Remembering names of exons and introns.

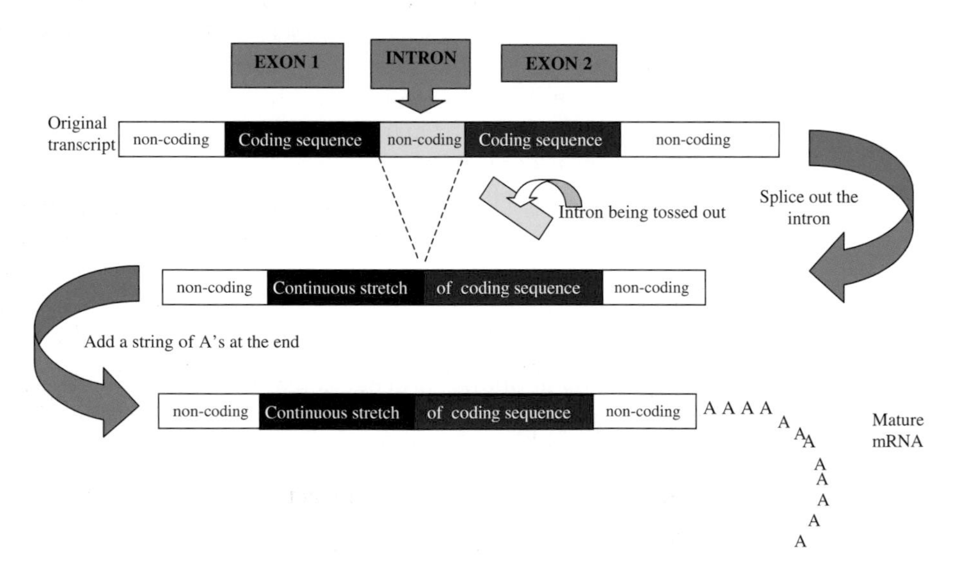

FIGURE 8.2 Two key steps in producing a mature mRNA—removing introns and adding a polyA tail. The splicing step removes parts of the transcript that don't contain useful information, and the polyA tail serves as a tag that tells the cell that this is a messenger RNA that should be translated into protein. Introns removed within the transcript contain areas that do not encode protein sequence, but other non-coding regions at the very beginning and end of the transcript do not get removed. Notice that exon 1 contains both non-coding information (at the very beginning) and coding information used in the making of the protein. Similarly, exon 2 has both coding information and the non-coding information at the very end.

TABLE 8.1 Different types of RNA and the role they play in the cell

TYPE OF RNA	ALSO CALLED	LOCATION	FUNCTION
Messenger RNA	mRNA	Nucleus and ribosomes	Carries information from the nucleus to the cytoplasm
Ribosomal RNA	rRNA	Ribosomes	Part of the ribosome structure that translates the message
Transfer RNA	tRNA	Cytoplasm and ribosomes	Translates genetic code to amino acid sequence
Heterogeneous nuclear RNA	hnRNA	Inside the nucleus	Partially processed transcripts not yet ready for use, also discarded RNA
Small Nuclear RNA	snRA	Inside the nucleus	Part of the nuclear machinery that helps splice the mRNA
Small interfering RNA	siRNA	Nucleus	Interfere with expression of a gene

The other thing that happens to the transcript, whether or not any sections get spliced out, is that a *polyA tail* gets added to the end of the transcript. This marks the transcript as being mature (ready to be used by the ribosomes) and also marks it as being an mRNA that should be translated. This is an important distinction since there are other kinds of RNA molecules that don't get translated and that do not get polyA tails (Table 8.1). We have already heard about tRNA molecules that help the ribosomes carry out translation. In addition, there are RNA molecules that form part of the ribosome structure (ribosomal RNA) and other RNA molecules that help carry out splicing steps.

THE MIX-AND-MATCH GENE PRODUCT

What happens in the case in which alternative splicing takes place? Imagine that we have a protein that can carry out a particular biochemical reaction if it comes together with another protein to form one larger protein complex. Now imagine that we want that protein complex to carry out its reaction while stuck to the surface of the cell that made the protein, and then imagine that there are situations in which we want that protein to be able to leave that cell and go to another part of the body. What does our gene need to have to fulfill all of the above requirments? It needs:

- One module to make enzymatically active part of the protein
- One module to anchor the protein to the cell
- One module that functions as a vehicle to carry the protein elsewhere in the body
- One pairing module that lets two copies of the protein stick together

Once we know about splicing, it is easy to conceive of mixing and matching the modules (Figure 8.3). Enzyme plus anchor plus binding module gives us a version of the enzyme that binds the protein complex and stays with the

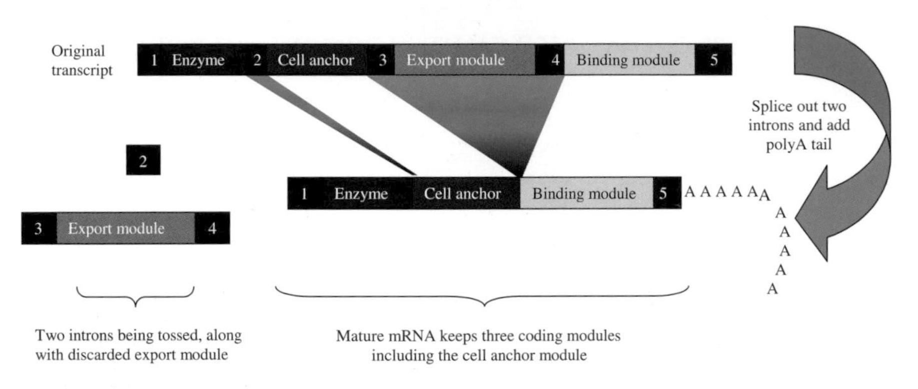

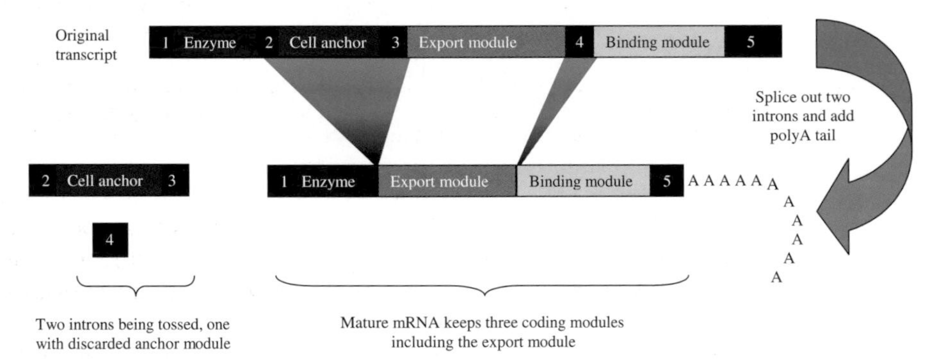

FIGURE 8.3 Alternative splicing allows modular use of different parts of a gene to allow the gene to accomplish some functions under multiple circumstances while allowing other parts of the gene to be used optionally according to whether they are needed at that time, under those conditions, or in that type of cell.

cell. Enzyme plus vehicle plus binding module gets us a version of the enzyme that travels and binds the protein complex.

Why not just make two different genes, one gene that makes an anchored version of the enzyme and a different gene that makes a traveling version of the enzyme? In fact, sometimes the body accomplishes these multiple goals by having different genes that share some functions but not others. However, sometimes the genome is especially efficient and uses the modular approach to get more uses out of one gene, so that the same gene can serve a local purpose, such as putting a particular enzyme function on the surface of the cell that makes the protein or exporting that enzyme function to the bloodstream to carry out its function elsewhere in the body (Figure 8.4).

THE IMPLICATIONS OF MODULAR GENES

Some genes are not spliced. What you see in the DNA tells you what you get in the mRNA and the protein. Some genes are spliced, but the gene struc-

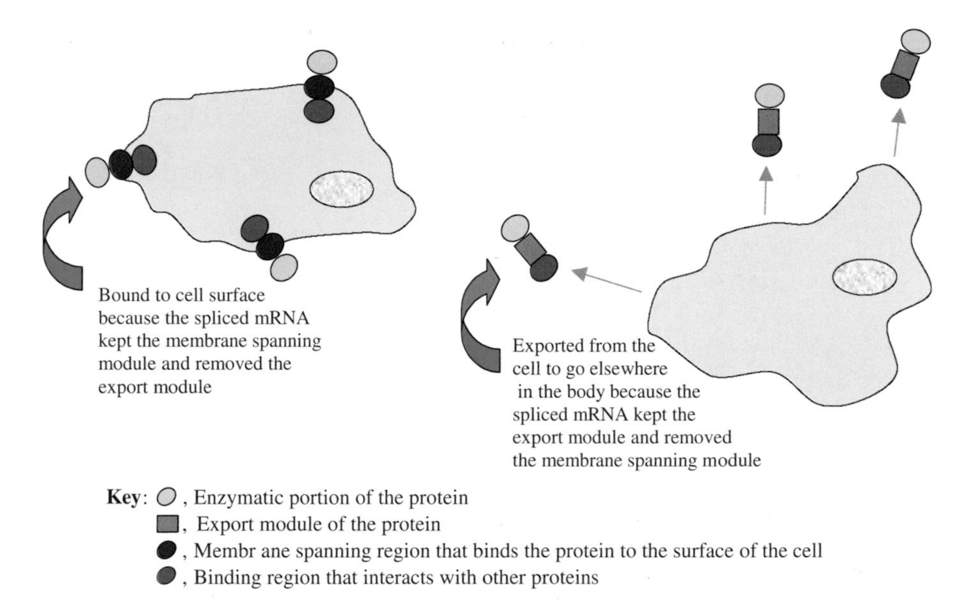

FIGURE 8.4 The gene splicing events in Figure 8.3 can result in the gene product being bound to the cell surface (on the left) or exported from the cell to carry out its function elsewhere (on the right).

ture is simple and the final mRNA made from that gene will always be the same. Some genes are alternatively spliced, with some genes using two alternatively spliced forms and some using many. These alternatively spliced genes are the modular genes.

What are the implications of modular genes? First, the total number of proteins that the human organism can make is effectively much larger than the total number of genes. Second, because sometimes the alternative splicing takes place differently in different cells, this means that cells with different needs can make different uses of the same genes. Third, it adds another level of complexity to the cellular processes for converting information in the DNA into a final set of proteins and functions. Here we considered what happens when we can use different parts of genes in different combinations with each other. In Chapter 9, we take a look at the next level of complexity that comes from using different combinations of genes in different types of cells and how the human genome is orchestrated.

ORCHESTRATING THE HUMAN GENOME

9

As a whiff of ether sent the swarming cloud of fruit flies drifting off to sleep they settled onto the bed of beige "fly food" at the bottom of the clear glass bottle. Stacy peered at the label to confirm that she had the right bottle, then gently tipped the flies out onto a white plate and adjusted the microscope to bring the sleeping flies into focus. The eyes of fifteen flies gleamed back at her from their ruby-red, multifaceted surfaces. She briefly scanned the thirty eyes and thirty little antennae that perched on the fifteen normal little fly heads to confirm that nothing was out of the ordinary. Then her eyes turned to the other dozen flies under the scope, the dozen flies of interest to her project, the dozen flies that had developed without eyes. After carefully counting each group and recording the numbers in her notebook, she scooted the flies back into the fly bottle and reinserted the stopper before they could awaken. She repeated the operation with the next bottle of flies after checking the label on the bottle. Once again she saw a group of normal flies with jewel-like eyes and tiny antennae sticking out of the tops of their heads. As expected, some of the flies from this bottle had tiny legs sticking out of the tops of their heads where their antennae should have been (Figure 9.1).

As expected? Legs on their heads? Is this a joke? Why would Stacy expect to find that her fly bottles held flies with no eyes or flies with legs growing out of their heads? How did they end up like that? And can this kind of thing

FIGURE 9.1 On the left is an image of a normal fly head seen straight on. On the right is the same view of a head from an *Antennapedia* mutant in which a change in gene expression causes production of legs where the antennae should have been. See page 79 for more about this mutation. (From The Interactive Fly 1995, 1996 Thomas B. Brody, Ph.D. Photos by Anthony Mahowald and Rudy Turner.)

happen to humans? To understand what Stacy was looking at and what this has to do with the human genome, let's talk about gene expression, the selective use of different genes by different cells in the body.

Each cell in the body carries the same genetic blueprint, the same nucleus full of information, yet liver cells and blood cells and brain cells are amazingly different in both form and function. If all of the cells have the same genes, how can they be so different? The answer is that not every gene is always expressed, that is to say, transcripts do not get made from every gene in every cell. Some genes that are expressed in all cells at all times are called *housekeeping* genes. Each type of cell also expresses a distinct *cell-type specific pattern* of genes that gives that cell its special properties and functions. Expression of some *developmental regulatory* genes is specific to certain stages during the growth and development of the person. *Inducible* genes are genes that are not normally expressed but that can be expressed in response to something to which the cell is exposed.

Let's look at some examples. All cells need to produce the enzymes that generate energy to run the cell. However, only red blood cells need to produce hemoglobins, the proteins that carry oxygen from the lungs to the tissues and return carbon dioxide from the tissues to the lungs. Only the cells of the retina in the eye need to produce the light-sensitive proteins, such as rhodopsin, that permit us to see. This is not to say that liver cells lack the genes for hemoglobin or rhodopsin; rather, liver cells simply do not express these genes (i.e., they do not make RNA transcripts from those genes). Instead, the liver cell expresses its own specific repertoire of genes needed to make the proteins that carry out the specific functions of the liver.

Indeed, the development of a human being from conception to death is the result of a complex, preset program of expression of genes that get used in some cell types and not in others or get used only at specific times during development of the fertilized egg into a living, breathing baby. The cell selectively accesses the information it needs while ignoring the information that is meant for some other cell or situation.

The effect is similar to that achieved when an orchestra uses the same set of about eighty notes to generate vastly different pieces of music (Figure 9.2). The same set of instruments achieves effects as different as Beethoven's Fifth Symphony or Classical Gas depending on which of the approximately eighty available notes get used and in what order. Similarly, orchestration of the use of the complete set of genes (the *genome*) can achieve the profound differences between a muscle cell and a nerve cell (patterns of cell-type specific gene expression) depending on which genes get used, the order in which the genes are transcribed, how often they are transcribed, how long the mRNA endures once it is produced, simultaneous expression of multiple genes, and coordinated regulation of expression of those genes. When the wrong notes play in an orchestra, the melody is changed or dissonance occurs; when the wrong notes play in a genetic symphony, the results can be as profound and perplexing as having legs appear in place of antennae on the head of a fly. Just as the orchestra dropping a few bars of Mozart into a Beethoven piece could greatly change the pattern, so would expressing leg genes where antenna genes should be expressed.

The ability of a cell to control *gene expression* depends on sequences within the DNA itself known as promoters and regulatory elements and on a

FIGURE 9.2 An orchestra offers an interesting model for the combination of spatial and temporal differences in gene expression that take place at different stages in life in different types of cells in the body. This picture shows the University of Michigan Life Sciences Orchestra, which is made up of members of the life sciences community from throughout the campus. (Courtesy of the University of Michigan Health System Gifts of Art program. Photo by Lan Chang.)

set of regulatory proteins that bind to those DNA sequences to control the expression of each gene. The *promoter* lies at the beginning of the transcribed region for each gene and defines the site at which transcription is begun. It is the binding site for RNA polymerase. *Regulatory elements* may lie upstream, within, or down stream of a gene, and they determine how accessible the promoter is to RNA polymerase (and thus the extent of transcription). Genes are switched on or off by the binding of proteins known as *transcription factors* to these regulatory elements. The combined action of these proteins and DNA sequences allows each cell to express a specific subset of genes at various times in the life of the organism. This exquisite control of gene expression allows the development of complex life forms such as ourselves by facilitating the development of thousands of different types of cells. To understand this mechanism we need to now consider the biology of each of these players in transcriptional regulation more carefully.

THE CONCEPT OF A PROMOTER

As shown in Figure 9.3, transcription begins by the binding of a very large enzyme called the *RNA polymerase* to a site on the DNA next to where transcript of RNA will begin. The region from which the RNA polymerase initiates transcription is called the *promoter*. The promoter is usually located close to the beginning of the RNA transcript. Once the polymerase is bound to the promoter, it moves along the DNA. As it moves, it makes a single RNA copy

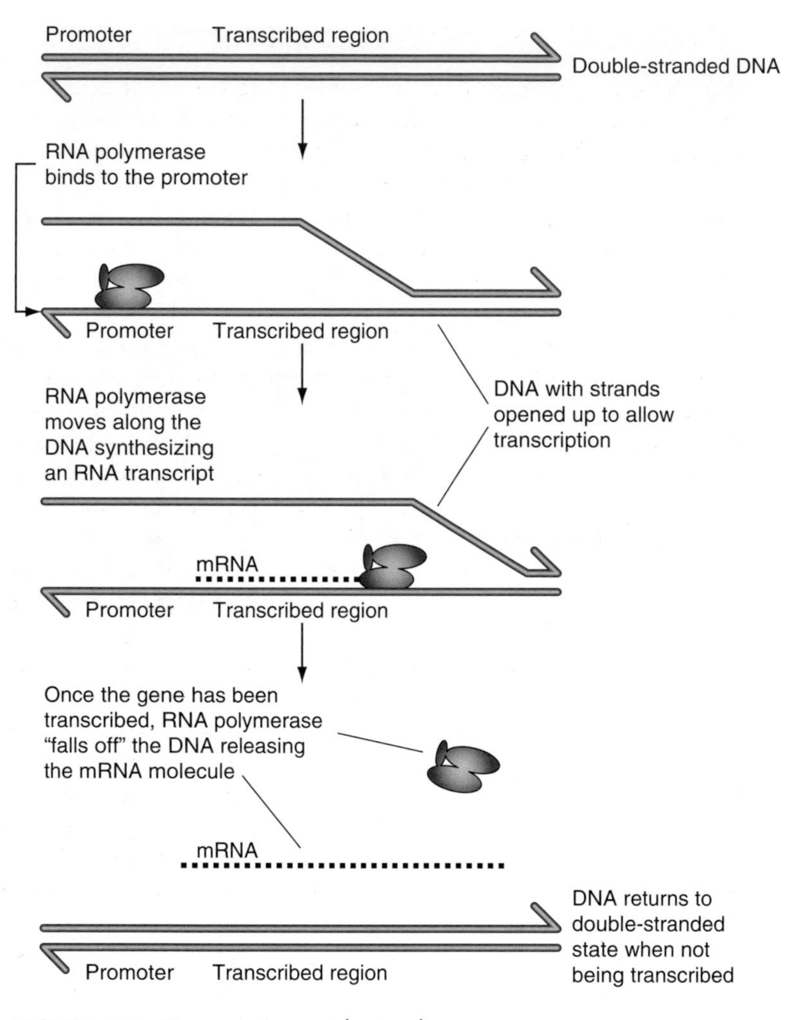

FIGURE 9.3 Transcription made simple.

from only one strand of the DNA double helix. It adds new bases to the growing RNA strand by using the rules of base pairing to insert the bases that complement the bases present on the DNA. When the RNA polymerase reaches the end of the DNA that comprises the gene, it detaches from the DNA and releases the newly made RNA molecule to have a polyA tail added and be spliced if it is a spliced gene.

Regulation of gene expression happens mostly at the level of transcription by controlling whether or not an RNA copy of the DNA sequence gets made. Regulation can also be imposed by modulating how often an active gene is transcribed (i.e., how many RNA copies of that gene are made in a given interval of time) or how stable the RNA is (how long the RNA copies stick around to be reused). So when we think about regulating transcription, we can think about whether or not the polymerase has access to the DNA that is to be transcribed, and we can think about what affects how frequently or rapidly transcripts get made.

CONTROL "SWITCHES" AND "REGULATORS"

As a metaphor, we can think in terms of two different things that affect transcription: switches that can control transcription (which take the form of DNA sequence) and regulators that operate the switches (which take the form of proteins that bind to the DNA). Control of gene expression may sometimes function like a light switch (expression is either on or off), but at other times it may act like a rheostat (sometimes few copies are made and sometimes many copies are made). The points in the DNA sequence that can act as "on" switches, "off" switches, "dimmer" switches, or boosters of expression are called *regulatory elements* (Figure 9.4). A gene normally has a control panel of regulatory elements adjoining the transcription start site and can sometimes have additional control elements located elsewhere in the vicinity of the gene. Thus control of whether the gene is transcribed and how much RNA gets made is normally not controlled by a single, simple switch; rather, control of transcription of any given gene is normally the product of the combined effects of multiple regulatory elements.

If the amount of transcription that takes place is the product of action of multiple regulatory elements in the DNA sequence, what controls whether they are on, off, up, or down? The answer is that regulatory proteins called *transcription factors* control the action of the regulatory elements by binding to them (Figure 9.5). When a regulatory protein binds to a regulatory element, it effectively changes the setting of the regulatory element's "switch." For some regulatory elements, binding of regulatory proteins can change the setting to "on." For other regulatory elements, binding of the regulatory protein changes the setting to "off." Similarly, if a regulatory protein is normally bound to the regulatory element but then stops binding the regulatory element and comes off the DNA, that can change what is happening to transcription, too.

As with locks and keys, the regulatory proteins are very specific to the switches. Thus there is not one type of protein in the cell that regulates all of the "on" switches for all of the genes. In fact, a particular DNA sequence that acts as a regulatory element may actually be found in the vicinity of the promoter region of may different genes. When the cell makes copies of the regulatory protein that binds to that regulatory element, that regulatory protein

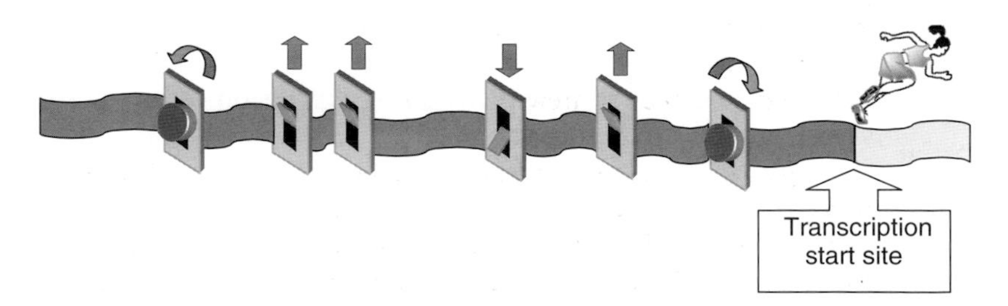

FIGURE 9.4 Level of expression of a gene may be affected by regulatory proteins called transcription factors that bind to regulatory elements in the DNA sequence to affect binding of the polymerase to the DNA and the subsequent transcription event. These act like on/off switches and rheostats that affect how much transcription can take off from the nearby transcription start site.

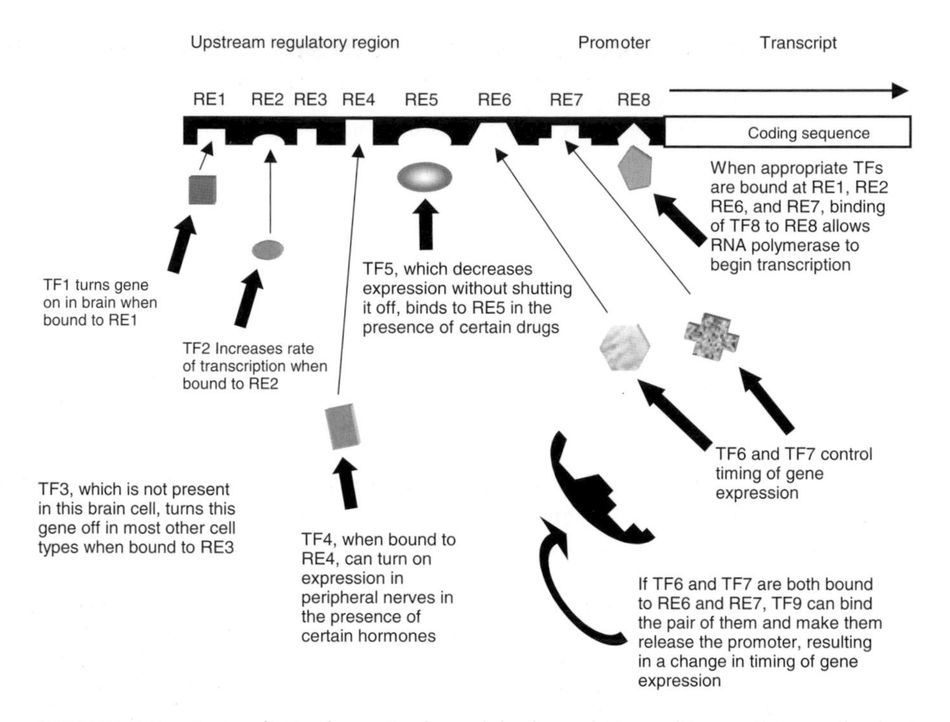

FIGURE 9.5 A simplistic theoretical model of regulation of transcription of a brain specific transcript would work. Regulatory elements (specific bits of sequence in the regulatory region before the beginning of the transcribed sequence) and transcription factors (proteins that direct the level and timing of transcription of the gene) can act in concert by binding to the DNA and/or each other to turn gene expression on, off, up, or down in a way that may differ in different cell types or at different points in the growth and development of the individual. A schematic diagram of the upstream regulatory region for a brain specific transcript is provided. *TF,* Transcription factor; *RE,* regulatory element to which TFs bind.

will bind to and change the state of not one but many genes. Thus some regulatory proteins that are present in the eye can turn on (or off, up, or down) multiple different genes expressed in the eye if they all have the same regulatory element to which that regulatory protein binds. If that regulatory protein is not also found in red blood cells, those same eye genes might not be expressed in the red blood cells. However, some other regulatory proteins found in the eye are also found in other cell types.

So patterns of tissue-specific gene expression are the result of complex combinations of regulatory proteins acting to turn the switches next to the genes into their correct positions of on, off, up, or down. If a gene is being expressed in a cell that only makes regulatory proteins that bind to the positive switches for that gene, the gene will be transcribed. That same gene will not be transcribed in a cell that is making only the regulatory proteins that bind to the negative switches for that gene but none of the regulators of the positive switches. In many cases, some combination of negative and positive regulation may be going on, or interactions of two proteins may enhance expression over what either of those proteins would bring about on their own (Figure 9.6).

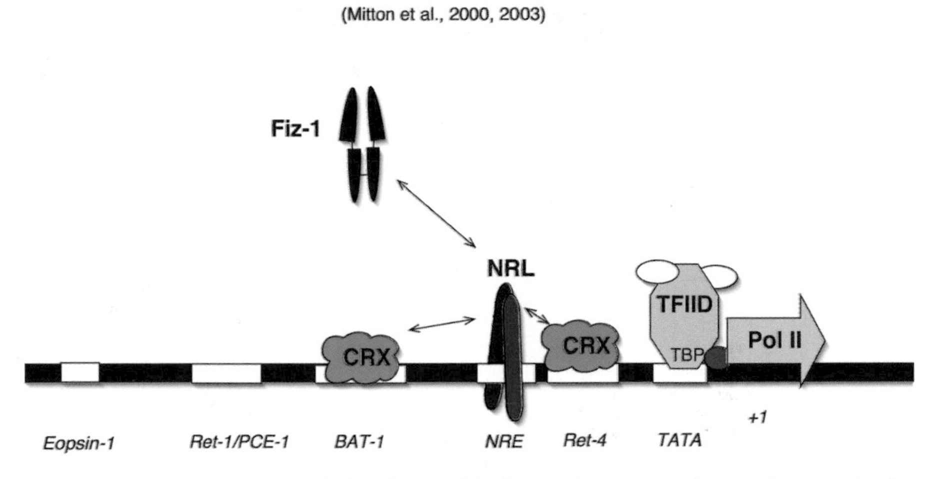

FIGURE 9.6 Known transcription factors binding to known regulatory elements in the well-studied rhodopsin promoter. Binding of transcription fractor NRL to this region can activate transcription even if transcription fractor CRX is not there. Similarly, binding of CRX can activate transcription even if NRL is not present. However, binding of both NRL and CRX at the same time results in synergistic interaction that substantially increases expression over and above what either of them bring about separately. (Courtesy of Ken Mitten, Oakland University Eye Research Institute, Rochester, MI).

Having multiple switches and multiple regulatory proteins for a gene might seem excessively complex. However, this pattern of gene regulation allows for groups of genes that use the same switch and regulator to be coordinately regulated—turned on and off under the same circumstances. It allows for a gene to be "on" in the cell that has the right set of regulators and "off" in the cell that doesn't. Also, use of combinations of regulators to achieve specific regulation allows the complex expression of many genes in many different cell types without having to have one or more regulatory proteins per gene.

OUTSIDE INFLUENCES

Like an electrical appliance in your home that is not plugged in, some regulatory proteins may be present in a cell but not in an active state. What turns on, or activates, these regulatory proteins is not electricity but rather binding to another molecule, such as a hormone. For example, hormones are small proteins that are made by some cell types to be used as signals to send messages to other cell types. A hormone receptor may sit inactively on the surface of the cell, waiting until the hormone that can activate it comes along. Once the right hormone comes along and binds to the hormone's receptor, some kinds of hormone/receptor complexes become regulatory proteins that leave the cell surface, go to the nucleus, and bind to switches in the promoter regions of genes (Figure 9.7). This allows the cell to activate a specific set of genes in the presence of the hormone and to leave them inactive if the

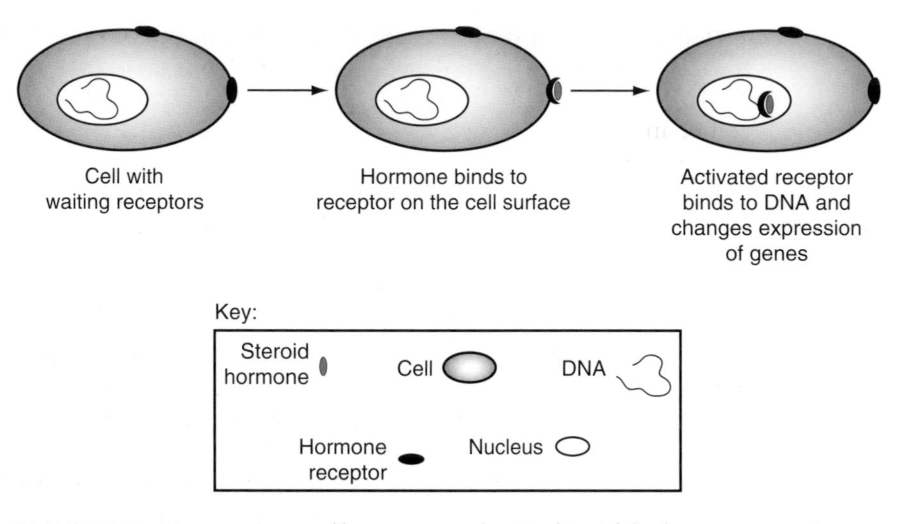

FIGURE 9.7 How one type of hormone works. Binding of the hormone to the receptor on the surface of the cell changes the properties of the receptor, and the receptor/hormone complex travels from the cell surface to the nucleus, binds to DNA in the vicinity of the promoters of multiple genes, and causes changes in levels of transcription of those genes.

hormone is not there. In other cases, binding of the hormone to the receptor causes the receptor to send a signal into the physiological machinery of the cell without physically travelling into the cell, but the result can be similarly profound changes on events taking place within the cell. Many genes change expression when a new hormone is introduced or the amount of it is increased or decreased. And expression of those genes can change again if that hormone decreases or goes away. Think about how this can help explain things such as growth spurts and changes in the human body during puberty!

Many other external factors can affect expression of human genes, including infections, inflammation, allergies, injury, nutrition, medications, temperature, and emotional reactions. In many cases, several of these effects may be going on at once.

Many of the changes in gene expression that take place in response to external factors are designed to help us heal or adapt to our environments. During wound healing, expression of collagen genes produces collagen proteins that become part of the scar that eventually seals off the wound site. However, some changes in gene expression can turn out to be maladaptive, such as those that take place during inflammatory processes that can cause further damage after some kinds of injury.

HOW WE BECOME HUMAN

Embryonic development starts with the fertilized egg that is nothing like a finished human body. There follows an amazing progression not only in the number of cells present but also in the shape of the embryo and the gradual addition of new organs and cell types. Rapid and dramatic changes in gene expression accompany the changes in the embryo. The actions of transcription factors such as those we have been describing gradually distinguish cells that will become the outsides of our bodies from cells that will make up our "innards." As the combinations of transcription factors that are present in a cell change during embryogenesis, cells that initially are specified only as to general regions of the body—inside or outside, towards the top or towards the bottom—gain more and more specific instruction on what they will become. Thus changes in transcription factors bring about the gradual differentiation of cells. A cell that has initially been "told" by a combination of transcription factors only that it will be internal to the body then gets a set of signals telling it to become part of the top of the organism. Later, as it continues to divide and make more cells, some of them get signals indicating that they are now fated to be something neuronal, then that they are to be part of the brain, then that they have become the type of brain cell that receives signals that come from the eye.

If genes regulating gene expression during development are not expressed exactly when they are supposed to be expressed, the embryo may not be able to go on to later stages of development and the effects can be lethal. Some of the genes expressed early on determine profound things such as which end of the organism will be the head rather than the feet. Major errors in laying out the basic pattern of the embryo tend to be lethal quite early.

Study of *developmental regulatory* genes tells us that specific gene regulation events can set off a cascade of changes in gene expression that correspond with changes in the developmental program for a set of cells, taking cells that were all destined to become parts of the front end of the organism and committing some of those cells to become parts of the eye while committing other cells to become parts of the brain. Some of the most dramatic lessons on this subject come from the study of fruit fly mutants like the *Antennapedia* mutant in Figure 9.1. These types of mutants, called homeotic mutants, result in changing the developmental fate of a set of cells, so that cells that were destined to become one body part become a different body part instead. It is now known that in humans as well as flies, the *homeotic* genes responsible for some of these cell fate commitments are transcription factors that bind to the DNA to regulate the expression of many other genes. Interestingly, mutations that cause expression of a homeotic gene in the wrong place with the resulting production of an organ not normally seen in that location (*ectopic expression*) can cause a very different set of characteristics than loss-of-function mutations in that same gene. The dominant *Antennapedia* mutation alters gene regulation in such a way that cells that should turn on the battery of genes whose functions lead to antenna development, instead activate a set of leg-building genes.

Some genes are responsible for left-right asymmetry, such as specifying the internal asymmetry of the human organs—heart on the left, liver on the right, and so on. Some people who have their hearts on the right instead of the left have two damaged copies of a gene called the *situs inversus* gene (Figure 9.8). You might think that this means that the damaged copy of the *situs inversus* gene makes the heart end up in the wrong position. You might also think that hearts end up on the right unless the *situs inversus* gene is working correctly. The story is more complicated and interesting than that. What actually happens is that when the *situs inversus* gene product is missing,

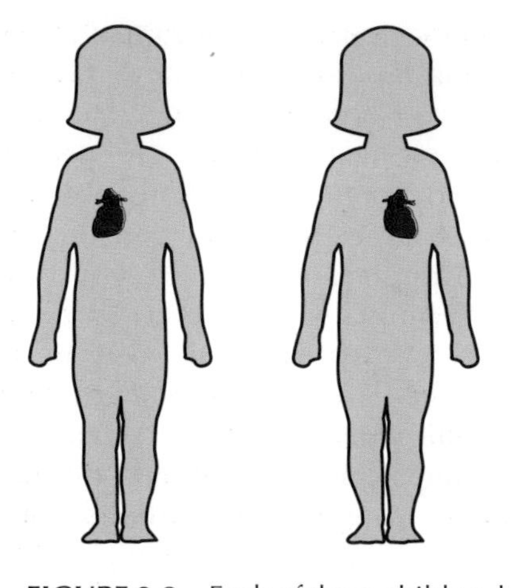

FIGURE 9.8 Each of these children has two damaged copies of the *situs inversus* gene. The lack of the *situs inversus* gene product removed a regulatory step that normally would tell the body which side to put the heart on, so in each case the laterality got assigned randomly. In one case, one time the heart happened to get assigned to the left where it belonged, and in the other case it happened to get assigned to the right. In fact, in these people, more organs than just the heart are affected. Imagine how confusing it would be to sort out the mode of inheritance for this gene for which half of the affected individuals with the gene defect have the normal phenotype. It is an example of a situation in which studying animal models of disease can greatly simplify our understanding of what is going on.

the developing body does not know whether to put the heart on the right or on the left and so assigns the sidedness randomly. So half the people with two damaged copies of the *situs inversus* gene have their hearts on the right, and half the people with two damaged copies of the *situs inversus* gene have the hearts in the normal position on the left. Lack of *situs inversus* protein does not cause right-sided placement of the heart; it causes a random decision to be made as to whether to go to the right or the left.

As the embryo develops, changes in gene expression gradually cause more and more specialized events. The very first sets of gene expression in the egg direct cells to divide and migrate and establish specialized pools of precursor cells. Very early actions by transcription factors set off cascades of gene expression changes downstream of the initial transciption event, including production of other transcription factors that are each followed by their own complex array of gene expression changes. Genes specify where the organs are located, what form they take, and what functions they can carry out. For example, regulatory proteins that affect expression of genes involved in formation of the eyes must be expressed in just the right amount and at just the right time, or the baby might end up with damaged eyes or even no eyes at all (similar to the flies without eyes at the beginning of the chapter).

So we see that the human body is made up of highly specialized organs and cells, each with their own specific patterns of gene expression running chronically or sometimes in response to some stimulus from the environment. Liver cells produce certain key enzymes in response to transcription factors binding to the regulatory elements in the genes that produce the enzymes. Similarly, signaling molecules in nerves, ion channels in kidneys, hormones carrying messages from one cell type to another, and structural molecules making up muscles and bone all originate in the activities of transcription factors interacting with regulatory sequences for those genes. However, there is another layer of complexity to the makeup of a human being, and that is the cascade of gene regulation events that causes some cells to take on the structure of a nerve and gain the ability to express neurotransmitters even as a cell in a different developmental lineage is taking on the ability to make immunoglobulin molecules that help protect us against infection. The orchestration of this complex series of gene expression events is initiated when the sperm fertilizes the egg, but the end of this orchestrated piece is not birth or even completion of adolescence because, frankly, we continue changing our gene expression capabilities throughout the aging process. If we look at our makeup at the level of gene expression, we are truly in a process of "becoming" throughout the full lengths of our lives.

EYE-BUILDING GENES IN FLIES AND HUMANS

But what about the missing eyes in the flies? Flies homozygous for loss of function mutations in the PAX6 gene arrive in this world with no eyes. The PAX6 gene encodes a transcription factor (or regulatory protein) that binds to regulatory elements in multiple different genes that are important for eye development. In the absense of *PAX6*, some of the switches to which this regulatory protein should bind are left empty. For genes that need *PAX6* to operate their switches, regulation of transcription changes when *PAX6* is missing or if there is not enough of it.

Does this really have anything to do with humans? Sometimes babies are born with no eyes or very tiny, nonfunctional eyes. Since it turns out that humans have a version of the *PAX6* gene, we have to wonder whether damage to the *PAX6* gene could be the cause of the lack of eyes in these babies. For more than a decade, it has been known that human beings who have one damaged copy of *PAX6* make about half the usual amount of *PAX6* transcript. They often suffer from a disease called aniridia, in which part or all of the iris of the eye is missing but most of their ocular structures and functions are normal. Does this mean that *PAX6* has nothing to do with missing eyes in human beings? No. In fact, one child has been found in whom both copies of the *PAX6* gene were damaged so that the child could make no normal *PAX6* protein. This child was born too ill to survive long, with head and face deformities, central nervous system problems, and no eyes. Such extensive problems were not the result of any massive trauma or infection before he was born. Rather, the problem was an error in gene expression due to the lack of one specific regulatory protein that was needed to operate some critical regulatory element switches. It seems a terrible thing that a change in the blueprint too small to see even under the most high-powered microscope should be able to have such devastating consequences in this newborn child.

BEYOND REGULATORY SWITCHES

Are we trying to say that the entire process of regulating where a gene is used, when it is used, and how much gene product is made comes down to a matter of on and off switches in the promoter that lead to increases and decreases in the amount of mRNA being used to make critical proteins?

Actually, it is not that simple. There are quite a number of other levels at which things can be regulated, including not only the rate at which mRNA is produced but also the rate at which it is gotten rid of. Since mRNA molecules are the temporary messengers rather than the permanent information repository, the cell has a regular process of mRNA breakdown going on to get rid of messages that are no longer needed. Some mRNA molecules are used briefly before being discarded, but others are more stable, that is, they stay around in the cell for a longer period of time before the cell breaks them down and gets rid of them. If the developmental stage advances or something changes in the cell's environment so that the cell finds that it has a lot of mRNA present from a gene it no longer wants to use, the cell might just wait for the natural decay rate for that mRNA to take its course.

Sometimes, however, the cell may need to be able to get rid of that mRNA and stop making that protein rapidly without waiting for the gradual loss of the mRNA. One of the ways that cells cope with the need to reduce the amount of an mRNA that has already been made is by making small intefering RNAs (siRNAs) that contain a sequence complementary to the sequence in the mRNA to be eliminated. Pairing of the siRNA with the mRNA signals to the cell that this is an mRNA to be eliminated without waiting out the normal life span of the mRNA.

Similarly, sometimes the cell needs to be able to get rid of a protein that has already been made from the gene's information. In some cases, a protein to be eliminated may be digested by a *protease* that the cell makes for just that purpose. In other cases, the cell may conserve its resources and keep the protein around in an inactive form so it can use it again later. One of the ways it does this is by sticking a chemical tag onto the protein that is supposed to be active and then removing that chemical tag when it wants the protein to shut down for a while.

Thus the cell is actually orchestrating a very complex array of events that control the production of the mRNAs, the persistence and reuse of those mRNAs, the amount of the protein present in the cell, and the active status of proteins that are kept around even though they are temporarily not needed.

SUMMARY

Regulation of gene expression is carried out by the coordinated efforts of the regulatory proteins and the regulatory element switches that they control. Changes in regulation of gene expression not only determines specific differences between cell types in the body but also between different stages of life. Thus the human genome is carefully orchestrated, with a symphony that begins with a frenzy of fluctuating gene expression throughout fetal development and childhood and then settles into a slightly more stately adagio in which gene expression continues to evolve as we age. This cascade of gene expression changes, which seemingly starts with fertilizaton of the egg, really begins a step before that, at the point when it is determined what genetic material goes into the egg and into the sperm. To begin exploring the process by which the genetic blueprint specific to one human being comes about, we will look through the microscope into the nucleus in Chapter 10.

Section 3

HOW CHROMOSOMES MOVE

Here we present a microscopic view of human chromosomes, discuss how the cell passes the right number of chromosomes along to new cells, and discuss how we pass the right number of chromosomes along from one generation to the next.

SO WHAT ARE CHROMOSOMES ANYWAY? 10

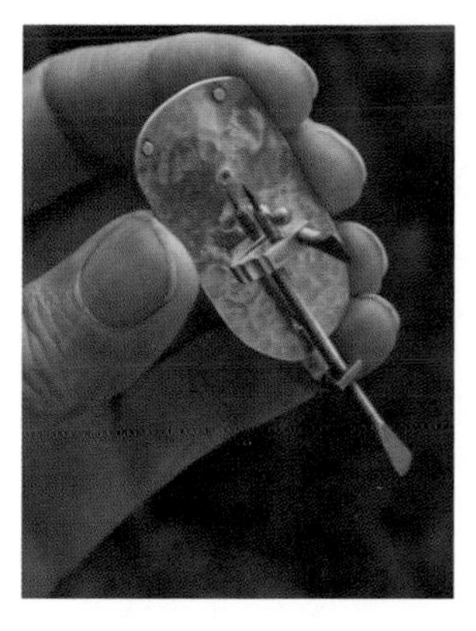

(Photo courtesy of Alan Shinn)

Antonie van Leeuwenhoek (1632–1723) was a Dutch tradesman who invented the first microscope capable of viewing individual cells and microscopic life forms. He was the first to discover bacteria, microscopic parasites, and sperm. His microscopes were tiny and quite different from modern instruments found in genetics labs. Alan Shinn has created a set of directions for construction of a replica of van Leeuwenhoek's microscope, which, as you can see in the picture shown here, is small enough to hold in one hand.[1] Some types of cells are large enough to be distinguished with this early technology, but it will not let you see individual chromosomes. His microscopes, small enough to be held in the palm of the hand, were not powerful enough to allow visualization of DNA, but his breakthroughs provided foundations for more advanced microscopy used today to view chromosomes.

We talk about the genetic information inside of our cells being like a blueprint, but it may be more useful to think of our genetic information being stored in a 23-volume set of encyclopedias, with two copies of each volume present in each cell. These volumes, called *chromosomes*, each exist as a long string of DNA made up of millions of genetic letters spelling out thousands of genes along the length of each chromosome. Thus, when the cell does something to a chromosome, it is actually acting on a large number of genes all at once.

VIEWING THE NUCLEUS THROUGH A MICROSCOPE

Here is an amazing concept—we can see these long strings of DNA through a microscope! If we start out looking at a magnification of 200 or more, we find that many cells have a nucleus that is filled with indistinct material that offers no clues about what is in there, something that is typical of nondividing cells in a metabolic resting state (Figure 10.1).

[1] Instructions for making this microscope can be found at www.mindspring.com/~alshinn or by searching on Shinn and van Leeuwenhoek.

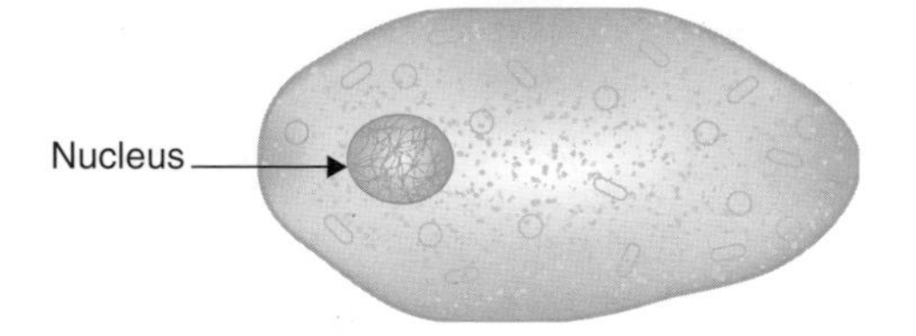

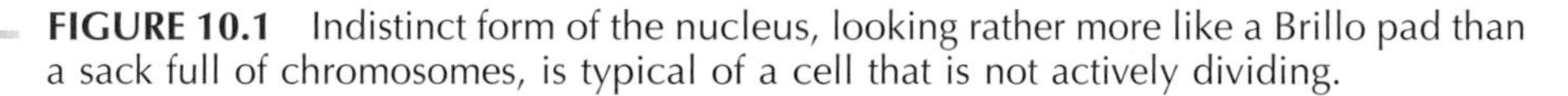

FIGURE 10.1 Indistinct form of the nucleus, looking rather more like a Brillo pad than a sack full of chromosomes, is typical of a cell that is not actively dividing.

If we increase the magnification, we find that some actively growing and dividing cells show distinct features in the nucleus—long, thin, threadlike structures. These are the chromosomes, each one a piece of DNA that is millions of genetic letters in length. If we spread them out so that we can count them, we find that there are forty-six chromosomes in most human cells.

THE VISIBLE LANDMARKS ON THE CHROMOSOMES

A first look at chromosomes under the microscope can seem fairly confusing. However, there are several physical features that let us identify the different chromosomes and tell them apart. First, the chromosomes come in different sizes, with the longest chromosome being more than five times the length of the smallest. Second, each chromosome has a constriction called a centromere, which can be found near the middle of some chromosomes and near the end of others, that divides the chromosome into longer and shorter arms (Box 10.1).

However, some chromosomes are very similar in size, and this overall shape—length plus centromere position—does not let us tell all of the chro-

BOX 10.1 CHROMOSOME ARMS: THE LONG AND THE SHORT OF IT

Length, centromere position, and banding pattern all help distinguish different chromosomes from each other. Chromosomes are numbered according to their size. Chromosome 1 is the longest and chromosome 22 is the smallest. Two chromosomes are left out of this numbering scheme—the X and Y chromosomes. On some chromosomes the constriction called the centromere is near the middle, and on some it is near the end. Noting the long arm (called the p arm) and the short arm (called the q arm) of a chromosome can help in telling one chromosome from another, especially in cases in which sizes are so similar that they are hard to tell apart. Thus, if someone says that a gene is located on chromosome 10p, they mean that the gene is on the short arm of chromosome 10. There is a tale that says that the names for the chromosome arms were originally supposed to be p (as in petite) and g (as in grande) but that a printer misread the g as a q, leaving us with a naming system that is not so obvious.

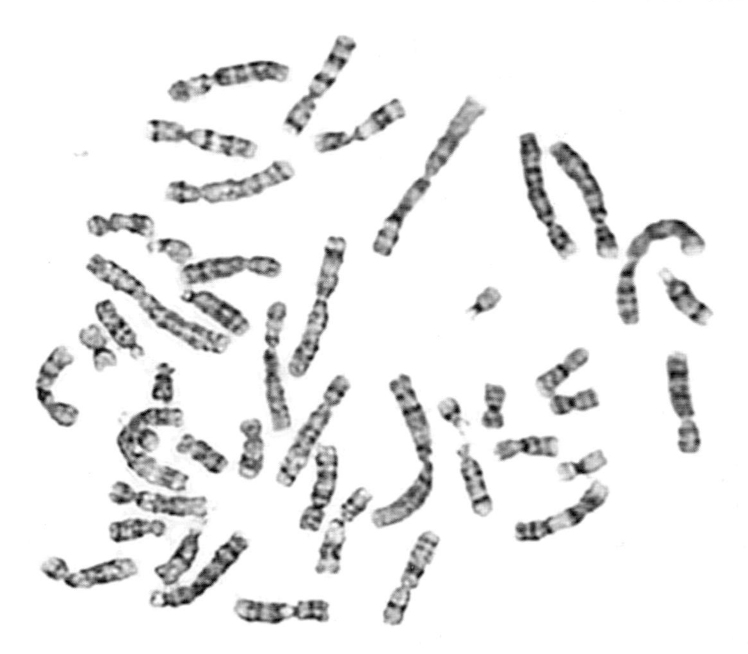

FIGURE 10.2 Intact metaphase cell. (Courtesy of the Clinical Cytogenetics Laboratory, University of Michigan, Ann Arbor, Michigan; Diane Roulston, Ph.D., Director.)

mosomes apart uniquely. An additional trick, the use of dyes that stain some parts of a chromosome more darkly than others, produces a pattern of light and dark bands rather like a bar code specific to each of the individual chromosomes. Once the *banding pattern* is combined with other information on size and shape of the chromosome, we can tell all of the chromosomes apart from each other, even chromosomes such as 11 and 12 that are very similar in size and centromere position. In Figure 10.2, it may not be obvious that the banding pattern is all that helpful, but if we cut out the pictures of the chromosomes and arrange them systematically, the banding pattern becomes a great aid (Figure 10.3).

The usefulness of banding patterns becomes a bit more apparent when the pictures of the chromosomes are cut out and arranged by size, centromere position, and banding pattern to produce this karyotype picture. For instance, looking at Figure 10.2 did not make it immediately obvious whether the individual was male or female, but once the chromosomes are lined up in pairs and by size as in Figure 10.3, it is much easier to tell that this cell has two X chromosomes and thus comes from a female. Figure 10.4 shows an idealized image of the human karyotype—the set of chromosomes arranged by size, centromere position, and banding pattern. Because the real pictures are less clear than this, karyotyping normally involves photographing and studying chromosomes from multiple cells to be sure that clear enough images have been obtained for each chromosome. To help people communicate about features they observe concerning the chromosomes, there is a system of naming for the pattern of bands that lets someone publish a geographic designation such as Xp21 and have others know exactly which band on the short arm of the X chromosome is being discussed (Figure 10.5).

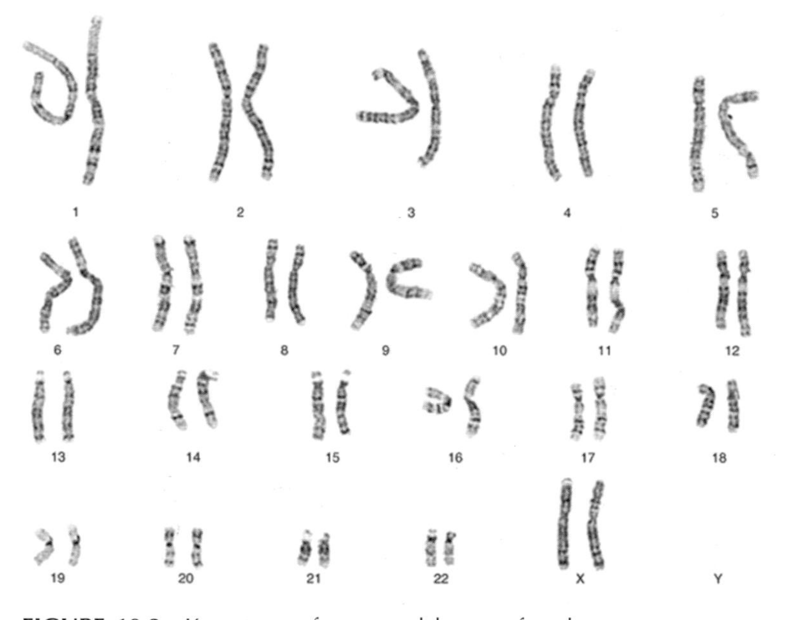

FIGURE 10.3 Karyotype of a normal human female. (Courtesy of the Clinical Cytogenetics Laboratory, University of Michigan, Ann Arbor, Michigan; Diane Roulston, Ph.D., Director.)

CHROMOSOMES COME IN PAIRS—MOSTLY

One of the very striking things about the carefully arrayed chromosomes of a woman shown in Figure 10.3 is the way in which the forty-six chromosomes can be grouped (by size, centromere position, and bar code banding pattern) into twenty-three pairs. The chromosomes that have been paired because they look alike are effectively two copies of the same chromosome, also known as *homologues*, with the same order of genes arranged along the length of the chromosome and almost exactly the same DNA sequence. Normally, the individual's mother donated one member of a pair of homologues and the father donated the other member of the pair. There are usually at least some differences between the two homologues at the level of the DNA sequence, but the gross structure of the two homologues is the same.

BUT MEN HAVE TWO CHROMOSOMES THAT DON'T MATCH

When researchers looked at cells from a man, they saw that only twenty-two pairs of visually similar chromosomes could be formed (Figure 10.6). This set of "matchable" chromosomes, which looks the same in both sexes, is referred to as the *autosomes* (chromosome pairs 1–22).

After matching those first twenty-two pairs, there were two chromosomes left over in the male cells that did not look alike. One of the medium-sized chromosomes, called the *X chromosome*, matches one of the chromosomes present in the twenty-third female chromosome pair. However, the unmatched chromosome was a very small novel chromosome (the Y *chromosome*) not present in female cells. As we will see later, the structure, function, and behavior of these chromosomes help us to understand sex

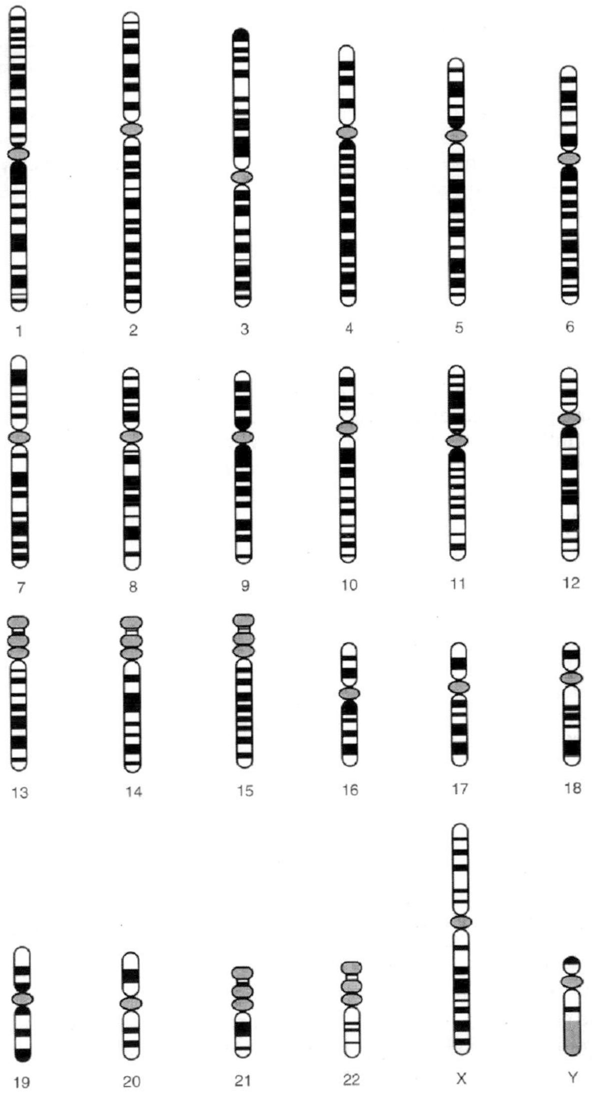

FIGURE 10.4 Idealized diagram of a chromosome banding pattern that helps with identification of the images that are harder to distinguish in a real microscopic chromosome spread.

determination, the process that decides whether the baby will be male or female.

FISH AND CHROMOSOME PAINTING

Only very large changes in a chromosome can be detected using normal karyotypes. The more recent use of brilliantly colored fluorescent dyes makes many things about the microscopic chromosomes easier to identify and distinguish. A process called *fluorescence in situ hybridization* (*FISH*) allows a brightly dyed piece of DNA to find and attach to a spot on a chromosome

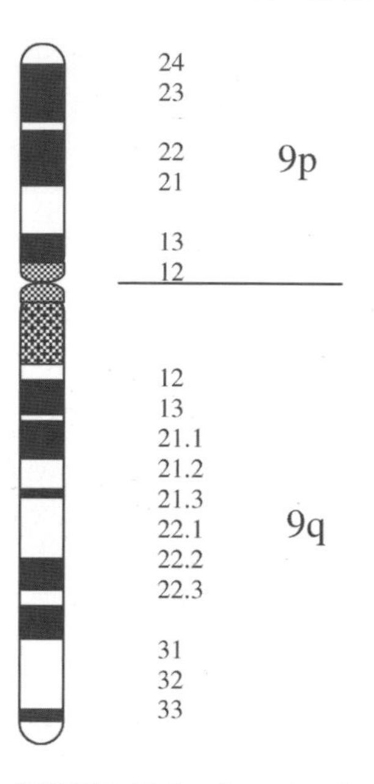

FIGURE 10.5 Counting bands in the chromosomal bar code. Bands on the cytogenetic bar code are numbered outwards from the centromere. As shown in this diagram of chromosome 9, a gene located on band 9p21 is on the short arm, and a gene located on band 9q31 is on the long arm. A gene located on band 9p23 is on the short arm of chromosome 9 and is farther from the centomere than a gene located at band 9p22. If a gene is located on band 9q21.3 its position on the long arm of chromosome 9 is farther from the centromere than a gene at band 9q21.1.

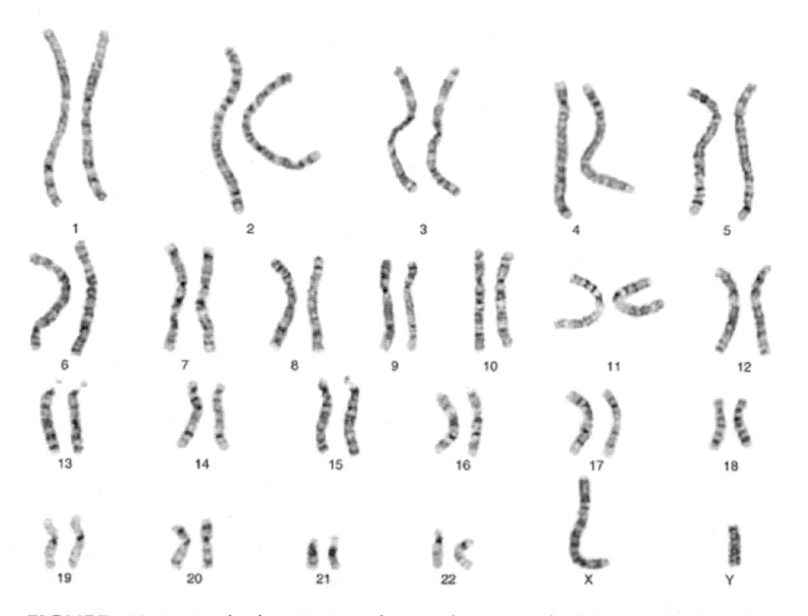

FIGURE 10.6 Male karyotype shows the two chromosomes that are not matched in males, the X and the Y, which are not included in the size-based chromosome numbering scheme. (Courtesy of the Clinical Cytogenetics Laboratory, University of Michigan, Ann Arbor, Michigan; Diane Roulston, Ph.D., Director.)

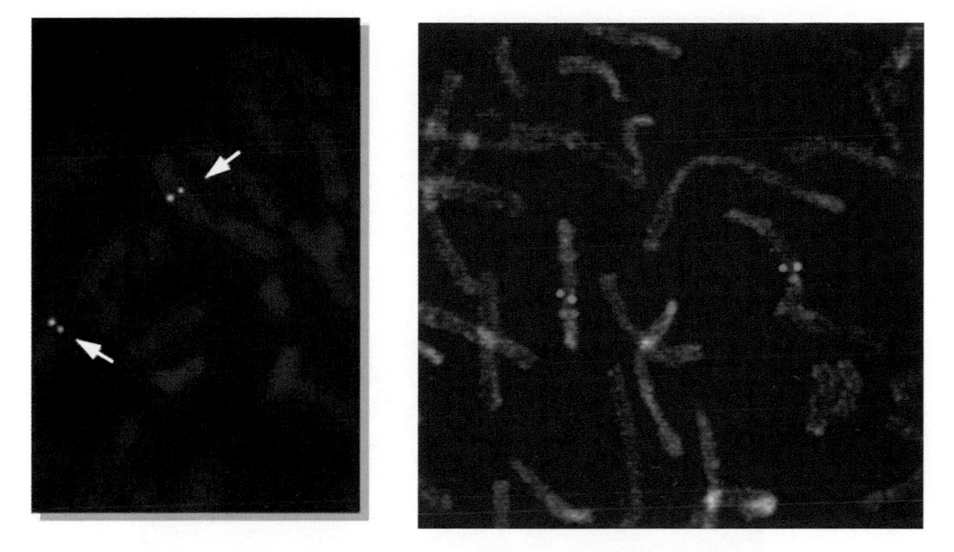

FIGURE 10.7 FISH lets us attach a brightly dyed piece of DNA to the point on a chromosome that contains the same sequences present on the dyed probe. In this picture, we can tell that the gene corresponding to this particular probe is not deleted because it lights up on both copies of the chromosome on which this gene is normally found. (Single-color FISH courtesy of Thomas Glover and multi-color FISH courtesy of Octavian Henegariu.)

that has the same sequence as the fluorescent DNA probe. In some cases, this has been used to let investigators find out where a particular gene is located (Figure 10.7). In other cases, it provides a sensitive test of whether a particular gene is missing from a chromosome that lets researchers detect deletions that are much smaller than anything detectable in a standard karyotype. Normally, spots on two chromosomes would light up for most genes, but if one copy of the gene were deleted, spots would appear on only one chromosome instead.

Very recently, it has even become possible to stain each chromosome a different color in an impressive process called *chromosome painting* (Figure 10.8). This provides a very powerful way to distinguish different chromosomes that in some ways seems like a major improvement over the tiny black-and-white banding patterns, although some of these new color-based technologies can be problematic for researchers who are color-blind. Later on, when we talk about damaged chromosomes and some kinds of cancer, we will tell you more about ways this new imaging approach can be used to study medically important processes.

This ability to see chromosomes through a microscope lens and distinguish the chromosomes from each other offers powerful opportunities to answer important questions. As we saw here, it demonstrates a fundamental genetic difference between males and females (XY vs. XX). In Chapters 11 to 13, we will show you how this kind of microscopy lets us tell a great deal about how cells move chromosomes around and pass them from one generation to the next. As you will see, understanding what happens to the chromosomes when they are passed to new cells or new generations in a human family shows us some important things about the relationship between chromosomes and Mendel's units of heredity.

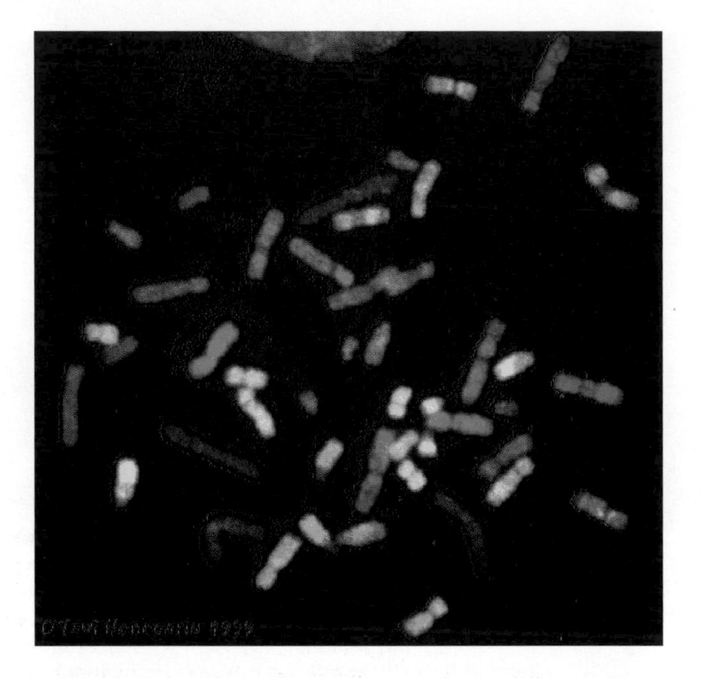

FIGURE 10.8 Chromosome painting shows different chromosomes in distinct colors. (Courtesy of Octavian Henegariu.)

We cannot amplify the magnification far enough to read the order in which the As, Cs, Gs, and Ts are arranged, even if we use an electron microscope; we have to use other biochemical tricks to decipher the actual order of the genetic letters and "read" the blueprint. Even if we can't read the blueprint through the microscope, we can distinguish the different volumes of the encyclopedia from each other and track them as the cell moves them around. As we will see in the next several chapters, being able to see the chromosomes lets us look at some profoundly important processes that go on in the cell.

HOW CELLS MOVE YOUR GENES AROUND

11

Several years ago, a brief visit to the dermatologist left a precise surgical hole in Julia's thumb about the depth of a dime and half as wide. Over the course of several weeks, a pinkish mist of cells gradually spread inwards from the edges to fill the hole with solid skin and scar tissue. This migration of cells across the open space represented not just movement but cells growing and making new copies of themselves at a frantic pace. With each round of cell growth, the genetic blueprint in those cells was being copied and passed along to new cells with a level of speed, efficiency, and precision that human industry has never matched.

Have you had a rug burn lately? Or perhaps cut yourself? Have you ever wondered about the processes that go on as such damage is repaired? You spend your whole life replacing and repairing losses due to erosion and injury. In fact, cell types such as skin cells divide many times as we age in an effort to keep up with the rate at which we are losing cells. We have all repeatedly used this process of *cell division* to turn a small number of cells into a larger number of cells. It is how we developed into the large, complex animals we are from the single-celled *zygotes* that were created many years ago when *that* sperm met *that* egg and we came into being.

It seems a simple enough thing to imagine how a skin cell would duplicate itself as it joins the rush of cell growth that will fill in a damaged area. The cell is basically a sack full of organelles, the little biological engines and factories that run different functions within the cell. If this sack gets bigger while making extra copies of everything inside of it, there will be enough extras to make up two complete cells identical to the original cell. Once enough of the cell's innards have been duplicated, the cell divides down the middle to make two new cells.

For some organelles in the cell, such as the mitochondrial energy factories and the ribosomal protein-synthesis machinery, there are so many copies that both daughter cells are bound to get their fair share. If there are millions of ribosomes in a cell, there is no great concern that a dividing cell would put all of them into one daughter cell and leave none in the other (Figure 11.1). It is very simple: duplicate the contents, pinch the growing cell into two new cells, and there will be plenty of organelles in both cells. Besides that, the cell can make more of these plentiful organelles if it needs them because each daughter cell has the genetic blueprint that gives directions for their synthesis.

PASSING MOST ORGANELLES ALONG IS EASY

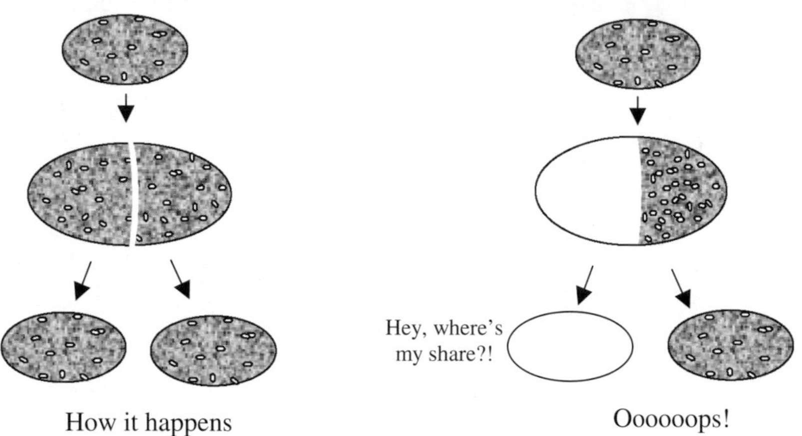

FIGURE 11.1 Dividing the resources during cell division. No, it's not a pizza being cut, it's a cell undergoing division. This cell needs to be sure that both daughter cells end up getting their share of the cellular resources. Because there are so many organelles such as ribosomes (black dots) and mitochondria (white spots), they are spread around the cell uniformly enough so that, when the cell divides, both halves have some of the organelles as shown on the left. Just as we don't expect all of the air in the room to up and move over to the right hand half of the room, we do not expect a fairly uniform distribution of millions of tiny organelles to all run to one side of the cell before it is cleaved down the middle. So the cell doesn't need a mechanism to actively put the right number of ribosomes and mitochondria into the two separate daughter cells, in part because it does not care whether the two cells have exactly the same number of organelles. They will end up with approximately the same number, so each cell ends up with enough and the exact number does not matter. This might seem obvious, but there are a lot of other ways this could be done that could have caused the cell to have to really work at dividing up these resources. We have left the nucleus out of this picture because this mechanism of approximate division of broadly dispersed organelles does not apply to it.

Passing the *genetic blueprint* along to new cells during cell growth is not nearly as simple as passing along the other organelles. The cell needs to be very careful about passing along copies of the genetic blueprint into the daughter cells. It actually really matters (in fact, it matters a lot!) whether a cell has exactly the right number of copies of the genetic blueprint. Remember that each cell has only two copies of the blueprint in the form of two copies of each volume of the genetic encyclopedia, or, as we were able to see through the microscope, two copies of each chromosome. So the cell cannot just split down the middle and hope that exactly two copies of each chromosome happen to be sitting to the left of the dividing line and exactly two copies happen to be sitting to the right of the dividing line. It would be too easy to end up with three copies over here and one over there, with disastrous consequences to both daughter cells (Figure 11.2).

So although the cell can fairly passively divide most of the types of organelles that exist in large numbers, it has to take active steps to be sure that partitioning of the small numbers of copies of the blueprint come out correctly by actively moving one copy to the right of the dividing line and the

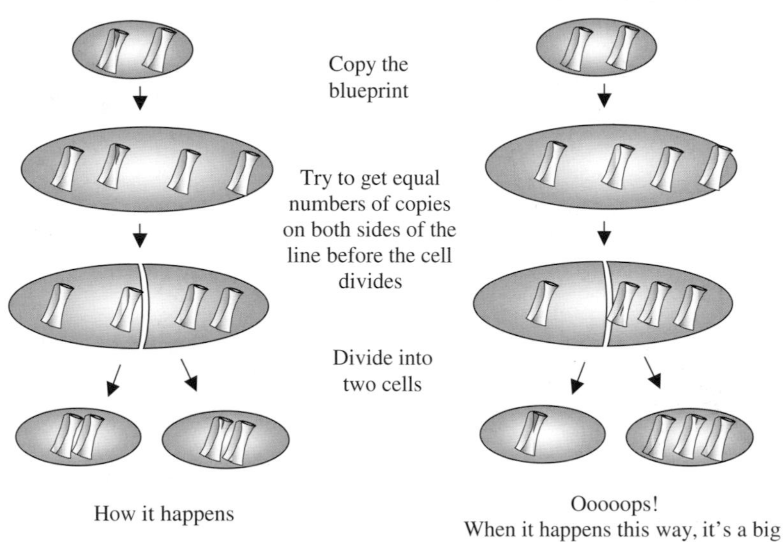

FIGURE 11.2 The cell has to work actively to make each daughter cell end up with exactly, not approximately, the same number of copies of the genetic blueprint, or else it can end up with the scene shown on the right, with some cells getting too many copies and some not getting enough. As we will discuss in Chapter 22, having the wrong number of copies of all or part of the blueprint can cause major health problems. Since there are actually forty-six chromosomes, or volumes, to this blueprint, the whole process is much more complex than this, as we will show in the rest of the chapter.

other copy to the left. Since we are talking about a blueprint made up of twenty-three pairs of chromosomes, each of which must separately be passed along in correct numbers, the situation is far more complex than what is shown in Figure 11.2.

Fortunately, we can use the microscope as we did in Chapter 10 to look at what happens to the encyclopedia volumes (chromosomes) as the cell copies its contents and then splits into two daughter cells. We can start with cheek cells, white blood cells, or other sources of cells. The important thing is that we want to be looking at actively dividing cells so that we can watch the transfer of chromosomes from parent cell to daughter cells.

THE CELL CYCLE

First, we need to understand the basic process of cell division that creates identical cells. The process of cell division involves the completion of a series of cellular events known as a cell cycle, which includes duplication of the cellular resources and *mitosis,* a crucial process that moves copies of the genetic blueprint into the daughter cells (Figure 11.3).

Through most of the cell's life the DNA molecules are loosely entwined with each other in the cell nucleus, going about the gentle business of running various aspects of metabolism and growth. During this time, the chromosomes are not visible as separate entities; rather, the nucleus looks like an

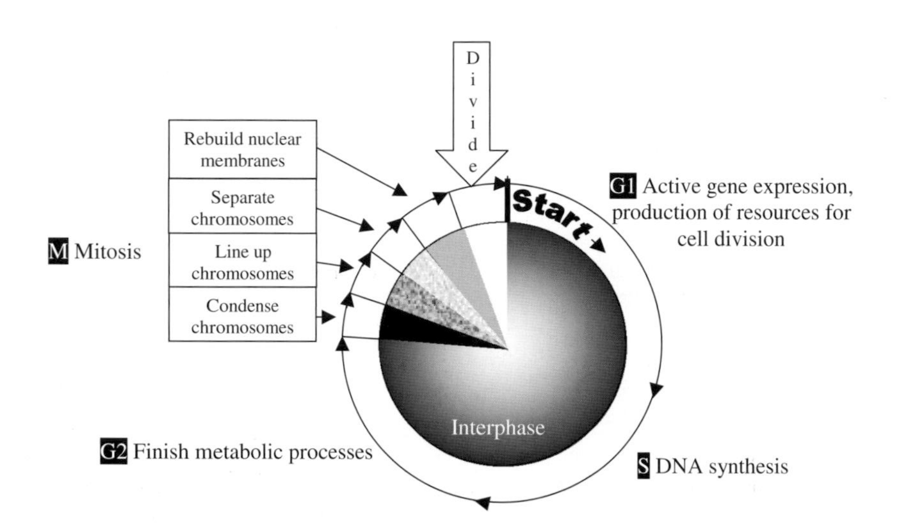

FIGURE 11.3 The cell cycle. Every dividing cell goes through the series of steps shown here. It starts with G1, the period at the beginning when the cell accumulates resources needed for the next round of cell division. During the next step, S, DNA synthesis copies the chromosomes. During G2, the cell finishes off any remaining metabolic processes needed for cell division. Interphase consists of G1 plus S plus G2, a period during which the cell looks pretty much the same under the microscope. The microscopic view starts to change during M phase, or mitosis, when the chromosomes are condensed (prophase), line up (metaphase), moved to the separate ends of the cell (anaphase), and packaged back into a nucleus as the cell prepares to divide (telophase) in an order manner. Cytokinesis is the actual cell division step that separates the two new daughter cells. Cells that are growing slowly spend a lot of time in G1. This pie chart shows an average representation of amount of time in the cell cycle spent in each of these stages. It also shows that mitosis is a very brief part of the cell cycle. If cells are truly inactive and not dividing, they go into a metabolic resting state called G0 instead of going into the metabolically active state of G1. *Take-home message: During interphase, when the nucleus looks like a Brillo pad, the cell makes copies of everything and gets ready for cell division. During the visibly distinct stages of mitosis, the cell carts the chromosomes around to where they should be (a process we can see under the microscope), and cytokinesis completes the separation into two cells.*

old Brillo pad. Only once the cell starts the process of mitosis do we begin to see distinct structures within the nucleus. So let's take a look at mitosis and see how it works.

MITOSIS PUTS CHROMOSOMES INTO DAUGHTER CELLS

Let us start our examination of mitosis, the process of getting the right chromosomes to end up in the right copy numbers in the right cells, by looking at a simplified case. Let's imagine a very simple fictitious organism *Organisma hypothetica*, otherwise referred to as *O. hypothetica.* This imaginary beastie consists of a small number of cells whose genes are arranged on only a single pair of homologous (which is to say, essentially identical) chromosomes (Figure

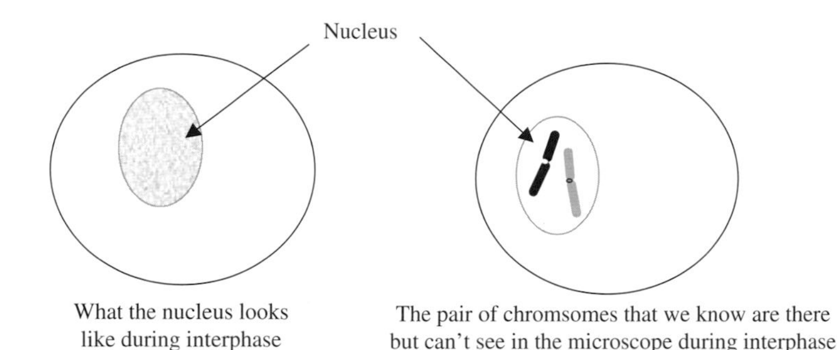

FIGURE 11.4 A very simplified diagram of the original fictitious *O. hypothetica* cell that we are going to follow through mitosis. In a real cell, if we looked at the nucleus, it would look indistinct like the cell on the left, but since we know the pair of chromosomes is in there, we show a picture of the pair of chromosomes so that you can begin to follow what happens. We are leaving out the other organelles because we have decided that they are not having a problem getting transferred to the daughter cells. To help keep track of this pair of chromosomes as we go, we are showing the chromosome that came from the mother in black and the chromosome that came from the father in gray. These are *homologous* chromosomes, with the same genes in the same order arranged along the length of the chromosome.

11.4). As you may recall, each of these chromosomes consists of one very long piece of double-stranded DNA. Now suppose that one cell in this organism needs to divide in order to form some necessary structure consisting of two or more cells.

At the beginning of the cell cycle of *O. hypothetica*, there are two DNA molecules and thus two chromosomes, each consisting of a long double-stranded piece of DNA (see Figure 11.4). Since these are two copies of the same chromosome, we will call it a pair of chromosomes. The cell copies these two DNA molecules to create four complete copies of the chromosome and then puts two copies into one daughter cell and two copies into the other when it divides. Let's take a more detailed look at this process of copying the chromosomes and moving them into the new cells.

When a chromosome is sitting in a nondividing cell prior to its replication, it is composed of only one long, double-stranded DNA molecule. Once the cell cycle starts, the cell gets to S phase, the DNA synthesis phase. At this point, it replicates the one, double-stranded DNA molecule in each of the two chromosomes, which results in a more complex structure consisting of two complete copies of the DNA molecule held together at the constricted point called the centromere (Figure 11.5).

Each complete copy of the whole DNA molecule in this X-shaped chromosome is called a *sister chromatid* (Figure 11.6). Each single chromosome in a nondividing cell contains thousands of genes arranged in a linear array along the length of the chromosomes. Thus each X-shaped chromosome found in dividing cells contains two identical linear arrays of genes running in parallel, in exactly the same order along the length of the sister chromatids.

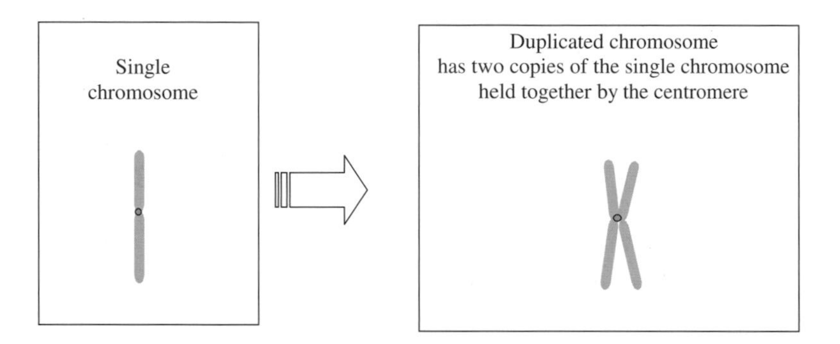

FIGURE 11.5 *Interphase—S phase* copies the chromosomes. Two forms in which chromosomes exist. In a resting cell that is not actively growing and dividing, a chromosome is a long, single chain of complementary double-stranded DNA with a constriction called a centromere, but we usually can't see individual chromosomes when we look at resting cells. Usually, we look at actively dividing cells to see the X-like structure of the replicated chromosome shown on the right, which actually contains two copies of the chromosome from the left, joined at the centromere so that the cell can move them around together until ready to separate them.

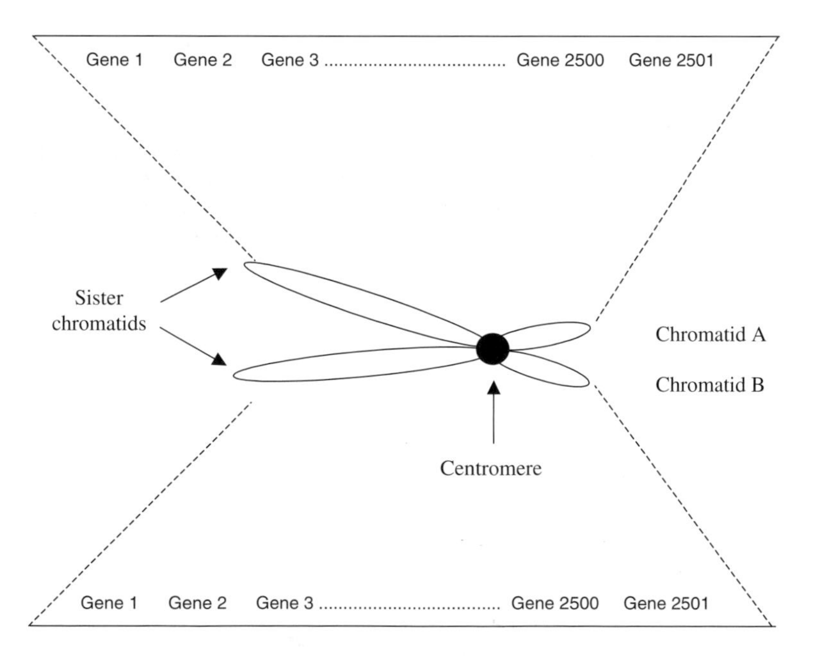

FIGURE 11.6 The replicated chromosome. Diagram of a replicated chromosome before cell division. Chromatids A and B are copies of each other. Thousands of genes are arranged along the DNA in the chromosome rather like beads on a string, one after another. The order of the genes along the two chromatids is the same.

Once copying of the DNA is complete, the chromosomes begin to condense and become visible under the microscope as distinct entities. The stage is called *prophase*. It is easy to remember this term if you think of *pro*-as meaning "before," as in before the chromosomes start moving around within the cell. As prophase continues, the cell begins to assemble a scaffold with two poles called a *spindle apparatus* around the nucleus. This structure is comprised of protein assemblies called microtubules that will facilitate the process of chromosome movement. The end of prophase is signalled by the breakdown of the nuclear membrane that surrounds the chromosomes. As the membrane breaks down, the centromeres of each chromosome attach to microtubules emanating from each pole, such that one chromatid is attached to each pole at its centromere (Figures 11.7 and 11.8). These microtubules,

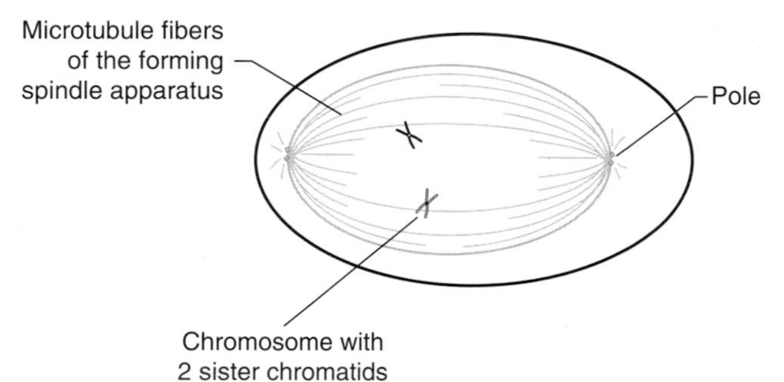

FIGURE 11.7 *Nuclear breakdown* at the end of prophrase—the nucleus opens and chromosomes attach to spindle fibers. Fragments of the nuclear membrane appear as fragments of dashed lines. X-shaped replicated chromosomes have condensed into a form that makes them easier to visualize under a microscope. A protein scaffolding called a spindle aparatus forms threads that run between the two poles of the cell and attach to the chromosomes.

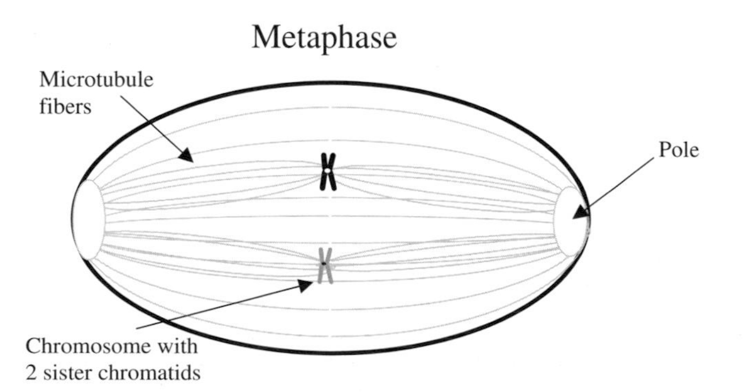

FIGURE 11.8 *Metaphase*—the chromosomes are pulled to the center of the cell in a schematic diagram of chromosomes on the spindle at metaphase in our fictitious cell with only one pair of chromosomes. In a real human cell, there would be forty-six chromosomes, each attached to the microtubule fibers of the spindle, and the spindle apparatus would contain a vastly greater number of the threadlike microtubule fibers. Notice that fibers of the spindle attach to the chromosomes at one end and the opposite poles of the cell at the other end. Note the spindle apparatus attaches to two sides of the same centromere.

which are the protein "train tracks" along which the centromeres can move chromosomes, run from the centromeres to the ends or poles of the spindle. Complex protein structures, known as kinetochores, which are assembled at each sister centromere, actually connect the sister centromere of each chromosome to microtubule fibers. Each replicated chromosome ends up bound to its own set of tracks within the spindle apparatus, such that microtubules connect the centromere of one sister chromatid to one pole while other microtubules connect the other sister chromatid to the opposite pole.

By the time the cell cycle advances to the next stage called *metaphase*, the chromosomes have moved to the center of the cell, midway between the poles, and lined up on a "plate" that is the cross section through the center of the cell. Again, kinetochores at the centromeres of each sister chromatid are connected by microtubule fibers to poles at both ends of the cell, such that one sister chromatid of the X-shaped structure is oriented toward each pole (Figure 11.8). As a result of these attachments to the poles at opposite ends of the cell, chromosomes have lined up along the equator or midpoint of the cell (also known as the *metaphase plate*). In this case, we can think of *meta-* in metaphase as meaning "between" or "among" because it takes place right in the thick of things, after the chromosomes have been copied and before the cell divides. You might think of metaphase as the "middle stage" because of when it happens, or the "middle place" because of where it happens, at the metaphase plate in the middle of the cell.

The next step in cell division is known as *anaphase* (Figure 11.9). This step depends on the fact that the kintechores bound to the centromeres contain motor proteins that act to move the chromosome along the spindle fibers during cell division. All of a sudden, the two sister chromatids completely split, right at their connection point at the centromere, and the centromere motors then move the separated sister chromatids rapidly to opposite poles of the cell by pulling them along the tracks of the spindle fibers. We are just now beginning to figure out how this terribly complex process works, and many

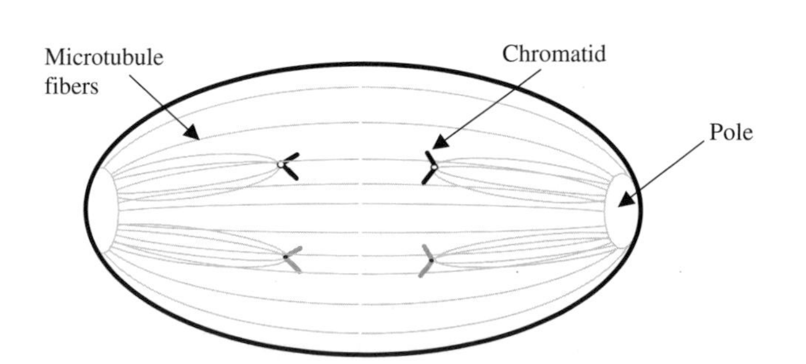

FIGURE 11.9 Anaphase—the sister chromatids of each chromosome split apart and move to opposite poles of the cell. The same cell seen in Figure 11.5 has now advanced to anaphase, in which the motors in the centromeres move the divided chromosomes along the tracks of the spindle apparatus towards the opposite poles of the cell. Notice that the two halves of the X have split and one half (one chromatid) is moving towards the right side of the cell while the other half is moving towards the left side of the cell.

but not all of the proteins involved have been identified. The key concept to understand here is that there are proteins at the centromere that function as motors that pull the chromatids along the spindle fiber tracks toward the opposite poles of the cell.

The phase of the cell cycle that occurs once the chromatids have reached the poles of the spindle is called *telophase.* At this point, the membrane around the nucleus reforms and we begin to see where the cell will split into two parts (Figure 11.10).

After telophase, actual cell division, called *cytokinesis,* occurs (Figure 11.11). During metaphase, each of the pair of replicated chromosomes had two chromatids, which means that the cell momentarily had four copies of each gene instead of the two copies normally present in a resting cell (Figure 11.12). Now, after cytokinesis, there are two daughter cells whose genotype and DNA content are identical to the original cell.

Telophase

Beginning signs of where cell will divide

Nuclear membrane reforming to enclose the chromosomes

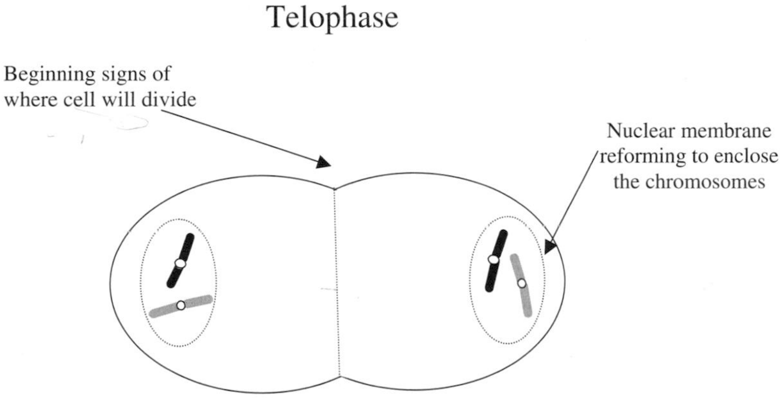

FIGURE 11.10 Telophase—the nuclear membrane begins to reform and the cell prepares to divide. The identical, duplicated copies of each chromosome have been pulled to opposite sides of the cell by the motor apparatus and are now being set off from the surrounding cytoplasm as the nuclear membrane begins to form again.

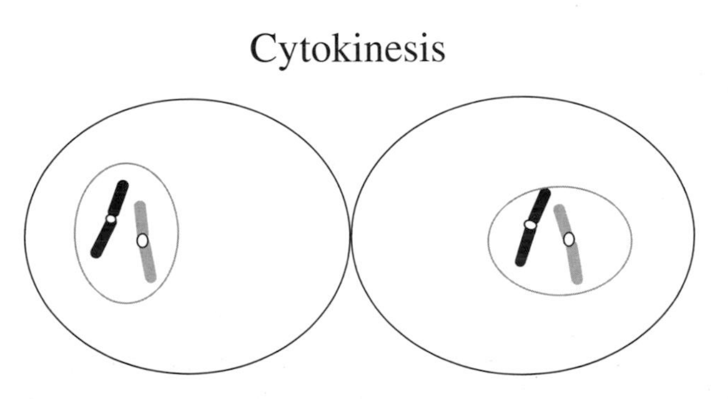

FIGURE 11.11 Cytokinesis—physical separation into two cells. The separation into two separate cells is completed, and the nucleus has completely reformed. The result is two daughter cells that are identical to the original cell, with identical copies of the pair of chromosomes that the original cell started out with.

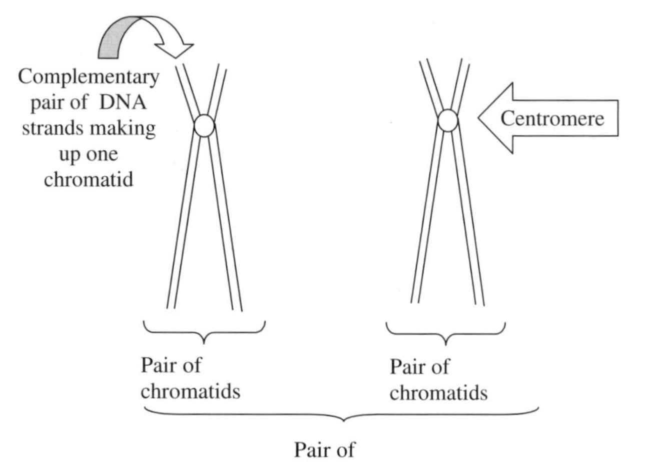

FIGURE 11.12 Pairs of pairs of pairs. When we look at the chromosomes in an actively dividing cell, we end up with a lot of pairs of things. First, we find that we have pairs of chromosomes. Second, each of these chromosomes passes through a stage right after it is replicated but before the cell divides when the chromosome has a pair of chromatids. Finally, each chromatid consists of a pair of complementary single strands joined into a double-stranded structure. It is important to keep track of whether we are talking about pairs of strands, pairs of chromosomes, or pairs of chromatids within the X-shaped replicated chromosome structure of a dividing cell.

Remember, when we started out with our fictitious cell, we said that the black chromosome came from one parent and the gray chromosome came from the other. Notice that each cell has ended up with one black and one gray chromosome, not two gray chromosomes in one cell and two black chromosomes in the other cell. As we will discuss later in Chapter 23, it is important that each new daughter cell gets one copy of the pair from mom and the other copy of the pair from dad. That is, it really matters that we keep track of the black copies and the gray copies so that we can be sure that the final cells end up with the combinations black-gray and black-gray, not black-black and gray-gray.

It is this process of mitosis that allows individuals to develop from a single-celled zygote (the product of sperm and egg fusion) to a complex organism with "gazillions" of cells, all of which are genetically the same. When there are more chromosomes, the process can be more complicated (Figure 11.13).

In a human cell, there are forty-six chromosomes lined up in the center of the cell at metaphase. Each of these 46 chromosomes is attached to the spindle apparatus and needs to have the centromere motors pull the two halves of the chromosome apart and carry them to opposite ends of the cell. In fact, the gathering of replicated chromosomes at the metaphase plate during mitosis in a human cell is a terribly crowded and complex event aimed not at passing along one pair of chromosomes but rather at seeing to it that a copy of each and every one of the forty-six chromosomes ends up in each of the daughter cells at the end of mitosis.

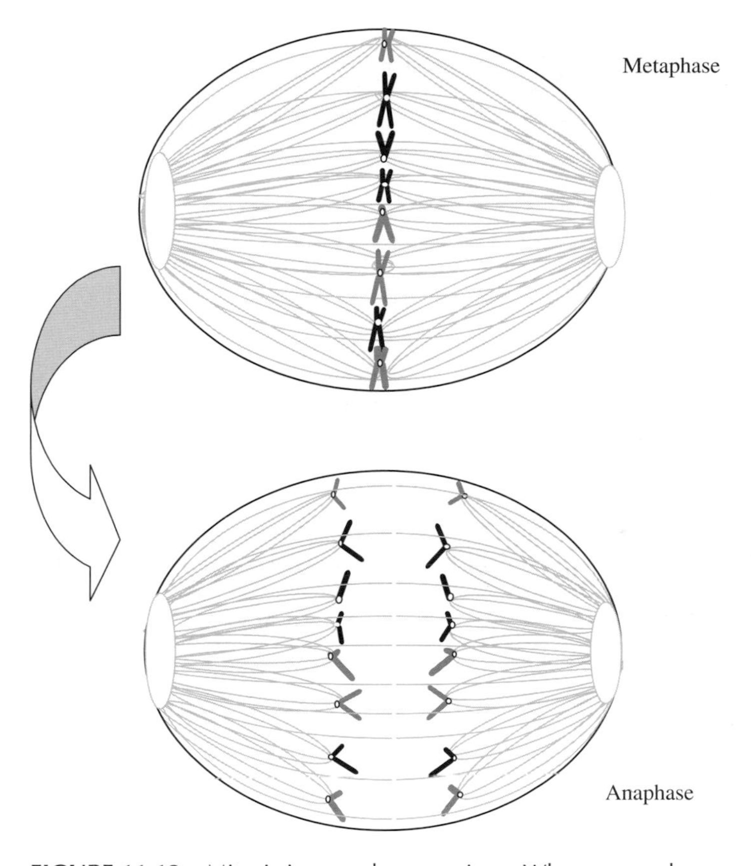

FIGURE 11.13 Mitosis in complex organisms. When many chromosomes are present in the cell, correct segregation into the separate cells still happens correctly because each chromosome gets moved around individually. Each replicated chromosome gets aligned at the metaphase plate and attached to the spindle apparatus separately. So, even though there are many chromosomes involved, the cell handles each one as an individual problem. Here we see the cell handling eight chromosomes (four pairs of homologous chromosomes, with a black copy and a grey copy of each chromosome) as it advances from metaphase to anaphase.

The basic pattern of events in the cell cycle is the same as in our hypothetical organism. In Figure 11.14 the whole series of steps in mitosis is shown in photographs of real cells. Green and red dyes show the locations of proteins involved, such as those that form the spindle apparatus, and blue dyes show where the chromosomes are.

So the process of getting the right number of copies of the genetic blueprint into the daughter cells doesn't seem that tough. During the cell cycle, the cell copies everything in it, including the chromosomes. Replicated chromosomes stay attached to each other while they line up at the center of the cell and become attached to "tracks" that connect to the poles at opposite ends of the cell. Motor proteins in the centromeres separate the X-shaped chromosome back into two single chromosomes that get pulled to the

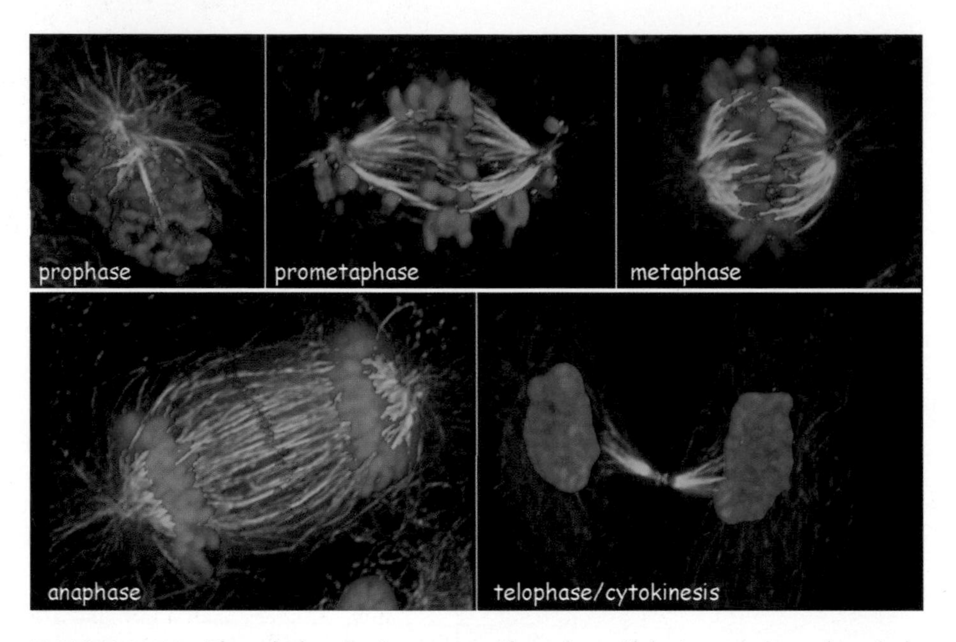

FIGURE 11.14 The whole mitotic process. These beautiful pictures show the whole process, from prophase, prometaphase, through metaphase, anaphase, telophase, and cytokinesis. DNA is labelled in blue, and protein machinery involved in putting the chromosomes through their paces are labelled in green and red. (Courtesy of William C. Ernshaw.)

opposite poles of the cell along the protein tracks. The key to this successful allocation of copies into the new cells is the X-shaped chromosome in which the duplicated copies are kept locked together and moved around as a unit until the cell is ready to send them to opposite ends of the cell.

It turns out that things get even more complicated when the blueprint gets handed down from parent to child. Chapter 12 shows how chromosomes get moved around when making sperm and eggs.

PASSING GENES BETWEEN GENERATIONS 12

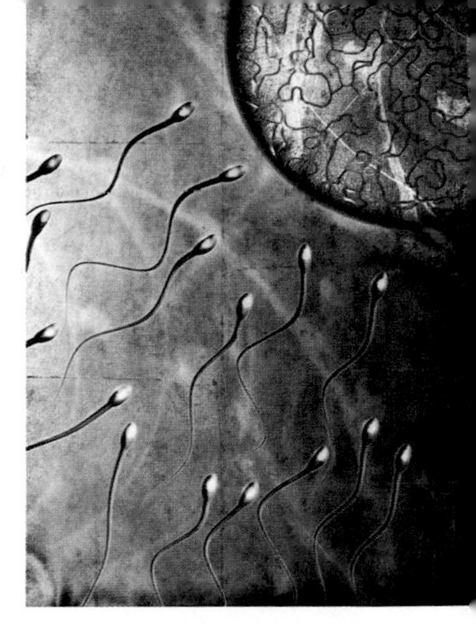

The earliest known member of the family suffered from a blinding eye disease, a very early-onset form of glaucoma that was untreatable in the early years of the nineteenth century. In 1834, he passed glaucoma along to one of his two daughters, and his affected daughter in her turn bore nine children. By the turn of the twenty-first century, she had more than 700 descendants, including more than seventy who inherited juvenile-onset glaucoma. This form of glaucoma is much more treatable than it was eight generations ago. The gene that causes it is now known, and it is even possible to test for mutations in this gene so that children at high risk can be identified and begin frequent eye exams to ensure that treatment will begin at the earliest possible moment to prevent vision loss. But the question remains, as each at-risk child is born into this family: what is it that decided that some of them would inherit the gene that causes the disease while others did not?

Traits get passed from one generation to the next when *that* sperm meets *that* egg and a zygote is formed. If we look at the many descendants of the young woman with juvenile-onset glaucoma, we find that even knowing which gene has a defect does not tell us the mechanism by which some of her children inherited a defective copy while others inherited an undamaged copy. Each affected member of the juvenile-onset glaucoma family has two copies of the genetic blueprint, but each of them makes sperm or eggs that have only one copy of the blueprint, that is to say, only one copy of each chromosome. So how does this happen? How does only half of the blueprint get transferred into any one sperm or egg? Clearly, the processes of mitosis that are designed to preserve the full number of chromosomes won't work here.

To see what happens to the chromosomes during the creation of germline cells, let's return to our fictitious friend *O. hypothetica* with its one pair of chromosomes (Figure 12.1). It has gotten the urge to mate. Now it needs to make a gamete. In order to do things the Mendelian way, it needs to produce a sperm or egg with only one copy of each chromosome pair or, in this particular case, one chromosome. As with mitosis and cell division, getting the right number of chromosomes into the gamete is going to be something difficult that cell is going to have to actively orchestrate. The process that the cell will use to accomplish this is called *meiosis.*

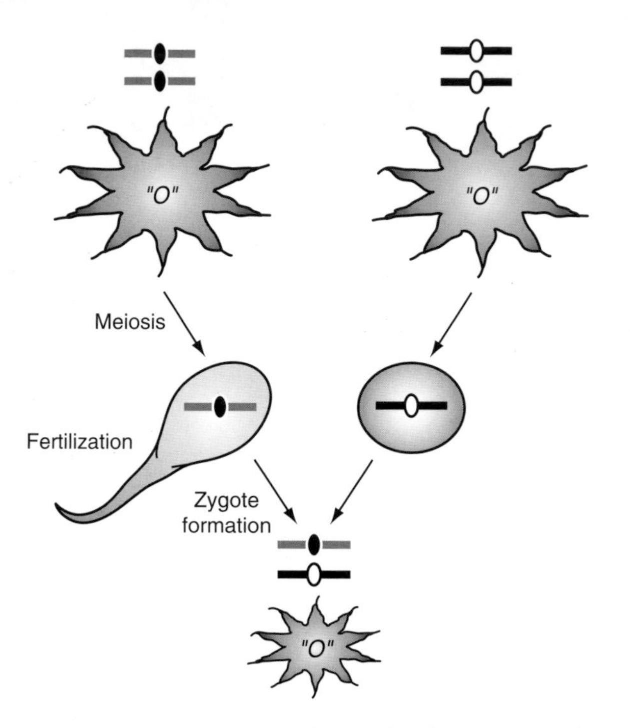

FIGURE 12.1 Meiosis made simple. This oversimplification shows how two *O. hypothetica* produce offspring. Papa *hypothetica* and Mama *hypothetica* each give a chromosome to the sperm and egg that fuse to make Baby *hypothetica*.

MEIOSIS MADE SIMPLE

Meiosis actually encompasses two cell division events, (cleverly called the first meiotic cell division and second meiotic cell division!). The purpose here is simple: get single copies of the blueprint into each germline cell. To do this, the cell carries out one round of DNA replication (just like in mitosis) but then carries out two rounds of cell division instead of the one cell division found in mitosis. (And yes, as the math wizards can tell us, one round of chromosome replication accompanied by two cell divisions will in fact cut the chromosome number in half). Each parental *O. hypothetica* has two copies of the chromosome, but a sperm or egg produced by *O. hypothetica* will have but one chromosome. Thus, when the sperm and egg come together, the new organism will once again have two chromosomes (see Figure 12.1).

Of course, it is all much more complex than just replicate-divide-divide, because once again we have to be sure that things end up in the right place at the end of this. So first, let's take a pictorial overview of meiosis (Figure 12.2) in terms of where the chromosomes are and how they get moved around and a description of those steps (Box 12.1). Once we have seen how the chromosomes get moved around, then we can move on to discuss more details about some of the critical steps, especially steps 2 and 3, first meiotic prophase and first meiotic metaphase.

Remember, the two chromosomes that *O. hypothetica* starts out with are homologous chromosomes; that is, they are effectively the same chromosome, with the same genes in the same order along the whole chromosome. To keep track of the two copies separately, we will color them red and blue.

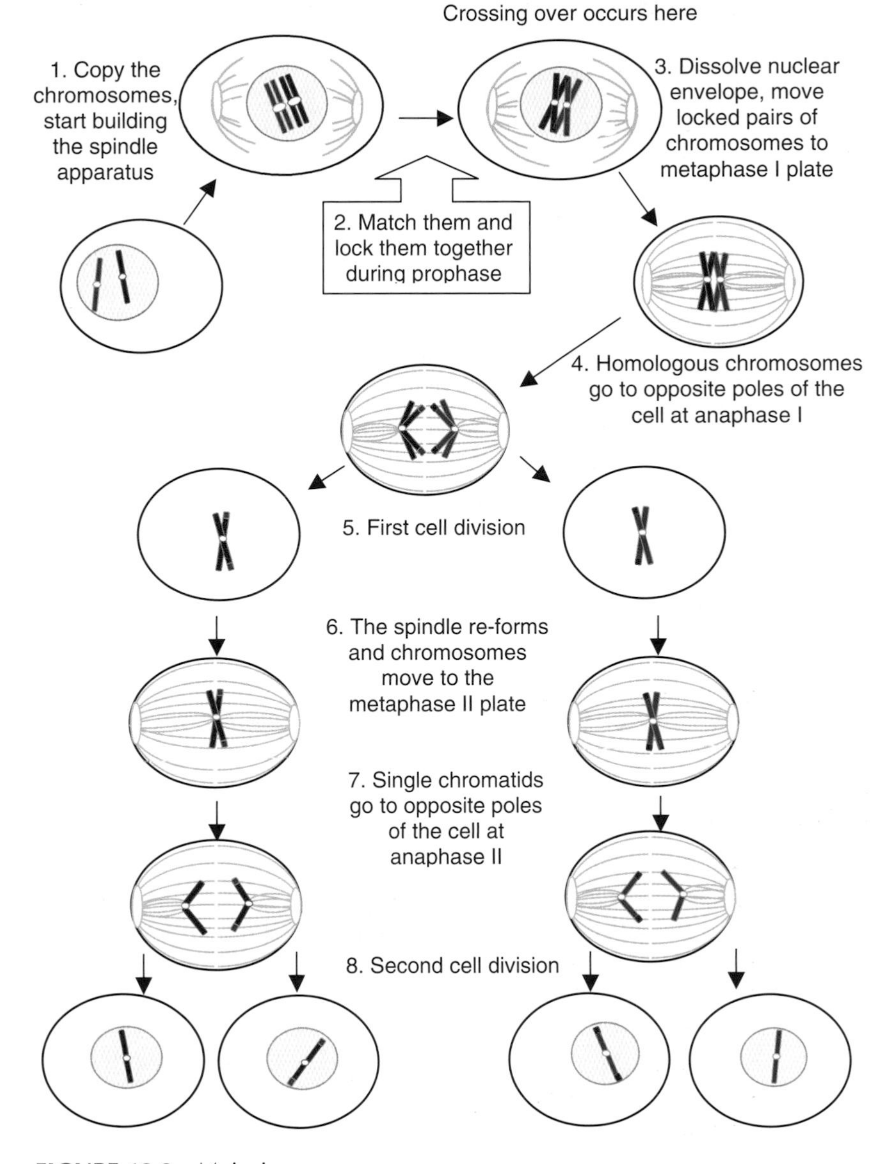

FIGURE 12.2 Meiosis.

RECOMBINATION PROCESSES LET CHROMOSOMES STAY PAIRED

Now let's go back and take a closer look at several steps in meiosis. We are only going to give more detailed discussion to a few of the steps that show critical events in meiosis.

First, consider Step 2 (also known as first meiotic prophase). As indicated in Fig. 12.2 and Box 12.1, this is the point at which the homologous pairs of replicated chromosomes find each other and link-up to each other before moving. During this step, homologous chromosomes pair along their entire length as part of locking the two chromosomes together. This is a step not seen in mitotic cells. In the mitotic cells, the replicated chromosomes stay

BOX 12.1 THE STAGES OF MEIOSIS

Meiosis is the process used to create the germ cells—sperm and eggs. Two homologous chromosomes replicate to become four copies that eventually separate to end up as four single chromosomes in the four separate sperm that are created. (In contrast, only one of the four cells at the end of oogenesis becomes a viable egg.) This list of stages from Figure 12.1 has been greatly simplified. For instance, first meiotic prophase alone is usually divided into five stages with unlikely-sounding names such as zygotene and diakinesis, but we have left out much of that. You don't have to struggle with distinguishing things such as different levels of condensation of the chromosomes to be able to follow the critical steps we present here and in Fig. 12.2. In fact, if you compare this list to the description we gave when we talked about mitosis, you will see that we have even condensed some of the steps we talked about in Chapter 11. For instance, we have grouped the telophase step when the nuclear envelope reforms together with the cytokinesis step in which the cells divide. This streamlined version of meiosis shows the essential steps that get the chromosomes replicated and then reduced in number so that each germ cell holds one unreplicated copy of the chromosome instead of two.

Meiosis I

Step 1. During interphase, the cell copied everything including the DNA, resulting in the X-shaped replicated chromosomes.

Step 2. Each replicated chromosome finds its homologue and the homologues lock together. At this point, a critical event happens—DNA gets exchanged between the replicated chromosomes, something we will talk about in more detail during this chapter.

Step 3. After the nuclear envelope breaks down, the matched, locked-together chromosome complex—called a bivalent—moves to the metaphase plate, where the spindle apparatus attaches the centromere of one replicated chromosome to one pole of the cell, and the centromere of the other replicated chromosome to the other pole of the cell.

Step 4. The two chromosomes that make up each bivalent get pulled to opposite ends of the spindle. Chromatids in the replicated chromosome stay together because the centromere was only attached to one end of the cell, not both.

Step 5. Cytokinesis divides the cell into two cells, each with one bivalent.

Meiosis II

Step 6. In each cell, replicated chromosomes move to the metaphase plate. The spindle apparatus now attaches to the centromeres of both sister chromatids so that two sister centromeres are attached to opposite poles of the spindle.

Step 7. As the motors pull towards the opposite ends of the cell, the two separated sister chromatids move towards the opposite poles of the cell.

Step 8. The nuclear membrane reforms and cytokinesis separates the cell into two separate cells, each one containing a single unreplicated chromosome.

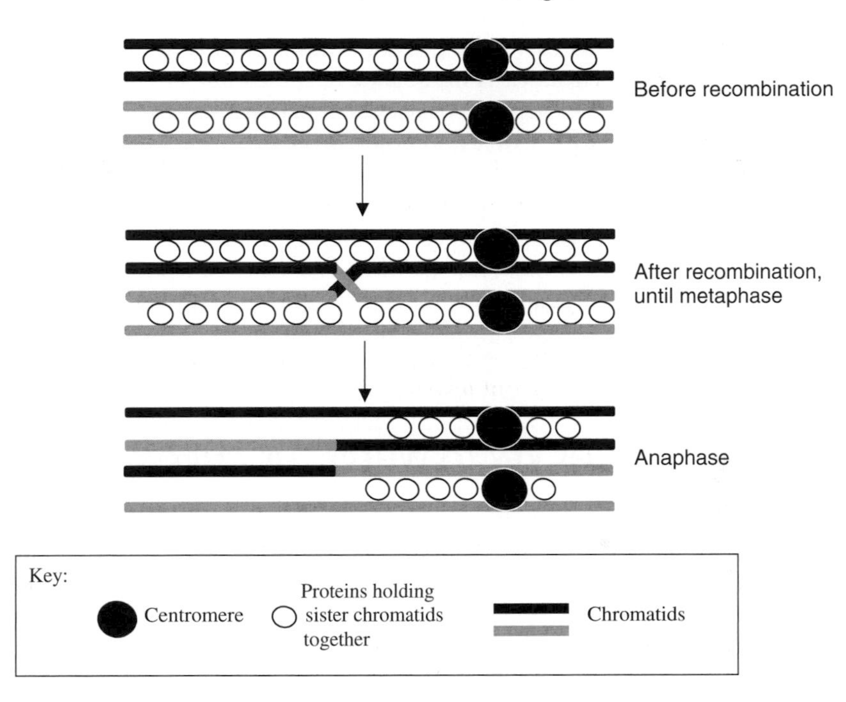

FIGURE 12.3 Meiotic recombination. Replicated chromosomes pair, and sections of sister chromatids are traded between them during first meiosis. This is normally a process of exact exchange in which any region that is "given away" by one chromosome gets replaced by "taking back" exactly that same region from the other chromosome. At the beginning of anaphase, sister chromatid cohesion is released along the arms, except near the centromeres, allowing the homologues to separate from each other.

apart from each other and get moved around as independent units. These paired chromosomes in meiosis, known as *bivalents*, are then locked together. They are allowed to exchange large regions of homologous DNA by using a process known as *recombination*, or *crossing-over* (Figure 12.3). We tend to think of meiotic prophase as consisting of three functional steps:

Match them (pairing)	Lock them (recombination)	Move them (towards the metaphase plate)

This process of hooking the chromosomes together has a purpose of much greater importance than trading DNA: these recombination events serve to link homologous chromosomes together to ensure that they go where they are supposed to at the end of the first meiotic division—to opposite poles of the cell before it divides. Think of it this way: pairing does not take place to allow recombination; rather, recombination takes place as a by-product of the pairing process that holds the chromosomes together at a critical point when the cell needs to handle them as a single unit while moving them around.

Those who study human populations may sometimes be seduced by the view that the critical point of recombination is to generate diversity among the progeny. It is an especially attractive view because the existence of such diversity lets us carry out genetic studies. However, those who actually study meiosis know that cells don't lock chromosomes together to achieve recombination and diversity. Cells use the mechanisms that produce recombination as an engineering process to hold chromosomes together so that they will end up going where they are supposed to. Where such exchange or recombination events don't take place, the homologous chromosomes may fail to go to the opposite poles of the cell at the first meiotic division. In such cases, the chromosomes don't end up where they are supposed to be by the end of meiosis. So the point of recombination is not diversity, however beneficial that side effect may be. The point of recombination is getting the right chromosomes to show up at the right place at the right time.

CENTROMERES ARE THE KEY TO WHERE THE CHROMOSOMES GO

When we look at Step 3 (first meiotic metaphase) in detail, we see one of the other points that is critically different from the steps of mitosis. Where mitosis moves individual duplicated chromosomes to the metaphase plate, meiosis I moves bivalents, the locked complex of two duplicated chromosomes. At this point in meiosis, the centromeres of the two sister chromatids do not attach to opposite poles. Rather, each chromosome has a centromere that is attached to only one pole by microtubule fibers (Figure 12.4). Thus, at this point in the meiotic cycle, the centromere is doing something fundamentally differ-

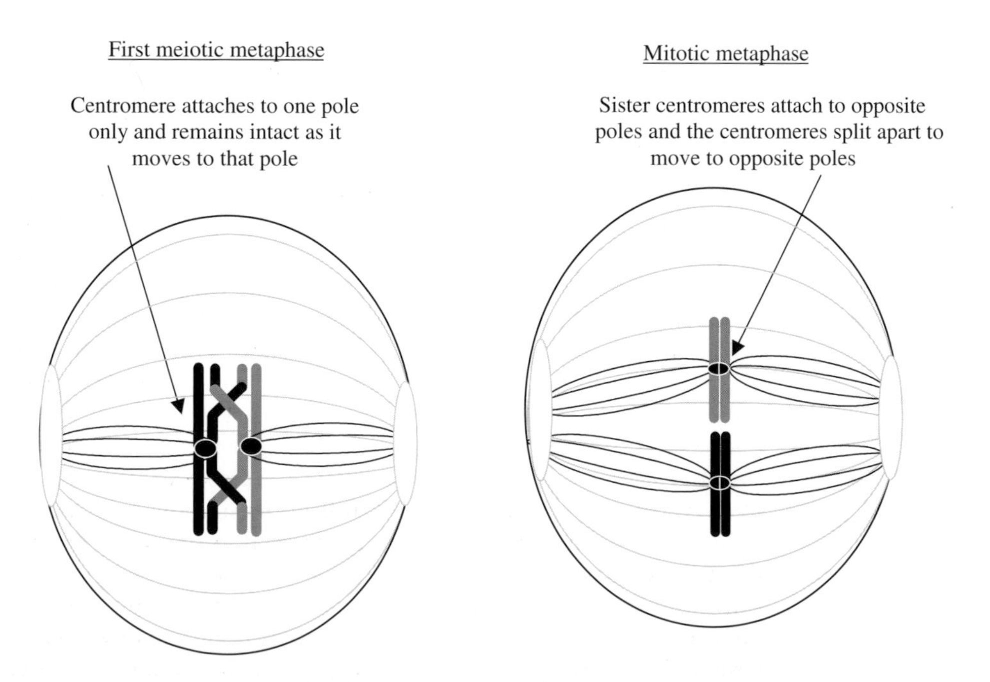

FIGURE 12.4 Comparison of first meiotic metaphase and mitotic metaphase. On the left, where the centromeres are not attached to both poles at once, notice that the crossovers hold the bivalents together until the cell is ready to move the replicated chromosomes to the opposite poles of the cell. On the right, attachment of one centromere to both poles is all that is needed to allow the cell to move things to where they should be in a mitotic division.

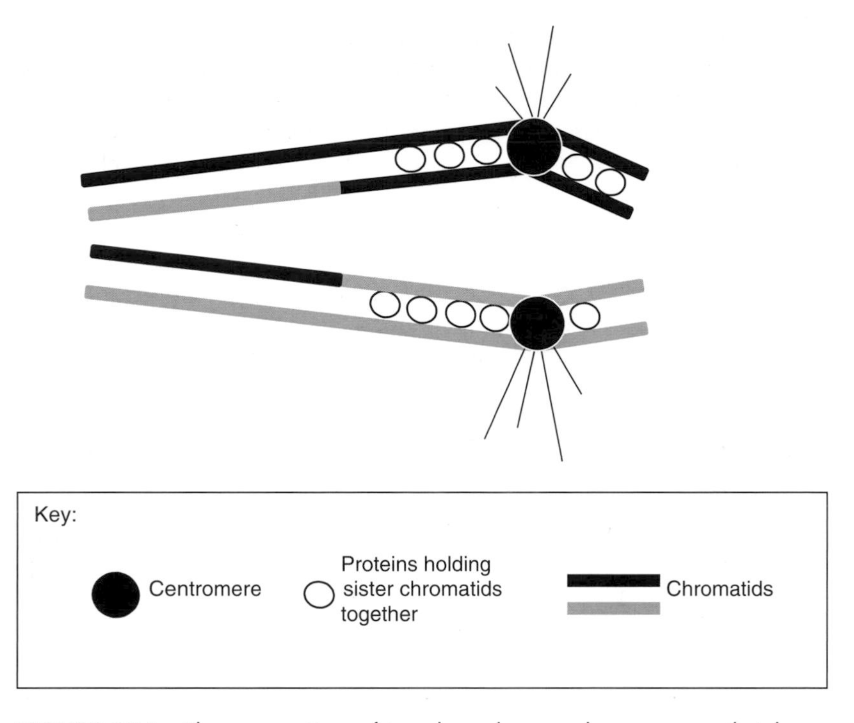

FIGURE 12.5 The segregation of two homologous chromosomes that have recombined to opposite poles. Note the change in the pattern of *sister chromatid cohesion*.

ent from what it does during mitosis: namely, the two sister centromeres of the mitotic centromere attach to two opposite poles, which, following centromere separation, leads to the movement of two sister centromeres to opposite poles. However, at the first meiotic division the centromere of each replicated chromosome attaches to only a single pole, so the meiotic centromere travels intact to just one of the two poles.

Another critical event in meiosis happens at Step 4 (first meiotic anaphase) when the two replicated chromosomes that comprise each bivalent separate and move to opposite poles of the spindle (Figures 12.1 and 12.5). This is the crucial meiotic event. Movement of the two homologous replicated chromosomes to opposite poles explains Mendel's observation that only one copy of a given pair of alleles will be included in a gamete—the other allele just went to the opposite pole at anaphase I and will go into a separate cell at the end of meiosis. Or, as we see in the pictures, the pair of black chromatids went to one pole and the pair of gray chromatids went to the other. The cells then proceed to meiosis II, which uses processes resembling mitosis to separate the two sister chromatids into separate cells, which each end up with one copy of the chromosome in place of the two copies the cell started with.

If we ask more about the replicated chromosomes that are being moved to the opposite poles during the first meiotic division, we find that each chromosome has recombined (traded material between the two locked chromosomes) and consists of some DNA from each of the two replicated chromosomes that were held together in the bivalent. Also, we find that some of the proteins that held the sister chromatids together are gone.

So meiosis, when we reduce it to its simplest elements amounts to this:

- *Replicate the chromosomes, so each chromosome consists of two chromatids*
- Pair the replicated homologs
- Allow the paired chromosomes to recombine, thus locking them together
- Pull the two homologous chromosomes (each still possessing two sister chromatids) to opposite poles at meiosis I.
- Divide into two daughter cells, each has half the original chromosome number
- Now, without any more replication, *line the chromosomes up on a new spindle*
- *Split the sister chromatids, with one chromatid going to each pole*
- *Complete cell division*

We can compare this to a similarly reduced version of mitosis, which amounts to:

- Replicate the chromosomes, so each chromosome consists of two chromatids
- Line the chromosomes up on the metaphase plate of a spindle
- Split the sister chromatids, with one chromatid going to each pole
- Complete cell division

As you can see, the italicized steps for meiosis look an awful lot like the steps for mitosis. When we delve into the details, there are some differences there, but at the level of understanding how the cell moves the chromosomes around, it is quite striking that meiosis II operates much like mitosis. One of the key points here is that there is no round of DNA duplication after meiosis I and before meiosis II.

We can see some key differences between mitosis and meiosis. Chromosomes do not pair or recombine during mitosis, or the second meiotic division that resembles mitosis; chromosomes only pair and recombine during the first meiotic division. Mitosis produces two identical daughter cells, each with two copies of each homologous chromosome; meiosis produces four gametes, each with one copy of each homologous chromosome. Mitosis takes places whenever cells divide, especially in cells that comprise the skin, the bone marrow, and the inside of the gut; meiosis only takes place in ovaries and testicles, with the objective of producing sperm or eggs.

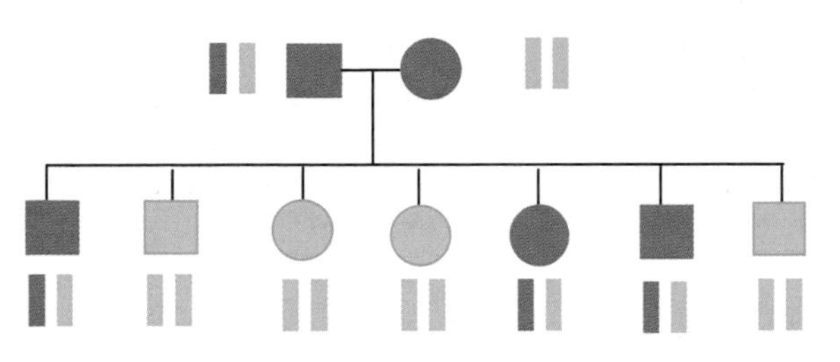

FIGURE 12.6 Brown eye color is dominant over blue eye color. Circles and squares are males and females. Brown symbols mark brown eyes, blue symbols mark blue eyes, black chromosomes carry brown alleles, blue chromosomes carry blue alleles. Thus someone with brown eye color who marries someone with blue eye color might produce 100% brown-eyed children if both of his eye color alleles are brown, but will produce 50% brown-eyed children if he carries a brown allele on one copy of the BEY2 locus on chromosome 15 and a blue allele on the homologous copy of BEY2. In actual fact, there are other genes that can also affect eye color so the situation is not always this simple.

MITOSIS AND MEIOSIS IN THE CONTEXT OF AN EYE COLOR GENE

The process by which the iris in a brown eye was formed went through many rounds of cell division to make all of the cells in the iris. As each cell containing the genes encoding eye color divided and passed along its genetic blueprint, every cell that was produced received the same eye color alleles. When our brown-eyed hero married someone with blue eyes and had children, three of them turned out to have brown eyes and four of them turned out to have blue (Figure 12.6). The fact that half of his children are blue-eyed suggests several things. First, that brown eye color is dominant over blue, and second, that he is carrying one brown allele and one blue allele. Because meiosis passes only one of his two copies of chromosome 15 along to each child, each one receives either a chromosome 15 homologue bearing a brown allele or a chromosome 15 homologue bearing a blue allele (see Figure 12.6).

The situation is actually more complicated than this since this is not the only locus that can affect eye color. The choices of eye color as we all know are more complex than just blue, green, and brown, including mixed colors (hazel) and different shades of any one color. While it is common for two brown-eyed parents to have blue-eyed children because of the recessive blue alleles they carry, the fact that blue-eyed parents do also rarely produce brown-eyed children simply reinforces the view that the genetics of eye color is complex and involves genes that are being moved around on multiple different chromosomes.

SUMMARY

For each stage of meiosis or mitosis, we see one common theme: when the cell wants to move chromosomes somewhere before dividing, it first sets up the move by placing the items to be moved at the metaphase plate and attaching the items to the spindle apparatus that will move the chromosomes to the poles of the cells. At that point, whether items go to the same place or to different places is all a matter of whether the microtubules connect the centromere to one pole or both poles.

In this dive down through the looking glass of the microscope, it may seem as if we have gotten rather far removed from the sorts of genetic issues we started out with early in the book—issues of human characteristics. In fact, the processes of chromosome mechanics discussed here sit at the very heart of genetics. If we take the same processes we just looked at and reexamine them in terms of multiple chromosomes and multiple genes per chromosome, we begin to understand the processes by which Mendel's peas could pass along pea pod color separately from plant height or flower color. In Chapter 13, we look at the chromosomal basis of heredity.

THE CHROMOSOMAL BASIS OF HEREDITY

13

And so we arrive at a junction of the different ideas we have been discussing. The genes Mendel was talking about, pieces of information connected to phenotypic characteristics, are in fact the bits of DNA sequence along the chromosomes, and meiosis is the mechanism that causes only one copy of a gene to be passed along because only one copy of the chromosome carrying that gene gets passed along.

So how does knowing that genes are being carried along on the chromosomes and segregated into separate germ cells during meiosis help us understand genetic and phenotypic variation between people, even people within the same family?

DOING MEIOSIS WITH TWO PAIRS OF CHROMOSOMES

Let's first take a look in Figure 13.2 to see what happens when two pairs of chromosomes segregate in our new fictitious friend, *O. complexica.* This more

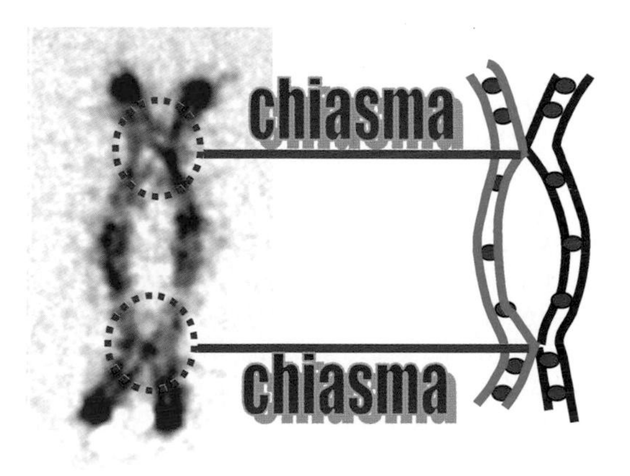

FIGURE 13.1 This picture and the diagram next to it show the structure that exists as chromosomes pair and exchange material. The chiasma at the point of exchange separates groups of genes above it from groups of genes below it when it results in trading of DNA between the two chromosomes. (After Petronczki *et al.*, *Cell*, 2003; 112(4):423–40. Photo courtesy of Jasna Puizina.)

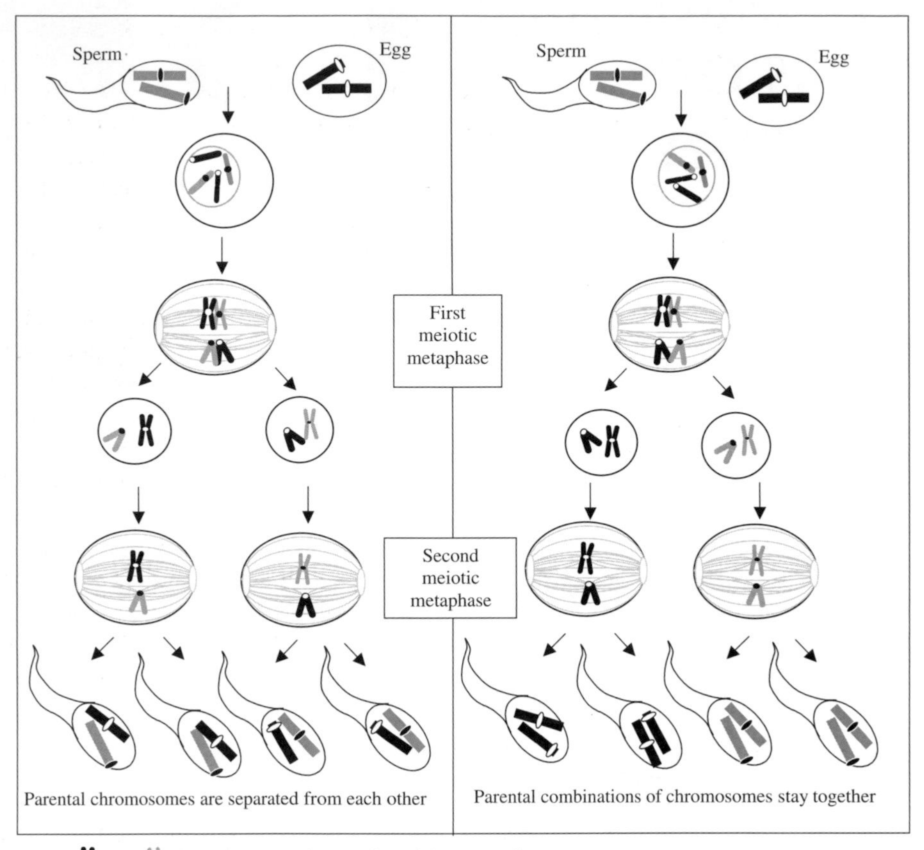

Key: and , homologous pairs, copies of the same chromosome.
and , homologous pairs.

FIGURE 13.2 There are two different possible outcomes from meiosis in the same individual, in this case a male. Here we show only the key steps needed to trace where the different copies of chromosomes are going so we have not included recombination in this figure. Since this individual makes more than one set of gametes in a lifetime, sometimes the outcome will be that shown on the left, where the parental chromosomes segregate apart from each other (black plus gray), and sometimes it will be what is shown on the right, where parental chromosomes stay together (black-black or gray-gray combinations). Notice that the key steps in deciding whether the parental combinations stay together or separate occur at first meiotic metaphase.

complex species has two pairs of chromosomes, one pair of metacentric chromosomes (having the centromere near the middle) and one pair of telocentric chromosomes (having the centromere near the end). If we color the chromosomes from his mom black and the chromosomes from his dad gray, we can keep track of what happens to the two forms of chromosomes and we can also keep track of the chromosomes from each of the parents.

Figure 13.2 shows that there are two different kinds of results that come out of meiosis in *O. complexica.* One possible result is shown in the right-hand panel, where we can see that the chromosomes that came in from the mother both end up going into the same cell after the first meiotic metaphase to produce a chromosome combination like that donated by the organism's mother. Chromosomes that came in from the father both end up going into the same cell after the first meiotic metaphase, which results in sperm with the same chromosomes donated by the father.

However, if we look at the left-hand panel, we see the other possible outcome. Instead of having the chromosomes that came in together stay together when they get passed along, the alternative outcome sees the chromosomes that came in together leaving in separate cells. That is, the chromosomes that came in to this organism together segregate apart from each other at the first meiotic metaphase, resulting in gametes that have a different combination of chromosomes from that donated by either of the parents. Each of these outcomes is about equally likely. In fact, for each "black-black" sperm the organism makes, we expect it to also make a "gray-gray" sperm and two "gray-black" sperm.

So, what does this mean? It means that the cell does not care whether DNA that came in together goes out together. The two gray chromosomes that came from dad can go into the same sperm cells or different sperm cells. However, it does care that one copy of each chromosome pair goes to each pole, so there must always be one copy of the chromosome with the centromere in the middle and one copy of the chromosome with the centromere near the end. Thus we do not end up seeing one sperm get all metacentric chromosomes and the other sperm get only acrocentric chromosomes; each sperm always gets one acrocentric and one metacentric.

THINKING ABOUT MEIOSIS IN TERMS OF GENES

We can now consider the meiotic process in terms of two or more pairs of genes and in terms of organisms like us that have more than two chromosomes. Minimally there are two cases we need to consider: (1) when two genes are located on different pairs of chromosomes, and (2) when both pairs of genes are located on the same pair of homologous chromosomes. Because the case in which the genes to be considered fall on different chromosomes turns out to be both simpler and more common, we will consider it first.

Gene Pairs Located on Different Chromosomes Segregate at Random

Mendel asserted that the alleles of one gene will assort independently of the alleles of another gene. That is, an individual of the genotype AaBb, where A and a are alleles of one gene and B and b are alleles of a *different* gene, will produce four types of gametes (AB, Ab, aB, and ab) with equal frequency (Figure 13.3). By now, we hope you know why we do not end up with genotype combinations like AA in one sperm and Bb in the other.

Note that a gamete carrying the A allele is as likely to carry the b allele as it is to carry the B allele. The same is true for gametes carrying the a allele; about half of them will carry B and half will carry b. Mendel stated that the two gene pairs segregate independently such that there is no preference for a gamete to carry a particular combination of alleles. He referred to this rule of segregation as *independent assortment.*

As shown in Figure 13.2, independent assortment results from the fact that two bivalents will orient at random on the metaphase plate with respect to each other, so half the time the black chromosomes are connected to the same pole and go to the same end of the cell, and half the time the black

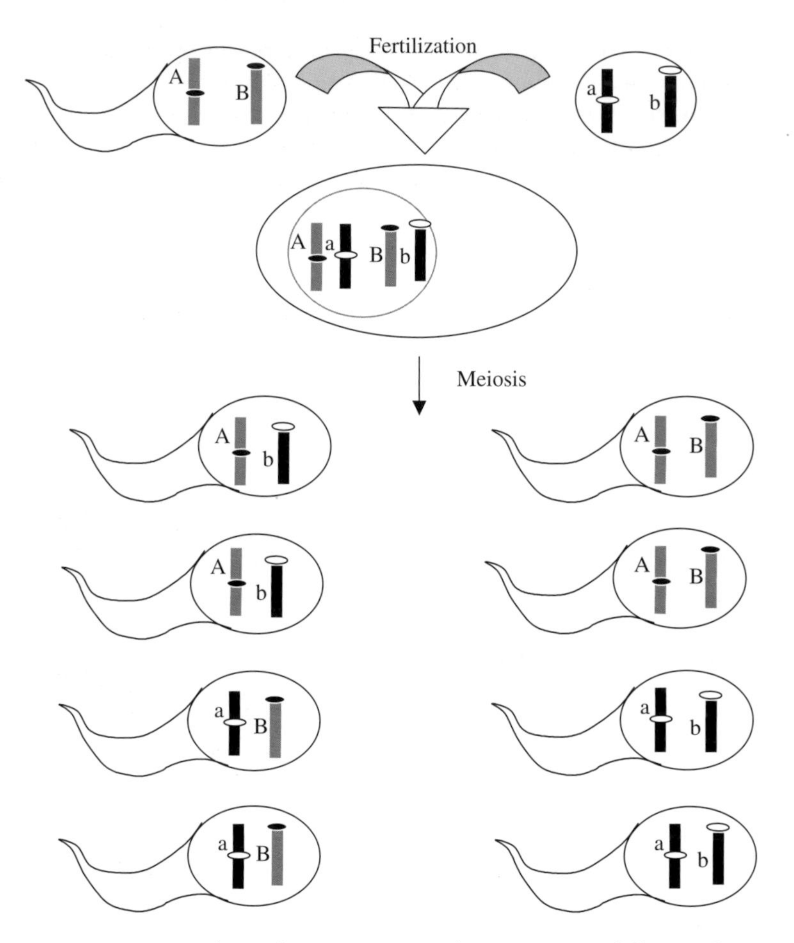

FIGURE 13.3 Independent assortment of genes on two different chromosomes. A can be found with either B or b, and B can be found with either A or a.

chromosomes are connected to opposite ends of the cell and go to separate germ cells in the end.

Recombination and Pairs of Genes Located on the Same Chromosome

Although you might expect alleles of genes on the same chromosome to travel together in moving between generations, alletes of two different genes that come in located on the same chromosome can actually leave on different chromosomes. Let's take a look and see how this happens.

As shown in Figure 13.3, the rule of independent assortment does not apply when two genes, R and S, are located on the same chromosome. Indeed, in the simplest case, the previous pictures of meiosis we have looked at would suggest that an individual of the genotype RrSs, in which R and S alleles are on one homologue and the r and s alleles are on the other, would only produce RS or rs gametes. This exception to Mendel's rule of independent assortment is called *linkage*. This idea, linkage of things located on the same chromosome, makes intuitive sense because the two genes are located on the same physical entity that is one chromosome, but the situation is more complicated than what we see in Figure 13.4.

As shown in Figure 13.5, the rearranged combination of Rs or rS in the gametes can only be produced when crossing over occurs in the region of the chromosome that is located between the two genes.

Recombination, or crossing over, events are, however, relatively frequent during human meiosis. (Note that there are two synonyms for *crossing over*,

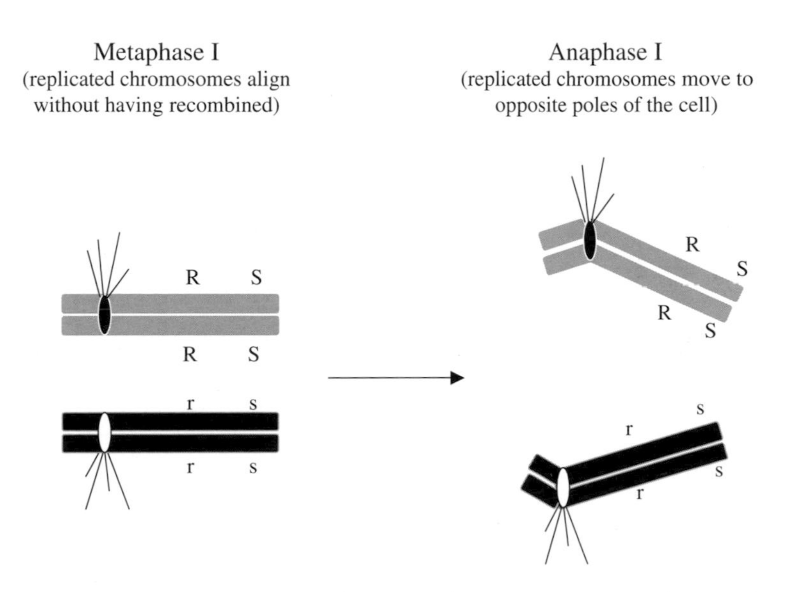

FIGURE 13.4 Two genes on the same chromosome at metaphase and anaphase I.

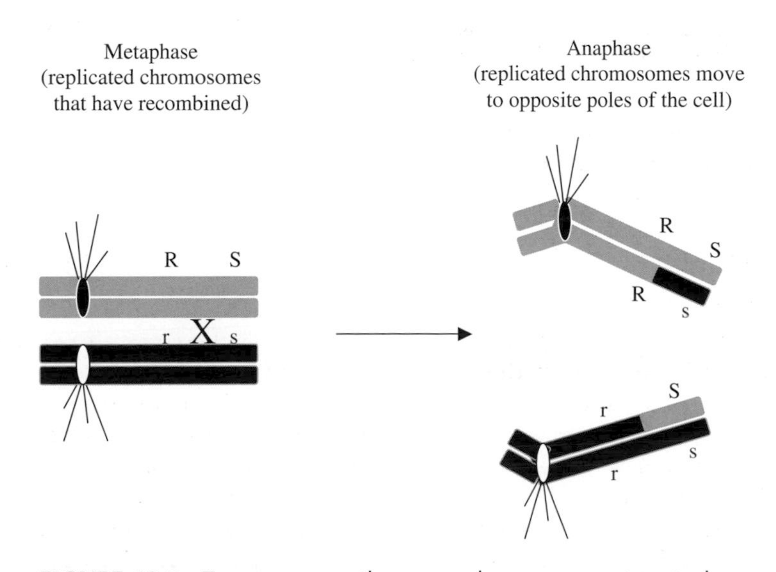

FIGURE 13.5 Two genes on the same chromosome, at metaphase and anaphase I, with a crossover. This view shows only the chromosomes and leaves out the cell and its structures. Notice that the black chromosome now contains a region of DNA with an S allele that used to be the gray chromosome, and the gray chromosome now contains a region of DNA with an s allele that used to be on the black chromosome.

namely *recombination* and *exchange*. All three terms will be used interchangeably). There is usually at least one such recombination event each time a pair of chromosomes come together to form a bivalent. In the case of large chromosomes, recombination may be more frequent. This is especially true for large chromosomes with the centromere near the middle, in which recombination events will likely occur on both arms of the bivalent.

Because recombination can occur at most sites along the chromosome, the probability that a recombination event will occur between two genes is dependent on the distance between those two genes on the chromosome. Thus, if two genes map at opposite ends of a chromosome, the probability of a recombination event occurring between them is high. If the genes are very close to each other, the chance that the recombination event will fall in between them is smaller than the chance that it will fall outside of the area containing the pair of genes. In fact, an approximation (but only an approximation) would be this: if the length of DNA on the chromosome outside of the pair of markers is about 10 times as long as the length of DNA between the markers, the chance of the recombination event falling outside the markers is about 10 times the chance of the recombination event falling between the markers.

Of course, the situation becomes more complex as we start trying to follow more and more genes. In looking at just three genes on the same chromosome, Figure 13.6 shows six exchange events involving a bivalent marked with Ee, Ff, and Gg. All the exchange events fell between the E and the G genes. Within that area between E and G, five crossovers fell between E and F, but only one of the exchanges fell between F and G. Thus the frequency of exchange events between any two markers is approximately proportional to the physical distance between them.

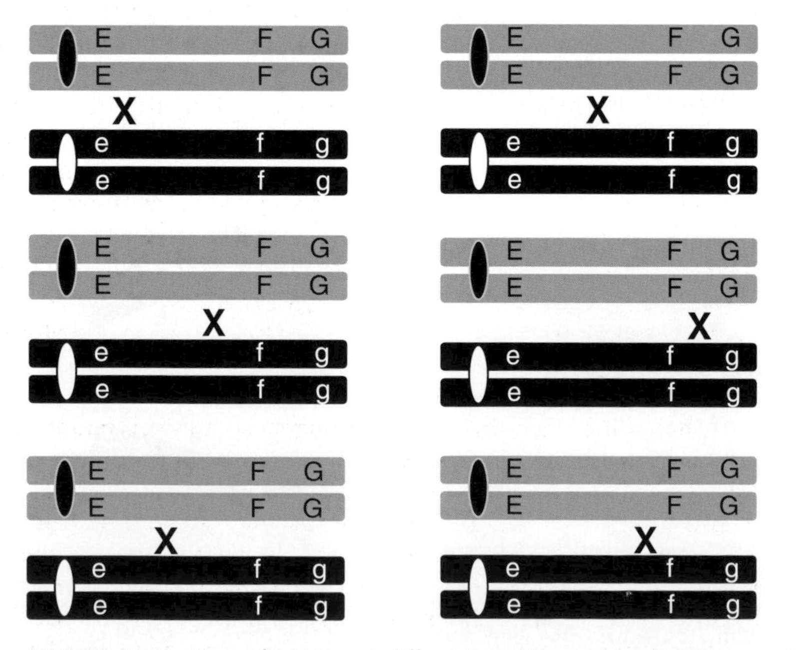

FIGURE 13.6 Recombination at different positions on a chromosome. X marks the point of recombination in each of six different rounds of meiosis for the same pair of chromosomes locked into a bivalent structure.

Meiosis is Executed Quite Differently in Human Males and Females

Given the importance of this process, it is surprising that it takes place in such different fashions and at different times in men and women, but it does, and these differences are truly impressive.

The most notable sex difference is in the timing of meiosis. In human females, meiosis begins during fetal development. Most if not all of the oocytes (eggs) that a human female will possess in her lifetime are produced while she is still *in utero* (in the uterus of her mother). These oocytes all begin the meiotic process during fetal development but arrest at the end of pachytene (the period of early meiosis during which chromosomes are observed to be fully synapsed along their length and are known to have completed meiotic recombination). Thus all of the meiotic recombination a human female will ever do is completed before she is born. These arrested oocytes remain quiescent until the girl enters puberty. At that point, a few oocytes are allowed to begin the maturation process during each menstrual cycle. Usually only a single oocyte is ovulated per cycle. It proceeds through the first meiotic division to metaphase of the second division, at which time it "arrests." Because completion of the second meiotic division is triggered only by *fertilization*, the number of completed meioses experienced by a human female roughly equals the number of conceptions.

Thus, in human females, recombination events must ensure chromosome segregation events that will occur decades later! This long delay in completing meiosis in females may well underlie the observation that the frequency of birth defects due to meiotic errors increases dramatically with advancing maternal age after age 35.

Male meiosis begins at puberty and continues uninterrupted throughout the life of the male. Male meiotic cells, known as *spermatocytes*, are continuously reproducing cells known as stem cells. Thus, unlike oogenesis, in which all the oocytes exist at birth, spermatocytes are constantly produced throughout the life of the male. Once a spermatocyte initiates the meiotic process, it takes less than seventy-five days to produce mature sperm. (Compare this with the process of oogenesis that must span decades!) Thus, in human males, the meiotic process is basically free running, with cells usually progressing through the meiotic process in an uninterrupted fashion. It is perhaps not surprising that geneticists have observed subtle differences in the patterns of recombination in the two sexes. Perhaps the temporal differences in meiotic prophase and the different requirements for ensuring chromosome segregation have imposed different pressures on the evolution of recombination patterns in males and females.

These differences in biology of oogenesis and spermatogenesis result in some rather large differences in the number of meiotic cells and of the number of *gametes* produced by the two sexes. Each female carries some 2 to 3 million oocytes at birth, but usually less than 400 of these oocytes will eventually mature during her life. However, the production of spermatocytes and the subsequent process of male meiosis occurs at a rate sufficient to produce the roughly 200 million sperm present in each ejaculate (approximately 1 trillion sperm during the life of the average male). The most important numerical difference is this: each female meiosis produces only a single oocyte; the remaining products of meiosis become nonfunctional cells called *polar bodies.* However, each male meiosis produces four functional sperm.

The actual molecular mechanisms that ensure meiotic segregation appear to be different, as well. In human males, the meiotic spindle is organized by cytoplasmic structures called *centrosomes.* The chromosomes then attach to the developing spindle. In females, the chromosomes themselves bind to the microtubules and build the spindle from the inside out without the assistance of centrosomes. Moreover, whereas human female meiosis includes frequent preprogrammed stops and selection appears to act at multiple points in the process, male meiosis appears to run uninterrupted once initiated. However, there do appear to be multiple checkpoints or control points in male meiosis that allow a spermatocyte that has made errors in meiosis to abort the meiotic process. Whether such checkpoints exist in female meiosis is a hotly debated issue.

It may be surprising to realize that meiosis is so different in the two sexes, but try to think about what the organism needs to accomplish. A sperm and an egg are very different cells. A sperm is basically a genetic torpedo. It has a payload (twenty-three chromosomes), a motor, and a rudder. Its function is to survive for a day or so and to swim to an egg. Once the sperm nucleus (called a *pronucleus*) is delivered to the cytoplasm of the egg, the rest of the sperm cells are destroyed. An egg, however, must possess all the supplies and determinants required to support embryonic development until the embryo can attach to the endometrium of the uterus and access the mother's blood supply. These two roles call for very different cellular machinery, and the process of human reproduction requires that a vast excess of sperm be produced for every egg, since the probability of any one sperm finding the one egg is very low!

SUMMARY

As we have seen in the last several chapters, several different factors contribute towards a child having different combinations of genetic information than were present in the child's parents and grandparents. First, each parent passed only half of their genetic information along to their child. Second, through independent assortment of chromosomes, alleles carried on those different chromosomes can pass independently down through multiple generations so that alleles of two different genes that were present in someone's grandmother may no longer be present together in the same germ cell that produces the grandchild. Third, even when specific alleles of two genes are located on the same chromosome together in the grandmother, recombination can exchange material between chromosomes in such a way that a different combination of alleles will be present on that chromosome in the grandchild. In Chapter 20, we will see that transmission of the X and Y chromosomes between generations leads to an unusual mode of inheritance and poses special problems in gene dosage for male and female organisms.

Section 4

MUTATION

There are a surprising number of truly different ways in which changes to the genetic blueprint can affect the traits we pass along to our children. In this section, we describe a variety of different kinds of mutations and discuss some very basic things about how mutations bring about their effects on human traits.

ABSENT ESSENTIALS AND MONKEY WRENCHES

14

"Now I get to grow up."
—Thank you note from a child with cystic fibrosis to researchers who helped find the CF gene.

When the gene responsible for cystic fibrosis was cloned, it was a landmark event in molecular genetics. Researchers all over the world had inched their way towards an answer to what was causing this killer disease. Finally, an international collaboration of doctors and molecular geneticists used cutting-edge technologies, traded resources, shared information, and pooled ideas to make the breakthrough and find the gene. At that time, many of today's advanced technologies and resources were not available, and some steps that are now done in a few hours with computers took years of laborious slogging through experiments at a lab bench.

Many people worked long, hard hours to clone this gene. Among them was a college student with cystic fibrosis who worked in one of the labs that made the breakthrough. Medical advances in treatment of cystic fibrosis had made it possible for him to grow up and attend college, but as the other scientists working on the project watched him bravely alternate between attending classes, working in the lab, and suffering repeated illnesses, it was clear that much more medical advancement was needed.

The tale of the cloning of the cystic fibrosis gene was really one of determination, heroism, and hope, as shown in the flood of mail that the researchers received after the announcement that the gene had been found: letters of congratulations, letters of hope, and letters from small children writing to thank them for being given the opportunity to grow up.

As more and more of humanity's disease genes are identified, people become more and more blasé about the process. Types of findings that previously made headlines may not even rate a news article these days. However, for every gene that is cloned, there are people like those young letter writers—children with other diseases yet unsolved, children who have been waiting for someone to come up with their particular breakthrough, to bring them, if not a cure, then at least the hope of one—the hope that they will get to grow up.

When Mendel conceived of different alleles of a gene being responsible for the differences in the observed traits, he had no model for the form the differences in information might take nor for why some forms of the information would manifest themselves as dominant traits while others would appear as recessive traits. Recessive inheritance can be seen in many diseases with serious or even potentially lethal consequences, such as cystic fibrosis, phenylketonuria, and Lesch-Nyhan syndrome. Dominant disorders include comparably severe illnesses such as Huntington's disease, Lou Gehrig's disease (amyotropic lateral sclerosis), and Marfan's syndrome, which is believed to have affected Abraham Lincoln. Why do changes in some genes cause a dominant problem when the information in that gene is altered, when changes in other genes cause a recessive problem? In the next several chapters, we will talk about the specifics of some different kinds of mutations, the changes in information that constitute the different alleles of a gene, but we want to start out here by talking about some general principles of how a change in information in a gene can affect the cell and the organ and the body in which the genetic defect exists.

THE RELATIONSHIP BETWEEN THE NATURE OF THE MUTATION AND THE RESULTING PHENOTPYE

On some fundamental level, mutations can be divided into two classes:

- *Absent essentials*: Mutations that result in some necessary function not being carried out. This may be due to the protein being missing, or it may be because the protein does not work even though it is there; either way, the result is a *loss of function.*
- *Monkey wrenches*: Some mutations produce an abnormal protein that actively does something wrong, and in doing so disrupt an essential cellular function. In some cases the abnormal proteins are essentially poisonous, and in other cases they exert their effects by doing the normal things they are supposed to do, but doing them at a time when or in a place where those functions should not be taking place. Some monkey wrenches that actively harm the cell are called *gain-of-function* mutations. That gain of function can be something very specific, such as binding to a different hormone than the one to which the protein usually binds. It can also be as nonspecific as poisoning the cell. Some monkey wrenches that act by interfering with the normal protein's ability to do their job are called *dominant negative* mutations.

Classically, people have described such mutations in terms of convenient terms, such as dominant or recessive. When people generalize, they often say that absent essential mutations show recessive inheritance and that monkey wrenches show dominant inheritance. Actually, that is not exactly what they say because "absent essentials" and "monkey wrenches" are our terms for these mutations. Mostly researchers refer to them with names like "loss-of-function" mutations or "gain-of-function" mutations. As shown below, although many absent essentials pass through families with a recessive pattern of inheritance and many monkey wrenches turn up in dominant pedigrees, there are no simple correlations between the nature of a mutation and its behavior in pedigrees (i.e., whether it is dominant or recessive). However, we begin to understand a lot about a disease if we separately identify whether a genetic defect has removed or added something functional in addition to determining whether the mode of inheritance is dominant or recessive. In cases in which an absent essential is not inherited in the expected recessive manner or the monkey wrench does not turn up with dominant inheritance, the dichotomy is usually telling us important things about the underlying mechanisms of the disease.

THE EXPECTED CASE: A LOSS-OF-FUNCTION MUTATION PRODUCES A RECESSIVE TRAIT

Let's start out by talking about what is going on in *cystic fibrosis* (Box 14.1), a disease caused by a gene in which mutations act as expected—loss of function leads to recessive inheritance. The cystic fibrosis gene makes a protein called CFTR that has to be present and working correctly for the lungs and other organs to stay healthy. The CFTR protein is an enzyme that transports salt (technically chloride and bicarbonate ions) across the membranes of several tissues, most notably the lungs and pancreas. If salt that is supposed to leave the cells of the lungs instead stays inside them, there will not be enough fluid outside the cells. Without enough fluid, the mucus in the lungs gets too thick, which in turn leads to inflammation and the possibility of chronic respiratory infections. These repetitive infections can be fatal, although improved treatments are allowing more and more children with cystic fibrosis to grow up.

Let's imagine a child, May, with moderately severe cystic fibrosis. When we examine her genetic blueprint, we find that both of her copies of the cystic fibrosis gene have a mutation that creates a stop codon, in place of amino acid 553, the fourth most common cystic fibrosis mutation in the world. May is missing more than 60% of the length of the protein, including regions of the protein that carry out important functions. Thus she appears to be lacking functional CFTR protein. This child's recurrent bouts of illness do not really occur because a change in a couple of bases of the sequence in her blueprint causes a problem, but rather on a more immediate causative level she is ill because lack of CFTR function is caused by the absence of the CFTR protein

BOX 14.1 CYSTIC FIBROSIS—A RECESSIVE LOSS-OF-FUNCTION DISEASE

According to the Cystic Fibrosis Foundation, one in every 30,000 people in the United States has cystic fibrosis, and more than ten million people are carriers. Vigorous percussion of back and chest, accompanied by regimens of antibiotics, antiinflammatories, and mucous thinning drugs are all ways to try to help cut down on infections that can threaten the patient's life. Enzymes, vitamins, and diet all help with digestive problems that affect some individuals with cystic fibrosis. With many medical advances in recent years, more and more children with cystic fibrosis are going on to become adults. Current research on treatments includes nutritional studies and investigations of antibiotics, drugs to change salt transport in the lungs, and even gene therapy trials.[1]

[1] For more information about cystic fibrosis, check out the Cystic Fibrosis Foundation at www.cff.org/clinical.htm.

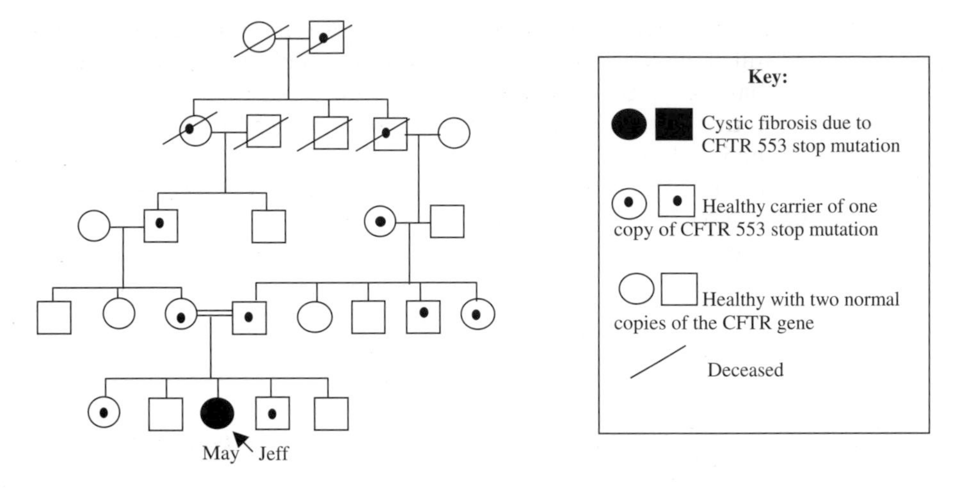

FIGURE 14.1 Classical situation in which a loss-of-function mutation, in this case in the cystic fibrosis gene, shows a recessive pattern of inheritance. In this family, we have left off the spouses and children of aunts and uncles and just show the immediate line of descent from the shared great-great grandfather who passed the mutation along to both sides of the family. Also notice that since no one else in this family developed cystic fibrosis, no one realized that this mutation was present in the family until May was born.

itself brought about by the typo in her blueprint. On the most fundamental level, she is ill because the functions that the CFTR protein should be carrying out are not being carried out. A look at her family finds that she is the only one who has cystic fibrosis but that several of her relatives are carriers who have one defective copy of the gene and one normal copy of the gene (Figure 14.1). Looking at her family structure, with her parents sharing a great grandfather who carried the mutation, also shows us how she might have come to have both copies of her CFTR gene share the same mistake.

Now consider this: what if her brother Jeff carries both a normal copy of the CFTR gene and a defective copy? Most texts and even some professors will glibly tell you that Jeff will be fine simply because the CFTR mutation is a loss-of-function mutation and that loss-of-function mutations are recessive (and thus the normal allele is dominant). We have to wonder, though: Why is Jeff fine? Does one normal copy of a gene per cell produce enough good protein to carry out the needed function and prevent damage or illness when the second copy is defective? The implication is that cells normally produce extra CFTR, so they don't mind losing some of the protein, or that CFTR is such an efficient protein that one half the normal levels of this protein can apparently manage enough salt and fluid transport to maintain health. Hence the oft-stated assertion that loss-of-function mutations will be recessive and that recessive mutations involved loss of function. Alternatively, think back to Chapter 9 and consider the idea that a cell that senses that levels of a protein are too low might have the ability to turn up the rheostat in the promoter region of a gene to increase expression of the normal copy of the gene.

For different genes, there may be different mechanisms by which one good copy can compensate for a damaged copy. The important point is that having just one good copy left works fine for many genes. After all, isn't that an important attraction of being diploid, being sure that needed processes still take place even if there is a problem with one copy of a gene? So we expect that for some recessive diseases, the gene already makes enough protein so that a reduction to half the normal level still leaves enough protein to get the job done. However, we expect that, in other cases, the body has the ability to compensate for a reduced amount of protein by putting in an order for increased production of protein from the normal copy of the gene. This is a system that works well until both copies of the gene are gone, and then there is no backup copy on which to turn up the rheostat. It's a system that only works if one or another of those mechanisms is at play for the particular gene in question.

THE SURPRISING CASE: A LOSS-OF-FUNCTION MUTATION PRODUCES A DOMINANT TRAIT

Although it is true that absent essentials often result in recessive traits, there are important examples of loss-of-function mutations that create severe phenotypes even in the presence of a normal allele. Some interesting examples of this unexpected situation include genes that produce structural proteins rather than enzymes, and genes that produce certain kinds of regulatory proteins that control the expression of other genes. Part of why loss-of-function mutations in these genes turn out to be dominant is because these are the genes for which the amount of gene product in the cell is critical. The cell apparently cannot tolerate a reduced amount of protein, and the cell has no ability to get the one normal copy to up-regulate its expression enough to compensate for what is missing.

Many structural proteins, such as type I collagen, are required in large amounts. Moreover, the demand for these building blocks of biological structures is usually so high that having half as much as normal is insufficient. When just one normal copy is present along with a damaged copy, the presence of one such nonfunctional allele results in a severe disease called type I osteogeneis imperfecta, a disease that causes brittle bones and early onset deafness (Box 14.2). So it is a loss-of-function mutation—the cell makes less of a structural protein than it needs. However, presence of disease in heterozygotes shows us that the disease allele predominates in the presence of the normal allele, so we consider this case of loss of function to be dominant.

THE EXPECTED CASE: THE PROVERBIAL MONKEY WRENCH RESULTS IN A DOMINANT TRAIT

Like ballet dancers and bank robbers, very few proteins truly act alone! Instead, many function either as *dimers*, an associated pair of identical proteins, or as parts of large macromolecular assemblies that are composed of many proteins. Imagine then a gene whose protein product is assembled into such a large structure. Now imagine that when a missense mutation occurs, a misfolded version of the protein cripples the structure (kind of like the proverbial weak link in a chain). Such mutations can and do create a cellular disorder/defect, even in the presence of a normal gene product.

BOX 14.2 OSTEOGENESIS IMPERFECTA I—DOMINANT LOSS-OF-FUNCTION ALLELES

Osteogenesis imperfecta is often inherited as an autosomal dominant disorder. Although some children may inherit it from one of their parents, about 25% of osteogenesis imperfecta cases appear to be new mutations. Types II, III, and IV osteogenesis imperfecta may involve an abnormal form of collagen being made, but in the case of type I, the problem is that not enough collagen is made. Type I osteogenesis imperfecta is considered on average the least severe form of the disease. Altogether, the Osteogenesis Imperfecta Foundation estimates that there are more than twenty thousand people in the United States with some form of osteogenesis imperfecta, and type I is the most common of the four forms. Children with severe osteogenesis imperfecta may be born with multiple fractures and suffer numerous additional broken bones in the course of growing up. One of the most heartbreaking commentaries on this disease is the fact that the Osteogenesis Imperfecta Foundation includes a whole section on their web page about child abuse and the problems of good, loving parents who are mistakenly arrested for child abuse. In the cases of children who turn up with broken bones in situations in which the injury that is described does not match the severity of the damage that results, the answer might actually be child abuse, but the answer can also be that nonabusive parents are dealing with a child with bones that are too fragile for even gentle, loving care by parents who do not realize that their child has such fragile bones. Fortunately, there are doctors who can do genetic and biochemical tests that can help distinguish a child with a genetic disorder from a child who has been injured by an abusive parent, but we have to wonder how many parents in earlier times protested their innocence to deaf ears because they and their accusers both did not know that this disease existed.[2]

Let's consider Marfan syndrome (Box 14.3), which is inherited as an autosomal dominant trait. As you will recall, the fact that this trait is dominant means that only one defective copy need be present to cause disease. Skeletal features of Marfan syndrome include unusual height, long limbs, and joint laxity. Some of these individuals have spinal deformities such as scoliosis, eye problems that include myopia, or even life-threatening cardiovascular defects. Individuals with Marfan syndrome have been shown to have a defect in the gene that makes *fibrillin*, a protein that combines with itself and with other proteins to form microfibrils. According to the National Marfan Foundation, there may be 200,000 people or more with Marfan's syndrome or something related to it.[3]

This broad array of problems—skeletal, cardiac, and ocular—results when defective fibrillin causes abnormal protein complexes in cartilage, tendons, blood vessels, muscles, skin, and various organs. It has been proposed that the defective fibrillin has a dominant negative effect, interfering with the forma-

[2] More information about osteogenesis imperfecta can be found at www.oif.org.

[3] For more information about Marfan syndrome, check out the Foundation's web site at www.marfan.org.

BOX 14.3 MARFAN SYNDROME—WAS DEFECTIVE FIBRILLIN LINCOLN'S MONKEY WRENCH?

Some people believe that Marfan syndrome was the cause of Abraham Lincoln's lanky build, although finding out for sure has been held up by controversy about whether testing available samples of Lincoln's DNA would constitute violation of his privacy. On the one side is the important consideration that dying does not mean that you have lost your privacy. On the other hand, it has been argued that Lincoln would not have minded having this question asked. He was, after all, a public figure who was not particularly protective of his personal privacy. It has also been argued that knowing that he had Marfan syndrome would not hurt him, but would likely help the cause of people with genetic defects in general and Marfan syndrome specifically by demonstrating that someone with such a mutation or disease can be so successful and admired. What a strange concept, that in an era when cardiac surgery and modern heart medicines had not yet been heard of, a cardiac time bomb could have been waiting to fell him even if an assassin's bullet had not. Perhaps we will never know, if the ethical issues regarding this DNA test are never resolved. Even if the test showed that he had the defect, we still would not know if he would have faced heart failure, since severity of genetic diseases can so often vary from one person to another. More on this subject and other issues arising from modern genetics can be found in *Abraham Lincoln's DNA and Other Adventures in Genetics* by Phillip R. Reilly.

tion of correct microfibrillar structures even though there is also normal fibrillin present.

THE SURPRISING CASE: A GAIN OF FUNCTION RESULTS IN A RECESSIVE TRAIT

The discussion above made sense, so aren't mutations that create disruptive or poisonous proteins always going to be dominant? Once again, things are not that simple. The answer is simply no, they won't always be dominant. In fact, while a mutant protein that is actively doing something new will normally act as a monkey wrench and cause problems, sometimes the new function of the protein can even be beneficial.

This point is made most clearly by considering the case of a disorder that can be serious or even lethal in homozygotes but can actually be beneficial or even life-saving in heterozygotes. The disease in question is *sickle cell anemia*, a severe blood disorder that can cause serious illness or even death (Box 14.4). The genetic basis of this disease is the presence of two defective copies of the gene that makes a *hemoglobin* protein essential for adult red blood cells to successfully carry out their role of transporting oxygen. A *missense mutation* causes the production of hemoglobin S (also called HbS), which has the wrong amino acid, valine, at position 6, which contains a glutamic acid in the normal variant, hemoglobin A (also called HbA). Some other genetic variants

BOX 14.4 MANIPULATING GENE EXPRESSION TO TREAT SICKLE CELL ANEMIA

Hemoglobins are the proteins in the red blood cells that carry oxygen to the tissues and carry carbon dioxide back to the lungs to be breathed out of the body. HbS is a damaged version of the HbA hemoglobin present in adult red blood cells. By itself, the sickle cell form of hemoglobin (HbS) forms abnormal biochemical strctures that cause the cell to sickle and become rigid. In heterozygotes, where HbA is also present, the abnormal hemoglobin structures that cause sickling don't take place. Fetal hemoglobin (HbF) is a form of hemoglobin that is mainly made before a baby is born. Expression of the HbF gene normally shuts down by the time a baby is born, but some expression may continue. HbF can provide a similar protection against sickling. Since the patient with the sickle cell anemia does not a have a gene present that can make HbA, you might think that manipulating gene expression would not help, since you can't turn on expression of something the patient does not have. However, since HbF is made by a different gene, it is available to be tapped for service when HbS is making the patient ill. By using a medication called hydroxyurea, doctors are able to turn on an increased expression of HbF. Presence of HbF reduces the amount of sickling and decreases the problems with bouts of pain caused by red cell breakage in the capillaries. It will be a long time before we know how beneficial this medicine is. It does not turn up expression of HbF in everyone who takes it, and it can have side effects. We will talk in later chapters about gene therapy in terms of actual changes to the DNA in the cell being treated, such as adding back a good copy of HbA into cells that cannot make HbA. Hydroxyurea, which has been in use since the mid-1990s, constitutes a different form of gene therapy that simply makes use of one of the patient's own endogenous genes to provide a substitute gene product that can take over at least some of the needed function. This is not an effective enough process to outright cure the patient, but it apparently does provide a level of remedy that can make a real difference for at least some patients. Unfortunately, not enough is known about the effects of this drug on children, so it is not available to them even though more than 10% of children with sickle cell anemia will have a stroke or other major problems before they are adults.[4]

are also known that can cause sickling of red cells, but HbS is the most common cause.

Sickle cell anemia is a recessive disease that occurs when both copies of the gene are defective. The HbS protein becomes insoluble and forms aggregates. Cells that contain only the abnormal HbS tend to become rigid and deformed, taking on a sickle shape that tends to get stuck in the capillaries and break. Among the complications that can result are severe pain, infections, leg ulcers, delayed growth, and eye damage. Some of the more severe

[4] More information can be found by going to the New York Online Access to Health, which offers a long list of links to sources of infomraiton about sickle cell anemia and hydroxyurea treatments at www.noah-health.org/english/illness/genetic_diseases/sickle.html.

complications can include strokes, lung congestion, and pneumonia. The consequence of this incorrect amino acid is an abnormal hemoglobin molecule that causes red blood cells to become rigid and deformed (shaped like a sickle) and to block the capillaries. Over time, lung and kidney damage can accumulate. Treatments include antibiotics, vitamins, avoiding dehydration, carrying out transfusions, and, in rare cases, even bone marrow transplants.

Red cells do not become rigid and sickle-shaped in a fetus with sickle cell anemia. The gene that makes fetal hemoglobin (HbF) is a different gene from the one that makes HbS, and normally the job of fetal hemoglobin gets taken over by HbA in an adult. How far "off" the HbF expression gets turned after birth apparently varies from one person to the next. And how far "off" the HbF gene gets turned also apparently affects how severe a case of sickle cell anemia will be. So one of the most interesting treatments recently developed involves manipulating gene expression to get expression of the fetal hemoglobin gene HbF to partially compensate for the defective hemoglobin in the red cells (see Box 14.4).

Interestingly, there is a different kind of gain of function associated with this same sickle cell mutation that is considered to be dominant. Although people with two copies of HbS struggle with pain and illness, people with one copy of HbA and one copy of HbS are actually sometimes better off than people who only have the normal sequence. Specifically, people who are heterozygotes are better off if they live in areas where malaria occurs. The heterozygotes, with one normal allele, and one "sickle" allele, are less likely to be infected by the parasite that causes malaria. In fact, the frequency of the sickle cell mutation is higher in areas with endemic malaria, and lower in areas where malaria is rare or nonexistent. Thus the sickle cell mutation is a recessive gain of function in one sense (the disease called sickle cell anemia) but a dominant gain of function in another respect (the beneficial trait that causes resistance to malaria).

The incidence of this disease is the highest in African populations and approaches one in twenty-five births in some parts of equatorial Africa. (The incidence of sickle cell anemia among African Americans is approximately one in five hundred.) Under normal conditions, people who have one good copy of the gene are fine (although they may exhibit some symptoms at very high altitudes, where the oxygen pressure is low). Although we think of sickle cell anemia as affecting Africans and African Americans, sickle cell anemia can also be found among people who live in other parts of the world where malaria is present, including among some Mediterranean populations and in India. What we are looking at is a kind of trade-off between the optimal genotype for a malaria-free environment vs. the optimal genotype for an environment in which malaria is endemic. Because disease resistance occurs in the heterozygotes and is especially frequent in regions endemic for malaria, the individuals with sickle cell anemia pay the price for a mutation that benefits the population overall while harming them as individuals.

There are a number of genes in the human body that affect our abilities to resist various kinds of infections, and the outcome of an infection can depend in part on how well our bodies are prepared to cope with the particular invader causing an illness (Box 14.5). Some of the mechanisms by which we protect ourselves seem fairly obvious. If an infectious organism has

BOX 14.5 WHO WILL BE AROUND AFTER THE NEXT EPIDEMIC?

The processes that generate genetic diversity, while sometimes causing problems, can also provide advantages under the right circumstances. Diversity turns out to be of especially great importance in terms of responses to infectious disease. There are a large number of genetic differences between individuals that affect their resistance to different diseases. Some of these differences actually involve molecules of the immune system but can also affect genes involved in many other host defense systems. The result is that people who survive a polio epidemic may be a different subset of the population than those who would have survived if it had been a smallpox epidemic instead. If we built a population of clones derived from one individual, they would share similar (though not necessarily identical) fates in the next epidemic. If it happened that the population had been built from an individual with good defensive mutations against that particular disease, they would fare well, but what is the chance they would all be similarly genetically prepared for the next infection, and the next, and the one after that? This is a problem already faced in agriculture, where trends towards growing certain popular genetically identical strains puts crops at risk of an all-or-nothing fate depending on whether they are or are not resistant to the next pest or bug that comes through the region. Imagine the hazard to a human community, or humanity as a whole, if we all shared identical sequences in the genes that affect resistance to infectious diseases. Even if we could engineer it so that we all started out resistant to the known diseases, new diseases and new strains keep coming along. Influenza, and the need to keep getting new flu vaccinations each year to keep up with the constant trickle of new antigenic types, offers one of the strongest lessons in the rate at which infectious diseases can keep changing almost faster than our ability to cope with them. Our diversity, then, is one of our greatest protections; not in the sense of protecting any one individual, but rather in the sense of protecting populations overall so that there is someone left to carry on after the epidemic is over, and someone to take care of those who are ill while the epidemic is ongoing.

to attach to a particular protein or receptor to enter a cell, a human being may be protected if she is lacking that protein or if she makes a variant form of the protein that the "bug" doesn't recognize. If the invading organism causes damage when a protease cleaves an important human protein, protection can come from a change in the protein sequence that eliminates the cleavage site. Just one type of bacterium involved in a single infection in a human being may be using quite a diverse arsenal of biochemical tricks to assist in establishing a connection with the host, facilitating invasion to arrive at its favorite target cells, diminishing host defenses (immune and otherwise), and causing damage to the cells and/or surrounding tissues. Every single point of interaction of that bacterial cell with the human body represents a possible point of susceptibility or resistance to the invader, depending on whether mutations in human genes have produced altered forms of the human proteins with which the bacterium interacts. In many cases, even if someone has a sequence change in one of those key proteins, it may not be a change that affects the critical points of interaction between human and bacterial proteins. In a large population, with many different mutations having occurred over long periods of time, it is likely that some people will have different reactions to any given pathogenic factor in the bacterium's arsenal.

SUMMARY

There would appear to be some standard correlations, that absent essentials (loss of function) usually will be recessive and monkey wrenches (gain of function) usually will be dominant simply because that is often what happens. In reality, there are no absolute correlations between the actual nature of mutations, in terms of their effect on gene function and their phenotype when a bad copy and a good copy are both present. The relationship between a given form of a gene and its phenotype depends on the nature of the encoded protein, its biological function, the cell type in which it acts, and the environmental factors that influence expression.

For these reasons, we prefer to couple the terms *dominant* and *recessive* with a separate description of mutations in terms of the gene's ability to synthesize functional or poisonous proteins. Thus we often couple terms together, referring to a recessive loss-of-function allele or a dominant gain-of-function allele to give the combined information about how the mutation acts in a pedigree and what it actually does in terms of protein production. However, sometimes we do not have enough information to know which molecular mechanism is involved if we are dealing with a phenotype and a mode of inheritance for which the actual gene or biochemical pathway remains unknown.

HOW WE DETECT MUTATIONS

15

"You take a clear solution, you add a clear solution, you get a clear solution, and you call that a result!"
—*Spouse of a genetics graduate student*

The processes by which we look at DNA can be quite different, depending on what we are looking for. Earlier in this book, when talking about what chromosomes are, we showed pictures of chromosomes as seen under a microscope. However, even the highest-powered microscope will not allow us to read the order of As, Cs, Gs, and Ts in the DNA. Yet clearly in a world where headlines trumpet news about the human genome sequence, someone is managing to read the order of genetic letters contained on those chromosomes we were looking at.

Almost 30 years ago, when the perplexed nonscientist wife of a genetics graduate student made her declaration about our ability to see results in a test tube containing what looks like a drop of water, reading even a small bit of DNA sequence was technically difficult and required weeks or months of laborious effort. Some of the earliest experiments to read the sequence of DNA used enzyme-based technologies similar to what we use today, but Julia spent her graduate school days in the late 1970s and early 1980s using a rather terrible and fascinating mix of rocket fuel and carcinogens to pry the genetic spelling out of the pieces of DNA she was studying. It remains true to this day that when we want to read the sequence of a piece of DNA, we in fact start with a clear solution, add a clear solution, and end up with a clear solution. Fortunately, we now have instruments that will let us "read" the order of genetic letters present on that piece of DNA (in place of radioactivity and x-ray films used in the first versions of DNA sequencing, and still used in some places to this day) after we have completed the enzymatic reactions that go on in that clear drop of liquid.

Sequencing technologies have evolved in a long, gradual manner. Rather than spin lengthy tales of the technical hardships of an earlier era, we want to tell you how we currently find out the sequence of a piece of DNA, something that is critical to our ability to tell whether someone has a mutation in a particular gene.

For years the hardest part of getting at the information contained in any given gene has been the problem of getting our hands on the gene we want to know about. With the information content of the genome spanning billions of base pairs and our target being one out of many tens of thousands of genes, the process of gene discovery has traditionally been lengthy, complex, and expensive. However, once a gene has been discovered and its sequence is known, it is relatively simple to look at that gene in many different people to see whether any of them have a copy of that gene in which the sequence is different. Whenever we reach the point of doing genetic testing in a clinical context, that is, doing mutation screening in patients who want to know whether a typo in the blueprint is causing their problems, we are always talking about screening genes for which we already know the sequence. Sometimes that sequence resulted from long years on the trail of a specific gene, such as the Huntington disease gene, but in many cases, the sequence of a gene we want to study is the gift of the Human Genome Project, which we will discuss in Chapter 25.

MAKING COPIES OF DNA

The biggest problem we face when we want to read the sequence of a particular gene is the same problem we face when listening to someone talk in a crowded, noisy room—the signal we want to detect is there but is surrounded by too much other information that is very similar, and we can't distinguish the real signal from the background noise. With more than three billion base pairs of sequence in the human genome, if we try to read one piece of sequence while all of the rest of the genome is also present, we cannot detect the sequence we want to read even though the signal is present. So one of the most important steps in reading the sequence of a particular gene is to get it separated away from the rest of the genes in the genome, to bring it out of its "noisy" background into a quiet, separate place where it is the only thing present that our sequencing technology can detect.

We can get our hands on a gene that is already known by making copies of the DNA we want through the use of an *in vitro* (Box 15.1) process known as polymerase chain reaction (PCR). The use of PCR lets us make billions of copies of a single gene or piece of DNA in a matter of hours at a cost of less than a dollar per sample. This effectively amplifies the signal we want (the copied piece of DNA) billions of times compared to the background noise (the rest of the genome that did not get copied).

BOX 15.1 *IN VITRO* AND *IN VIVO*

According to the Miriam-Webster Dictionary, the term *in vitro,* which comes from Latin and means literally "in glass," dates to the year 1894. *In vitro* actually refers to processes that can be carried on outside of the living cell, with the concept "in glass" referring to the glass flasks and glass test tubes in which so much early biological work was carried out. As science has evolved, many processes are now carried out in tubes and flasks made of plastic yet are still referred to as being done *in vitro*. The counterpart term, *in vivo,* which they trace back to 1901, also comes from Latin and means literally "in the living." Thus an enzymatic reaction taking place in the cells of your body are going on *in vivo,* but if we duplicate that enzymatic reaction in a test tube by isolating the enzyme away from any living cells, then it is an *in vitro* process.

HOW PCR WORKS

PCR copies DNA using the same systems a living cell uses to copy DNA. As you may recall from Chapter 6, the secret to the replication of DNA is that the DNA polymerase enzyme copies a single strand of DNA by putting in place bases on the second strand that are complementary to the sequence of the first strand. One of the biggest problems with trying to use this approach is that we don't want to copy all of the DNA in the genome (or in the test tube in front of us on the lab bench!); we want to copy one small stretch of DNA that is perhaps only a few hundred bases in length. So if we were to just add DNA polymerase and copy everything in the test tube, we would be amplifying our background noise at the same rate we amplify the signal we are after.

One of the secrets to making PCR specifically copy only the piece of DNA we want, rather than copying all of the DNA in the test tube, is based on one of the biggest limitations of DNA replication. *DNA polymerase will only copy a single-stranded piece of DNA if there is a double-stranded section from which to begin adding bases* (Figure 15.1). So if we can create a double-stranded piece of DNA right next to the single-stranded piece of DNA we want to copy while avoiding having double-stranded DNA in other regions of the genome, we can make DNA polymerase do its job exactly where we want it.

We can create the conditions that DNA polymerase needs to carry out its copying function. First we need a source of single-stranded DNA that we will copy, which means we need to isolate DNA from cells of the person whose DNA we are going to examine (Box 15.2).

BOX 15.2 DNA FOR GENETIC TESTING

Easy sources of DNA include white blood cells from a blood sample or perhaps buccal cells that can be obtained by rubbing the inside of the cheek with a sterile swab. In some cases a sperm sample may be used. If we want to look at the DNA of a deceased person, it is often possible to get DNA from cells attached to hairs on a hair brush or even from pathology samples left over from a biopsy sample taken years ago. PCR works even if there is very little DNA in the sample to be tested, and it works well even if the cells have not been stored under ideal conditions. In forensic cases, DNA can be obtained from samples that have been in storage for decades, and in anthropology studies, DNA can sometimes be obtained from very old tissue samples.

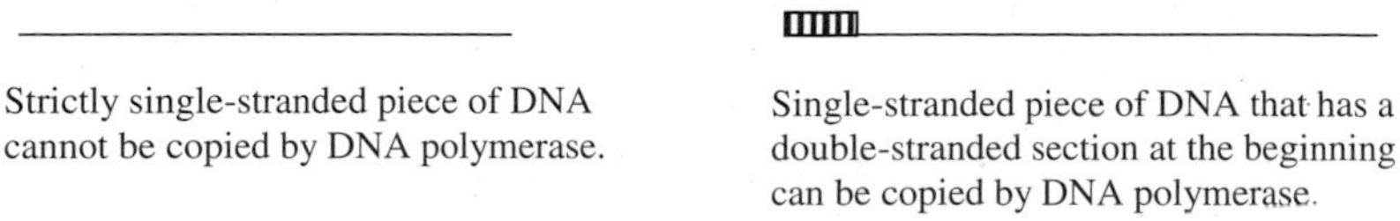

Strictly single-stranded piece of DNA cannot be copied by DNA polymerase.

Single-stranded piece of DNA that has a double-stranded section at the beginning can be copied by DNA polymerase.

FIGURE 15.1 Copying DNA. We cannot simply make huge numbers of copies of someone's genome by putting in DNA polymerase because it requires very special circumstances to carry out its copying function. It needs a region in which two strands of DNA are paired based on complementary base pairing as a starting point for its copying operation. Living cells have mechanisms for creating this condition when copying chromosomes, but it does not just happen naturally in the test tube.

Then we need to pull the strands of the DNA apart so that they are single-stranded, not double-stranded, DNA (Figure 15.2). This can be accomplished simply by boiling the DNA to break the chemical bonds holding the strands together.

There are more than three billion base pairs of human DNA sequence floating around in the test tube on the lab bench. We only want to copy the second exon of gene M, which is only 200 bases long and is the region of DNA in which mutations have previously been seen to cause the disease we are studying.

We use DNA polymerase's need for double-stranded DNA to force it to copy only the DNA containing that 200 base chunk of gene M. How? If we already know the sequence of gene M, we can call up a commercial DNA synthesis lab, and for about ten dollars, they will make us a tube filled with copies of a short piece of DNA (about eighteen base pairs in length) that matches a piece of sequence in or next to gene M (Figure 15.3).

If we add this short synthetic piece of DNA, called a *primer*, to a test tube containing single-stranded human DNA, the primer will find the piece of human DNA that contains its complement and will bind to (hybridize with) that piece of human DNA using the base-pairing mechanisms we talked about in Chapter 6. This creates the essential structure: a single-stranded region that we want to copy connected to a double-stranded region that gives polymerase its starting point. DNA polymerase will then start at the edge of the double-stranded piece and beginning adding new bases complementary to the adjoining region, including bases complementary to the gene M exon that we want to test for mutations (Figure 15.4).

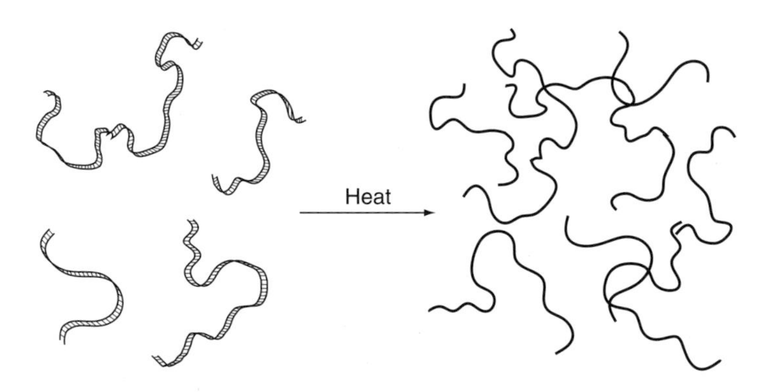

FIGURE 15.2 Heating makes double-stranded DNA come apart into single strands.

FIGURE 15.3 Targeting gene M. If we pick the sequence of our primer to match a spot in the intron right next to the exon we want to sequence, we can force DNA polymerase to copy the DNA right next to it by creating a hybrid that is double-stranded in the region of the primer and single-stranded in the region right next to it. DNA polymerase will begin inserting new bases right next to the arrowhead on the primer in this picture.

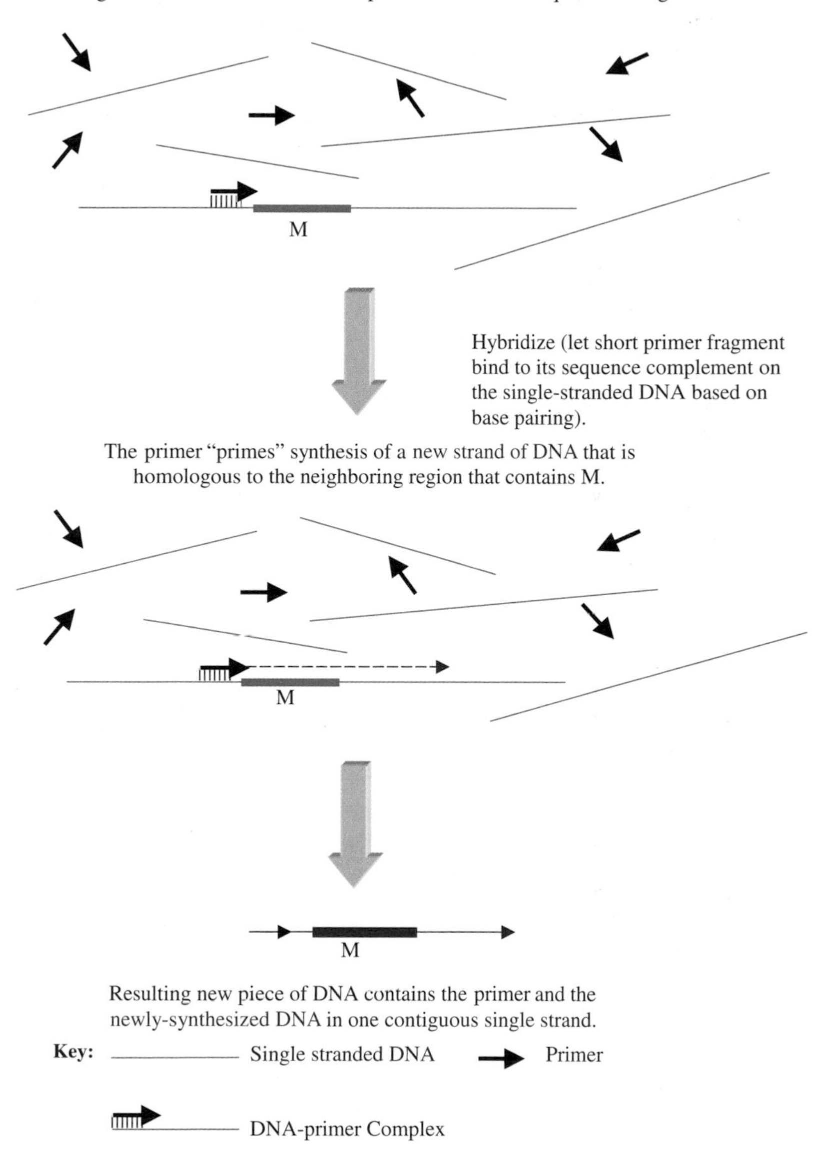

FIGURE 15.4 Newly synthesized strand of DNA containing the primer sequence at the beginning and a copy of the gene M exon in the middle of the newly synthesized strand. This strand was produced by mixing primer plus "target" DNA and letting DNA polymerase copy the area next to the spot where the primer hybridized.

FROM REACTION TO CHAIN REACTION

Now we know how to make new DNA complementary to exactly one place in the genome that we are interested in, by sticking a primer "tag" onto the human DNA at a location right next to what we want to copy. This lets us bypass copying the rest of the genome.

However, Figure 15.4 shows that we made one copy. Now, instead of having one copy of gene M among tens of thousands of other genes, we have

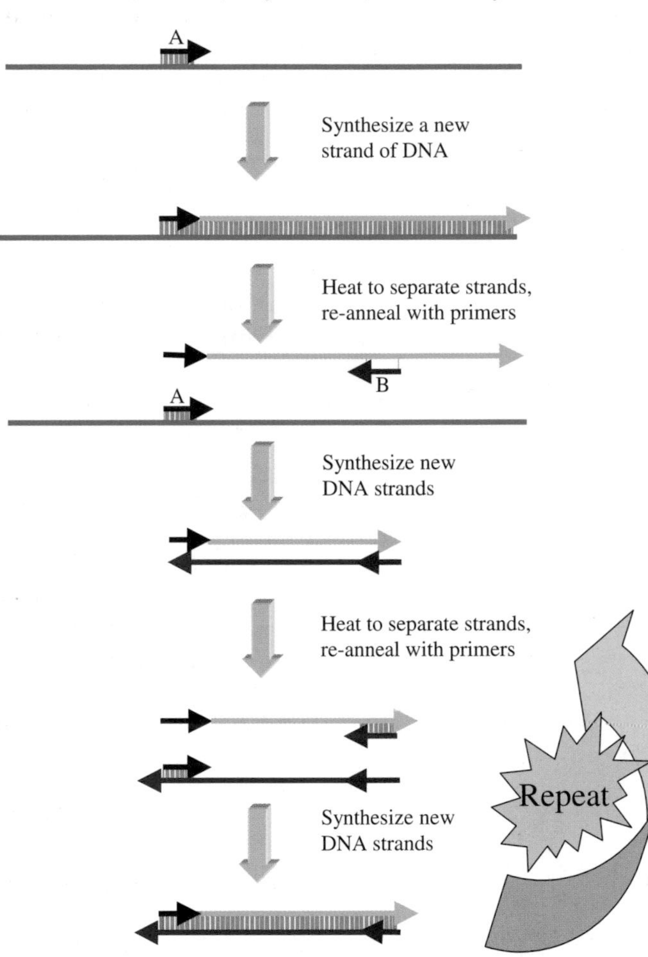

FIGURE 15.5 Polymerase chain reaction. After the first several rounds of DNA synthesis, PCR generates the simple double-stranded structure at the bottom of the figure, which can now be copied over and over through the "repeat" step that separates the DNA into single strands, attaches the primers, and synthesizes DNA again. The number of copies you can obtain is limited mostly by how many reagents (primers, individual nucleotides, buffers, and DNA polymerase) you want to expend, and frankly more often the limiting issue is how many copies you even need to make. Usually, one PCR reaction gives all the DNA needed for a sequencing reaction to read the DNA sequence of the PCR-generated DNA fragment.

two copies. This gains us almost nothing. We need a way to make a lot of copies. This is where the chain reaction part of PCR comes in. The real secret to PCR is not one primer that tags the spot you want to copy, but rather two primers that flank the place you want to copy that let you repeatedly copy exactly the same spot in the genome. So at the end of Step 1 in Figure 15.4, we had one new strand. If we have a second primer also present that can bind to the new strand, it can now copy back across the region containing M, but now it is copying on the other strand (Figure 15.5).

There is a second secret to PCR: the kind of DNA polymerase used in the reaction is stable at very high temperatures. In the procedure, every round of PCR calls for heating the DNA to separate the double strands back into single strands that can bind the primer. There is a problem here, which is that if you heat the reaction mixture to separate the DNA strands, you will kill most types of DNA polymerase, exactly the same enzyme that is needed in the next step to carry out the DNA synthesis step. A very clever solution to this problem was to go in search of organisms that live at very, very high temperatures. The logic is that, if an organism can live in a hot springs or along the edges of the hot thermal currents of the deep ocean vents, their enzymes must all be capable of surviving at temperatures close to boiling. By isolating DNA polymerase from organisms living in hot environments, scientists made PCR something practical and useful instead of a theoretical curiosity.

If you spend the time to follow Figure 15.5 through its steps, you can see how the use of two primers flanking a gene can rapidly isolate that gene (or other sequence of interest) onto a double-stranded fragment that has the primer sequences at its ends. Once this double-stranded sequence exists, it can go through the same loop—separate strands, bind to primers, synthesize new DNA, separate strands, bind to primers, synthesize new DNA—over and over.

You don't need to spend a lot of time contemplating the mechanisms in Figure 15.5 to be able to get the main point here: if you can make primers that flank the piece of DNA you want, you can make vast numbers of copies of the piece of DNA that lies in between the sequences where the primers bind based on their complementary base pairing (Figure 15.6). One double-stranded structure of this kind is copied to become two copies of the double-stranded structure after one round of PCR. After two rounds it has become four copies, after three rounds it has become eight copies, then sixteen copies, and so on. After twenty rounds of PCR, we have more than a million copies.

To truly appreciate the power of PCR, consider several things: After thirty rounds of PCR we have more than a billion copies of that one little double-stranded piece of DNA we started with. (In contrast, if we had used only one primer, we would have thirty new copies at the end of thirty rounds of DNA synthesis!) Usually, when we do a PCR reaction, we start out with more than one copy of the genome sitting in the test tube. Depending on how much DNA you need to end up with, a PCR reaction may often cost less than a dollar to carry out. Also, PCR is fast, with some reactions taking less than a minute per round (although sometimes it takes longer). It's no wonder that PCR is felt to have revolutionized modern biology and genetics!

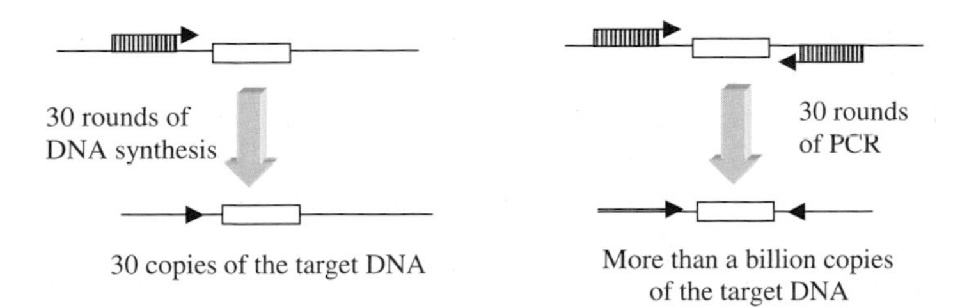

FIGURE 15.6 PCR can generate billions of copies of a DNA fragment in the time it takes a single-primer reaction to make tens of copies.

SEQUENCING

The next step is to determine the sequence of the DNA in the PCR-amplified DNA fragment. A set of technical tricks come together to make sequencing possible, including the ability to synthesize DNA in vitro, to label DNA with various dyes and tags, and to separate DNA fragments that differ in length by as little as one base.

Sequencing is basically a set of biochemical reactions that let us determine where the bases are located along a DNA chain by counting the number of letters away from the beginning of the sequence and asking what letter is at that position (see Figure 7.2). For instance, if our DNA fragment has the sequence GCTACCGCTTTCGACTGATGGCAT, we can perform a sequencing reaction that tests for where the Ts are in this sequence. The biochemical reaction we carry out will generate copies of the sequence that stop wherever there is a T in the sequence. So our sequencing reaction that reads the Ts in the sequence will make a lot of copies of out target fragment, some stopping at the first T, some stopping at the second, and so on. The T sequencing reaction will produce all of the fragments shown in Table 15.1.

In the test tube, where the biochemical reactions go on (in our clear drop of liquid), many copies of each of the above fragments are floating around in the solution. So we have carried out a biochemical step that creates fragments that end wherever there is a T in the sequence, but all we see is something that looks like a drop of water. How do we get useful information out of the fact that these different DNA fragments, these different "relatives" of the target fragment, are floating in solution? We use a very important technical trick: our ability to separate DNA fragments based on how big they are.

We create a gel matrix (that looks a bit like a giant, square, unflavored Jell-O Jiggler!) made up of cross-linked molecules with spaces between them through which the DNA can run. Because DNA has a lot of negative charges on it, it will move towards the positive pole of a battery or power supply. If we cut a hole (called a well) into the gel, put in our DNA samples with all of the different sizes of fragments, and turn on the electrical current, the DNA will move through the pores of the gel. The important part is this: the smallest fragments will move the fastest, and the largest fragments will move more slowly.

The use of dyes, in some cases fluorescent dyes, lets us tell where the DNA is located in the gel. If there were only one copy of each length of fragment,

TABLE 15.1 The T sequence reaction creates fragments that end wherever the sequence contains a T

Sequence of fragments generated by the "T" sequencing reaction	Fragment size
GCT	3
GCTACCGCT	9
GCTACCGCTT	10
GCTACCGCTTT	11
GCTACCGCTTTCGACT	16
GCTACCGCTTTCGACTGAT	19
GCTACCGCTTTCGACTGATGGCAT	24

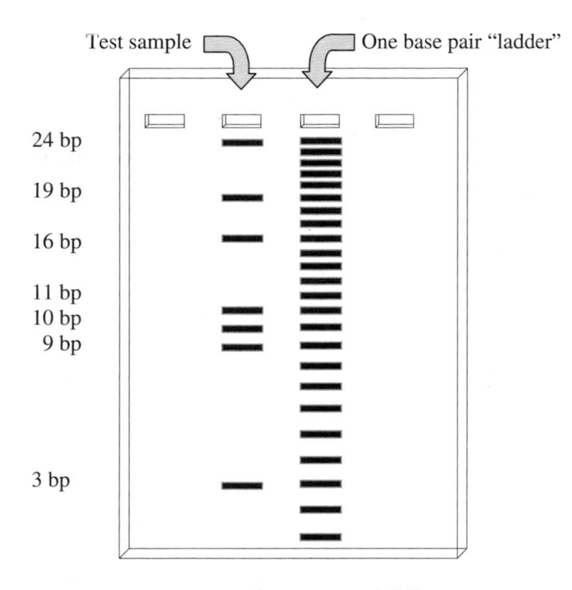

FIGURE 15.7 DNA fragments of different sizes run through gels at different rates. The smallest fragments run the fastest, and the larger fragments move more slowly. We can figure out the size of our DNA fragments by comparing our test DNA to fragments of known sizes (in this case, a one-base-pair ladder of fragments that differ in size by one base). Although we use this method to size many DNA fragments, this is not actually how we do it when we are sequencing DNA, as we will show you in Figure 15.8. (If this were a real gel, the gel itself would be almost transparent, and the DNA would be a fluorescent orangey-pink color, but clear gels don't make good images because they are hard to see.)

we would not be able to see the DNA, but since there are many copies of each length fragment, and since things the same size all run at the same position in the gel, use of dyes that detect DNA let us see where the different-sized fragments are in the gel (Figure 15.7). At a position where a group of same-sized DNA fragments runs together, they create a signal that we can visualize with the help of lab instruments, which looks like one of the bars in a bar code (although we actually call them bands).

Of course, we don't just want to know where the Ts are in the sequence; we want to know the order of all four bases. So if we want the order of all four bases, we can do four separate reactions: an A reaction that creates fragments that stop at all of the As, a C reaction that stops at the Cs, a G reaction that stops at Gs, and a T reaction just like the one we showed above that stops at the Ts. If we make a gel just like the one in Figure 15.7, we can load the A reaction in the first well, the C reaction in the second well, the G reaction in the third well, and the T reaction in the fourth well, then turn on the electrical current and let the small fragments outrace the larger fragments. A mock-up of such a gel in Figure 15.8 shows us how easy it is to read the sequence. We see that the G lane contains the smallest fragment, so G is the first genetic letter in this genetic word we are trying to read. The next-smallest fragment is in the C lane, so C must be the second letter, and the T lane holds the third-smallest fragment, making T the third letter in what we are trying to spell.

Real sequences of this kind are done on very large gels with as many as ninety-six wells allowing for sequencing of many samples in one experiment.

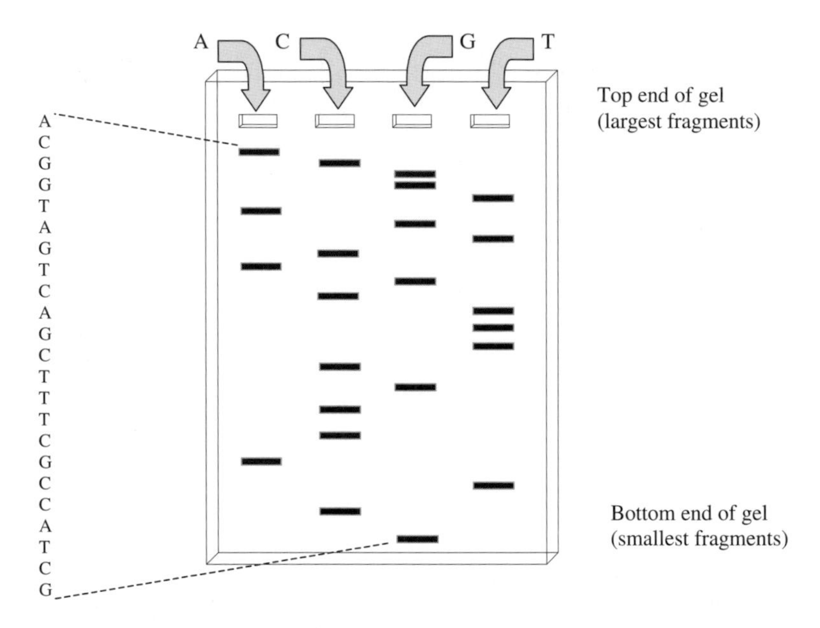

FIGURE 15.8 Diagram of a sequence gel showing the positions of DNA bands created by carrying out A, C, G, and T sequencing reactions on a piece of DNA with the sequence GCTACCGCTTTCGACTGATGGCA. The band closest to the bottom of the gel represents the smallest fragment created, so the first base in the sequence must be the base that corresponds to the reaction that created it (in this case a G). By starting at the bottom of the gel and reading upwards, calling off each band by the name of the reaction it came from (A,C,G, or T) results in reading of the order of bases in the DNA fragment.

Instead of staining the DNA with dyes, the DNA is tagged with radioactivity so that the image of where the DNA is on the gel can be captured on a piece of x-ray film (Figure 15.9).

The versions of sequencing we just showed are cumbersome, but they are very good for showing how sequencing works. Do four sequencing reactions that create fragments that end with the four bases, then separate the fragments by size, and start "reading" the sequence by seeing which lane contains the smallest fragment, then seeing which lane holds the next smallest and so on.

This type of sequence reaction and display via x-ray film can be used to detect mutations in human DNA samples. If someone is heterozygous for one of the bases in the region being sequenced, two different fragments will be generated that correspond to the two different "base terminating" reactions (Figure 15.10).

Although some sequencing is still done this way, more recently developed technologies that use fluorescent dyes read by lasers have revolutionized sequencing. The actual sequence reactions, the biochemical processes occurring in the test tube, are quite similar, except that they result in putting four different colors of fluorescent dyes onto the fragments generated by the sequencing reactions that end fragments at the four different bases, A, C, G, or T. The combined batch of all four colors of fragments are combined together and placed in an electrical field where the fragments begin moving,

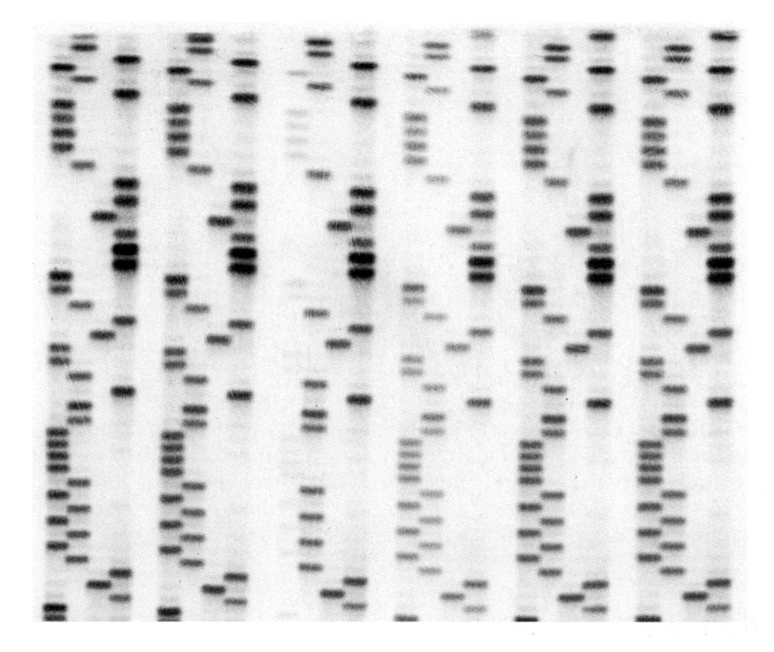

FIGURE 15.9 Picture of an x-ray film image of the same region of DNA sequenced from six different samples. Notice how the pattern is the same in all six samples. Some differences, such as fainter bands in some of the lanes, are due to technical differences in handling the samples and do not reflect any difference in the actual DNA sequence. (Courtesy Kathleen Scott.)

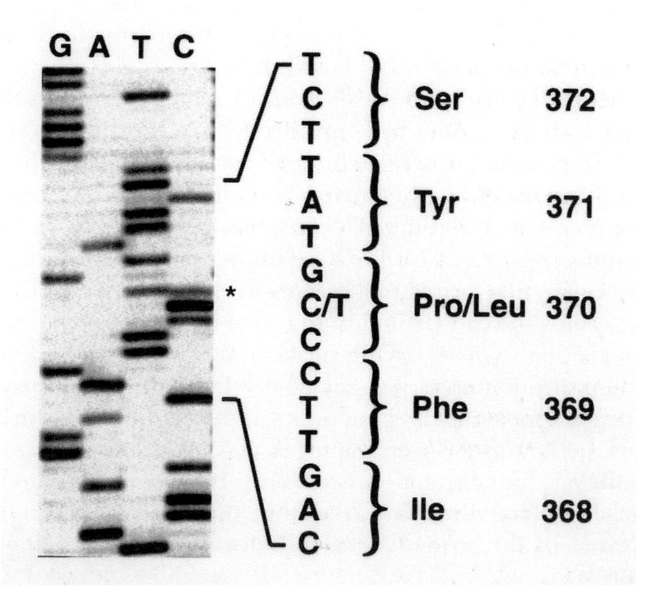

FIGURE 15.10 Sequence of exon three of the MYOC glaucoma gene, showing a mutation that changes amino acid 370 from a proline, in the normal protein, to a leucine in an individual who is affected. The mutation is detected as a point on the gel at which a band in the T lane appears to be the same size as a band in the C lane. (Courtesy of Frank Rozsa.)

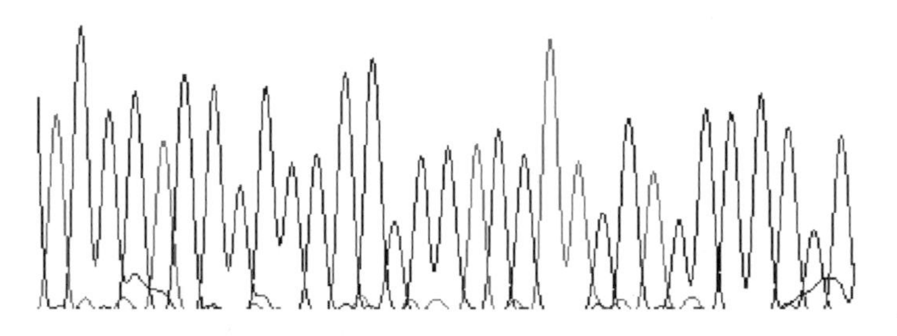

FIGURE 15.11 DNA sequence. This sequence, which was "read" by a laser-operated system that entered the data directly into a computer, is read from left to right. A green, C blue, G black, and T red. The letters across the top indicate which bases the computer thinks belong at that position in the sequence. Notice that there is one dominant color of peak at each position. (Courtesy of R. Ayala-Lugo.)

and as above, they move according to size with the smallest fragments moving the fastest and farthest.

As the DNA fragments migrate through the electrical field, they pass a point at which a laser "reads" the color of the dye passing in front of it. So if green is the dye that tags fragments ending in A, and the first dye to move past the laser is green, then we know that the first base in the sequence must be A. If T is tagged in red and the next two fragments to flow past the laser are red, the next two letters are T and then T.

Fragments that end with A will be dyed green, fragments that end with T are dyed red, those ending in G are dyed black, and those ending in C are dyed blue. By knowing which DNA bases correspond with which dye tags, we know that if the order in which dyed DNA moves past the laser is green-red-red-black-green-blue-red-red-black-black-blue, then the first part of the sequence must be ATTGACTTGGC. The information from the laser is read into a computer that prints out images like the one in Figure 15.11, which make it very easy to read (from left to right) the order of the bases. The computer also reads the order of the peaks and prints out the letters above the peaks. When a mutation is present in the DNA of a heterozgous individual and two peaks appear at the same point in the sequence (as in Figure 15.12), sometimes the computer even manages to "call" the presence of the double peak, signifying that a mutation is present. However, there are enough types of technical artifacts that can arise that mutation screening also calls for human review of the sequence traces before deciding whether an apparent mutation is real.

Because there are technical limitations to the sizes of DNA fragments we can measure accurately, sequence reactions often yield hundreds of base pairs of sequence, but not millions. So the sequence of a whole gene is often assembled by sequencing lots of separate pieces and then assembling those partial sequences back together into the completed large sequence.

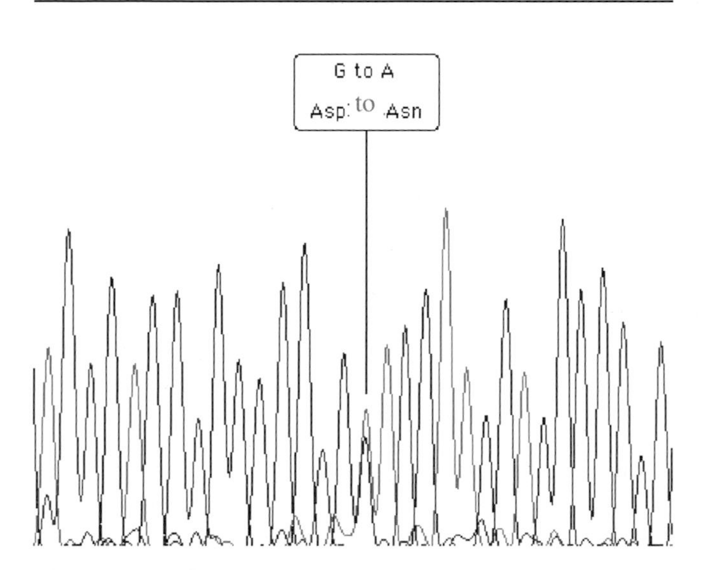

FIGURE 15.12 DNA sequence showing a mutation. This image shows a sequence of the same region of the same gene shown in Figure 15.10. However, at the position marked, there are two different peaks, a green peak indicating A and a black peak indicating G. This means that we are seeing both copies of this gene in someone who has G on one copy at the same point in the gene that holds an A on the other copy. This change in the DNA causes some proteins made from this gene to have asparagines in the protein at a position usually occupied by aspartic acid. (Courtesy of R. Ayala-Lugo.)

SEQUENCE TAGGED SITES AND THE BIOINFORMATICS REVOLUTION

The combination of PCR and sequencing brought about a massive *bioinformatics* revolution in biology. To carry out PCR, you need two primers flanking the item you want, each primer usually about eighteen bases long. So if you want to study a gene, you do not even have to start out knowing its sequence—all you need to know is the sequence of two primers that flank what you want. Usually, that means you need to know less than forty bases of sequence total to be able to get your hands on the DNA you are after.

So the big information revolution that went along with PCR was the development of the concept of the *sequence tagged site* (*STS*), a small piece of sequence that will let you tag the piece of DNA you want even if you don't yet know the sequence of the entire piece. When the concept of the STS was refined to its most efficient usage, researchers studying a particular gene no longer had to send a test tube containing a particular gene to another lab that also wanted to study that same gene; all they had to do was send them the sequence of the gene. In fact, even just the sequence of the primers that flank the gene that can be used to copy it via PCR (early on by e-mail, and later by posting the information in databases so that the researchers did not even have to spend time talking to each other to bring about the information transfer).

The really revolutionary part of the STS concept was the development of databases containing sequence information that anyone can access. Thus, if someone had identified a gene and wanted others to be able to study it as well, they did not have to send tubes of DNA to all of the other researchers in the world; they just had to put the sequence into the database. If someone wanted to study that gene, they could turn on their computer, use the Internet to access the database, look up the sequence of the primers specific for the gene they want, call up a DNA synthesis facility to order some primers, and do a PCR reaction. Some labs that do a lot of this even keep their own machinery for the synthesis of primers.

The result is that experiments that used to take weeks, months, or years at the lab bench may now occupy a few hours of computer time that may or may not require some follow-up lab work at the bench. In fact, some researchers now do "virtual PCR," running their tests electronically in the databases containing the human genome sequence instead of (or before) doing experiments with tubes and chemicals. Researchers now refer to these computer-based experiments as *in silico* experiments.

This has not eliminated real lab work, but it has increased the power of what can be accomplished for the same amount of lab work. It means that more time can be spent on the meaningful parts of the experiments, such as finding out what a gene does, and less time gets spent on trying to come up with the raw materials for the experiments.

WE ARE ALL MUTANTS

16

We are going to make a daring assertion here: we are all mutants, every single one of us. Those of us who struggle with ailments and curse our appearance in mirrors are mutants. Those of us who beam with pride over our fine health, good looks, or talents are mutants. Every human is a mutant. Actually, every cat, dog, horse, chicken, armadillo, dolphin, lobster, gecko, fruit fly, slime mold, corn plant, piranha, wildflower, and redwood tree is in some sense a mutant. In fact, each of us carries a surprising load of mutations scattered throughout our DNA, some benign, some beneficial, some potentially harmful. Some of those mutations combine to give each of us our unique characteristics, contributing to our virtues and our flaws, be they medical, cosmetic, or behavioral. Other mutations are carried silently by every single one of us, some having no potential to cause harm and others waiting unrecognized to be manifested only if they come together with another allele of that same recessive locus. It is estimated that each of us is a carrier for more than one mutation that would cause severe problems if both copies were knocked out. Fortunately for many of us, many of those harmful recessive mutations are incredibly rare, and we never find out that we have them or what they are because we don't end up having children with someone else who is a carrier for that same defect.

Sometimes, someone gets a bad roll of the dice or deal off the deck, whatever your favorite imagery is for manifestations of luck. We see this happen when two happy new parents are told that their apparently healthy child will be mentally retarded unless the child spends the first decade of life on a low-phenylalanine diet. We see this happen when parents with four healthy children are completely unprepared for the news that their fifth child has muscular dystrophy. We see this happen when a defect in nitrogen metabolism kills an infant whose parents have never even heard of nitrogen metabolism.

"Mirror, mirror on the wall,
Who is fairest of us all?"
—Snow White's Stepmother
(Courtesy of Sophia Tapio.)

Every individual who looks into the mirror on the wall saying mentally to themselves, "I am the fairest of them all, I am better than others who suffer from inborn maladies," needs to realize that we all carry changes in our genomes that could have been a problem under other circumstances. Everyone who responds to someone else's genetic tragedy by saying, "That's really too bad," while smugly thinking that their child is okay needs to remember the saying, "There but for the grace of God go I." Each of us with a healthy child could still be a genetic carrier for some other genetic lightning bolt that did not happen to hit us; that lighting strike might have been prevented because the person we had children with carried a different set of hidden genetic flaws than our own, or because the particular sperm and egg that met up did not happen to contain a harmful genetic combination we did not realize we harbored, or because our child was not exposed to some environmental situation that would allow their problem to manifest itself.

But hold on just a minute! We do not mean to cry doom and gloom. Let's be frank here. *Most babies are born healthy and go on to have healthy children of their own* (Box 16.1). The fact that we all carry some potentially detrimental mutations should not send any of us into fear and trembling at the idea of reproducing. Rather, it should send us in search of understanding. Being aware of your own genetic background, through understanding your genetic family history and being aware of the risks associated with dietary choices, drinking, smoking, and other behaviors that alter risk of birth defects, can help you make informed reproductive choices. Being aware that we all carry flawed genomes should help avoid some of the prejudiced reactions that have too often been manifested in the past towards those who are blind, crippled, or developmentally delayed or who in some other way wear public evidence of flawed genomes that the rest of us carry unknowingly.

JUST WHAT IS A MUTATION ANYWAY?

Back in Section I, we talked about Mendel's observations that true breeding strains of a plant would continue to breed true. One of the reasons we can follow the transmission of a trait from one generation to the next is because the associated allele acts as a constant "immutable" entity. In reality, genes can change in a stable and heritable fashion called *mutation.* In fact, mutation is

BOX 16.1 BIRTH DEFECTS

Talk of the many mutations present in the DNA of any one individual should not strike fear in the hearts of those who are planning families. The vast majority of babies are born healthy. In populations in which the parents are not related to each other, more than 97% of the children can be expected to be born without birth defects, and more than 99% will be free of major birth defects. This reflects, at least in part, the rarity of dominant alleles that can cause birth defects, but it also reflects just how often the hidden genetic defects (the recessive alleles) in one person are different from the hidden genetic defects in another. So what happens when people who are related have children together? They have an increased chance of sharing defective alleles, and the rate of birth defects may double. These numbers tell us that, even in the case of first-cousin marriages, most babies are born healthy. According to the March of Dimes, some of the most common birth defects include cerebral palsy, spina bifida, cleft lip/palate, lack of one or both kidneys, obstruction of the small intestine or urine passage, diaphragmatic hernia or abdominal wall malformations, chromosomal anomalies of which Down's syndrome is the most common, and limb malformations. Although there are genetic components to some birth defects, many people born with birth defects go on to have normal children, and good prenatal care reduces the chance of some types of birth defects. One of the most noticeable advances came about when it was discovered that addition of folic acid to the diet of pregnant women reduces the frequency of spina bifida.

the process that originally generated each of the alleles that we follow from one generation to the next when we look at inheritance of two different versions of a trait. Perhaps that mutation happened thousands of years ago and has been handed down to many descendants who all hold that mutation as something that makes them different from the rest of the human race. Or perhaps the mutation responsible for a particular trait happened in the sperm or egg that created you.

A mutation is a stable and heritable change in the genome of an organism, which in the case of a human being means a change in the sequence of the DNA that spells out the genetic blueprint. Not only is mutation a genetic process that actually alters the base sequence of the DNA, it is also the term used for the result of the process—a new sequence that is different from what was present in the DNA of that individual's parents. In the case of a human, a mutation can be either a change in the sequence of the DNA of the chromosomes in the nucleus or a change in the sequence of the mitochondrial chromosome. Most of the mutations are on the chromosomes in the nucleus, since the vast majority of the DNA is there. Mutations include changes from one base pair to another (for example, A-T to G-C), deletion of one or more base pairs, or insertion of one or more bases. In Chapter 14, we talked about the basic concept of how mutations affect us—through the production or lack of production of a gene product or altered gene product. However, we have not yet talked about the many different kinds of mutations or how they come about.

MISSENSE MUTATIONS CHANGE THE PROTEIN SEQUENCE

One type of mutation that is easiest to understand is the *missense mutation* (Figure 16.1), which is usually a change of one base in the DNA sequence but can involve more than one base. A missense mutation changes the amino acid specified to be used at that point in the protein. Since the amino acids have different properties—different size, different shape, different charge, different polarity—a change in amino acid can have a wide range of effects. Some changes will alter the gross structure and folding of the protein. Some changes will change more local properties, such as charge or the ability of that point on the protein to interact with some target molecule. Many missense mutations are what we call point mutants because only one point (or genetic letter) in the sequence was changed.

A disease that often results from missense mutations is phenylketonuria (PKU). Babies born with PKU used to be doomed to profound mental retardation and in some cases experienced additional problems, such as epilepsy. Their inability to convert phenylalanine to tyrosine would lead to accumulation of excess phenylalanine, with especially drastic consequences to the development of the brain in infants and children. Once the cause of PKU was identified, it became treatable by dietary approaches. However, the answer is not as simple as cutting out all of the phenylalanine, since damage will also result if the phenylalanine levels get too low. According to the National Newborn Screening and Genetics Resource Center, newborn screening for PKU now takes place in all fifty states of the United States so that infants can be placed on a low-phenylalanine diet as promptly as possible to minimize nervous system damage. Unfortunately, there are still many other places in this world where such children will have suffered irreparable damage before anyone figures out that they need to avoid phenylalanine. It may sound easy to say that putting these children on a low-phenylalanine diet fixes the problem, but it is incredibly difficult to keep the phenylalanine levels low enough, and all too often there is still some damage in the course of growing up. In addition, a woman with PKU has to go back on the severely restricted diet during pregnancy or her child will be harmed, an outcome that is far too common because of how hard it is to keep the phenylalanine levels low enough to protect the baby.

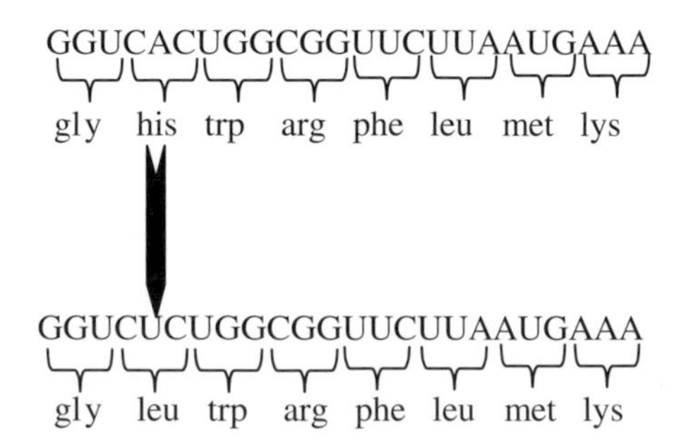

FIGURE 16.1 A missense mutation. A single base change in the mRNA due to a base change mutation in the DNA results in the incorporation of a different amino acid, leucine in place of histidine.

The common cause of PKU is defective phenylalanine hydroxylase, the enzyme that carries out the conversion of the amino acid phenylalanine to a different amino acid called tyrosine. Out of more than sixty different PKU alleles listed by Online Mendelian Inheritance in Man, more than three quarters of them are missense mutations and many of the others either delete an exon or affect a splice site. Also, there are many different missense mutations, hitting different locations spread across hundreds of amino acids, rather than one or a few specific mutations found repeatedly in many different individuals. The frequency of PKU varies from one population to another, and in some populations the existence of a very small number of mutations in all of the PKU cases suggests the existence of founder effects in which all cases alive today descend from a very small number of shared ancestors. In other populations, there are many different mutations indicating more heterogeneity.

NONSENSE MUTATIONS TRUNCATE THE PROTEIN

In other cases, a point mutation has a very different effect. Three of the sixty-four codons are stop codons that instruct the ribosome to stop adding amino acids to the growing protein chain. If the codon designating an amino acid is mutated to designate a stop codon instead, it is called a *nonsense mutation* (Figure 16.2). These mutations have the effect of truncating the protein, leaving a shorter protein that is missing its tail end. In some cases, if the nonsense mutation happens very early in the gene, most of the protein might be missing. In some cases this means that the cell has a bunch of truncated protein to deal with, but in some cases the cellular mechanisms for dealing with problematic proteins kick in and the "bad" protein is either gotten rid of or in some cases not made at all.

For which cases would we predict that missense and nonsense mutations would have very similar phenotypes? Both types of mutations can result in a completely inactive protein product, either through changing a critical amino acid or causing truncation that results in inactivity. We would predict similar phenotypes for cases in which missense and nonsense mutations are both

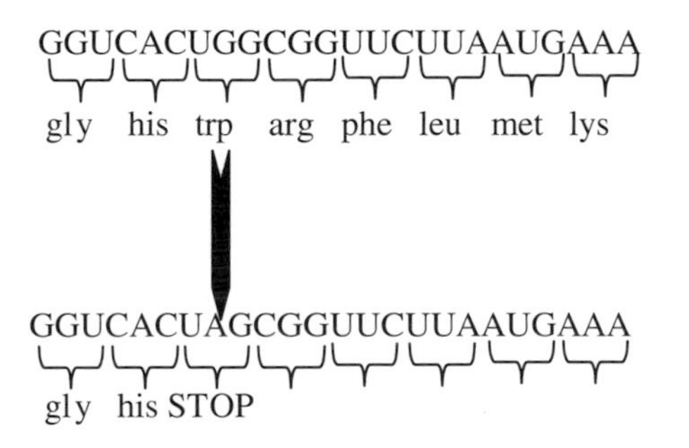

FIGURE 16.2 A nonsense mutation. A single base change in the mRNA due to a base change mutation in the DNA results in production of a STOP signal that tells the ribosome to stop adding amino acids to the growing protein chain.

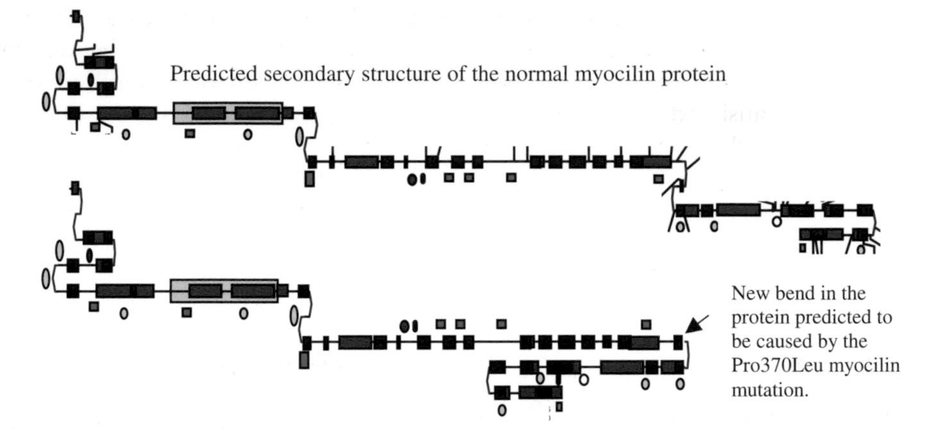

FIGURE 16.3 Predicted effect of a mutation on protein folding compares the secondary structure of the normal myocilin protein to the secondary structure of the myocilin protein that contains a leucine in place of a proline at position 370. The upper picture shows a computer prediction of some structures that might be found in the myocilin protein made by the GLC1A glaucoma gene. Some of the boxed areas show features such as regions of random coiling or pleated sheets. Symbols along the protein show where some types of chemical modifications could be added to the protein after it is made. The black lines show positions of amino acids changed by different mutations that have been found. The lower picture shows the same protein produced by a mutant copy of the gene. The arrow points to a new fold introduced into this protein by a mutation that causes an early-onset form of glaucoma. The real three-dimensional structure of the protein would be much more complex than this and involve interactions between different parts of the protein, but this kind of computer projection of structural changes can help us understand why a missense mutation is having an affect on the protein function. (Courtesy of Frank Rozsa.)

causing the disease via the same mechanism—loss-of-function. Thus, for many loss-of-function diseases, such as cystic fibrosis, we can see missense or nonsense mutations causing disease of comparable severity, and we see many mutations in each category.

It is important to note that missense mutations and nonsense mutations in the same gene will not always cause the same phenotype. This will happen in cases where the nonsense mutation causes loss of function but the missense mutation causes a gain of function.

For many disease genes, both missense and nonsense mutations are found. In the case of the gene that makes the *myocilin* protein, missense mutations towards the end of the gene most often cause an early-onset, severe form of *glaucoma* (Figure 16.3). As children, teenagers, or young adults, the patients experience a great increase in pressure inside of the eye, followed by gradual death of the nerves in the retina that carry visual signals back through the optic nerve to the brain. The nerve cells apparently die in response to the pressure, and in many cases the pace of nerve death can be relatively rapid and cause substantial visual deficits within a few years. In contrast to these missense mutations, there is a nonsense mutation that is actually the most common disease-causing mutation in the myocilin gene. It also causes glaucoma, but the glaucoma that results is usually a much later-onset disease, usually starting in middle age or later, involving much less pressure elevation in the eye and a much longer time period over which damage to nerves manifests.

So why would a nonsense mutation in myocilin cause a different phenotype than what we see for the missense mutations? It has been suggested that myocilin missense mutations that cause severe disease are monkey wrenches, causing disease through actively causing a problem rather than through a lack of the gene product or function. So for myocilin, does less severe glaucoma result from a nonsense mutation because it is operating through a different loss-of-function mechanism instead of acting as a monkey wrench? That remains to be seen, and some studies suggest that the difference may not be anything quite that simple.

MUTATIONS DON'T ALWAYS CHANGE THE PROTEIN

Notice that point mutations are not always missense mutations or nonsense mutations—that is, sometimes you can change a single base within the coding sequence and get no change in the amino acid or the protein. Why? Remember that there are sixty-four different codons that designate which amino acid will be used, and there are only twenty amino acids. This means that there is *redundancy* in the code. So some changes in the DNA sequence will replace one of the codons specifying leucine with a different codon specifying leucine. We joke sometimes about finding a leucine-to-leucine mutation, meaning a change in the sequence that does not affect the protein. These are called *silent mutations* (Figure 16.4).

Even when a missense mutation does change the protein sequence, sometimes this results in no functional change and no change in phenotype. This is especially easy to understand in cases in which an amino acid of very similar size, shape, charge, and polarity replaces the original amino acid. So missense mutations often cause functional problems, but just knowing that there is a change in amino acid is not enough to tell you that this change is necessarily bad enough to affect the phenotype. In one disease gene with more than fifty known missense mutations, only about half of them seem to cause the disease. The other half appear to be what we call *benign polymorphisms.*

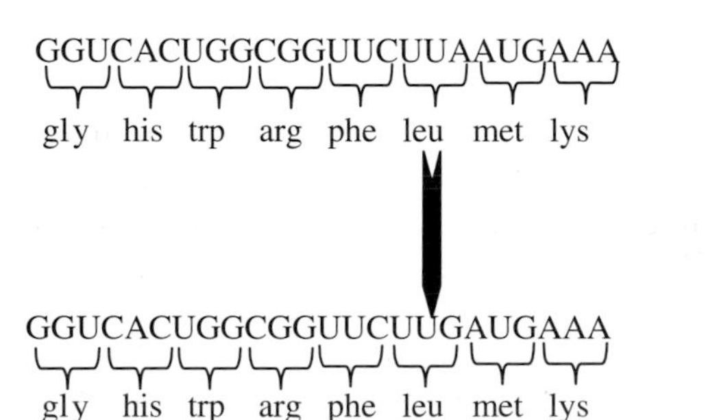

FIGURE 16.4 A silent mutation. A single base change in the mRNA due to a base change mutation in the DNA results in no change in the amino acid incorporated. Because a lot of the redundancy in the code occurs in the third position of a codon (in this case UUA goes to UUG), it is most common to see silent mutations "hit" the third position. However, arginine can be encoded by codons that begin with either C or A, and a number of other amino acids can be encoded by alternative codons differing at either position one or position two.

BOX 16.2 BENIGN POLYMORPHISMS

One of the things that we see in a lot of genes is that there are some categories of sequence change that are considered benign sequence variants—points in the sequence of that gene that differ between individuals in the population without showing any association with a trait. Sometimes these are also referred to as *polymorphisms*, but there is disagreement in the field about how the word is or should be defined; some see it as meaning any sequence change present in more than 1% of the population, and others see it as meaning only sequence changes that do not cause a trait, no matter what the frequency. Similarly, the term *mutation* does not mean the same thing to everyone. In some branches of study, a mutation is any change to the order of bases in the sequence; to other branches of study, mutation tends to be used to refer only to causative changes, that is to say, changes that result in a trait. We find the use of terms *causative mutation* and *benign sequence variant* as ways to bypass some of the conflicts in definition, but we can often be found falling back on the use of the easier and more familiar, but ambiguous, terms.

"SILENT" MUTATIONS THAT CAN HAVE AN EFFECT!

Most molecular biologists will glance at a newly discovered silent mutation in scorn or disappointment and set it aside as not being a causative mutation. After all, if mutations have their effects through changing something about the protein, and a silent mutation does not change the amino acid sequence, then how could it possibly have any effect and why would anyone spend time even thinking about it?

In the case of one silent mutation in the gene that causes Marfan's Syndrome, a silent mutation *did* turn out to be the cause of the disease. Does this overthrow the things we have told you about absent essentials and monkey wrenches? No, but it turns out that this silent mutation does have its effect through a mechanism other than causing a stop or an altered amino acid at that codon. In this case, the silent point mutation turns out to affect the splicing of the gene (Figure 16.5). Even though this sequence change is nowhere near the splice boundaries, it causes a new alternative splicing pattern that results in leaving out one of the exons that is normally always kept in the transcript.

Not surprisingly, changes in splicing of a gene can also result from changes in the sequence of the splice sites. If the cell recognizes a particular sequence within a gene as indicating the presence of a splice site, putting a mutation into that sequence should block the cell from recognizing that splice site. What happens then? Most of the time, the cell ends up bypassing the exon that that splice site goes with, jumping past it to splice to another exon. However, the result could also be a finished transcript that still has an intron present in the RNA. You can imagine what a mess that would create when the ribosome goes to translate the message.

In other cases, a mutation may create a new splice site. If the new splice site is present within an intron, this could have the effect of putting a shorter

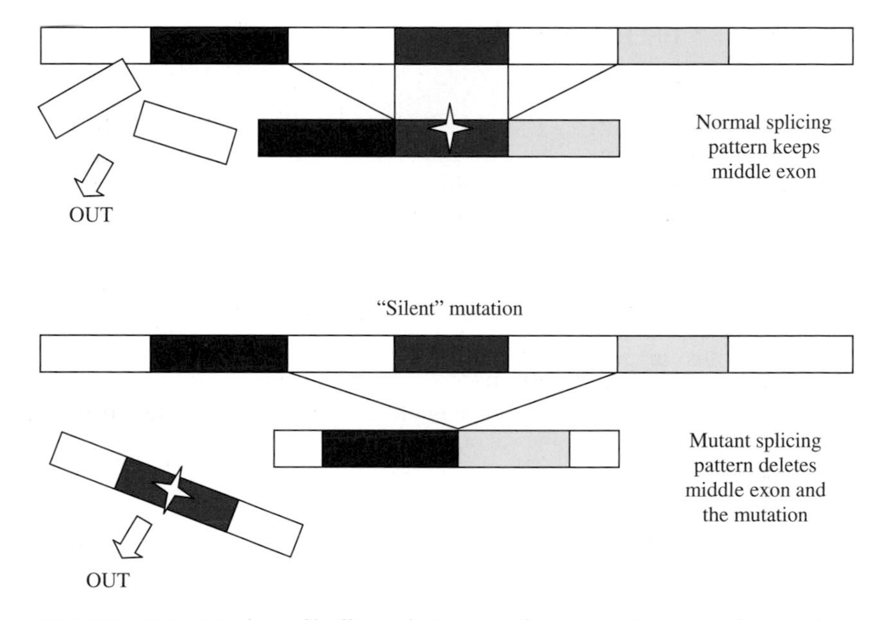

FIGURE 16.5 Marfan's fibrillin splicing. A silent mutation can change the protein sequence by altering the splicing of the gene.

version of that exon into the final transcript. In some cases, the new splice site might engage in alternative splicing to a different set of exons than the old splice site connected to. Sometimes the cell is actually smart enough to ignore the new splice site and still use the old one. At this point, it is still difficult to predict what will happen when a mutation affects a sequence involved in splicing.

BIGGER CHANGES

Sometimes, instead of changing one base in the sequence for another, a mutation adds (inserts) or removes (deletes) bases. Because the DNA code is read three letters at a time, inserting or deleting a number of bases that is a multiple of three will cause the simple addition or removal of one or more amino acids corresponding to those triplets (Figure 16.6). Whether that causes a problem will then depend on how the size, shape, charge, and polarity of that amino acid affect the protein's folding and functions.

Insertions and deletions can cause much larger problems if they affect numbers of bases that are not multiples of three. Why? Because the genetic code is read in units of three, any change involving three, six, or nine bases will only affect the one, two, or three amino acids encoded by those codons. On the other hand, a change involving one, two, four, or five bases will not only affect the codon being altered but will also throw off the frame in which the code is being read beyond the deletion. The resulting mutation, called a *frameshift mutation*, can not only change the local sequence plus the reading frame but also often results in the use of a stop codon at a different point in the sequence, which can cause the new form of the protein to be either longer or shorter than the normal protein (Figure 16.7). Thus a frameshift mutation

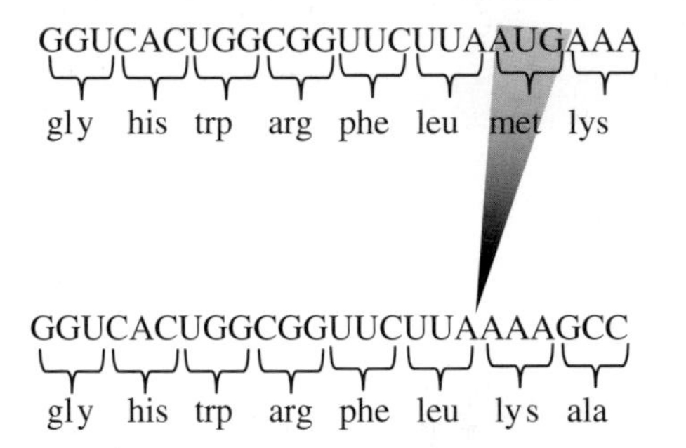

FIGURE 16.6 A deletion mutation. Removal of three bases (a codon) or a multiple of three bases results in removal of the accompanying amino acid(s), but the reading frame is not thrown off so that the sequence beyond the deletion is preserved and results in the incorporation of the same amino acids farther downstream even though the protein is now shorter. An insertion of three bases (or a multiple of three bases) will similarly affect only the sequence right at the point of the insertion but leave the sequence farther along the protein unaltered.

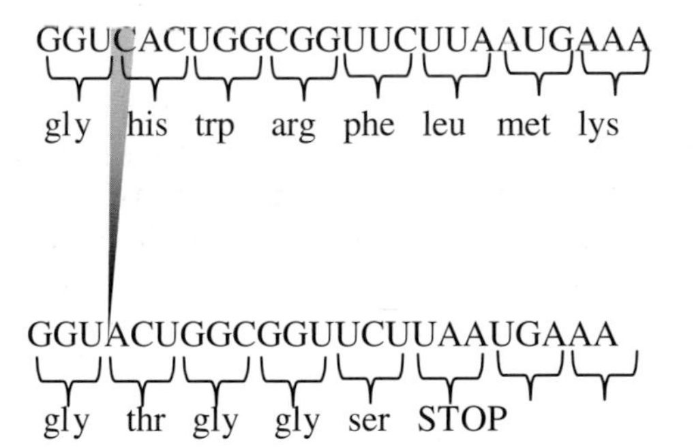

FIGURE 16.7 Frameshift mutation. Through adding or deleting a number of bases that is not a multiple of three, a frameshift mutation throws off the reading frame, rather as if spacing in a document joined the last half of a word to the first half of the next word so that the new information makes no sense. Notice in this case that the frame shift not only changed the sequence of amino acids beyond the point of the shift, but it also resulted in a STOP codon farther downstream in the reading frame, with no translation shown beyond that point because the ribosomes stop adding amino acids after a STOP codon.

(from either deletion or insertion) often results in a protein that is a different length than the original protein, with a new section of seemingly random amino acids attached to the end of the protein that have nothing to do with the sequence of amino acids that was there before. Thus the effects of a frameshift mutation seem as if they ought to be even more substantial than the effects of a simple nonsense mutation.

Some insertions or deletions can be quite large (Figure 16.8). In some cases, a deletion may take out a region of thousands or even millions of bases of sequence along a chromosome, in some cases causing such a large loss that

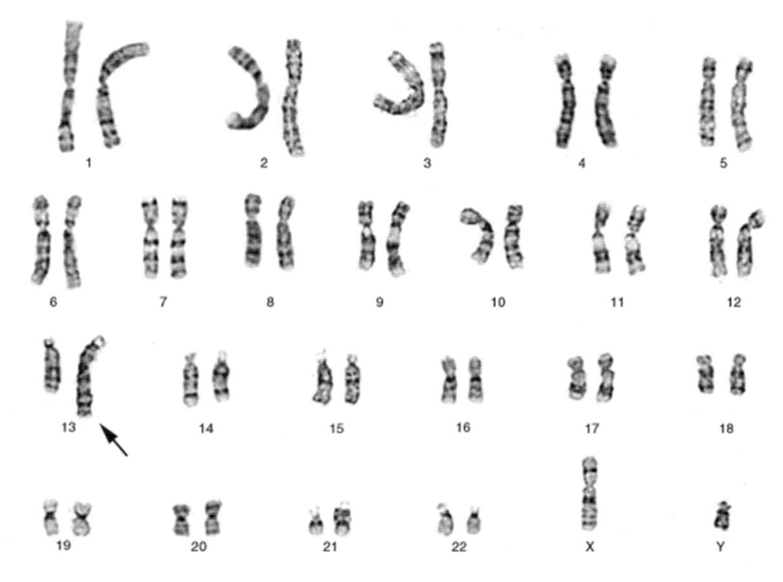

FIGURE 16.8 Duplication of a large section of the long arm of chromosome 13 becomes obvious when the chromosome images are cut out and placed next to each other in pairs. Notice the large difference between the two copies of chromosome 13 and the way in which the banding patterns are used to assist in correctly pairing up the different images. (Courtesy of the Clinical Cytogenetics Laboratory, University of Michigan, Ann Arbor, Michigan; Diane Roulston, Ph.D., Director.)

the effect is visible under the microscope. Not surprisingly, such a large deletion can sometimes include the loss of hundreds or thousands of genes. Deletions that large often are either lethal or cause problems that affect many different organ systems in the body. Many deletions that are viable (lead to a live birth) are hard to see or too small to see in a karyotype but may be detectable with the use of FISH.

WHERE DO MUTATIONS COME FROM?

After all is said and done, if you compare the sequence of any two human beings, you will find large numbers of differences between them. However, if you consider what fraction of the whole DNA sequence shows differences, the number is tiny. In fact, any two human beings share more than 99% identity in their DNA sequence. What a remarkable and wonderful thing, this great degree of similarity among human beings whose attention mostly seems to focus on the differences. But where do all of these differences come from?

Many different things cause mutations. In some cases, a mutation may result because the DNA has been broken, perhaps by radiation, and rearranged by the time it is repaired (Box 16.3). Although we most often hear about radiation as something that comes from such things as nuclear reactors, we are in fact naturally exposed to the chronic low levels of radiation produced by the environment in which we live. So the question is not so much one of whether radiation can cause mutations but rather one of how much radiation is a problem.

BOX 16.3 RADIATION CAN CAUSE MUTATIONS

In 1986 an accident destroyed one of the nuclear reactors at Chernobyl in Russia with a resulting release of tons of radioactive materials into the air. Thirty-one people are reported to have died from radiation exposure, and millions of people were exposed to radiation dispersed across the surrounding countryside. Increased thyroid cancer and detectable DNA alterations in children of parents who were exposed to the radiation, as well as reports of increased mutation rates in plants in the region, all support the idea that chronic exposure to low levels of radiation can result in mutations at a level that should be of concern to us. Surprisingly, long-term follow-up of the children of the survivors exposed briefly to high levels of radiation when nuclear weapons were dropped on Hiroshima and Nagasaki, Japan, in 1945 showed an apparently lower increase in the mutation rate than was expected. However, there is more than one way to look at mutation rates, and the studies in question only looked at some kinds of information on the subject. Frankly, increased cancer risk among the atomic bomb survivors themselves, rather than their children, suggests that mutation did take place, even if it did not pass along to the next generation in a way that was detectable by the studies carried out. The increased cancer risk also raises questions about what would be seen if changes throughout the entire genome could be looked at instead of very specific changes in individual genes, the only way that was technically feasible to investigate mutations during the original studies on this subject. Results on mutations resulting from radiation exposure have often been regarded as controversial as researchers struggle to sort out just how much radiation under what conditions will cause a problem. *The bottom line is: radiation can cause permanent changes in the sequence of DNA if someone is exposed to enough of it for a long enough time, but fear of any and all radiation may not be warranted.* Some kinds of exposure, such as medical x-rays, are purposefully designed to use low enough levels to be safe. We are all exposed regularly to background levels of naturally occurring radiation that cannot be avoided, but our bodies are adapted to live with this. There are some situations involving radiation that we should take steps to become informed about before putting our genetic blueprints in harm's way through exposure. Problematic situations to be concerned about include living around radon, smoking, excessive exposure to solar radiation via sunburns, or (in case it ever comes up in your neighborhood) joining in a cleanup crew after a radioactive accident if we are not trained to work with radiation and do not have appropriate protective gear. If we are well-informed about which radioactive situations are potentially harmful, we can help minimize risk of mutation from radiation.

Interestingly, our cells actually possess natural mechanisms for fixing some kinds of damage, such as certain types of breaks and unusual structures that radiation can cause in DNA. Do these exist to cope with natural background levels of radiation? Perhaps, but they also exist as part of the normal mechanisms for carrying out functions such as proofreading by which the cell evaluates and fixes sequence as it copies in an effort to maintain a low error rate during DNA replication, and other functions such as repair of single-strand breaks in DNA that are a natural part of the process of DNA replication and recombinational exchange of DNA between chromosomes.

In addition to radiation, mutations can result from exposure to certain chemicals and some subtances that occur naturally in foods (Box 16.4). Things that can cause mutations can be as different as pesticides, industrial chemicals, food additives, or naturally occurring ingredients in foods. Although people worry about pesticides and food additives with complex chemical names, we drink coffee and eat other foods that contain mutagens we ignore because they have not been labeled as "chemicals." We have to

BOX 16.4 CHEMICALS CAN CAUSE MUTATIONS

A test called the *Ames test* is designed to identify things that have the potential to cause mutations in human beings. While some chemicals can cause mutations, others don't, and it is important to be able to know the difference. The Ames test uses bacteria as a very sensitive biological indicator of whether or not a substance can cause a change in DNA sequence. Dr. Bruce Ames started out with a bacterial strain that has a mutation in a gene required to make the amino acid histidine. Because of the defect in this gene, the bacteria can only grow on food that provides histidine. When he exposed the bacteria to a disc of filter paper containing a *mutagen*—something that can cause changes in the DNA sequence—some of the bacteria ended up with a new mutation in the gene required for making histidine. This new mutation—called a back mutation—restores the ability of the bacteria to grow on food lacking histidine. If he exposed the bacteria to a disc containing something that is not a mutagen, no change took place in the histidine production gene and the bacteria did not grow on the histidine-deficient food. We think of the Ames test as a way to check out "chemicals," as if knowing the chemical structure and name of something makes it potentially more harmful; in fact, some chemicals are mutagens and some are not. There are many naturally occurring compounds that also test positive as mutagens in the Ames test, including substances found in moldy peanuts and overcooked hamburgers, so just asking whether something is tagged with the dread word "chemical" or "additive" or asking whether it comes from some safe-sounding source such as an herbal compound or a health-food store does not tell us whether it is in fact safe to consume.

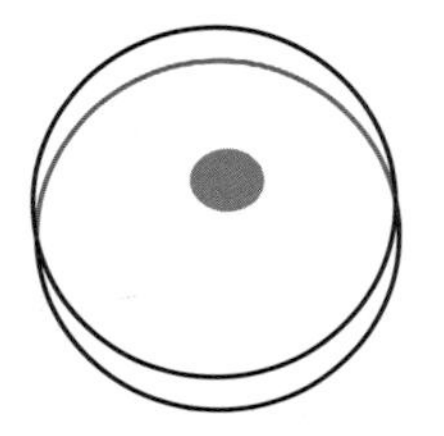

Disc with harmless substance placed in petri dish of His- bacteria causes no mutations so bacteria do not grow.

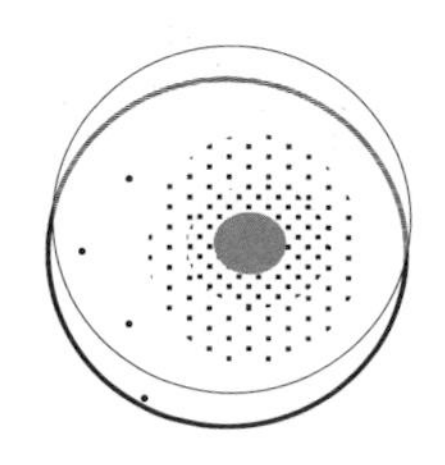

Disc with mutagen placed in petri dish of His- bacteria causes mutant bacteria to grow closer to the disc where the concentration of the mutagen is the highest.

BOX 16.5 MUTATIONS CAN HAPPEN WITHOUT ANY HELP FROM ENVIRONMENTAL EXPOSURE

We cannot simply blame our load of mutations on exposures to things around us. A certain amount of mutation goes on no matter what we eat or drink, apparently as a simple result of the rate of errors made by the machinery of the cell as it copies the DNA. Even the most pristine lifestyle will not protect against the fact that the natural machinery by which the genetic blueprint replicates has built into it limits on how perfectly it can carry out its copying functions. As the polymerase moves along the DNA making a new copy, it must correctly read a base and put its correct complement into place, every time, over and over again, for more than a billion bases of sequence if it is going to correctly replicate the entire genome once without making any mistakes. Frankly, without any exposure to chemicals or radiation at all, sometimes the polymerase gets it wrong. At some points in the DNA, the rate at which polymerase makes errors is increased because a naturally occurring chemical modification (that only gets made to some of the bases in the DNA) makes the base "look" like a different base to the polymerase as it comes through making copies. When you take antioxidant vitamins, you are helping your cells to repair oxidative damage to your DNA that your cell machinery must work constantly to repair. So there is already a base-line rate of mutation going on before we ever add in additional insults such as smoking to further aggravate the ability of the cell to get it right, every time, over and over and over.

wonder whether people would react differently to the mutagens and carginogens in cigarettes if they were labeled on the packet with the sinister sounds of their chemical names. The fact that something is herbal does not mean that it is inherently either safer or more dangerous—it only means that it was synthesized by a plant instead of a factory and likely contains a relatively complex mixture of biochemicals. Within that herbal preparation are biochemicals that have exactly the same chemical structures and names that they would have if they had been synthesized by a pharmaceutical company. The fact that something has been chemically synthesized does not necessarily mean that it is either more or less likely to be a mutagen—it only means that it is a relatively better characterized item that is often less biochemically complex than the same item derived from an herbal source. So the real key to whether something is more or less likely to be harmful or mutagenic depends less on whether you got it from a pharmacy or a health food store and more on what the particular biochemical is (Box 16.5). And some amount of mutation will happen without any chemical exposures at all.

HOW OFTEN DO MUTATIONS HAPPEN?

To some degree, mutation can be considered a spontaneous process. DNA polymerase, the enzyme that executes DNA replication, is an unbelievably accurate enzyme. Still, it inserts the wrong base at a frequency of about one error in every 10 billion bases replicated. That number may seem small, but remember that every time the human genome is replicated, DNA polymerase must copy approximately 6 billion base pairs. Thus, on average, slightly less than one new base pair mutation will occur every time the human DNA complement is replicated. If that number seems small to you, remember that a human being will go through more than 1 quadrillion complete replications of his or her DNA (cell division) in his or her lifetime. Thus a very large fraction of the cells in our bodies might be expected to carry one or more base change mutations, but as we have discussed, only some of those changes will actually make any functional difference to the gene product and cellular functions.

Realize that the vast majority of new mutations will occur in somatic cells and that even those deleterious mutations that do occur will likely have little or no effect because they only exist in the single cell in which the mutation event occurred. In most cases, the resulting impairment in gene function will be "covered" or "masked" by the unmutated copy on the normal homologue. Moreover, even if such mutations were to result in the death or impairment of a single cell and its somatic descendants, it is unlikely that the loss of a single cell or cluster of cells would be terribly deleterious to the organism. (We will, however, consider a rather dramatic exception to this generalization when we discuss the genetics of cancer in Chapter 33).

Perhaps of more interest to us is the frequency of mutations in the germline, mutations that get passed along to a child who then carries the mutation in every cell in the body. What fraction of human gametes might be expected to carry a new mutation that will impair or prevent the proper function of a given gene? Based on assaying the frequency of those new mutations in known genes that have phenotypic consequences, scientists have concluded that each gene in the human genome will be mutated (that is to say functionally altered, not just changed silently) only once in every 100,000 gametes. By this measure, Mendel does not seem to have been so far off—mutation is a very rare process indeed. Remember, however, that the actual frequency of DNA changes in a germline are much higher because of the types of silent mutations mentioned above. The observed mutation rate measures only those changes that dramatically alter gene function. Accordingly, whenever you think about mutation rate, stop and ask yourself whether you are looking at all heritable changes happening to the DNA (many) or whether you are looking at changes that are detectable as a phenotypic change in a living person (a much smaller number). And as we have mentioned, some agents in the environment can greatly increase the mutation rate above the baseline level of errors made by the polymerase during normal copying of the DNA (see Boxes 16.3 to 16.5).

Efforts to discuss functional *mutation rates*—the rate at which a mutation can cause a change in a protein that will alter some function or structure in the cell—are complicated. The mutation rate relative to a particular disease or gene depends on many variables. It depends on what size of gene we are talking about, what type of mutation we are talking about, and what region of the genome we are looking at. It also depends on whether we are asking about mutations with detectable biochemical effects, mutations that alter the amino acid sequence, or mutations that simply change the DNA sequence. The mutation frequency for a gene that is 900 base pairs long is likely to be quite different from the mutation frequency found for a gene that is 9,000 base pairs in length. The mutation frequency will be different for deletions vs. point mutations. Even different types of point mutations happen at rates that can vary more than tenfold, with an occasional hot spot in the genome showing an unusually high mutation rate that might be a thousand times that of some other bases in the genome.

One mutation in any given gene per 100,000 people might seem like a very rare event, but stop to think about what that number means. If you only had one gene, that mutation rate would mean that your chances of passing a new mutation along to your child would be about the same as the chance of winning one of the big lotteries. But if you had 30,000 genes, suddenly the 1 in 100,000 chance of passing along a mutation in one of those genes somewhere in your genome would not look so rare.

So even in a perfect world with no mutagens at all, we would all still have some amount of mutation going on in our cells over the course of a lifetime. Fortunately, most of the time, most of those mutations fall between genes, or within introns, or bring about silent changes, or happen in a skin cell that then dies and is sloughed off. Much of the time, even if a mutation takes place in one of our cells, we will not pass it along to our children unless the mutation happens to take place in the germline—in the lineage of cells that produces the eggs or sperm that carry the blueprint along to the next generation.

Most of these differences in the DNA sequence have been handed along to us by our parents, but new mutations can arise and be passed along to our children. Although we cannot control what we received from our parents, we can affect the chances of creating a new genetic problem that will plague our descendants. So you are going to be stuck with some level of mutation going on no matter how pristine an existence you live, but working around radiation without taking appropriate protective precautions (see Box 16.3) or living on a toxic waste dump site (see Box 16.4) can actually increase the chances that you will pass a new mutation along to your children. So the next time you find yourself wanting to roll your eyes at what some environmentalist is saying, stop and ask yourself how much you know about what causes mutations and what effects those mutations can have. In some cases the answer will be that the particular environmental situation is actually already safe enough, but sometimes the answer will be that there is something that needs to be cleaner, not only for our protection but also for the protection of future generations.

WHAT CONSTITUTES NORMAL?

17

As usual, the school morning started out with crabby protests of, "I don't feel good, Mom." The cheerful camaraderie that Ari shared with her mother in the early evenings had disappeared the night before, around bedtime, with Ari's first complaints of a stomachache. Now the downhill spiral of their morning interactions progressed, as always, as Ari objected to the clothes her mother had laid out and glared at her mother's efforts to bring some kind of order to her mass of curls. And then, as usual, it was time to leave and the crisis erupted. "Mom, I don't feel good; can't I stay home?" Her mother frowned and said, "Not again. Do you have a test today? Is there some problem at school?" Ari hunched over the thin arms that crossed her stomach protectively and said in a small, quiet voice, "No, I don't have a test, nothing is wrong at school, I like school, I just have a stomachache, OK?" Her mother shook her head in aggravation, breaking out the antacids and the analgesics and wondering what she was supposed to do now. Psychologists, doctors, nurses, and teachers over the course of several years had all been mystified at what to do about this seemingly healthy, self-confident, smart, academically successful child for whom bedtime and leaving for school regularly turned into a stomachache. Weekends and early evenings she seemed happy and well adjusted. Bedtimes and school mornings, she felt ill and asked to stay home. "Separation anxiety," they said, "and distress about going to school." It all seemed like a mystery until her mom had a guest to dinner who declined the ice cream, saying, "I can't eat dairy products. Milk gives me a stomachache because I have lactose intolerance." Milk. Stomachaches. Lactose intolerance. She thought about Ari's weekday routine, including a great big glass of milk at dinner, several hours before the stomachaches began. After several doctor's visits and some tests, a diagnosis of lactose intolerance revolutionized Ari's life by letting her avoid milk or use enzymes in pills to break down the milk sugar called lactose. Ari's stomachaches were banished, and so were the questions about whether she harbored secret anxieties about separating from her mother or going to school. Interestingly, the story of lactose intolerance offers us two different fascinating lessons, one genetic and one societal. So let's find out what is causing Ari's lactose intolerance, and then let's consider what an understanding of lactose intolerance can tell us about the concept of being "normal."

Ari's lactose intolerance results from her inability to make enough of an enzyme called lactase. According to the American Gastroenterological Association, as many as fifty million Americans may suffer from symptoms of lactose intolerance, including more than three quarters of those with African, Middle Eastern, or Native American ancestry and more than ninety percent of those with Asian ancestors. In populations around the world, babies

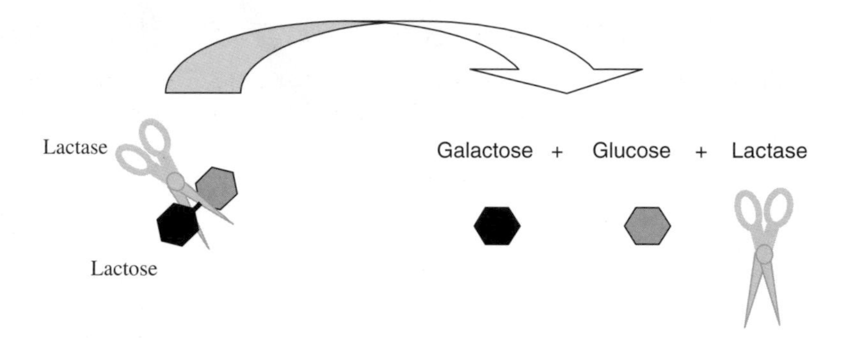

FIGURE 17.1 Metabolizing lactose. Lactose is made up of two other sugars that are joined together, galactose plus glucose. Notice that, after the enzymatic reaction, the substrate of the reaction (the sugar lactose) has been changed into two separate molecules, one of galactose and one of glucose, but that the enzyme itself is unchanged and ready to carry on the same reaction again if more lactose turns up. Lactose intolerance results when an individual does not make enough of the lactase enzyme, and thus ends up dealing with lactose instead of galactose and glucose.

routinely use an enzyme called *lactase* to break down the milk sugar *lactose* in their mother's milk into glucose plus galactose (Figure 17.1). They maintain the ability to make that enzyme and break down lactose until they are weaned. What happens next can be quite different depending on what population this child was born into. If he was born in Thailand, he probably lost the ability to break down lactose by the time he was two years old. If he was born into a Caucasian family in England, he might continue being able to use lactose for the rest of his life. Symptoms of lactose intolerance can include abdominal pain, gas, and diarrhea. Consuming milk can make a lactose-intolerant adult uncomfortable, but resulting dehydration can be a complication with consequences beyond discomfort in very young children. It is interesting that in each of these populations, some people retain the ability to use lactose throughout their lives because they keep making the full amount of the enzyme that digests the lactose. In Scandinavia, such individuals are quite common; in Southeast Asia, they are quite rare.

So why do people lose the ability to make the lactase enzyme that digests lactose? Babies need this enzyme because their milk-based diet contains high amounts of lactose. Consider this: if your diet beyond that point no longer included lactose, why would your body want to continue wasting energy making large amounts of lactase? In fact, even in lactose-intolerant individuals, some residual lactase is still present, just not the levels needed to cope with large amounts of milk. So this ability to regulate lactase production, and to stop making so much of it after weaning, would appear to result in savings of energy and resources for the cells that normally make the enzyme. Unless, of course, you live on a dairy farm in Wisconsin or England, in which case the savings in energy and resources would hardly be a reasonable trade-off for the losses in dietary advantage you would have if you couldn't drink milk.

THE PROBLEM WITH DIAGNOSING AN INDUCIBLE PHENOTYPE

In the case of Ari's family, what we see looks a lot like recessive inheritance of a disease gene (Figure 17.2), but it can sometimes be hard to tell if someone has lactose intolerance if they are not exposed to a lot of lactose. Her father's family, with western European ancestry and a standard American diet full of dairy products, appears to be full of lactose-persistent people. However, it is actually harder to tell about people on her mother's side of the family, who consume a more standard Middle Eastern diet. Although they also use dairy products, they eat recipes based on yogurt more often than milk, and drink tea or wine or water more often than milk. Lactose intolerance can be hard to diagnose even in someone who is exposed to a lot of milk, as happened with Ari; someone who is consuming a nearly lactose-free diet simply because of cultural context may not even know whether milk gives them problems. We know from many studies of other families and individuals that lactase persistence is dominant over lactose intolerance.

Telling just who is or isn't affected with a particular problem is a common problem in many genetic studies, but especially so with situations such as lactose intolerance that have an environmental component. A genetic disease called favism is only detected in individuals with the defect who eat fava beans. Normally, detection of malignant hyperthermia happens because someone undergoes general anaesthetic for surgery. Individuals susceptible to steroid glaucoma will only develop this potentially blinding eye disease if they are exposed to certain corticosteroid medications. So for many of these complex traits with both genetic and environmental elements, many of us have no idea what our phenotype or genotype might be because we have not encountered the conditions that would elicit the trait in someone with the predisposing genotype.

This process of bringing about the expression of a phenotype in response to exposure to something is called *induction*. Although many inherited phenotypes are congenital (present from birth) or developmental (develop at a particular stage in the course of development and aging), there are many phenotypes that are inducible. Tanning is the process of inducing more skin melanin in response to sunlight (or tanning beds). Certain allergies might be

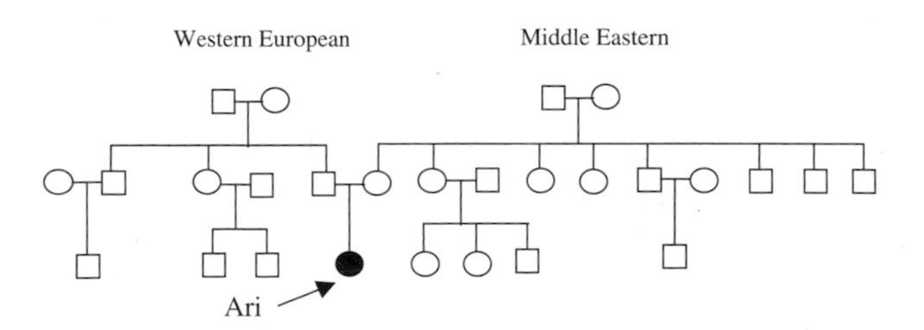

FIGURE 17.2 Recessive inheritance of lactose intolerance in Ari's family. The black symbol marks the one known case of lactose intolerance on both sides of the family. It is clear that her father's family has many individuals with lactase persistence, but the phenotype is unclear for most of her mother's side of the family because of their relatively low-lactose diet in which milk products tend to be yogurt based. We do know that Ari's mother and her father can both drink milk without problems, so they both appear to be carriers. This implies that at least one of her maternal grandparents should be lactase persistent.

said to be induced by exposure to the allergen, such as penicillin or a bee sting. In genetic terms, we usually use the term *induction* to refer to a process that brings about increased expression of a gene.

INDUCTION AND GENE REGULATION

In Chapter 14, when we talked about absent essentials and monkey wrenches, we talked about the various ways in which messing up a protein can cause a problem. When we discussed some of the common types of mutations, we focused on the coding sequence and mutations that change the resulting protein product. In the case of *hypolactasia* (a state of reduced lactase activity resulting in lactose intolerance) it looks as if we might be dealing with a different type of mutation, a mutation in the promoter region that affects regulation of expression of the gene.

So what is causing Ari's lactose intolerance? Theories have evolved over the years regarding the cause of lactose intolerance, but it was not until 2002 that a possible cause was identified. A group of researchers in Finland, where about 18% of the population is lactose intolerant, studied the gene responsible for making the lactase enzyme. The exons of the gene that makes the lactase enzyme (called lactase-phlorizin hydrolase) are spread over more than 30,000 base pairs of sequence on the long arm of chromosome 2. If you compare DNA from a lot of different people, you find that there are many differences in the DNA of the transcribed region that contains the coding sequence of the gene, but none of the differences anywhere in the transcript can account for the differences between people who keep being able to drink milk and those who can't. However, when researchers from Finland looked in the promoter region of this gene, at two places that are located thousands of bases before where the gene begins being transcribed, they found two different changes that seem to turn up repeatedly in people in Finland who can drink milk, but not in people with lactose intolerance (Figure 17.3).

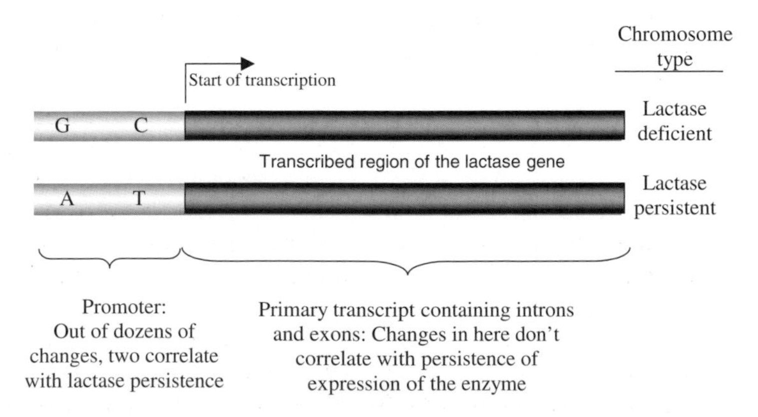

FIGURE 17.3 *Peristent* and *deficient* versions of the gene that makes the lactase enzyme. Persistence of expression of lactase in the Finnish population seems to correlate with two sequence changes in the promoter region that are many thousands of base pairs away from the transcription start site. Different changes in the promoter seem to be involved in other populations. Most of the people in the world have a G and a C at these positions rather than the T and A found in the lactase-persistent individuals.

So some people in this world have a G $_{22,018}$/C $_{13,910}$ genotype (G at a position 22,018 base pairs before the transcription start site, and C at a position 13,910 before the transcription start site) and others have an A $_{22,018}$/T $_{13,910}$ genotype. Those in the Finnish population with the G $_{22,018}$/C $_{13,910}$ combination lose the ability to make adequate levels of lactase, usually some time between the ages of ten and twenty years old. Those with the A $_{22,018}$/T $_{13,910}$ combination keep making lactase and continue being able to consume milk. If we look in other populations, the sequence that correlates with lactose intolerance is also G $_{22,018}$/C $_{13,910}$, but the sequence associated with continuing to make lactose may be different in other populations. So we arrive at a model that says that the A $_{22,018}$/T $_{13,910}$ sequence is in fact causing the change in gene expression and sugar utilization (Box 17.1), but it is really just a model, and more work will be needed to prove that regulation of the gene works just this way.

BOX 17.1 PROVING IT'S A MUTATION

In fact, the evidence offered so far strongly suggests that A $_{22,018}$/T $_{13,910}$ is causing some people in Finland to retain high levels of expression of the lactase gene. Does this kind of association (the genotype is there when the phenotype is there and missing when it is missing) actually prove that this is the cause? No. It is considered to be highly significant evidence, but you can have things co-occur without one causing the other. For instance, what if there is another mutation even farther away from the transcription start site that is the actual cause and they just haven't found it yet? If A $_{22,018}$ and T $_{13,910}$ are each part of the genetic fingerprint of the chromosome on which the causative mutations took place, we would expect that "affected" individuals who are still making lactase as adults would have the mutant alleles at all three positions in the sequence if they are not too far apart, even though only one of those changes is causing the phenotype. Why? Think back to our discussions of recombination in the meiosis in Chapters 12 and 13. The chance of a recombination event falling between two things is proportional to how big the distance between those two things is. In genetic terms, a few tens of thousands of bases is really a rather small chromosomal region that can sometimes be transmitted over a surprisingly large number of generations without recombining. So what can we do to tell whether A $_{22,018}$ or T $_{13,910}$ or the combination of the two are the cause of lactose persistence in Finland? The research to prove this model is actually likely to go in several different directions. Studies of additional populations, as well as studies of families, may help to extend the generalizations arising from the initial study, but a formal proof that these are causative mutations would be greatly helped if we could identify transcription factors that bind differently to a promoter with these mutations than to a promoter with A $_{22,018}$/T $_{13,910}$ than they do to a promoter with G $_{22,018}$/C $_{13,910}$. This is just one of the approaches that would work, and in the end such experiments are likely to support this current model that mutating the promoter sequence is what granted lactase persistence to a large number of Finns. But we won't know for sure until someone does the experiment.

SO WHAT OR WHO IS NORMAL ANYWAY?

So what is the "normal" genotype, or for that matter, what is the "normal" phenotype? There is a tendency to think that the genotype that makes you sick must be the mutant genotype, and that the phenotype that involves illness must be the mutant phenotype. Certainly, the way lactose intolerance and lactase persistence are talked about further contributes to this idea. People who get sick when they drink milk are said to have lactose intolerance; it is something they go to the doctor about, perhaps even something they use "medication" for (the enzyme tablets to digest the lactose). You will notice that, earlier, we asked what is causing Ari's lactose intolerance, and it seemed like a perfectly normal question. We would bet that you did not find yourself saying, "No, that's the wrong question. The question is: why there are people who don't get sick when they drink milk?"

In fact, the right question really is, "What makes some people persist in making high levels of lactase long after being weaned?" Why is that the right question? Because the "normal" state appears to be lactose intolerance. Studies of the genetic fingerprints of the region of chromosome 2 surrounding the lactase gene show that the lactose-intolerant genotype $G_{22,018}/C_{13,910}$ is a much older genotype that has been around long before the lactase-persistent $A_{22,018}/T_{13,910}$ arose.

More generally, if we look outside of the Finnish population, it has long been suspected that humankind started out lactose intolerant and over time natural selection favored an increased representation of lactase persistence in some populations, beginning about 10,000 years ago at the time of the introduction of dairy farming. The argument is that children in a dairy farming culture would experience improved nutrition, not only in terms of calories, protein, etc., but also in terms of vitamin D in northern climes, where scarcity of vitamin D in the diet can be a potential problem. This mechanistic model for how certain populations came to have a much larger representation of lactase-persistent individuals is not yet proven, but it makes sense and fits the information available.

So what we conclude at this point is that the original genotype in most of humanity was apparently the $G_{22,018}/C_{13,910}$ genotype, and that the normal nutritional state of the human race was use of milk in babies and toddlers followed by loss of the ability to use milk after weaning. Most people descended from those lactose-intolerant $G_{22,018}/C_{13,910}$ ancestors throughout much of Africa and Asia still have that genotype. The mutants, then, are the rare Africans and Asians, the more frequent Middle Easterners, and the very common Western European Caucasians who are able to continue metabolizing the lactose in milk throughout life.

You might think our point would then be that the lactose-intolerant $G_{22,018}/C_{13,910}$ genotypes must be the normal ones and the lactase-persistent $A_{22,018}/T_{13,910}$ genotype is abnormal. However, normal vs. abnormal is not actually what either the genotypes or phenotypes are about. The real point is that there is tremendous diversity among the different populations of the earth, and whether something is normal or not is really just a matter of whether it is more common or not, and not whether it is maladaptive or not. Whether a particular genotype is helpful or harmful depends on a lot of factors. If you were to ask Ari if she is ill, she might answer "yes" because she is surrounded by foods she must avoid, and even when trying to avoid things she knows about, she still periodically ends up with a stomachache if she eats crackers that she did not realize contained whey (a milk product in which lactose has been concentrated). However, if you went to Asia and tried to identify individuals who are lactose intolerant, might you have a hard time even telling who is and who is not? If you are surrounded by a culture in which rice and vegetables, meat and fish, and tea and juice make up the diet, a culture in which whey is not used in cooking, would you consider yourself to have an illness just because you don't happen to still be making an enzyme you have no use for? Under most circumstances, if you were lactase deficient in a lactose-less environment, you would never even find out which genotype or phenotype you have. Frankly, in that environment, your body would have the advantage (however minor that might or might not be) of avoiding wasting resources making a lot of enzyme that won't be used.

So the fact that we know which individuals are the mutants, that is, which individuals have the version of the sequence that is more recently arisen, does not actually mean that we know anything about what constitutes normal, about which individuals might be ill, or even whether the mutation has beneficial or negative impact on the lives of the people of any given genotype. Some mutations cause problems. Some mutations have so little effect that we can't even detect a phenotype that results from the mutation. And some mutations, like the ones we have just been talking about, can actually be beneficial. Knowing who is a mutant does not tell you who is normal or not normal, and it does not tell you who is ill or not ill. As we learn more and more about the correlations between genotype and phenotype, we see more and more that the word *normal* may not even be a useful term. If normal is simply whatever is usual, and relative advantages or disadvantages vary with the environment, we are eventually going to have to learn to remove the judgmental tone that goes with the word. Being normal might offer you some assistance under some circumstances and not under others, but there is nothing that says that having the most common genotype is inherently either better or worse.

18 MUTATIONS IN MAMMOTH GENES

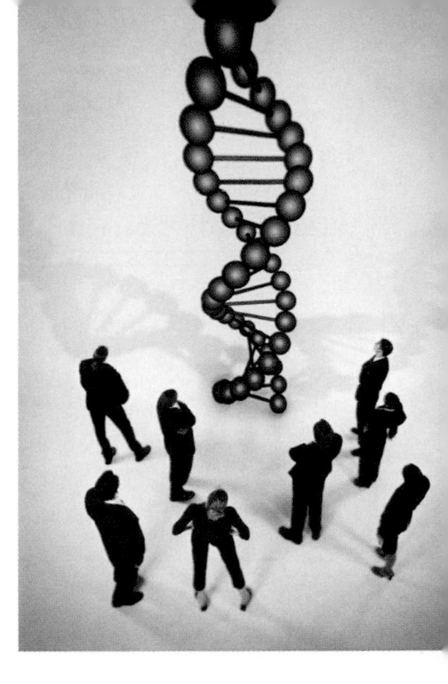

Duchenne's muscular dystrophy is an illness that forever changes anyone who comes in contact with it. You might expect that Scott's brief encounter with two boys dying of this disease would offer only lessons in stark reality, but his memories of this experience also retain impressions of hope and dignity. Back when he was a high school student in the late 1960s, Scott did a brief stint of volunteer work for the Muscular Dystrophy Association. He did what a high school student could do: he answered phones for the telethon and visited two brothers who were both suffering from this disease at a local convalescent home. The reality of their illness defies description; this is truly a terrible disease in which alert young people slowly waste away to death in their late teens. Amidst the realities of the prisons their bodies had become, those two boys wanted to talk of just one thing or, more correctly, of one person. They both idolized Elvis Presley. For his part, Mr. Presley had gone to some great lengths to return their affection. He had flown them to one of his concerts in Las Vegas and met with them before and after. There had been cards of best wishes and, as Scott recalls, a phone call or two from Mr. Presley. Despite what this disease was doing to their bodies, Mr. Presley's kindnesses had made these two boys feel quite special and, as one of them told Scott, even quite lucky.

Duchenne's muscular dystrophy (*DMD*) is a well-known fatal disorder that results in death in the late teens due to muscle wastage and deterioration. Because the DMD gene is X linked recessive and a normal copy of the DMD gene on the second copy of the X chromosome protects most female carriers from being affected, most of the affected individuals are males. These males are normal at birth but develop muscle weakness at age four to five years. These males are normally confined to a wheelchair by their early teens. By their late teens, these individuals usually succumb to either respiratory or cardiac failure. We now know that the DMD gene encodes a protein called *dystrophin* that is required for muscle maintenance and to prevent muscular atrophy. Affected males lack this protein and thus succumb to progressive muscular atrophy.

The DMD gene is so big that it is considered a mammoth gene. According to Victor McKusick, the author of *Mendelian Inheritance in Man*, human genes can be divided into roughly five size groups (Table 18.1). The groupings are based on the size of the region of chromosome covered by the introns, exons, and untranslated regions. *Small* genes, such as those for the blood proteins -globin and -globin, cover pieces of chromosomal DNA that range in size from 800 to approximately 4000 bp. McKusick's *medium* gene

TABLE 18.1 Characteristics of Several Human Genes

Class	Example	Genomic Size	Transcript Size	Number Introns	DNA Bases Per Intron	RNA Bases Per Intron
Small	-globin	1,500 bp	600 bases	2	750	300
Medium	Albumin	25,000 bp	2,100 bases	14	1,786	150
Large	Phenylalanine hydroxylase	90,000 bp	2,400 bases	12	7,500	200
Giant	Factor VIII	186,000 bp	9,000 bases	26	346	346
Mammoth	DMD	2,400,000 bp	14,000 bases	78	30,769	179

Modified from *Mendelian Inheritance in Man*, 1992, by Victor McKusick.

category includes genes ranging from 11,000 to 45,000 bp in length (11 to 45 kb). This medium class includes such genes as those encoding the collagen and albumin proteins. McKusick's next class, *large* genes, is represented by the 45,000 bp phenylalanine hydroxylase gene. The fourth class, which he refers to as *giant* genes, ranges from 160,000 to 250,000 bp and includes the cystic fibrosis gene discussed in the last chapter. Many of the genes we have talked about so far qualify as small, medium, or in some cases even large genes.

WHY ARE SOME HUMAN GENES SO BIG? (OR, REMEMBER "INTRONS?")

The gene responsible for DMD qualifies in the fifth class, what are called mammoth genes. This gene, which is approximately 2,500,000 bp long, covers about 1.5% of the length of the X chromosome. Many different mutations in the DMD gene can cause the disease known as Duchenne's muscular dystrophy. As researchers were eventually surprised to find, mutations in the DMD gene are also the cause of a different disorder called Becker's muscular dystrophy (BMD), as well as being responsible for about ten percent of cases with X-linked dilated cardiomyopathy. To give you some perspective on mammoth genes, this one mammoth gene is bigger than the entire genome of the bacterium *Haemophilus influenzae.*

By any standard of comparison, the term *mammoth* seems truly appropriate for the DMD gene. True, the DMD gene does encode a 14,000 base-pair messenger RNA, which is about ten times the size of average mRNAs. However, less than one percent of the gene can be accounted for by the size of the transcript. Similar inequalities can be seen for the smaller gene classes, as well. So why are these genes so big?

First, a gene that is giant or mammoth does not necessarily encode giant or mammoth messenger RNAs or proteins. Table 18.1 compares the gene size, the number of introns, and the size of the final mRNA for several human genes. As presaged in Chapter 8, much of the length of these larger genes is taken up by the noncoding introns that are interspersed with the coding exons.

GIANT GENES VS. GIANT TRANSCRIPTS

In fact, the *Titin* gene, which produces one of the biggest known mRNAs encoding one of the biggest known proteins, covers a stretch of chromosome that is perhaps only a tenth the size of the DMD gene (Box 18.1). Although some small genes have no introns at all, the Titin gene, which has a mammoth transcript of more than 80,000 nucleotides even if it is only a giant gene, has more than 360 introns (Figure 18.1). Because extensive alternative splicing

---

BOX 18.1 TITIN—A MAMMOTH TRANSCRIPT FROM A GIANT GENE

Although the DMD gene covers one of the largest stretches of chromosomal DNA, the Titin gene is sometimes mentioned as being the largest human gene. The chromosomal region occupied by DMD is almost ten times the size of that occupied by Titin, which covers 281,000 base pairs of sequence along chromosome 2. So why would anyone call Titin the largest gene? Because it produces such a large transcript and protein. One of many finished transcripts produced by this alternatively spliced 363-exon gene comes out to a final spliced length of 81,755 base pairs (almost six times the length of the spliced DMD mRNA), which makes a protein of 34,350 amino acids. That is just one of the alternatively spliced transcripts, so there are quite a variety of different transcipt lengths and protein sizes that come from this gene.

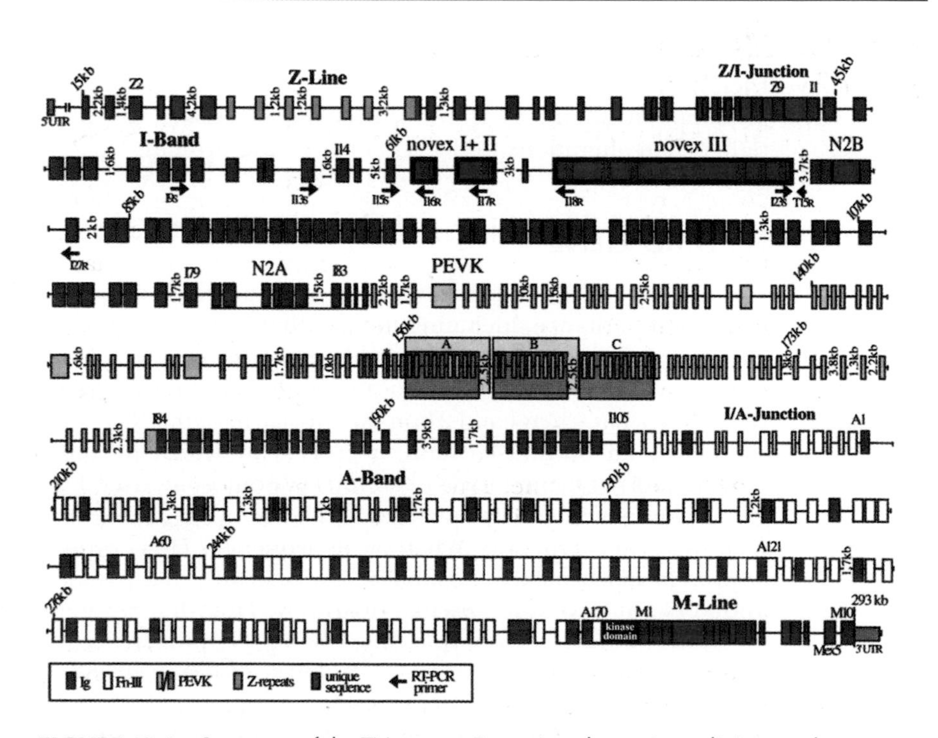

FIGURE 18.1 Structure of the Titin gene. Extensive alternative splicing produces many different transcripts and many different protein variants from this one gene. (From Bang et al., *Circ Res* 2001; 89:1065–1072.)

goes on in many of these biggest genes, many alternative smaller transcripts and proteins are also produced from the very complex primary structure of the gene.

DOES LARGE SIZE AFFECT THE GENETIC BEHAVIOR OF THESE GENES?

Imagine a gene like DMD that is so large that it occupies more than one percent of the length of the chromosome on which it resides, in this case the X chromosome. This represents a huge target for mutation. Indeed, such genes are extremely mutable and the types of mutations observed commonly include chromosome aberrations such as translocations, duplications, and deletions in addition to point mutations such as nonsense and missense mutations. At least one third of DMD males born into families where no one else has DMD are the result of newly arisen mutations. Indeed, the rate of mutation at the DMD gene is estimated to be one new mutation in every 10,000 gametes, a rate 10 to 100 times greater than that observed for most human genes.

Large deletions, ranging from those removing just one exon to cytologically visible deletions that remove the entire gene, account for more than sixty percent of the loss-of-function mutations that cause DMD. Most, if not all, of these deletions are sufficient either in their extent or structure to completely prevent the cell from making dystrophin, that is, the small deficiencies observed in DMD patients apparently either alter the reading frame of whatever message is produced or block proper splicing. In addition, nonsense mutations that result in truncation and insertions that throw off the reading frame can also cause disease. The one theme that seems to be missing from the population of DMD patients is disease resulting from missense mutation. Even in those cases that do not appear to delete most or all of the gene, the predominant theme in DMD is the failure to make the dystrophin protein rather than alterations to the sequence of the protein.

FINDING THE DMD GENE

The DMD gene was originally mapped to the X chromosome based on its mode of inheritance, including the fact that DMD occurs mostly in boys who have the defect passed to them by their carrier mothers. Some of the more precise mapping of the gene's location came about when it was realized that DMD can occur when band 21 on the short arm of the X chromosome (Xp21) was disrupted by translocation, a process by which the chromosome had broken and reattached itself to another chromosome (see Figure 18.2).

Once the gene was mapped to Xp21, the next problem was to figure out what the actual gene was. What was its sequence? How big was it? What kind of protein did it make, and what was wrong with that protein in individuals who were affected with muscular dystrophy? How did they get their hands on the gene to answer these questions? Some of the earliest technologies for studying DNA allowed researchers to detect or even retrieve a piece of DNA that matches any piece of DNA that they already had a copy of. Simply put, the piece of DNA you already have can serve as a probe that can detect other pieces of DNA that have the same sequence. How? By using the rules of base pairing to stick to the piece of DNA you want to find. (For instance, a probe

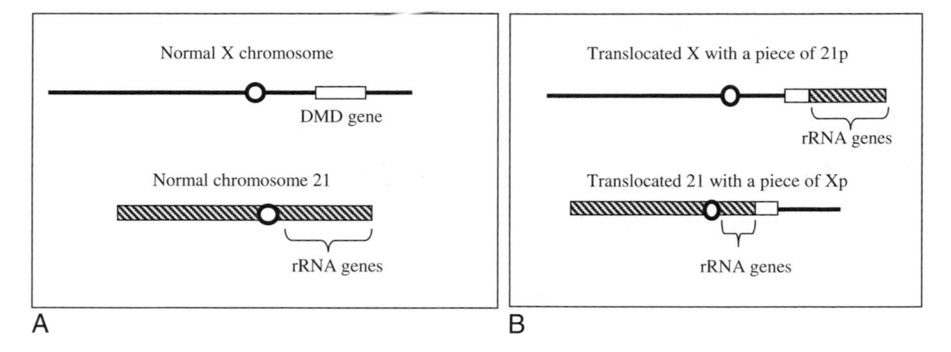

FIGURE 18.2 **A,** Normally, the DMD gene sits at band Xp21. The short arm of chromosome 21 contains a cluster of rRNA genes. One of the ways researchers found out where the DMD gene is located was by observing that translocations in women with DMD kept turning out to be broken in band Xp21. **B,** This translocation traded the tip of Xp for the tip of 21p. This broke the DMD gene into two pieces, one left on the X and the other now attached to chromosome 21. This translocation also broke the cluster of rRNA genes into two pieces, so that some rRNA genes stayed on 21 and some became attached to the X chromosome.

with sequence AAAAAAAAAAAAAAAA will stick to and identify a piece of DNA that is homologous to that sequence by pairing with the opposite strand that contains the sequence TTTTTTTTTTTTTTTT. Since a proble with a more complex or longer sequence can also find its match through the rules of base pairing, pieces of DNA hundreds or even thousands of bases long can be used as probes).

This might seem rather pointless: why would you need to get your hands on something you already have? The answer is that sometimes if you can pull out a piece of DNA that is a copy of something you already have, you can also obtain the adjoining DNA containing sequences you do not already have.

An excellent example of the usefulness of this technique can be seen in one of the ways the DMD gene was obtained. Researchers were able to use a piece of a known rRNA gene as a probe to isolate the piece of DNA containing the translocation break point from an individual with DMD that had a translocation between the X chromosome and chromosome 21 (see Figure 18.2). The new piece of DNA that they found with their rRNA probe was a chimeric piece that had rRNA sequences at one end and a piece of the broken DMD gene at the other end (Figure 18.3). Thus the rRNA gene became a kind of molecular fishhook that could be used to fish out a piece of the adjoining DMD gene that was fused to the rRNA gene via the chromosomal translocation.

Once they had their hands on the first piece of the DMD gene, the scientists were able to use a process called *chromosome walking* (see Figure 18.4), in which DNA from the first fragment they fished out could became a new *probe* with which to isolate the next overlapping piece of DNA farther down the chromosome. Each step in this chromosome walking process only allowed retrieval of a small region of DNA compared to the overall size of the DMD gene. Fortunately, the walking steps can be carried out over and over to move along a chromosome.

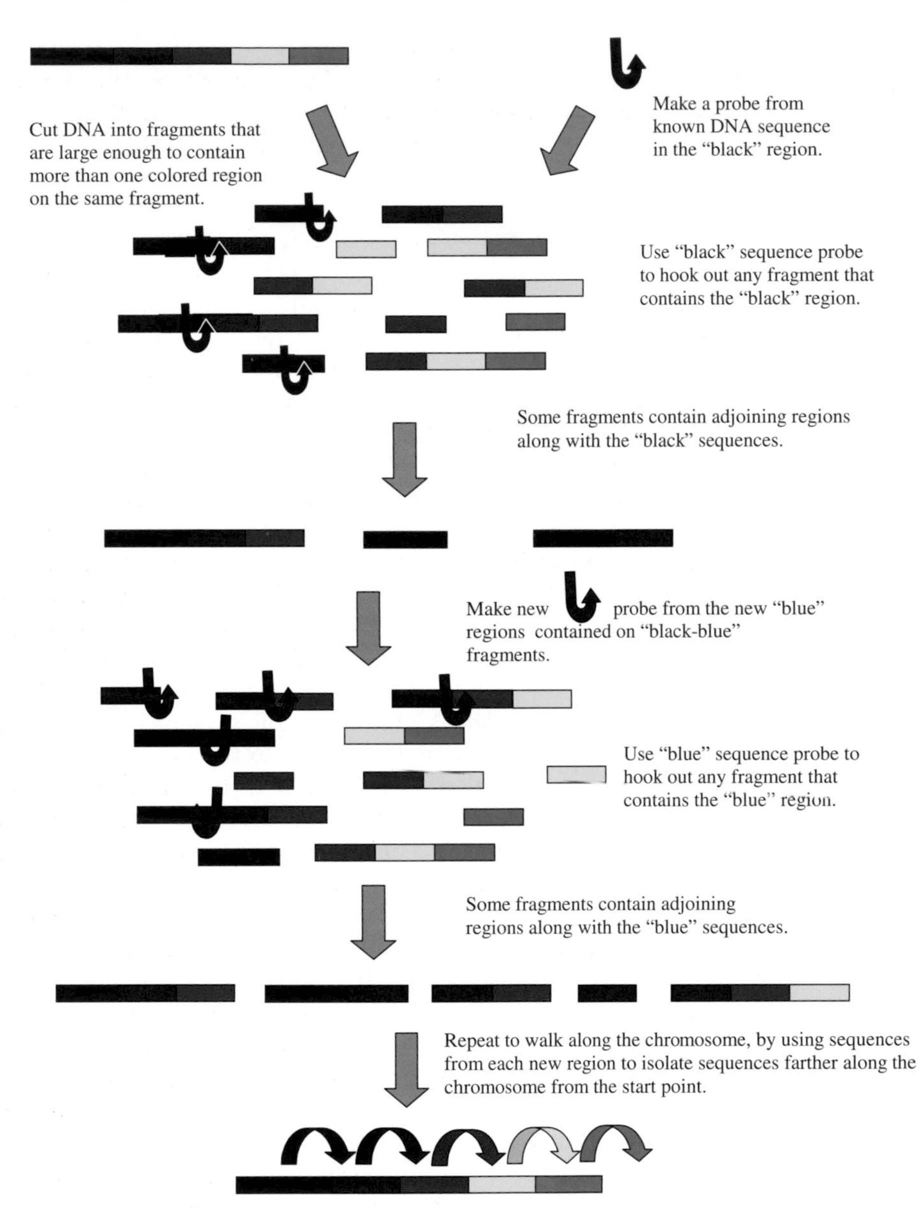

FIGURE 18.3 Chromosome walking. If we start out with a "probe" consisting of a known piece of sequence, we can then use it to isolate adjoining pieces of DNA so that we can then use them to isolate the next piece of DNA beyond that. If we keep repeating this operation, we can walk along a piece of DNA, isolating overlapping pieces until the whole region becomes known sequence. Many of the early successes in identification of human disease genes made use of this kind of approach and worked especially well in cases such as this, where a translocation or naturally occuring event placed a known piece of DNA next door to an unknown piece of DNA that needed to be studied.

Back at the time that the DMD gene was identified, many rounds of chromosomal walking were needed to obtain an overlapping set of clones that covered a region as large as the DMD gene. This is mostly not how we go after human genes these days, but such technologies are still used in some specialized cases, such as when the DNA adjoining a new translocation needs to be found.

There is a lesson here about the importance of very, very rare individuals with unusual mutations and/or phenotypes in doing genetics, human or otherwise. Females with full-blown DMD are very, very rare, yet the careful study of just these females, especially those with Xp21 translocations, made both mapping and cloning the DMD gene much easier. Frequently, one of the keys in the use of a rare individual to solve a genetic problem is an observant physician who realizes the unusual, important nature of this individual and arranges the patient's participation in a genetics study.

One of the greatest frustrations for the human geneticist is that we *know* that there are patients out there who have some unusual sets of characteristics that could provide a breakthrough, which could help many other patients in the long run, but who do not know about our studies and thus cannot volunteer to help in the search for the gene of greatest importance to them. Finding the rare breakthrough case in the office of some clinician in a small town in Texas or the middle of New York City can be difficult if the researcher has a lab in Kansas or Michigan or California and does not know either the rare individual or the physician working with that person. One of the ways that some genetic studies are starting to deal with locating unusual patients, or even just expanding the number of people participating in the studies, is through Web pages that provide information about the disease and invite individuals or families to volunteer to participate in the study.

An alternative approach was also successful in getting the DMD gene. In this case, the rare individual was a boy affected simultaneously with DMD and two other disorders associated with the loss or inactivation of genes known to be tightly linked to the X chromosomal DMD gene: retinitis pigmentosa and X-linked chronic granulomatous disease. Such cases, where a single deficiency removes several essential genes and thus produces a complex set of phenotypes, are often referred to as *contiguous gene syndromes* because they are presumed to reflect the simultaneous loss of two or more closely linked genes. In this case, the boy in question carried a small deletion that was visible under the microscope that removed band Xp2l. By using difficult trick called *subtractive hybridization*, researchers were able to identify DNA that is present in normal individuals but missing from the DNA of the person with the Xp21 deletion.

One piece of DNA they retrieved turned out to come from the deleted region on Xp21 and to also be missing from the DNA of several other children with DMD caused by deletions in the region. They were then able to perform a chromosomal walk like the one shown in Figure 18.3 to isolate adjoining pieces of DNA until they gradually obtained overlapping DNA containing not only the DMD gene but also the gene for chronic granulomatous disease, which is near the DMD gene.

MUTATIONS THAT DON'T ELIMINATE DYSTROPHIN

With a gene as big as DMD, might one also expect to find mutations that don't completely knock out the gene, mutations that might permit some level of dystrophin production? Indeed, one does find such mutations; they appear to produce a muscular weakness disorder with a very different phenotype known as *Becker's muscular dystrophy*. The symptoms of BMD are very similar to DMD, and the mutations that produce BMD map within the DMD gene. BMD is first picked up clinically at six to eighteen years of age, and patients often aren't confined to a wheelchair until twenty-five to thirty years old. They live to forty to fifty years of age and often produce children. Indeed, the ability of men with BMD to produce children is reflected in the fact that most cases of this disorder are inherited. Unlike DMD, where approximately one third of the cases are due to new mutations, less than ten percent of BMD cases result from new mutations.

Like DMD, BMD results from a lack of functional dystrophin. However, most mutations that lead to BMD arise from either single-base missense mutations or from small deletions that do not disrupt the reading frame of the protein. Thus, unlike DMD patients, they have a problem with dystrophin, but they do produce at least some dystrophin protein with at least some level of activity. Even this very small amount of dystrophin activity appears to significantly deter muscle wastage and thus greatly ameliorates the phenotype, at least in comparison with DMD. You might think of the difference here as being that a DMD individual makes *no* gene product, a BMD individual makes

BOX 18.2 DMD: ONE GENE, THREE DIFFERENT DISEASES

DMD is both more common and more severe than BMD, even though they both result from changes in the same gene. Boys with DMD are usually diagnosed during the preschool years. They are usually delayed in learning to walk, become wheelchair bound by middle-school age, and often die before twenty years of age even with the best medical care now available. Death is commonly the result of either breathing problems or dilated cardiomyopathy, which results from lack of the dystrophin protein in the heart and skeletal muscle. Boys with BMD become wheelchair bound as young adults and tend to die in middle age. Almost half of the heterozygous carriers for Duchenne's and Becker's muscular dystrophies show some cardiac involvement, including dilated cardiomyopathy in about a fifth of them. In addition, cases of X-linked dilated cardiomyopathy with no skeletal muscular dystrophy have been attributed to mutations in this gene. What makes the difference? The difference is that individuals with dilated cardiomyopathy make normal skeletal muscle dystrophin, with skeletal muscle mRNA not including the portion of the gene with the defect, which only turns up in transcripts made in heart muscle. Individuals with BMD make defective dystrophin protein. Individuals with DMD make effectively no dystrophin protein. Ongoing research is aimed with great hope towards development of gene therapy approaches to providing functional dystrophin to muscles that lack it.

a gene product that doesn't work very well, and an individual with dilated cardiomyopathy makes a completely normal gene product in muscle and only makes a protein with the altered sequence in cardiac tissue.

Please note the crucial lesson here: mutations at the same locus can produce different phenotypes, depending on the type of mutation and its position within the gene. In some cases the phenotype is not the consequence of the damaged gene but rather the result of the failure of that gene to make a functional product. However, the effect of producing an altered product may be quite different from the effect of producing no product at all, and the effect of having the altered part of the sequence used in some transcripts but not others can result in tissue-specific effects.

SUMMARY

Each of the known types of DMD mutations has major consequences for the individuals affected by the altered or missing protein. Because of the severity of DMD, the DMD gene is one of the important potential targets for genetic therapy. Efforts are being made to develop genetic therapies for DMD. These efforts are hindered by the size of the gene and complicated by the differences in manifestation of different mutation types. As with so many things in science, the development of these new therapeutic processes inch forward by taking baby step after baby step, with an occasional leap thrown in. We know there will be many baby steps in any research project, but we never know when the next leap will come along. It is sorrowfully too late for the two boys that Scott met and so many others, but we continue watching for new scientific leaps that we hope might make the current efforts relevant to children who are now alive.

EXPANDED REPEAT TRAITS

19

When Alex first started having major health problems in her late seventies, her children didn't know about it. She had gradually become more and more rigid about her daily routines, and she had developed a tendency to restate the same thing several times in a row, but her behavior changes were attributed to her advancing age. Although Alex had become a bit shaky as she aged and had had several falls, no one was terribly concerned because her sister had shown much more severe movement problems without anyone ever assigning it any medical significance. Besides, the movement problems seemed minor when compared to the disturbing problems that surfaced when Alex's daughter took a trip with her, including severe anorexia, obsessive compulsive behaviors, and hallucinations. Initial efforts to diagnose what was wrong lead to discussions of Alzheimer disease as one of the possibilities, but the big surprise was the final answer. Alex has Huntington disease, a disease that usually starts in early middle age and is most famous for the irregular jerking movements known as Huntington's chorea. Alex did go on to develop the characteristic movement problems, but the first signs that she had a problem were all mental. Alex's case is not so unusual, since some cases of Huntington disease manifest psychiatric and cognitive symptoms in addition to the movement disorder. Alex's case is quite unusual because her symptoms started so late in life. So what causes Huntington disease, how did the doctors figure out that that is what she has, and why did Alex remain free of symptoms long past the age at which many Huntington patients die?

Huntington disease (also sometimes referred to as *HD*) is autosomal dominant, which means that it only takes one defective copy of the gene to initiate the disease. Huntington disease involves a long, slow process of neurological degeneration. Onset is usually in the thirties or forties, although in rare cases the first signs of disease can show up in very young children or the elderly. Death occurs on average about seventeen years after the disease starts. Symptoms include the progressive development of uncontrolled or jerky movements. Although cognitive and psychiatric effects are quite common, some individuals and even whole families can remain free of the typical dementia even at late ages and stages of disease. HD is also known as Woody Guthrie disease, after one of its most famous victims.

Throughout most of the planet, perhaps one in 20,000 human beings have Huntington disease. The greatest concentration of Huntington disease cases found so far in the world occurs in some of the little fishing villages of Lake Maracaibo, Venezuela, where neurological manifestations that many of us have never witnessed are a commonplace part of everyday life. In the early 1800s, one Venezuelan woman with Huntington disease had ten children, and by the turn of the twenty-first century, a family tree of more than 17,000 of her descendants has been built. The pattern of inheritance seen in this family is called *autosomal dominant*, which means that the allele causing the trait is dominant and that the gene is located on one of the autosomes (chromosomes 1 through 22).

By the shores of Lake Maracaibo, thousands of members of this enormous family mark the days of their lives by the stages of the disease they call *El Mal de San Vito*. Autosomal dominant inheritance means that, on average, half of the descendants of anyone who is affected will also develop the disease. The half that are spared from developing the disease themselves often find their lives dominated by it anyway as their loved ones fall prey to the disease. Many of them struggle to wrest a marginal living from activities such as fishing in a community marked by the poverty that results when so many of the adults are too ill to work. Many of those who are not ill are overwhelmed with caring for those who are ill.

In 1979, an angel of mercy arrived in Venezuela to turn the local people with Huntington disease into partners in a search for the cause of the disease. Dr. Nancy Sabin Wexler (Figure 19.1), a psychologist on the faculty at Columbia University, had heard of the unusual concentration of Huntington disease in the area and came to survey the situation for herself. The expedition she led was the beginning of a series of annual expeditions that have taken doctors and researchers back to Venezuela in search of a cure.

Dr. Wexler's involvement with Huntington disease was not just professional. It was driven by her mother's death from Huntington disease. It was driven by the knowledge that she and her sister Alice each faced a fifty percent chance of sharing her mother's fate. It was motivated by her involvement with the Hereditary Disease Foundation, the organization that her father had founded for the purpose of beating this terrible disease before it could have a chance to fell one of his daughters. It was founded on her immense concern for all of the people whose lives are touched by the disease.

Her involvement with Huntington disease made use of her boundless energy and keen intelligence, tools that helped shepherd an international consortium of scientists through the process of searching for the gene. After scientists at Harvard University used a breakthrough concept in human genetic mapping to find the location of the gene on chromosome 4, members of the scientific consortium spent most of the next decade sifting through the genes in that region of chromosome 4 to finally find the culprit. The Venezuelan Huntington disease family, along with members of many other Huntington disease families from all over the world, contributed blood samples and medical information that were the keys to finding the gene.

There was a victory at the end of the search that began in 1979 when Nancy Wexler turned to this unlikely cluster of poverty-stricken fishing villages for help in solving one of the most puzzling neurological mysteries in modern medicine. What was it that they found at the end of one of the longest disease gene hunts of the late twentieth century? As you might guess, we are talking to you about Huntington disease not only because of its medical importance, but also because it offers us another set of genetic lessons. What they found in their search provides a framework with which to understand a whole family of genetic diseases. What unites these diseases is not any set of symptoms but rather the fact that they are all caused by the same type of mutation, an

FIGURE 19.1 Nancy Wexler—professor, researcher, philanthropist, and psychologist—gets a big hug from a Venezuelan child during one of Dr. Wexler's annual pilgrimages. Along with the sense of hope that she has brought to the shores of Lake Maracaibo with each of her visits, she has also brought these people medical help, basic necessities such as clothing, and a tremendous amount of compassion. (Courtesy of the Hereditary Disease Foundation. Photo by Peter Ginter.)

unusual type of mutation quite different from the other mutations we have been talking about.

THE EXPANDING AND CONTRACTING REPEATS

Simple *tandemly repeated sequences* occur throughout the genome, with more than 50,000 copies of some repeat categories found scattered around the genome. Most often, they are found between genes and in introns but can also

* This photo is from the web site for the Hereditary Disease Foundation, where more information is available about the disease, the research, and the Venezuelan kindred that was one of many HD families who helped researchers find the gene (www.hdfoundation.org). You might also want to read Mapping Fate by Nancy's sister, Alice Wexler. Helpful information can also be found at the Huntington Disease Society of America at www.hdsa.org, the Huntington Society of Canada www.hsc-ca.org/english/main.shtml, the National Institute of Neurological Disorders and Stroke at NIH www.ninds.nih.gov, Huntington Disease Society Online in the UK www.hda.org.uk, and an organization called We Move at www.wemove.org. (Photo courtesy of Peter Ginter and the Herediatry Disease Foundation.)

be in promoter regions, in untranslated sections of transcripts, and occasionally even in coding sequence. These tandemly repeated sequences can take the form of mononucleotide repeats (e.g., AAAAAAAAAAAAAAAAA), dinucleotide repeats (e.g., CACACACACACACACACACA), trinucleotide repeats (e.g., CATCATCATCATCATCATCATCATCATCATCATCATCATCAT), or repeats with a longer subunit length. They share the property that sometimes runs of a simple sequence repeat can change length—undergo expansion and contraction, with very different consequences to coding sequence depending on whether or not the repeat unit length is a multiple of three bases (the length of a codon) or some other length (Figure 19.2).

The repeat units in a *trinucleotide repeat* are the same length as the length of a codon—three bases. If a trinucleotide repeat within a coding sequence gains or loses one or two subunits, as commonly happens, it gains or loses one or two copies of an amino acid in the protein sequence. If a mononucleotide or a dincleotide repeat gains or loses a copy, it throws off the reading frame, causing a frame shift mutation that normally is expected to alter the amino acid sequence beyond the point of the mutation and will often result in picking up a stop codon in the out-of-frame reading of the sequence.

More than a half dozen different genetic diseases involve expansion of trinucleotide repeats located in coding sequence. Thus trinucleotide repeat expansions, if the change in number is small, have a small effect on the protein, whereas repeat unit lengths that are not a multiple of three can cause even a small change in DNA sequence to result in a major frame shift mutation with resulting major alterations to the sequence and structure of the protein. However, sometimes trinucleotide repeat expansions can be large, with truly drastic consequences, even though the resulting sequence has remained "in frame" and retains the normal sequence beyond the region of the repeat (Table 19.1). As we discuss below, the expansion of CAG repeats

TABLE 19.1 Examples of Disorders Caused by Expansion of Simple Sequence Repeats

Disorder	Gene	Repeat Unit	Location of the Repeat Within the Gene
Dentatorubral-pallidoluysian atrophy	DRPLA	CAG	Coding sequence
Huntington disease	HD	CAG	Coding sequence
Spinobulbar muscular atrophy	AR	CAG	Coding sequence
Spinocerebellar ataxia 1	SCA1	CAG	Coding sequence
Spinocerebellar ataxia 2	SCA2	CAG	Coding sequence
Spinocerebellar ataxia 3	SCA3	CAG	Coding sequence
Spinocerebellar ataxia 6	SCA6	CAG	Coding sequence
Spinocerebellar ataxia 7	SCA7	CAG	Coding sequence
Myotonic dystrophy 2	ZFN9	CCTG	Intron
Fredreich ataxia	X25	AAG	Intron
DM1-associated cataract	SIX5	CTG	Promoter
Progressive myoclonus epilepsy	Cystatin B	12 bases	Promoter
Myotonic dystrophy 1	DMPK	CTG	3 untranslated region
Spinocerebellar ataxia 8	SCA8	CTG	3 untranslated region
Fragile X	FRAXA	CCG	5 untranslated region
Fragile XE	FRAXE	CCG	5 untranslated region
Spinocerebellar ataxia 12	SCA12	CAG	5 untranslated region

A. MONONUCLEOTIDE REPEAT EXPANSION

Expansion by adding one, two, four or five units causes a frame shift.
Adding three units or any multiple of three units maintains the reading frame.

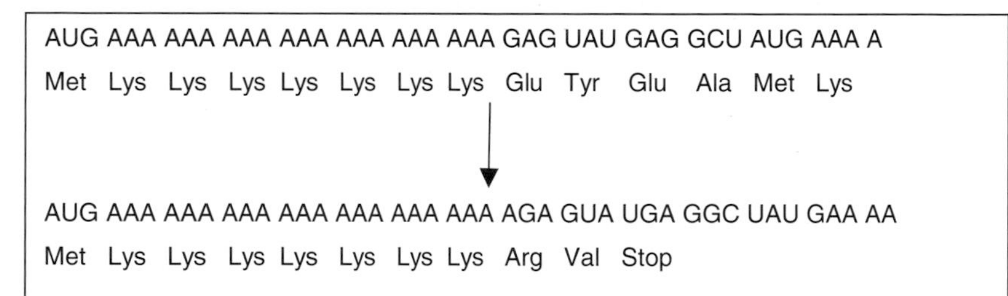

B. DINUCLEOTIDE REPEAT EXPANSION

Expansion by adding one, two, four or five units causes a frame shift.
Adding three units or multiples of three units maintains the reading frame.

AUG ACA CAC ACA CAC ACA CAC ACA GAG UAU GAG GCU AUG AAA A
Met Thr His Thr His Thr His Thr Glu Tyr Glu Ala Met Lys

↓

AUG ACA CAC ACA CAC ACA CAC ACA CAG AGU AUG AGG CUA UGA AAA
Met Thr His Thr His Thr His Thr Gln Ser Met Arg Leu Stop

C. TRINUCLEOTIDE REPEAT EXPANSION

Addition of any number of repeats maintains the reading frame.

AUG CAC CAC CAC CAC CAC CAC CAC GAG UAU GAG GCU AUG AAA A
Met His His His His His His His Glu Tyr Glu Ala Met Lys

↓

AUG CAC CAC CAC CAC CAC CAC CAC CAC CAC GAG UAU GAG GCU AUG AAA A
Met His His His His His His His His His Glu Tyr Glu Ala Met Lys

FIGURE 19.2 Small changes in number of repeats units is least likely to have any substantial impact on coding sequence if the repeat unit length is a multiple of three, the length of the codon. Otherwise, the reading frame will be shifted and the protein sequence beyond that point will be altered unless the number of units added or subtracted is a multiple of three.

leads to the production of an expanded stretch of glutamine amino acids at some point in the offending protein.

However, not all of the trinucleotide repeat expansions take place in coding sequence (see Table 19.1). In these noncoding cases, models for how the repeat expansion cause disease include effects on RNA metabolism and changes in gene expression; in one case, the final effect is accumulation of

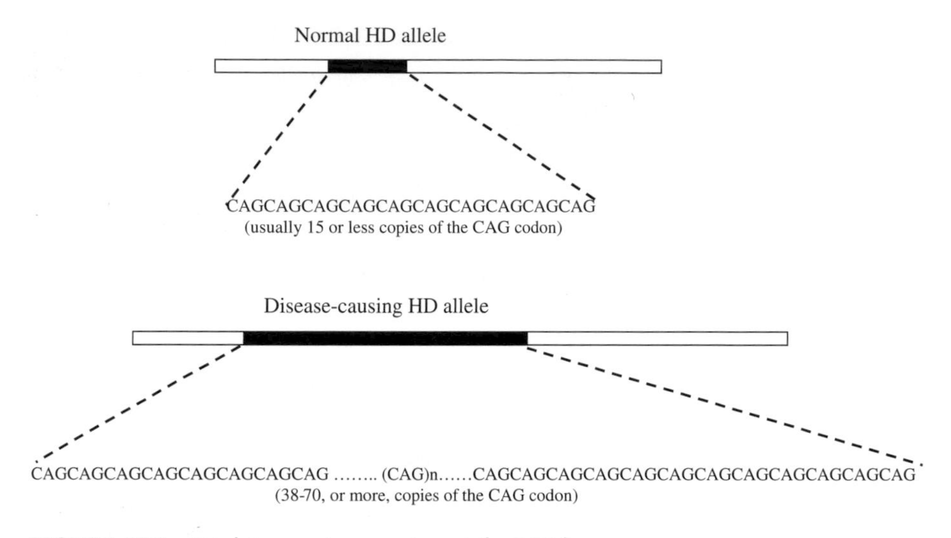

FIGURE 19.3 Triplet repeat expansion at the HD locus.

iron in the mitochondria. Much still needs to be learned about how these noncoding expansions play a role in disease.

Huntington disease is caused by one of the CAG repeat expansions in coding sequence. As shown in Figure 19.3, the mutation underlying Huntington disease is due to the amplification of the triplet codon repeat CAG (which encodes glutamine, as we discussed in Chapter 7). Normally, this gene includes a stretch of many copies of the glutamine codon arranged as one long, tandemly repeated array. If we looked at the sequence of the gene from a normal individual, we would find that different people have different numbers of glutamine codons at this point in the gene, anywhere from ten to thirty-five repeats of CAG in a row. In mutant HD alleles, the copy number has expanded to anywhere from thirty-six to about one hundred copies.

The mechanism by which new mutations arise in the HD gene remains unclear. As shown in Figure 19.4, an easy model to explain this kind of amplification involves the step at which DNA polymerase attempts to copy the tandem repeat of CAG codons in the normal gene. If it sometimes "slips," or "falls back," and then recopies the same set of codons or "falls forward" and then continues copying, this would increase or reduce the number of copies in the replicated strand. If such errors are left unrepaired, new mutations that are expansions or contractions are generated.

REPEAT LENGTH CHANGES

Regardless of the mechanism by which instability occurs, it is clearly present and it gets worse as the number of copies of the repeat rises. Using a technique called *polymerase chain reaction* (PCR), which was described in more detail in Chapter 15, the HD gene mutation rate in different individuals was measured in terms of the expansion or contraction of the repeat number for this gene by analyzing the genes in a single sperm! Normal- or average-sized HD alleles (fifteen to eighteen repeats) showed three contraction events (reduction in the number of repeats) among 175 sperm. Even at that low

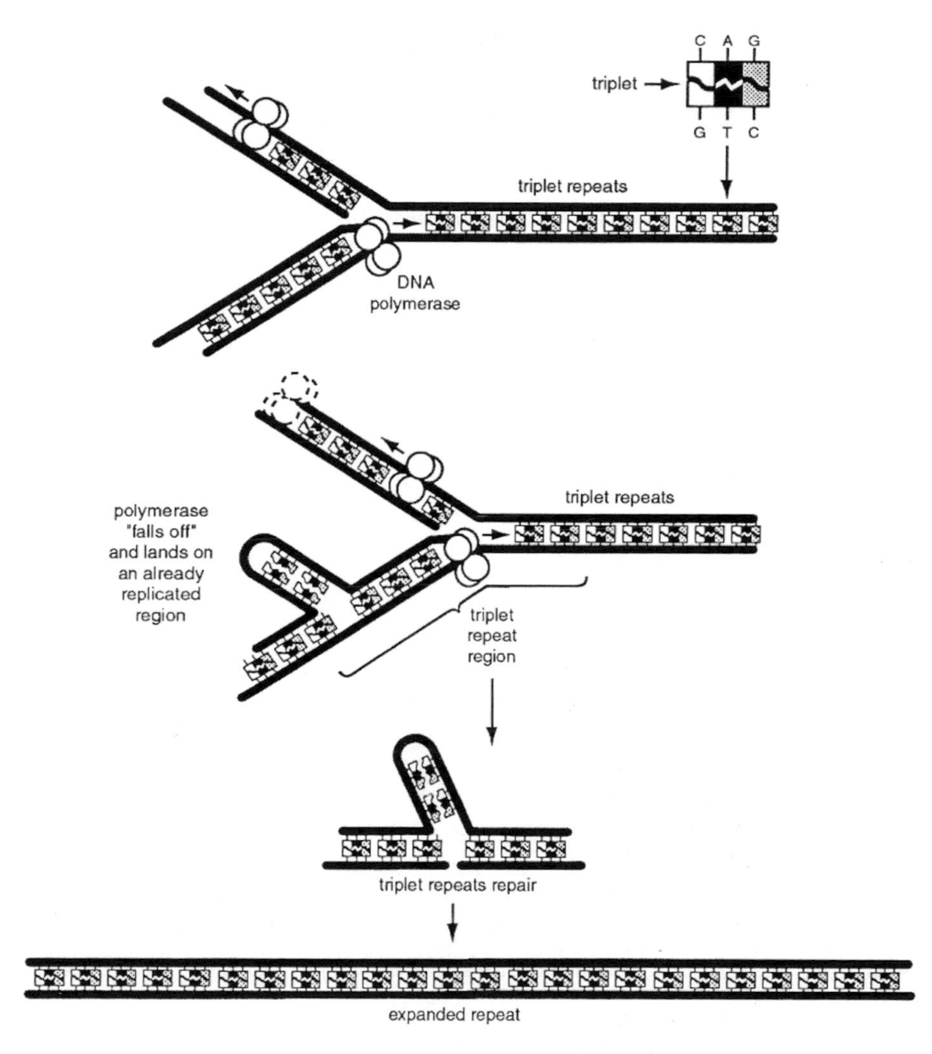

FIGURE 19.4 A possible molecular mechanism for triplet-repeat expansion.

level, the rate of expansion or contraction is an astonishing 0.6%. However, when they looked at a man bearing a normal allele with thirty triplet repeats, the mutation rate in terms of expansions and contractions went up 11%, that is, 11% of all of the sperm carrying this allele carried a variant copy of the HD gene. (Remember what was said earlier, the standard mutation rate is on the order of 1 in 100,000, so it would appear that the mutation rate here is 10,000 times higher than that!) An allele with thirty-six repeats showed a mutation rate of 53%, and in fact 8% of the sperm bearing this allele ended up with expansions so large that they would have caused disease. Disease-causing alleles, with thirty-eight to fifty-one repeats, showed expansions or contractions in more than 90% of the sperm carrying these alleles.

Simply put, as the number of repeats increases, so does the frequency of changes in repeat length, including expansions of the repeat. (Curiously, the frequency of contractions also increases up to thirty-six triplets but falls off as the copy number of the allele increases above thirty-six.) So one can imagine

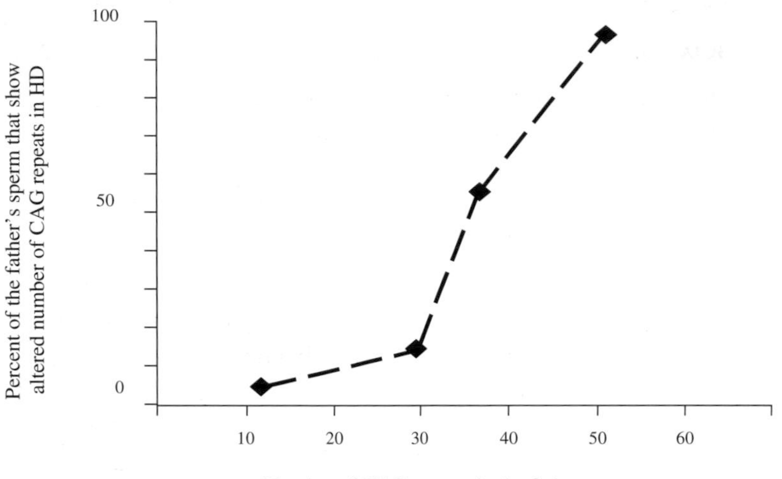

FIGURE 19.5 A man who has more repetitions of CAG in the HD gene sequence will have more sperm in which the length of that CAG repeat has changed, that is, mutated to a new length. Individuals with a higher number of copies within the normal length will produce some sperm in which the new CAG repeat length is long enough to cause disease.

how these mutations arise: one small increase makes a second increase more likely, and that increase further increases chances of an increase at the next generation, and so on. Realize that the mutation rate from a normal allele (fifteen to eighteen repeats) to an intermediate and unstable allele (say an increase from fifteen to thirty repeats) is quite low; however, once the repeat number gets above thirty repetitions, the mutation rate to a disease-causing allele is much higher. So someone with a repeat length of sixty repetitions is expected to transmit the "wrong" repeat length to the next generation more frequently than someone who starts out with sixteen copies.

THE EFFECT OF REPEAT LENGTH DIFFERENCES

Not only does the rate of mutation increase along with the increasing repeat length, so does the severity of the disease. For Huntington disease, the age at onset is related to the number of repeats, although the length of the repeat appears to account for only part of the variance in the phenotype. A larger number of copies not only predicts a higher mutation rater, but it also predicts an earlier age at onset of the disease.

A juvenile-onset form of the disease happens more often when the HD allele is passed on from the father. This phenomenon, in which a much younger age at onset seems to occur among the younger generation, is called *anticipation.* For many diseases, it can be hard to tell whether apparent anticipation is real or whether improvements in medical care, including more aggressive pursuit of diagnosis and better diagnostic tools, are actually responsible for earlier identification of affected people. Anticipation seems to be real

BOX 19.1 MYOTONIC DYSTROPHY—AMPLIFICATION IN THE EXTREME

More than a dozen different genes in the human genome can cause neurological damage as a result of the expansion of a region of trinucleotide repeats. Myotonic dystrophy, the most common form of muscular dystrophy, can be found in 1 per 8000 individuals and is inherited as an autosomal dominant trait. As for the HD gene, the mutations in the DMPK gene that cause myotonic dystrophy involve a triplet repeat expansion. However, the DMPK triplets are located in the noncoding part of the transcript beyond the stop codon, so the effect cannot be explained in terms of changing the number of repetitions of an amino acid in the protein. Although a normal individual only has five copies of the triplet, affected individuals can have one hundred or even one thousand or more copies of the repeat. Although expansions in the HD gene are more likely to come from the paternal line, expansions in the DMPK repeats seem to take place in the female germline.

in the case of Huntington disease. One dramatic example is that of a patient with onset around retirement age who had a child who developed symptoms in middle age and a grandchild who developed the disease as a very young adult. However, in many families the age at which the disease starts may appear to be fairly consistent, with middle age onset in most families and late age onset in some. Anticipation is not all that common, but it gets attention because it is so alarming to see a teenager developing disease symptoms at such a young age. This type of anticipation has been observed in others of the trinucleotide repeat expansion disorders and is not specific to Huntington disease (Box 19.1).

TREATING HUNTINGON DISEASE

Current options for treating Huntington disease are very limited. Antidopaminergic agents are used to try to treat the choreiform movements, and antidepressants and antipsychotics help treat the psychiatric manifestations of the disease. Elevated calorie intake helps fight the typical weight loss, and physical therapy provides further assistance with movement problems. However, none of the current treatments can stop the pathologic processes of the disease itself, which advances through an inevitable progression of cell death among neurons in specific regions of the brain that affect movement and cognitive functions. Past efforts to stem disease progression through use of antioxidant therapies have failed.

Recently, better understanding of the disease process has led researchers to begin exploring new therapies. Since the gene was first discovered, studies have continued using cells from human brains, human and animal cells grown in culture medium in the lab, and animal models of Huntington disease including both mouse and fruit fly models (Box 19.2). In the course of these studies, researchers have identified a variety of important proteins and biochemical pathways that are contributing to the disease process, including proteins involved in a process called *programmed cell death,* a set of excitotoxic

BOX 19.2 TRANSGENIC MOUSE MODELS OF HUMAN DISEASE

Researchers have made *transgenic animals*, animals that have been altered by putting in or taking out genetic material, of great importance to Huntington disease research. By constructing mice that have an expanded repeat in the mouse copy of the Huntington disease gene, they have made an animal model of the disease (the HD mouse) in which it is possible to study the same kinds of cellular and neurological processes seen in the Huntington disease patients. By making a different kind of transgenic animal called a "knockout" mouse that lacks the Huntington disease gene, researchers showed that animals cannot remain healthy without Huntington protein, so a gene therapy approach that simply removes the Huntington disease gene seems unlikely to work. Animals with altered or missing copies of a gene provide a valuable tool for the first stages in drug discovery. Although we most often think of transgenic animals as being mice, other kinds of organisms can be genetically modified through similar technologies. The most important other animal model of Huntington disease currently seems to be a transgenic fruit fly—the HD fly! The HD mouse and the HD fly have each been used in testing compounds that seem as if they have a chance of protecting against some aspect of the disease pathology. HD mice treated to make their Huntington less "gluey" lived longer and had fewer symptoms than the untreated mice. A compound that blocks HDAC deacteylase activity seemed to stop the disease process in the fly. Testing drugs in animal models allows researchers to identify which drugs and strategies are safe enough and work well enough to consider for testing in humans. Looking at the disease processes in the genetically modified animals helps researchers understand the basic underlying processes of the disease. However, since things that work in mice and flies do not always work the same way in humans, researchers need to take very careful steps as they work to find out whether a successful treatment of a transgenic animal is safe and effective and can be applied to human patients.

proteins that can overstimulate nerve cells even to the point of death, and an enzyme that chemically modifies important proteins in the cell. One of the most interesting observations is that many copies of glutamine in a row apparently have the ability to act like a kind of biological glue that sticks to certain other proteins in the cell and pulls them out of circulation so that they are not available to perform their functions (see below).

Two new therapies have clinical trials ongoing. One of the drugs being tested is minocycline, an antibiotic, that has the surprising side effect of blocking programmed cell death, the process by which the brain cells die in a Huntington disease patient. Riluzole, which blocks a process by which nerves can be overstimulated, is being looked at relative to Huntington disease and Parkinson disease. It is already in use for Lou Gehrig disease (amyotrophic lateral sclerosis, or ALS) and has been shown to allow only a small increase in life-span of the patients. Each of these therapies is based on general issues that potentially apply to a variety of neurodegenerative diseases whether or not they involve CAG repeat expansion or polyglutamine "glue" gumming up other proteins in the cell. At this stage, it is not yet known whether they will help patients with Huntington disease or whether any problems with their use will surface.

JUST AN IDEA? AN APPROACH TO POLYGLUTAMINE DISORDERS

In investigations of how polyglutamine proteins cause problems, researchers have discovered that, when cells contain either mutant Huntington protein or a synthetic protein made of nothing but glutamines, the polyglutamine acts like a glue that sticks to certain important proteins in the cell that get "pulled out" of use. One critical type of protein that sticks to the polyglutamine glue is called a transacetylase (TA), and the job of the TAs is to carry out a specific chemical modification that affects the level of activity of many other proteins. Normally, TAs add the chemical acetyl group to proteins and a different group of proteins called deacetylases (DAC) take the acetyl groups back off the proteins. The two processes need to be operating in balance with each other, and if there is too much or too little TA activity compared to the amount of DAC activity, the proteins in the cell will have too many or too few acetyl modifications (Figure 19.6). This would affect the activities of many different proteins in the cell.

A new treatment idea for Huntington disease arises from this model of the disease process and presumes that if the TA activity can be restored to be in balance with the DAC activity, the cells will be much healthier. Researchers

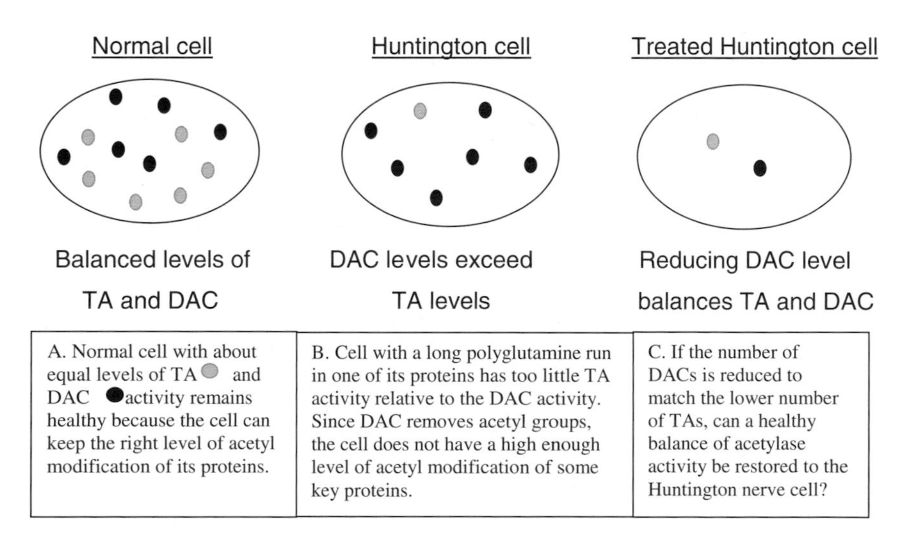

FIGURE 19.6 A model for how Huntington disease might result if the long polyglutamine string in the Huntingtin protein soaked up TA enzymes. Many proteins in the cell would have fewer acetyl groups attached to them and would not function correctly, in contrast with the normal cell that keeps a balance between adding and removing acetyl groups. The interesting point here is that this process of adding and removing acetyl groups may be key even though the Huntington protein itself plays no role in adding or removing acetyl groups. So is the answer to keeping the right level of acetylase activity as simple as a plumbing problem—if the drain is slow, turn down the rate the faucet drips to match the slower drain and thus manage to keep the same water level? Tests on the HD mouse suggest this might be at least part of the answer.

tested a DAC inhibitor in a most interesting animal model of Huntington disease—the HD fruit fly (see Box 19.2). The recently released amazing finding is that this medication appears to stop the disease in the HD fly! When this medication is tried on the HD mouse (see Box 19.2), we will watch eagerly for whether enough evidence develops to allow this medication to progress to clinical trials in humans. Although the fly studies are intriguing, they do not guarantee that the result will translate into viable use for human beings.

This model makes the assumption that the cell is very sensitive to the exact levels of the proteins that are sticking to the "glue." While we know that the cell is indeed very sensitive to the exact levels of some proteins, such as transcription factors or structural proteins, it is much more common for cells to be relatively unbothered by losing only one copy of a gene encoding an enzyme. Thus the need to precisely balance levels of TA and DAC proteins makes the assumption that the cell is much more sensitive to the exact levels of these proteins than it is to levels of most other enzymes. In the end, it may turn out that this balancing act is critical even though these are enzymatic activities we are dealing with, or it may turn out that the situation is much more complex than the "gluey" model.

Although we might someday see results from efforts to treat Huntington disease (or other CAG repeat expansion disorders) by targeting the activities of the "gluey protein" or proteins that interact with it, it has been suggested that the concept of protein glue as the primary mode of initiation of the disease may be inconsistent with the tissue-specific pathogenesis and with the autosomal dominant manner of inheritance. If the model for the primary pathology were as simple and nonspecific as a polyglutamine glue, we might expect to see expansion of polyglutamines in many different proteins leading to very similar phenotypes. However, what we see is actually some very different pathogenic events involving very different cell and tissue types resulting from CAG repeat expansions in different genes. While some of this may turn out to have to do with tissue specificity of expression of the different CAG-repeat genes, that is likely not enough for a generalized "glueyness" to account for the substantial differences in disease processes that are observed, given that in a number of cases the genes in question are not limited to expression in the cell type that ends up being damaged. However, the fact that the "gluey protein" may not be the key to the specific features of the disease does not keep biomedical researchers from looking for ways to take advantage of the fact that this generalized "glue" seems to be playing a role in the disease, even if it can't account for everything that is going on.

It sounds quite exciting to have a list of four possible new ways to treat Huntington disease:

- Prevention of programmed cell death
- Reduction of excitoxic stimulation of nerve cells
- Rebalancing the TA and DAC activity levels within the cell
- Reducing the glueyness of the polyglutamine glue

However, none of these treatments appears to be ready for prime time yet. Any of the new treatments being explored will have to go through careful processes of testing and evaluation before they become generally available

BOX 19.3 CLINICAL TRIALS TESTING OF POTENTIAL NEW TREATMENTS

The process for drug approval starts with Phase I trials that screen dozens of subjects to test for any major harmful effects. Phase II trials screen hundreds of subjects to test for whether the compound actually does any good. The process finishes up with Phase III trials that screen thousands of individuals to find out whether the drug is both safe and helpful when tested on large numbers of people, and to help pin down the optimal dosages and better understand the side effects, if any.[1]

from the medical community. Some drugs already in use in humans for other purposes may need a shorter testing period, but compounds that have not yet been approved for any use in humans will take longer to develop. We expect that, within a few years, at least some of the treatment models we have presented here will have been disproved, but if one of the current ideas does not turn out to be "it", researchers will have made further progress in developing additional treatment approaches by that time.

In the case of a disease like Huntington disease, someone who feels desperate for a cure may wonder why they can't just start taking medications such those discussed here. Unfortunately, long experience has taught the medical community that only a fraction of the brilliant breakthrough ideas turn out to work and that, in some cases, something that sounds like a great idea can end up causing more harm than good (Box 19.3). One of the most dramatic examples of this kind of unexpected negative outcome for a promising drug is thalidomide, a drug that helped prevent nausea in pregnant women with morning sickness but then caused some of their babies to be born with major limb deformities. So the medical system persists in requiring rigorous testing to keep the newest drugs from causing even greater harm than the disease they are supposed to cure.

Because there is not yet any cure available for Huntington disease, the discovery of a genetic marker that could be used accurately to identify people carrying HD mutations was a bit of a mixed blessing. For years before such testing was available, members of Huntington disease families asked if there wasn't some way to test and find out who would end up being affected. Once the test became available, many who had asked for the testing backpedaled and indicated that they didn't want to be tested yet or that they didn't want to be tested at all.

Dr. Nancy Wexler, who played a leadership role in the international consortium that cloned the HD gene, has applied the name *Tiresias complex* to the dilemma of making the choice regarding whether to be tested for something for which there is no cure. The name comes from the blind seer Tiresias who, in *Oedipus the King* by Sophocles, said, "It is but sorrow to be wise when wisdom profits not." In describing the Tiresias complex, Dr. Wexler asked, "Do you

[1] More information about treatment ideas being developed and clinical trials going on can be found at web sites for the Hereditary Disease Foundation (www.hdfoundation.org), the Huntington Disease Society of America (www.hdsa.org), and the National Institutes of Health of the United States (www.nih.gov). More Information on another HD clinical trial can be obtained from Huntington Disease Drug Works (HDDW) at http://www.hddrugworks.org.

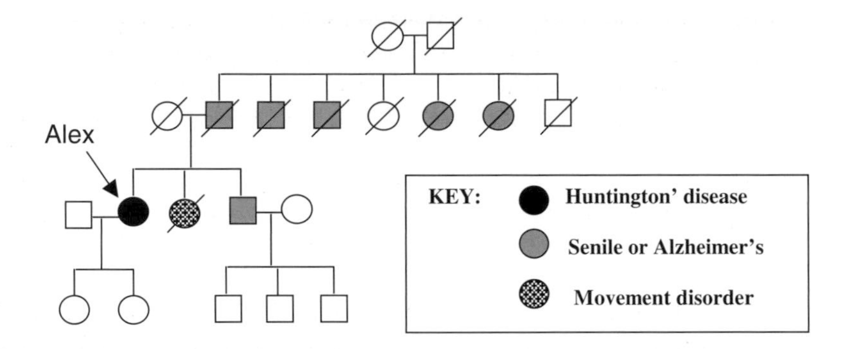

FIGURE 19.7 Alex's pedigree may actually be showing us a family history of Huntington disease that no one realized was going on because the onset of the disease was so late that most affected individuals had simply been considered "senile," including her brother and her father. Four of her father's siblings had been diagnosed with either senility or Alzheimer disease. One of her father's brothers has recently developed movement problems as he is approaching ninety years of age. Her sister also had not been diagnosed with Huntington disease, although she demonstrated fairly severe movement problems reminiscent of Huntington chorea. We have to wonder if the many cases of reported senility shown here represent a family history of undiagnosed Huntington disease in which the first symptoms were mental rather than physical.

want to know how and when you are going to die, especially if you have no power to change the outcome? Should such knowledge be made freely available? How does a person choose to learn this momentous information? How does one cope with the answer?"[2]

This is an issue that the human genetic community is struggling with for Huntington disease and many other fatal disorders that we can currently diagnose but not cure.

A RETURN TO ALEX'S STORY

So we now return to Alex's story. At the beginning of this chapter, we asked three questions. We have already answered the question regarding what causes Huntington disease: a CAG repeat expansion causes the Huntington protein to have too many glutamines, which then serve as a glue to remove other key proteins needed to keep the right balance of activities in the cell, which sends the nerve cells into a process of programmed cell death. Although this may be oversimplifying the overall pathogenic process, it seems to be at the core of what is going on.

So let's answer the next question we asked: How did they find out that Alex has Huntington disease? If we look at her family history, we do not see the history that we expect to see in a Huntington disease family (Figure 19.7). Usually, we would expect to see a family history covering multiple generations, with about fifty percent of the at-risk individuals affected. Instead, we see a

[2] Wexler, N. S., The Tiresias Complex: Huntington Disease as a Paradigm of Testing for Late-onset Disorders. *FASEB J*, 1992; 6, 2820–2825.

history of "senility" among Alex's father and his siblings, and no particular diagnosis at all for her sister, whose movement problems made it look like she was bowing and dancing when she tried to use her walker.

While they were considering the list of possible diagnoses (called the *differential diagnosis* list), in addition to Alzheimer disease and several other things, Huntington disease was among the possibilities. Since there is a very simple genetic test for CAG repeat expansions in the Huntington disease gene, the doctors arranged to have a sample of Alex's DNA tested for a CAG repeat expansion in her copies of the HD gene. A very clear answer came back with the genetic testing results that told the doctors that Alex has Huntington disease because of the presence of the expanded CAG repeat.

The test result also answered our last question: Why did Alex's disease start so late? The answer is that Alex has only forty copies of the CAG repeat, one of the shortest lengths of repeat that can cause disease (remembering that thirty-five copies appears to be short enough to leave someone free of the disease, at least within the length of a normal human life span). When we look at Alex's grandparents and find no reports of senility or Huntington disease, we wonder whether the one that passed the disease along did not live long enough to manifest the disease. On the other hand, we have to consider that they might never have been diagnosed even if they displayed symptoms because views of mental and physical incapacity in the elderly back then might have allowed for the dismissal of shakiness and some mental problems as the simple manifestations of aging without perhaps ever even consulting a doctor.

You might think that the main point of the diagnosis would be a medical issue focused on symptoms and treatment. Certainly knowing what is happening to Alex lets her doctors optimize their use of those treatments that are available. However, there are real limits to what they can do for her, and the biggest gains here appear to have been psychological. Her daughter reports that the whole family ended up with a sense of relief just to be able to understand what was happening to her mother and to have a better understanding of what this implies for the rest of them. Apparently, the uncertainties before the diagnosis were worse than the final certain answer.

Now Alex's children and grandchildren face decisions about whether to take the genetic test themselves. You might think that the Tiresias complex would loom large for them, but their biggest concerns are quite different. They worry about what kind of use could be made of their testing information. As Alex's daughter said, "If one of us is going to end up ill and in need of medical help at Mom's age, the last thing we want to do is take some genetic test now that will prevent us from having insurance protection when we need it the most." As we look at the amount of care that Alex needs now, and as we consider whether the eventual answers to Huntington disease might include any kind of expensive genetic therapies, we can sympathize with the family. Clearly, they feel that no one has been able to provide them with the levels of assurance they would need to make them comfortable about taking the test without risking loss of coverage. If they end up with the assurances they want and face going ahead with the test, we have to wonder whether the Tiresias complex will suddenly become a more important issue than it seems to them right now.

SUMMARY

Basically, the human genome is a pretty stable place. Errors in replication seem to occur rarely and, when they do occur, are rapidly corrected so that the overall mutation rate stays low. However, certain triplet repeats appear to represent an Achilles' heel for the replication and repair systems. Triplet repeat expansions occur more often than most of the other mutation types we have discussed, and the rate of triplet repeat expansion mutations increases after the first expansion mutation occurs. Some of these expansions, the ones that exhibit primarily paternal instability, appear to reflect errors in male meiosis. Others, such as the expansion arrays that underline fragile X syndrome and myotonic dystrophy, appear to reflect events occurring in the female germline.

Because diseases involving CAG repeats in the coding sequence may operate through the same polyglutamine-glue mechanism, there is some hope that treatments developed for one of these diseases will apply to the whole family of polyglutamine disorders. New treatment ideas have emerged from the study of this class of mutations, but it remains to be seen whether any of them will be the answer that is needed.

GENES, CHROMOSOMES, AND SEX

In this section, we present the sex chromosomes, explain the role of genes and chromosomes in determining sex, discuss distinctions between sex, gender, and orientation, and talk about why having extra copies of sex chromosomes may sometimes be more "allowable" than having extra copies of some of the autosomes.

THE X AND Y CHROMOSOMES: THE ODD COUPLE

20

"What color is the light, Linda?" Two days into their cross-country road trip, ten-year-old Linda had lost track of how many times her adored grandfather had asked her this. As they approached a red traffic light, she smiled at his little game and said, "red." He carefully slowed down and stopped to wait for the light to turn green again. At the next green light, feeling bored, she answered, "red," so he smiled indulgently in her direction and slowed to a stop. He waited patiently until she told him the light was green and he started up again. What child would not be charmed by a grandfather who would accommodate her whims and stop anyway when she answered with the wrong color? Later that day, as he made his same query at a red light, she said, "green," with a little secret smile and he pulled right on through the intersection, which was fortunately free of other cars. Linda was just sure that she had the coolest, funniest grandfather in the state of Ohio, until years later when someone mentioned something she had never realized: her grandfather was color blind. Can you imagine the shiver that ran up her spine as she mentally flashed back to the scene of their car confidently pulling through a red light into the fortuitously empty intersection? (Don't be alarmed at the idea that color-blind people are driving around trying to distinguish whether they are supposed to stop at the light at the intersection! These days, at least, positioning of the order of the light color is standard throughout the country, so someone who is red-green color blind can tell whether the light is red or green simply by telling whether it is the top or the bottom light that is lit. In fact, for color-blind individuals, knowing whether to stop at a stop-light can often be far easier than telling whether a pair of socks match.)

When we look at the story of Linda and her grandfather, it is not surprising that the person who could not tell the color of the lights is a man. About one in every ten men has some form of *color blindness*, but less than one percent of women are color blind. In previous chapters, inheritance seemed a simple matter bearing no relationship to the sex of the individual. For trait after trait, we have seen a kind of genetic equal rights movement: examples in which traits affect males and females equally, get passed along equally to sons and daughters, and pose the same risk whether coming from a mother or a father. Then we come across a trait such as color blindness, and we have to wonder how the rules of inheritance work in a family if the two sexes have different chances of being affected. Do the two sexes actually have a different chance of inheriting the genetic defect, or is there some difference such as hormones that affect whether the person with the defect manifests the trait? To better understand what is going on with traits that turn up with different frequencies in the two sexes, let's start by looking at the X and Y chromosomes, which are sometimes referred to as the *sex chromosomes*.

PASSING THE X AND Y CHROMOSOMES BETWEEN GENERATIONS

In previous chapters, we have talked about the fact that females have two X chromosomes and males have an X and a Y. This genetic asymmetry presents the cell with several problems, one of which is how to get two different chromosomes, the X and the Y, to go through meiosis when neither of them have a homologous copy of themselves to pair with. During the formation of eggs in a female, the two X chromosomes offer no problem and are handled basically the same way the autosomes are handled, through pairing of the homologues (Figure 20.1, **A**).

Surprisingly, during formation of sperm, the cell's meiotic machinery manages to treat the X and Y chromosomes as if they were a homologous pair, even though even the simplest visual examination demonstrates that they are significantly different from each other. During meiosis I, the replicated X and the replicated Y come together to form a bivalent that lines up on the metaphase plate along with the bivalents that are autosomal (replicated versions of any chromosome other than the X or Y) (Figure 20.1, **B**). At the end of meiosis I, we find the X in one daughter cell and the Y in the other daughter cell. At the end of meiosis II, there are four sperm, two containing the X and two containing the Y.

How does this happen? What allows the cell to treat the X and the Y as if they were homologous structures? Clearly, there is not enough DNA on the Y chromosome for it to contain homologues of all of the genes on the X. In fact, however, there are a small number of genes that occur on both the X and the Y. The X and the Y each contain a region near the tip of the shorter chromosomal arm called the pseudo-autosomal region because these regions contain homologous copies of the same genes and can pair based on homology as if they were autosomes (Figure 20.2). DNA in this region can recombine, exchanging material between the X and the Y. Because of the exchanges, the chromosomes are locked together during male meiosis, but such exchange of material is normally strictly contained within the pseudo-

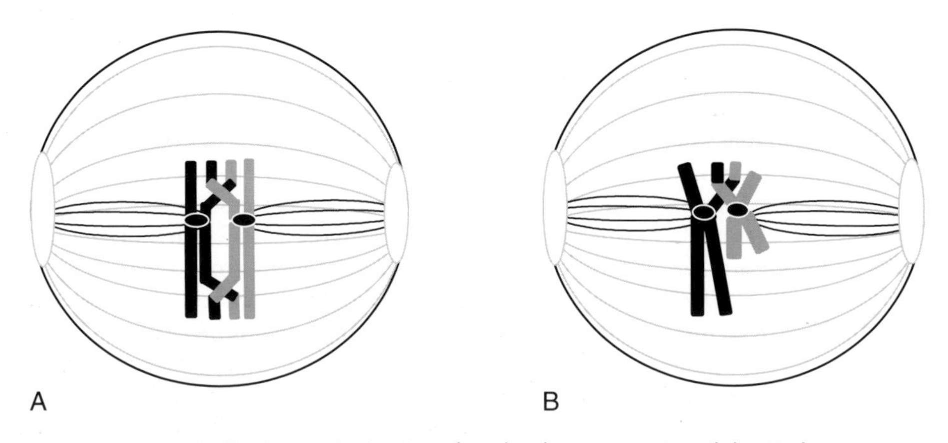

FIGURE 20.1 **A,** During meiosis I in a female, the two copies of the X chromosome pair along their lengths based on sequence homologues found all along the length of the chromosomes. **B,** During meiosis I in a male, only the very small regions of the X and Y that have the same genes can pair, and other regions that contain genes specific to just the X or just the Y cannot participate in aligning the chromosomes or holding them together at the metaphase plate.

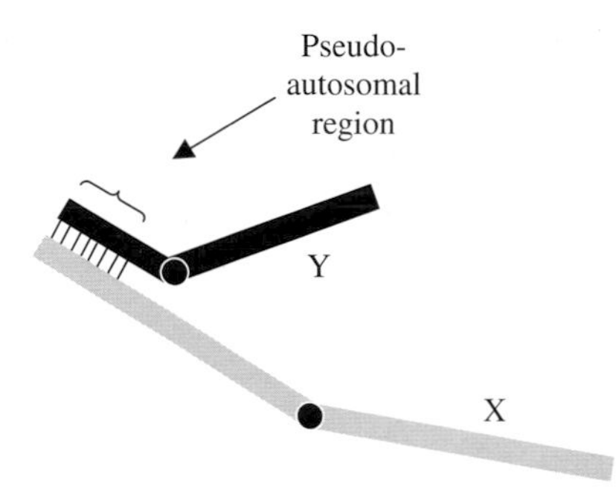

FIGURE 20.2 There is more than one region on the X and Y chromosomes that share genes, but there is a main region near the tip of the short arm of the X that is homologous to a region near the tip of the short arm of the Y chromosome that is called the pseudo-autosomal region, where pairing of the two chromosomes can be seen during meiosis.

autosomal region. Exchange of material farther out would put Y specific DNA onto the X chromosome, and vice versa.

COPY NUMBER

Another big problem faced by the human body is how to deal with the fact that there are genes on the X chromosome that are present in two copies in women but only one copy in men. In spite of the small number of pseudo-autosomal genes that allow the X and Y to pair, there are thousands of genes on the X chromosome that are not present on the Y, and a small number of genes on the Y not present on the X. There are many genes for which the number of copies of the gene affects the amount of gene product that is present in the cell. And although we can imagine that the amount of gene product might actually need to be different for the two sexes for some genes specific to sex, such as certain hormones, a large number of genes on the X chromosome have nothing to do with maleness or femaleness.

So it is possible to look at this from one of two perspectives: either the men have too few copies of the X-specific genes, or women have too many copies. We could also invoke much more complex regulatory models for what is going on, but it turns out that for many genes on the X, the cell needs exactly one copy rather than the two copies found for the autosomal genes. Thus the answer is that women have too many copies and the cell has to deal with that.

The way the cell deals with the extra copy of each X-chromosome gene present in a woman's cells is by inactivating the extra copy (Box 20.1). The cell turns the gene off so that it is not expressed: no transcription takes place and no RNA gets made. It accomplishes this by chemically modifying the copy of the X chromosome that is being turned "off." The cell packages the inactivated X chromosome into a little bundle that usually appears under the microscope as a little dot or blob called a Barr body (Figure 20.3). The result is that women have two copies of the X-specific genes for each copy that a man has, but a female cell uses only one of those copies. Thus equal levels of gene expression are achieved for men and women.

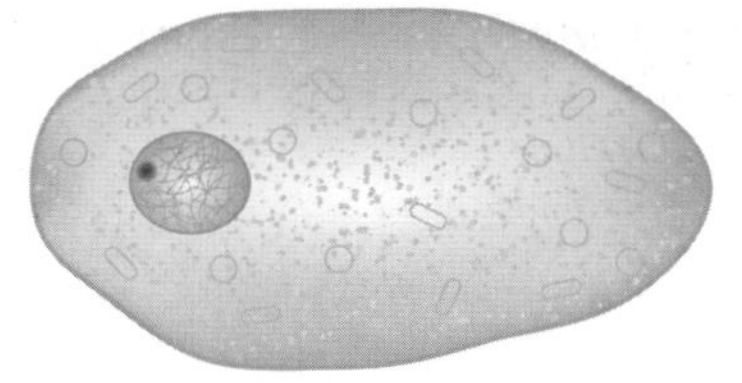

FIGURE 20.3 Picture of a cell with a Barr body appearing as a dark staining inclusion off to the side in the nucleus.

BOX 20.1 X INACTIVATION AND GENE DOSAGE

One X chromosome in each female cell has been inactivated, and the inactivated X appears as a spot in the nucleus that is structurally different from the X or any of the other chromosomes. The cell has two X chromosomes, and which one is inactivated in any given cell seems to be random. So if a carrier for an X-linked trait randomly turns off the normal copy in some cells and the altered copy in other cells, can X-linked recessive inheritance really work by the same mechanism as autosomal recessive inheritance, that is, by having the good copy make up for the defect? Apparently not, since cells in which the defective copy has remained active do not have another copy in the cell that can be used to produce the needed normal gene product. Apparently, for many of the genes on the X chromosome, having some of the cells making the normal gene product seems to be enough to compensate even though the defect is being expressed in some cells but not in others.

X inactivation occurs in humans quite early in human development, when the embryo consists of approximately 32 total cells. The individual cells are indifferent as to which X they will inactivate: it will be either the paternal X or the maternal X. Thus, this is a random event. However, in each individual cell once it has chosen whether to inactivate the maternal or the paternal copy of the X, all mitotic descendants of the cell will retain that committment to keep that same copy of the X inactivated. It is interesting to note that the paternal X is inactivated in extra-embryonic structures like placenta, and that the inactivated X is reactivated during female gametogenesis.

Because X inactivation occurs randomly and at an early stage in fetal development, it is possible for a tissue or organ to be comprised of cells that only have active paternal copies of X or that only have active maternal copies of X. Thus, in rare cases we may see an X-linked trait manifested in a female carrier who has inactivated the good copy of the X in all of the cells relevant to that trait.

However, if the women inactivate one of their two X chromosomes, they now have the reverse problem. For the pseudo-autosomal genes shared by the X and the Y, the men now have two copies in use and the women have only one. It turns out that the cell has an answer to this dilemma, too. On the inactivated X chromosome, only the X-specific genes get inactivated (Figure 20.4).

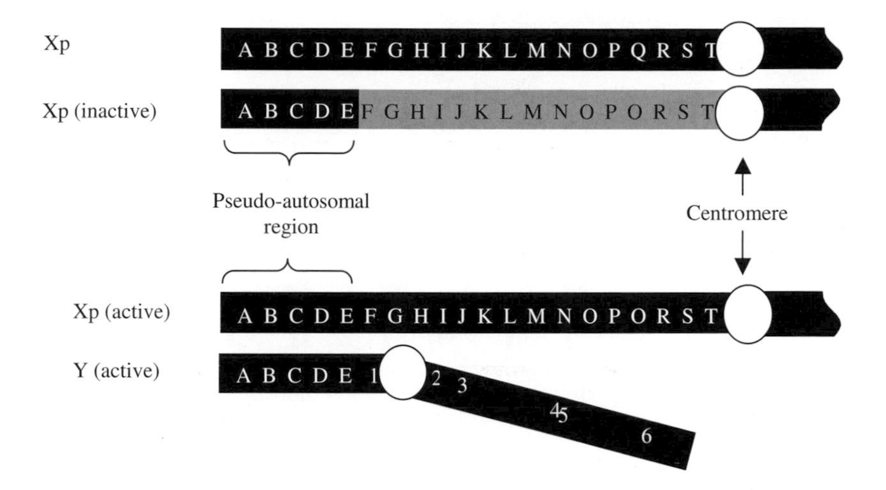

FIGURE 20.4 Active (white letters) and inactive (black letters) genes on the X and Y chromosomes. This cartoon of the Y chromosome and the short arm of the X chromosome shows how some genes in the pseudo-autosomal region (genes A, B, C, D, and E) are present in two active copies in both males and females, while other genes on the X (F through T) are present in one active copy. This oversimplification presents the general concept but misses the fact that there are additional unmodified areas outside of the pseudo-autosomal region where transcription still takes place on an inactive X, including some additional genes that are found on both the X and the Y. Notice that there are still some genes on the Y chromosome that are only present in males, a matter that we will tackle further in Chapter 21.

The pseudo-autosomal genes remain active. But wait a minute! Do we really know that this is completely true? Actually, we have not assayed every single gene, but as an approximation, this is a good model for what is going on with gene dosage on the X and Y chromosomes. How does the cell accomplish this? By turning off only the regions of the X that contain the X-specific genes. The psuedo-autosomal genes remain on, so they can still be transcribed. The result is that men and women both have one active copy of the X-specific genes in use, and both men and women have two copies of the pseudo-autosomal genes available for use. Thus a kind of transcriptional equality of the sexes is maintained.

X-LINKED RECESSIVE INHERITANCE

What does a family tree look like when we track a trait encoded by a gene on the X chromosome? If we look at inheritance of classical red-green color blindness, we are looking at the transmission of a genetic trait encoded near the bottom of the long arm of the X chromosome, a region that is nowhere near the pseudo-autosomal region. The genes in question—which make the proteins that detect red color and green color—do not have homologues on the Y chromosome.

Let's consider why Charles in Figure 20.5 would have only children with normal color vision if he is passing along a color vision defect on the X chromosome. Each child gets either the X or the Y from him, but not both. The color-blindness allele is on his only copy of the X chromosome. Its easy to see

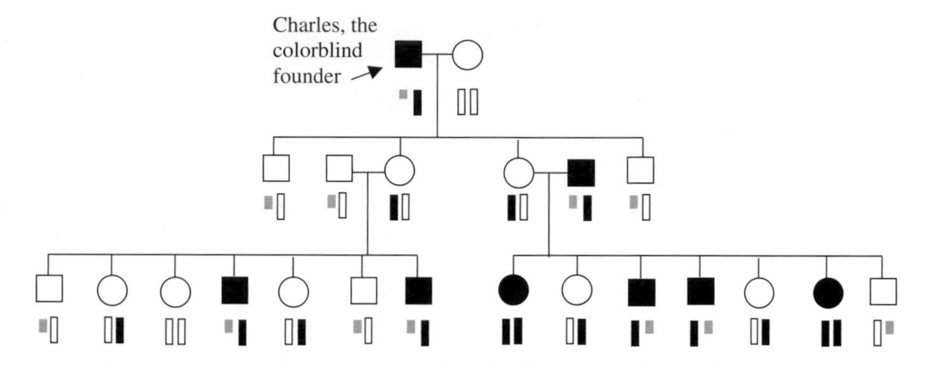

FIGURE 20.5 Inheritance of red-green color blindness in a theoretical kindred. Charles' sons are not affected because they did not get an X from him, and his daughters are not affected because their maternal X covers for the defect they inherited from him, but they are considered obligate carriers. Color vision deficits start to turn up among Charles' grandchildren, although in some families the X-linked trait may not turn up until later generations. The branch of the family on the left looks like a rather standard X-linked pedigree. The branch on the right is unusual because someone with the trait has married into the family, a not unlikely event with something as common as color blindness. The daughter on the left has sons who are color blind and some of her daughters are carriers, that is, they have the chromosome with the gene defect but they are not themselves affected. The daughter on the right has both daughters and sons with color vision defects because she is a carrier who married a man with a red-green genetic defect. Notice that the black chromosomes carry the color vision defect, but that the only females affected are the ones who have two black chromosomes, one from each parent.

that the boys will be unaffected, since he gave them a Y chromosome that does not carry the color-blindness allele. Since the girls get a good copy of the X from their mother along with Charles' X with the color-blindness allele, the fact that they are unaffected suggests that red-green color blindness is recessive to the normal version of the gene.

What happens when one of the carrier daughters in Figure 20.5 has children? The daughter on the left married a man with normal vision, so her sons, who could get either a normal or a mutation-bearing X from her, are each at fifty percent risk of being color blind. As predicted, half of her sons are color blind. The daughter on the right married a man who is red-green color blind. As happened in the branch of the family on the left, half of her sons are color blind. What is unusual is that each of her daughters is also at fifty percent risk of being color blind because half of the girls will end up with a mutation-bearing X from mom in addition to all of them getting a mutation-bearing X from dad.

When we look at Charles' grandchildren in Figure 20.5, we can tell which girls are carriers by looking at the chromosome diagram, but in real life, we normally have to function with much less information. Some X-linked recessive traits have a *carrier state*—a phenotype that is unusual but much less severe than the phenotype of people who are truly affected. The unaffected daughters of a red-green color blindness carrier usually cannot tell whether they received the normal copy of the gene or the defective copy.

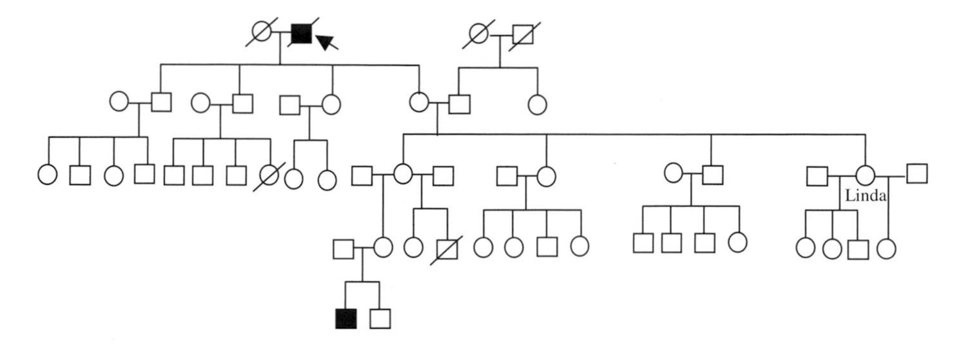

FIGURE 20.6 X-linked inheritance in Linda's family. Traits encoded on the X-specific parts of the X chromosome can sometimes turn up after many generations of seeming to disappear from a family. This family tree, which shows only some key parts of Linda's large family tree, presents a classical story of X-linked inheritance of color blindness. The X chromosome with Linda's grandfather's color vision mutation was passed to his daughter, granddaughter, and great-granddaughter before arriving at his great-great grandson (black square) and causing color blindness. By looking at this family, we cannot tell who all of the carriers are who are at risk of passing along the color vision defect, although in some cases, we can identify individuals who cannot have received the defective gene. For instance, from looking at this family tree, we cannot tell whether Linda's grandchildren or great grandchildren would be at risk of being color blind, although we could do a statistical calculation of their risk.

The result is that sometimes a mutant allele on the X chromosome can be passed down through multiple generations of female carriers without the family seeing any sign of someone with the trait. If we recall Linda's driving adventures and then look at part of Linda's family (Figure 20.6), we can see that her grandfather's color blindness seemed to disappear from the family. Imagine how surprised the family might have been when one young man in the youngest generation turned out to have a color vision defect that most of the family didn't even realize his great-great grandfather had.

If we did not already know that red-green color blindness results from changes in the blueprint in the vicinity of Xq28, the X-linked nature of the inheritance would only be obvious in some families. In the case of Linda's family, if this were some new trait with unknown mode of inheritance, the information in her family tree would not be enough to tell us that this is X-linked recessive rather than autosomal recessive inheritance, but we would know that X-linked inheritance is one of the likely possibilities. When trying to rule out X-linked inheritance, one of the key things we look for is whether or not the trait gets passed from a man to any of his sons. Since a man does not pass an X chromosome to his son, he cannot pass an X-linked trait to his son.

GIRLS WITH DUCHENNE'S MUSCULAR DYSTROPHY

More lessons about X inactivation come from the study of rare females affected with Duchenne muscular dystrophy (DMD). Most DMD carrier females (those with a mutation on one of their two X chromosomes) show no obvious symptoms, despite the fact that X inactivation causes half the nuclei in their muscles to express only the mutant allele. Why would they not develop illness? This is because muscles are composed of very large cells that arose by the fusion of many individual cells. These cells each contain hundreds to thousands of nuclei per cell. On average, half of these nuclei produce dystrophin (i.e., they have inactivated the mutant-bearing X), and this level of functional dystrophin production appears to be sufficient. Given that X inactivation is random, one might imagine, at least in some cases, that a female embryo might inactivate the normal X in a large fraction of those embryonic cells that will go on to produce muscles. Such a thing may not happen often, but it is theoretically possible, and indeed, some eight percent of carrier females show some detectable muscle weakness.

We can anticipate some scenarios that could lead to disease in a female. If there are so many new mutations, we can imagine that some of these could arise when a sperm with a new DMD mutation fertilizes an egg that received a previously existing DMD mutation from a carrier mother. We do not expect this to happen often, but then girls with DMD are rare. As we now understand from studies of aneuploidy, DMD might also be able to happen in an XO woman with Turner's syndrome if the X chromosome she received had a DMD mutation. However, there is another, more complicated situation that can arise in which disruption of the DMD gene by a translocation can cause DMD in a girl who has two X chromosomes, one of which has the normal sequence. Follow along with us as we explain the logic behind the seemingly unexpected event of a heterozygote affected with an X-linked recessive trait.

TRANSLOCATIONS INVOLVING THE DMD GENE

Recall that a translocation results from the breakage of two nonhomologous chromosomes (in this case, the X and an autosome) and subsequent rehealing by sticking the broken pieces back together incorrectly so that the broken end of one chromosome now caps the broken end of the other chromosome, and vice versa. Thus these females carry a normal X chromosome, a normal autosome, and the two rearranged chromosomes that resulted from the translocation. When the breakage events that created the translocation occur within a gene or genes, they can disrupt those genes and result in a loss-of-function mutation. RNA polymerase, the enzyme that carries out transcription, can do many neat tricks, but it can't jump between chromosomes. By splitting a gene into two parts and moving one part to a new chromosome, you have killed that gene. The RNA polymerase molecule simply has no way to leap to another site in the genome to complete transcibing this gene.

Women with balanced X-autosome translocations that disrupt the DMD gene end up affected, which at first seems counterintuitive. After all, they have a normal X chromosome in addition to the X involved in the translocation. The problem lies in the fact that all of the cells in the bodies of these females arose entirely from embryonic cells in which the normal X chromosome was inactivated (Figure 20.7). Why does that happen? It all comes down to a problem in gene dosage that kills off embryonic cells expressing the wrong number of copies of key genes.

In a normal female, half the cells will inactivate one copy of the X and half the cells will inactive the other copy. In a girl with a translocated X,

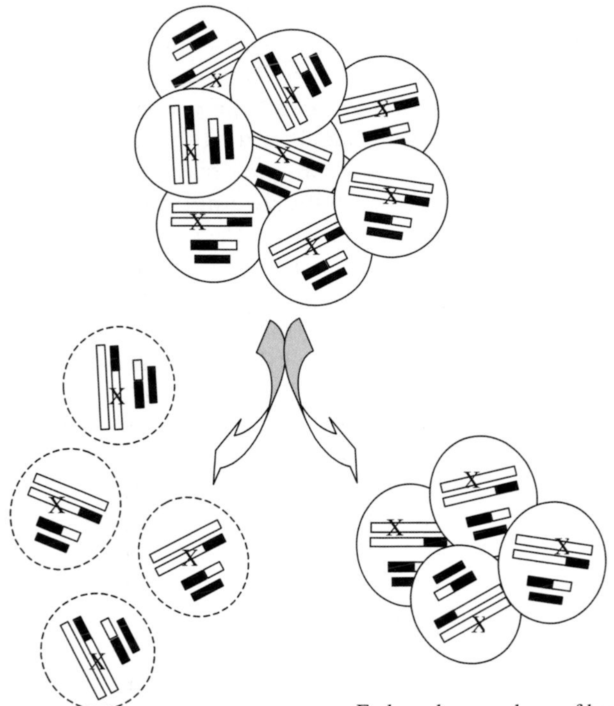

FIGURE 20.7 An embryo that starts out randomly inactivating the normal X in some cells and a translocated X in other cells ends up with all cells having the normal X inactivated because autosomal gene dosage problems kill the cells that inactivate the translocated X. If the translocation breaks the DMD gene, then each cell in this female embryo will have only one activated X and it will always have the broken copy of the gene, leading to a girl with DMD.

inactivation of the translocated X causes a problem. Even though the balanced translocation had originally left her with the right number of copies of the genes on the two copies of the autosome, inactivating the translocated X shuts off a lot of genes that would normally be transcriptionally active from their location on an autosome. So inactivating a translocated X is the equivalent of deleting one copy of the translocated autosomal region, an event that would normally be lethal when such a large chromosomal region is involved. So very early in embryogenesis, all of the cells that inactivated the translocated X die because of autosomal gene dosage effects. The embryo is then constructed from the remaining cells, all of which have the activated translocated X that can still express the attached autosomal genes. So females with a translocated X will all be born with the normal X inactivated because they have to keep the right gene dosage number for the cells to stay alive (Figure 20.7). If the translocation disrupted the DMD gene, they will have DMD because the only copy of the X chromosome from which they can transcribe the DMD gene is translocated. The lack of the DMD gene product, dystrophin, then causes DMD.

In talking about X-linked inheritance, we chose color vision and DMD to make it clear that many of the traits that turn up with sex-linked patterns of inheritance don't actually have anything to do with sexual traits. In fact, many traits encoded on the X chromosome have nothing to do with determining sex, such as hemophilia, muscular dystrophy, mental retardation, and more. But of course, the X and Y chromosomes are the fundamental determinants of whether we end up male or female. So it's not surprising to find that the X and the Y contain genes critical to the determination of whether we turn out to be male or female, a topic that we will take up again in Chapter 21.

GENETICS OF SEX, GENDER, AND ORIENTATION

21

With her full bust, small waist, broad hips, delicate facial features, and female genitalia, the Parisian fashion model fit a very classical view of femininity. She was a beautiful woman who was about to be married, a seemingly noncontroversial move for a woman at the beginning of the twentieth century. When she sought to have some "tumors" removed before the wedding and the doctors found them to be undescended testicles, they reclassified her as a male and informed her that her sexual attraction to men ("such as her fiancé?" we find ourselves musing) made her a male homosexual. This gonadal definition of sex has been only one of a variety of evolving medical views of *intersex* individuals who show some characteristics of each sex or intermediate development of external sexual anatomy. Although most individuals fall into one of two categories, clearly male or clearly female with complete consistency of genetic, gonadal, anatomic, and psychological aspects of sex within any one individual, there clearly is a complex gradient that runs from male to female occupied by many different varieties of people who do not fall neatly into one of the two standard sexual definitions. Whenever a child is born, it seems that there would be a simple answer to the question, Is the child a boy or a girl? However, for a surprising number of people in this world, the answer is unclear, or the answer may even change over the course of one's lifetime as new information comes to light or as medical views change. As we see in this tale of a woman who suddenly found herself being told that she was a man, sometimes efforts to answer the question are perplexing because the answer may be different depending on what aspect of the person you ask about. More details about the Parisian model and others with intersex phenotypes can be found in the writings of Alice Domurat Dreger, who offers many insights into the sexual complexity of people who occupy the gradient in the middle between the conventionally defined male and female. We in the field of genetics find that many of the cases that hold sociological and historical interest for Dr. Dreger also offer potential insights into the role of genes in the determination of different aspects of sex. In this chapter, we will tell you about some of the genes that determine whether we will look or feel female, including the gene that we know about today that might have led the doctors to tell this surprised woman that she was "really" a homosexual male, and we will talk about some of what is known (or mostly not known) about underlying genetic contributions to gender identity and sexual orientation.

As geneticists, we spend much of our time trying to understand how the cells of the developing organism make choices, such as whether or not to become a nerve cell or a muscle cell. One of perhaps the most fascinating processes that happen to the human embryo commits it to one pathway of sexual differentiation or the other. In this chapter, we are going to talk about the genes that control sexual differentiation, by which we mean several

TABLE 21.1 Genetic, Gonadal, Primary, and Secondary Characteristics

Sexual Features	Conventionally Defined Male	Conventionally Defined Female
Genotypic	TDF	No TDF
Karyotypic	XY	XX
Gonadal	Testes	Ovaries
Primary somatic	Penis, scrotum	Vagina, cervix, uterus, fallopian tubes, clitoris
Secondary somatic	Face and body hair, narrower hip structure, greater upper body strength, greater ability to rapidly add muscle mass	Breasts, little face and body hair, broader hip structure, less upper body strength, less ability to rapidly add muscle mass, increased body fat, menstrual cycle

different things: gonadal sex (whether you have ovaries, testes, or, in some cases, ovotestes), somatic sex (whether you have male or female body characteristics; Table 21.1), and sex role (gender) identification and sexual orientation.

If your *gonadal sex* is male, you have testes. *Somatic sex* characteristics are broken into *primary* and *secondary* characteristics. Male primary somatic sex characteristics are the penis and the scrotum. Secondary characteristics include facial and chest hair, increased body hair, pelvic build (lack of rounded hips), upper body muscular build, and the ability to generate muscle mass at a faster rate than the female.

If your gonadal sex is female, you have ovaries. Your primary sex characteristics are your vagina, uterus, fallopian tubes, clitoris, cervix, and the ability to bear children. Your secondary sex characteristics are your relative lack of body hair, thicker hair on your head (in some cases), rounded hips/figure, a decreased ability to generate muscle mass at a fast rate, decreased upper body strength, breasts, ability to nurse children, a menstrual cycle, and increased body fat composition. There are, of course, exceptions to any efforts to use a list of features to classify people into the conventionally defined sex categories. For instance, not all women succeed in beast-feeding their infants, even if they otherwise fit the conventional definition of female.

Sexual identification (*gender*) and *sexual orientation* define our sex roles and our choices in sexual partners. They are independent phenomena determined separately from whatever determines our gonadal and somatic sexual characteristics. Later in this chapter, we will talk about how sex, gender, and orientation are related to each other. Ultimately, in actual practice, sex categories usually end up being defined socially and not biologically.

The question we want to explore revolves around the degree to which each of these components of sex in human beings is genetically determined. Just how do our genes determine our sex, and to what extent do genes determine our sexual behaviors? To begin with, we will consider in detail some of the peculiar properties of the sex chromosomes in humans. Then we will address the more controversial issues of the role of genes in establishing sex roles or sexual orientation.

TABLE 21.2 Genetic, Gonadal, Primary, and Secondary Characteristics of Two X Chromosome Aneuploidies, Klinefelter's Syndrome (XXY) and Turner's Syndrome (XO)[1]

Sexual Features	XXY Males (Klinefelter's Syndrome)	XO Females (Turner's Syndrome)
Genotypic	TDF	No TDF
Karyotypic	XXY	XO
Gonadal	Testes often of reduced size after puberty —reduced levels of testosterone production and no sperm are produced[2]	Ovaries (greatly reduced in size due to loss of oocytes)
Primary somatic	Penis, scrotum	Vagina, cervix, uterus, fallopian tubes, clitoris
Secondary somatic	May show varying degrees of somatic "feminization," including breast development (gynomastia) and a more female-like pattern of hip development in a significant fraction of affected males[2]	May display reduced stature, less breast development

1. For further discussion of these syndromes, please see Chapter 22.
2. Administration of testosterone can ameliorate somatic feminization, especially if started early enough.

SEX CHROMOSOMES IN HUMAN BEINGS

As we noted in Chapter 20, the karyotype of a genetically normal human being contains twenty-three pairs of chromosomes, the twenty-two pairs called autosomes and one pair called sex chromosomes. A normal female possesses two X chromosomes, whereas a normal male possesses one X and one Y chromosome. The finding of XXY males and XO females (which we will talk about in Chapter 22) convinced geneticists that sex in humans was determined solely by the presence or absence of a Y chromosome (Table 21.2).

However, variant Y chromosomes have been found that were missing quite a bit of material but were still capable of determining maleness. All that seemed to matter in terms of being able to determine maleness was a small region on the short arm of the Y chromosome. These data demonstrated clearly that it is not simply the presence of the Y chromosome that creates a normal male but rather a small amount of genetic material now known to be a single gene located on the Y chromosome called the *testis determining factor* (TDF). The TDF gene promotes the body to develop male genitals.

TDF INITIATES MALE SEXUAL DIFFERENTIATION

Several lines of evidence argue that the TDF gene is both necessary and sufficient by itself to initiate male sexual differentiation. First, XX human beings are occasionally found in which a piece of the Y chromosome bearing the TDF gene has been appended (or translocated) onto the tip of one of the two X chromosomes, creating a chromosome that we will call X(TDF). Suppose a sperm bearing that translocated X(TDF) chromosome fertilizes an X-bearing egg?

Such XX(TDF) individuals will develop as a male but will suffer from testicular atrophy or, more simply put, small testes and thus sterility (Figure 21.1). (Why sterility? This is because it is not possible to have two X's present

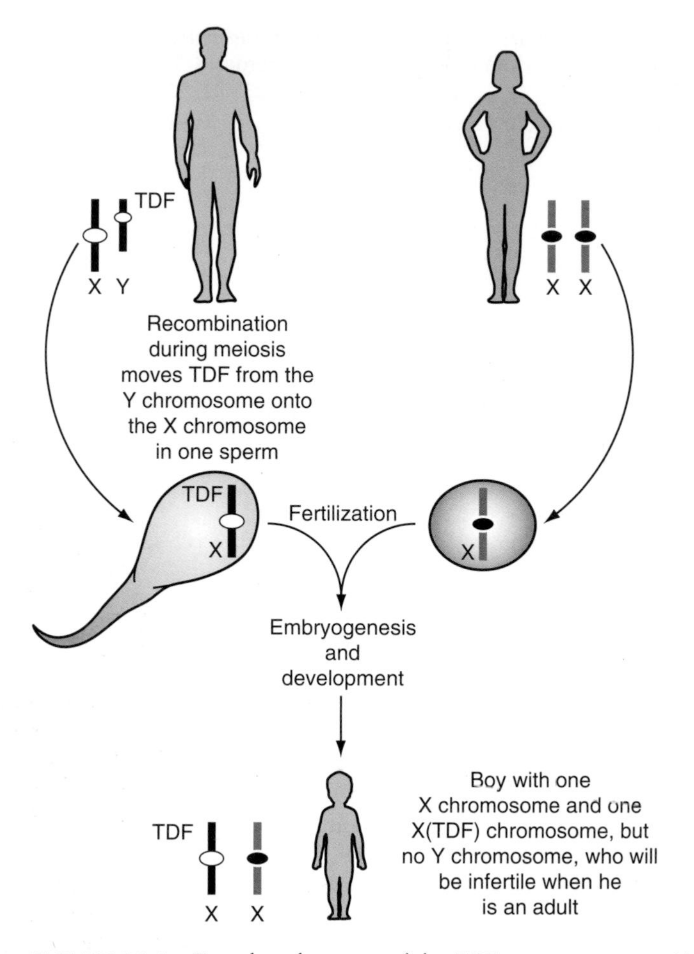

FIGURE 21.1 Transfer of a copy of the TDF gene creates an X chromosome that can make an XX individual be male.

in the male germline and still be fertile; the mere presence of another X chromosome acts almost like a poison to the germ cells and kills them during meiosis.) Hence, this individual is unable to produce healthy and happy living sperm. Nevertheless, regardless of the two XX's present, this individual is a male! He has male gonads, he has male genitals, and the rest of his primary and secondary sexual characteristics are male.

The second line of evidence that TDF causes an individual to become male comes from the finding that several XY *females* differ from normal males only by mutation of one base pair within the TDF gene (Figure 21.2). These women possess a normal or near-normal outward appearance, a cervix, a uterus, and normal vagina. However, because oocytes require two functional X chromosomes, oocyte death occurs during fetal development and, as a result, the ovaries are rather small and such women are sterile.

Finally, to prove that the TDF gene alone is responsible for male gonadal sex, researchers used some rather clever tricks of DNA manipulation to insert a mouse TDF gene, and just the TDF gene, into the genomes of XX mouse embryos. (This experiment works to answer this question because mice and

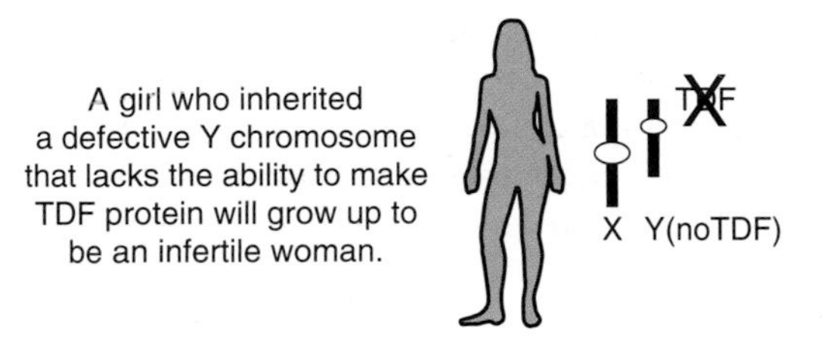

FIGURE 21.2 A defect in the TDF gene results in a girl with a male karyotype.

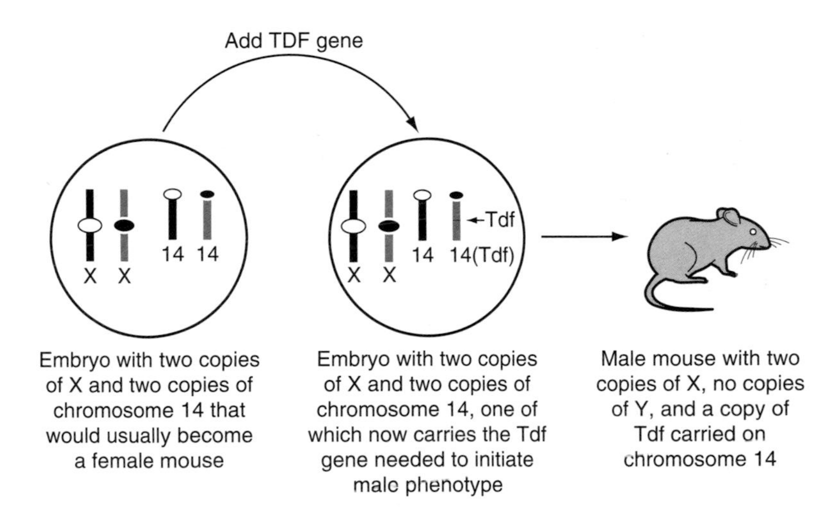

FIGURE 21.3 Addition of the TDF gene to an XX mouse embryo results in an XX with TDF associated with one of the chromosomes that does not have to be either the X or Y chromosome. Here we show it having become part of chromosome 14, but it could have joined with any of the chromosomes.

humans determine sex in exactly the same way.) These XX mouse embryos, which would have become female if they had not had a copy of the TDF gene added, developed into healthy but sterile male mice (Figure 21.3).

So the only thing that matters for gonadal sex determination is the TDF gene, but what does the TDF gene do, and how does it do it? We will explore that question next.

THE TDF GENE CAUSES THE INDIFFERENT GONADS TO DEVELOP AS TESTES

When you were first conceived, you began life with a pair of indifferent gonads. The term *indifferent gonads* is self-explanatory: the fetus' organs are literally "indifferent" to becoming either ovaries or testes. They are equally willing to become one or the other, depending on whether the fetus' germ cells do or do not carry the TDF gene. The presence of the TDF gene during the seventh to eighth week of fetal development gives the instruction to the indifferent gonad that it should develop into a male gonad. Note that the

TDF gene acts only during this brief moment in development and is inactive the rest of the time. Moreover, it acts only in a certain specific subset of the cells in the indifferent gonad. The expression of TDF in those cells is, however, sufficient to induce the indifferent gonads to become testes, which is the step that initiates all of the rest of the subsequent male development processes.

The TDF gene turns on for a brief period of time in a minor fraction of fetal cells and is then done for the rest of that individual's lifetime and not heard from again until the next generation. We are reminded of the lines of Shakespeare's *Macbeth*: "Life's but a walking shadow, a poor player that struts and frets his hour upon the stage and then is heard no more." Macbeth concludes that this moment on the stage produces a tale "signifying nothing," but in fact, TDF's brief turn on the biological stage to determine who among us will be male is truly significant.

WITHOUT TDF, THE INDIFFERENT GONAD BECOMES AN OVARY

So we arrive at a very important concept: if the TDF gene is not expressed, the cells of the indifferent gonad will follow a separate path and the indifferent gonad will develop as ovaries. Thus, although we might have imagined that the indifferent gonads would become testes if it received one signal and ovaries if it received a different signal, that is not how it works. There is a default state, the state that occurs if no signal is received, and that is proceeding along the developmental pathway to become an ovary (Figure 21.4).

GONADS DICTATE THE NEXT STEP IN DEVELOPMENT OF SOMATIC SEXUAL CHARACTERISTICS

Unlike gonadal sex, somatic sex (the sexual characteristics of the body) is independent of the presence or absence of the TDF gene and the Y chromosome. It is determined by the hormones that are produced by the developing gonads. You began life with two sets of reproductive "plumbing": the Müllerian ducts (female reproductive tract: uterus, primitive fallopian tubes, ovaries) and the Wolffian ducts (male reproductive tract: vas deferens, seminal vesicles). You also possess a small bud of tissue called a genital tubercule that will form either a penis or a clitoris. This is to say that, where the

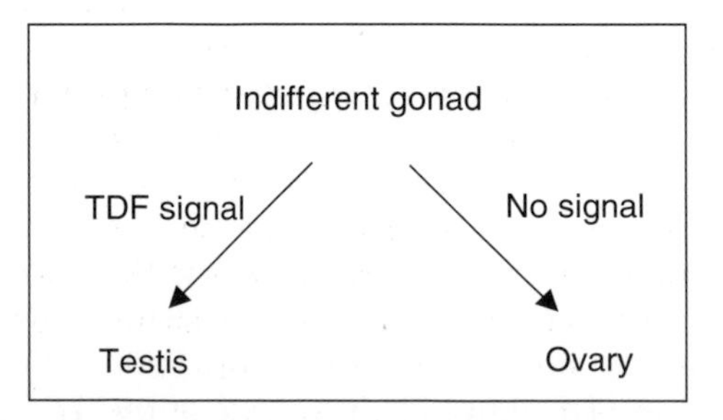

FIGURE 21.4 The default state for the indifferent gonad that receives no signal is to become an ovary. Only if the TDF signal is received will it become a testis.

indifferent gonad starts out as neither and then proceeds depending on whether or not it gets a signal, nature's first choice for the plumbing is to make both and then get rid of the one that it is not going to use!

Males possess a normal Y chromosome (or at least a TDF gene), and within the eighth week of development, the indifferent gonads became testes and began secreting androgen (testosterone) and the Müllerian inhibitory factor (MIF). The MIF causes the regression of the Müllerian ducts. In females, the indifferent gonads become ovaries and produced estrogen. During the thirteenth week of development, the Wolffian ducts degenerate and the Müllerian ducts develop. The relative levels of estrogen or testosterone also determine the development of the primary sexual sex characteristics. The high levels of testosterone produced by the testes cause the genital tubercule to develop into a penis, and a scrotum is formed. In the presence of high levels of estrogen, the same tissues will form a clitoris and a vagina. Notice that, once again, as with the gonadal differentiation, this takes place in response to a lack of signal.

Somatic sex manifestations can be altered in ways that are not the result of the infant's genes. Some developmental events are influenced by the uterine environment. For instance, if the mother has an adrenal tumor during pregnancy, her daughter might be born with masculinized genitalia.

CONGENITAL ADRENAL HYPERPLASIA AND AMBIGUOUS GENITALIA

The cells of the adrenal cortex (a part of your adrenal gland) also produce low levels of both estrogen and testosterone. Sometimes as a consequence of overactivity of the adrenal gland during development or of a defect in hormone synthesis, high levels of either estrogen or testosterone can be produced by the adrenal cortex. Congenital adrenal hyperplasia is not an intersex condition for males, but it is for females. Thus a developing female fetus could be exposed to high levels of both testosterone (from the adrenal cortex) and estrogen (from the ovaries). The result is a mixture or confusion of developmental processes, resulting in a newborn whose genitals seem to be "a little of both," or in some cases may even appear to be clearly male. These cases of ambiguous genitalia are quite disturbing to some parents and physicians, who may rapidly push to make the child's situation unambiguous at a very early age, before the child begins experiencing a wide array of sex-specific social interactions. Often surgeons seek to treat the situation as promptly as possible with plastic surgery, but some other medical specialties tend to prefer waiting before deciding about intervention. Treatment of these infants requires genetic evaluation to determine the sex chromosome composition and the presence or absence of ovaries and testis, surgical evaluation to determine the treatment most likely to produce a functional adult, and psychological evaluation and counseling of the parents.

The frequency of such congenital adrenal hyperplasia births (approximately 1 in 10,000) requires that we mention this disorder. We also mention ambiguous genitalia (which can be as frequent as 1 in 2000) because it vividly makes our point that genitals and other external features of sex are determined by hormonal messengers and not by the TDF gene on the Y chromosome. We are also well aware of the use of surgery and hormone treatment in the sexual reassignment of adult transsexual patients. Both of these cases should focus attention on the fact that the only step in primary or secondary sexual differentiation that is controlled by the TDF gene is the choice of testes or ovaries. The rest is determined through environmental events and a secondary set of steps directed by the hormones produced by testes or ovaries and other glands.

HOW HORMONES WORK

As puberty begins, these hormones will also determine the development of secondary sexual characteristics. The high levels of testosterone flowing through a male's body are responsible for his physically masculine appearance, whereas the high levels of estrogen flowing through a female's body are responsible for her physically feminine appearance. A general definition of the hormone is a chemical messenger that is produced by one cell type and released into the bloodstream and received by a target cell with the intention of altering this target cell's pattern of gene expression. The type of hormone considered here is the steroid hormone. Steroid hormones include testosterone and estrogen. Testosterone is excreted from the testes and the adrenal cortex in the male, whereas estrogen is excreted by the ovaries and the adrenal cortex in the female. Actually, both sexes produce both hormones. However, there is much more testosterone than estrogen in males and much more estrogen than testosterone in females.

When sex hormones are excreted into the bloodstream, they circulate until they encounter the target cell where they are needed to carry out their purpose, which is telling the target cell to alter its pattern of gene expression. These target cells have receptors that sit on the outside of the cell's membrane and wait for the needed hormone to float on by. When the receptors detect the presence of the hormone, they bind to the hormone and carry it through the plasma membrane of the cell to the awaiting nucleus. Once inside the nucleus, the hormone and the receptor complex bind to DNA regulatory elements and promote gene expression. The protein products of these testosterone- or estrogen-induced genes actually allow the cells and organs to execute sexual differentiation.

MUTATIONS IN THE GENE THAT ENCODES THE ANDROGEN RECEPTOR

Now imagine if a steroid hormone receptor in your body was not there or was not functional. Your hormones would continue to flow throughout your body, but when they arrived at the target cell, there would be no place on the cell surface for them to dock. If they don't dock with their receptor, the target cell cannot tell that the hormones are there and thus does not know that it needs to change which genes it is expressing and to change the levels of expression of some of the genes it is already using. In the case of sexual development, one of the key receptors is the *androgen receptor gene* (AR gene). It is encoded by a gene on the X chromosome, and loss-of-function alleles of the AR gene are referred to as *AIS mutations.* Because these mutations prevent the production of functional testosterone receptor, the phenotype of XY individuals is the result of a pattern of gene expression that has not been altered by signals from testosterone. The result is a disorder known as *androgen insensitivity syndrome* (AIS), sometimes also known as *testicular feminization* (TFM). AIS is seen in approximately 1 in 20,000 live births.

In XY embryos with an AIS mutation, the indifferent gonads receive the TDF signal and develop as testes while the Müllerian ducts regress in the presence of MIF. However, the cells of this embryo cannot sense the testosterone that is running around the body looking for androgen receptors. Instead, the somatic cells respond to the normal, low level of estrogen secreted by the adrenal cortex of both sexes, and the embryo develops along a female pathway (Figure 21.5). Consequently, the child at birth appears as a perfectly normal female. However, her vagina ends in a blind duct. The AIS female has no cervix, uterus, or fallopian tubes. Instead of fallopian tubes, there are two fully developed but undescended testes producing testosterone. These females are externally normal throughout childhood, puberty, and adult development, with the exception of a scarcity of underarm and pubic hair. Obviously, they will neither menstruate nor be able to bear children.

Given that such women are often detected as children or teenage girls, this is a serious issue in terms of how much information should be provided

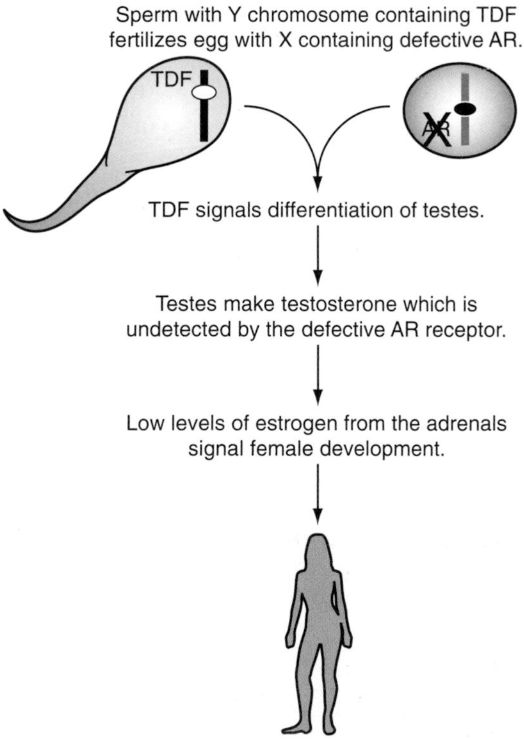

Sperm with Y chromosome containing TDF fertilizes egg with X containing defective AR.

TDF signals differentiation of testes.

Testes make testosterone which is undetected by the defective AR receptor.

Low levels of estrogen from the adrenals signal female development.

FIGURE 21.5 AIS syndrome. Individuals with a defective androgen receptor have an XY karyotype. Their gonads are undescended testes, and their sexual anatomy is female, except they have no ovaries, cervix, or uterus. The phenotype can be either complete or partial, with the latter resulting in some sexual ambiguity.

during diagnosis and counseling, how it should be provided, and who should receive the information. Some girls were not told that they had the AIS mutation, even though their doctors and parents knew. In such cases, the news can some as a shock if discovered later as an adult. However, there can also be serious health repercussions to not knowing if you have an AIS mutation, so withholding such information can be dangerous.

As noted above, the AIS female also possesses a fully developed set of testes that are located internally above where the scrotum would normally be. These testes reside inside the body, existing at a higher temperature inside the body than would normally exist for testes that have descended into the scrotum. It is recommended that such a female have her testes removed as a young adult because of an increased risk of testicular cancer that can develop later in life as a result of the elevated temperature.

AIS females are often considered quite attractive by contemporary standards, and they are often taller than the average woman. The health implications of AIS are risk of testicular cancer, infertility, gonadectomy, hormone replacement therapy, and, eventually, osteoporosis. Psychologically, they are as stable and happy (or not) as women with two X chromosomes who end up coping with fertility issues. They can be expected to live perfectly happy, normal lives and, when they so choose, become parents of adopted children or stepchildren.

Do AIS women (with XY karyotype and no functional androgen receptor) have the same characteristics as TDF-negative women (with XY karyotype and no functional TDF protein)? No. Recall that the TDF-negative woman, in the absence of the TDF signal, has produced female gonads, which provide an estrogen-dominated hormonal environment. Although they are infertile because their meiotic processes needed two copies of the X chromosome, they have a full set of female anatomy. The AIS woman, with an active TDF signal, has male gonads, only as much estrogen as the adrenal glands can supply, and no ovaries or uterus. Thus a TDF woman would be harder to identify without genetic testing, and an AIS woman has cancer risk to deal with in addition to infertility.

So what we see from this step-by-step walk through of the first several steps in sex determination in humans is the requirement for at least four elements: a Y chromosomal signal; a sensing mechanism in the indifferent gonad to respond to the Y chromosomal signal; a hormonal signal produced by the gonads (androgens or estrogens); and a set of sensors, androgen, and estrogen receptors, in the somatic tissues responding to the secondary signal coming from the gonads. In fact, there are other genes involved in sex determination that can affect a variety of the secondary steps that taking place in different cell types and tissues in different portions of the anatomy as the primary and secondary somatic characteristics emerge.

GENDER IDENTIFICATION, SEX ROLES, AND SEXUAL ORIENTATION

Genetic, gonadal, and somatic sex are consistent for most human beings. Similarly, an individual's sex is most often consistent with their *gender* (how individuals identify with male and female sex roles) and their *sexual orienta-*

tion (attraction to same, different, or both sexes); however, different combinations of sex, gender, and orientation can occur. Examples of the disconnect that can occur among these three traits can be found most noticeably among homosexuals attracted to individuals of the same sex and transgendered or transsexual individuals who grow up feeling as if they are trapped in a body of the wrong sex. The situation is further complicated by the existence of intersex individuals who have some or all of the physical characteristics of both sexes, some individuals who are bisexual (are attracted to both sexes), some individuals who self-identify with different gender roles at different times or under different circumstances, and some individuals who grow up to decide that they are a different gender than the gender they were raised as.

The fact that sex, gender, and orientation can occur in different combinations suggests that these three traits could have some different underlying determinants, whether genetic or environmental, just as some key determinants of gonadal and somatic sex are distinct. Although a number of the key elements leading to sex determination have been identified and turn out to be genetic, there is noticeably less known about the genetic components of gender or orientation, and some of what has been found is considered controversial.

MECHANISMS OF GENDER IDENTIFICATION

The controversial issue of whether gender is biological or acquired has been debated over the last century. For a long time, it was argued that the primary determinants of gender were environmental and that a child would acquire the sex roles with which they were raised. As we have already mentioned, some children are born with ambiguous genitalia or who are intersex individuals with some biological properties of each sex. Depending on the exact condition of the infant, the treatment of those children has often included "sexual correction," that is to say, surgical revision of the child's sexual anatomy, sometimes to recreate their anatomy to more closely resemble the anatomy usually expected for their genetic and gonadal sex, but sometimes instead to arrive at an external sexual anatomy that is different from their chromosomal sex. Part of the medical argument that such surgeries should be done, and done early, arises from reports that reassignment works well. Reports suggest that, where the parents are comfortable with the outcome of early revisions, the children will usually identify properly with the genders they have been assigned.

However, things may not be that simple. Two different schools of thought have developed—that a child will take on the sex roles and gender identity with which he or she was raised, or that there are biologically inherent determinants of sex roles and gender identity that cannot be reprogrammed by raising a child as if he or she were the opposite sex (Box 21.1). Perhaps the truth lies somewhere in between. If we look to other types of studies in the scientific literature, we find some evidence of both genetic and environmental components of gender identification. Twin studies of gender identity suggest that there is a strong genetic component to gender identity, but that

BOX 21.1 THE GIRL WHO WAS REALLY A BOY

A dramatic case that suggests a biological basis for gender is that of a male child who was "reassigned" as a female after irreversible genital damage during circumcision. During childhood, this child with a Y chromosome who dressed in dresses and had a collection of "girl" toys such as dolls was advanced as evidence that surgical reassignment of sex would result in the child's successful acceptance of his newly assigned gender role. Although reports in the literature repeatedly presented the view of a normally adjusted little girl, his real patterns of play as a child showed evidence of a taste for the toys and activities of the boys around him. In fact, this was a child struggling with a gender identity that did not fit. As a teenager, when he was finally told his medical history, he rejected the female identity that had been assigned to him and reembraced a male role in life. He took a male name and chose to live as a man. Because of cases like his, a number of workers believe that gender identification is biologically inborn and cannot simply be assigned (or reassigned) based on how the child is raised or what their external genitalia look like.

genetics cannot account for all of the determinants of gender. When animals of opposite sex develop together in the same womb, siblings may acquire sex-specific behaviors of the opposite sex, something possibly explained by exposure of one embryo to hormones being produced by another embryo sharing the same uterine environment; however, one study in human fraternal twins suggests that this might not be the case for human development.

Thus, although much on the subject remains confused, the overall picture we find is one of both genetic and environmental effects on gender identity. While the lack of a simple answer complicates efforts to make decisions about sex reassignment surgeries or to understand the processes that produce transgendered and transsexual individuals, it is perhaps not surprising if the real answer on such a complex subject is not a simple answer. Overall, it is rather surprising how little is known about biological determinants of gender in humans.

SEXUAL ORIENTATION

We will now turn our attention from the development of sexual or gender identities to the development of sexual orientation, another topic where not nearly enough is known about the real underlying determinants. We can only apologize in advance if our treatment of this topic (or anything else in this chapter) in any way fails to be adequately sensitive to the broad array of perspectives on such controversial topics.

To quote two major workers in this area, "Most men are sexually attracted to women, most women to men. To many people, this seems only the natural order of things, the appropriate manifestation of biological instinct, reinforced by education, religion, and law. Yet a significant minority of men and women, estimates range from 1 to 5%, are attracted exclusively to members of their own sex."[1] This statement raises some fascinating questions. First, just how is sexual attraction or orientation determined? Is it biological? Are there genes that direct males to be attracted to females and vice versa? Second, if sexual orientation is biologically programmed, how are we to understand the etiology of cases in which men choose men as lovers or women choose women? Could such people reflect genetic variation in genes for sexual orientation? If such genes and such variation do exist, what are those genes and what do they do? These questions will be our focus for the remainder of this chapter.

THE GENETICS OF SEXUAL ORIENTATION: POPULATION STUDIES

Sexual orientation is defined by the sex to which a given individual is sexually attracted. When, as is usually the case, a person is attracted to an individual of the opposite biological sex, that individual is referred to as *heterosexual.* In the case in which people are attracted to others of the same sex, they are referred to as *homosexual.* In some cases, in which an individual is attracted to both sexes, the term *bisexual* is used. Terms used in popular culture seem to keep changing, but in recent years, common parlance in the United States often refers to homosexuals as "gay" men and "lesbian" women.

Before the early 1990s, there were two lines of evidence to suggest that male homosexuality might be genetic. The first line of evidence came from studies of heritability, the measure of how often the trait is concordant or discordant in identical twins vs. fraternal twins. In the case of homosexuality, such heritability estimates are suggestive of an important role of genes in determining the phenotype (Box 21.2). For both gay males and lesbians, their homosexual orientation is found in more than half of their identical twins, compared to one sixth (lesbians) or one quarter (gay males) of their fraternal twins, and about one eighth of their non-twin siblings. Genetically identical individuals are more likely to be concordant than genetically different individuals if some aspect of the trait is genetic, so the fact that the identical twins show a much higher concordance for being gay or lesbian suggests a substantial genetic contribution to the trait. Also notice that brothers of gays tend to also be gay more frequently than expected than the one- to five-percent rate estimated for the American male population, another piece of information that helps support the view that there are genetic factors contributing to gay or lesbian phenotype.

However, these data also suggest that the determination of sexual orientation cannot be wholly genetic. If the gay or lesbian phenotypes were completely genetic, we might expect the concordance of identical twins to be 1.0, as it is for traits such as color blindness or cystic fibrosis. Clearly, genotype alone cannot account for those fifty percent of cases in which the twins were discordant.

[1] Levay S and Hamer DH. Scientific American 1994; 270: 44–49.

BOX 21.2 SOME LIMITATIONS OF HERITABILITY STUDIES

There are real limitations to what you can tell from measures of heritability. Any estimate of heritability is only good for that one particular population at the time that estimate is made and might or might not offer insights into other populations. Some of this is due to differences in the genetic composition of different populations. However, some of it is due to differences in exposure to environmental factors influencing manifestation of the trait. In addition, the accuracy of this method is limited by the accuracy with which the researcher can validly and accurately score people for the trait in question. If you want to compare levels of protein in urine, it may be possible to make simple quantitative assessments of whether the values are the same in two individuals. If you want to know whether or not two individuals are concordant for a trait that you cannot directly measure, the amount of nonconcordance in the test will be directly related to the chance that self-report of the trait is inaccurate for any one individual being questioned, either because the answer given is untrue or because the individual does not know the correct answer. Another factor that can confound heritability studies is something called age-related penetrance—the tendency for the same trait to develop at different ages in different individuals—which can make it hard to tell whether lack of concordance indicates that twins don't share the trait or whether it means that one of them simply has not yet developed a trait that will appear later in their development. In the case of homosexuality, the social environment could vary for different individuals in ways that not only influence the willingness of the study subject to self-identify as a homosexual, but that also influence the age at which individuals admit to themselves that they are homosexual. So studies of heritability in homosexual populations may well be confounded by a variety of factors—differential environmental influences on the development of the trait, differences in accuracy of self-report, differences in age at which the individual realizes they have the trait, and differences in study participation rates for some individuals depending on their attitudes towards their status and towards surrounding social reactions to their status. You might expect to get a much better assessment of heritability of homosexuality if you were trying to study it in a society that fully accepts it than in a society that is critical of it or seeks to suppress it.

How are we to explain these data? One explanation is that there may be genotypes that predispose individuals toward one's orientation to others (sexual orientation genes, if you will) but that these genotypes interact with the environment. These environmental influences may include obvious things such as family values, peer pressure, societal responses, personal relationships and specific sexual experiences or religious influences, but environmental effects could also be nonsocietal and could *theoretically* include things such as medical events or nutrition. The allelic differences, if indeed they exist at all, appear to not be fully penetrant, suggesting that they predispose rather than dictating a specific outcome.

An alternative explanation could be that the genetic components of homosexuality are even larger than they appear to be, with apparent cases of discordance representing underreporting of gay or lesbian status. Even in cases in which social pressures on the situation are not in evidence, self-reporting of medical status can often be inaccurate when self-reports are compared to medical records, so how much more of a problem could this be if there are societal or personal pressures against self-identifying as gay or lesbian?

THE GENETICS OF SEXUAL ORIENTATION: FAMILY STUDIES

A different form of support for a genetic basis for male homosexuality comes from studies of families. There are many pedigrees in which male homosexuality appears to segregate in a predictable and sex-linked fashion through a given kindred. The pedigree shown in Figure 21.6 is an example. Face it, if you didn't know the phenotype under consideration, you would have glanced at the pedigree, thought "sex-linked recessive inheritance", and moved on. It was pedigrees such as this that caused Dean Hamer and colleagues at the National Institutes of Health (NIH) to begin a careful study of the genetics of male homosexuality in the early 1990s.

The initial subjects in Hamer's study were seventy-six self-identified gay men and their relatives over the age of eighteen, as well as thirty-eight pairs of homosexual brothers and their relatives. The researchers recruited through an AIDS clinic, through local gay organizations in Washington, DC, and through advertisements in gay-oriented magazines and newsletters. Before we go any further, we need to think about this study group. This population consists of gay men open enough about their sexuality to agree to both participate in this study and involve their extended families. In other words, all of these men were fully "out," and functioning in the context of families that did not reject participation in such a study. (We are left with questions here about how studies of this population related to homosexuals who are not open about their status.)

After evaluating the initial group, assessing the phenotype among relatives can actually be complicated. Hamer and colleagues used two methods to ascertain the phenotype: self-assessment and a set of psychological tests known as the *Kinsey scales* (Figure 21.7).

Amazingly, both the self-assessment and the assessment by the original members of the study group (the probands) were remarkably concordant. To

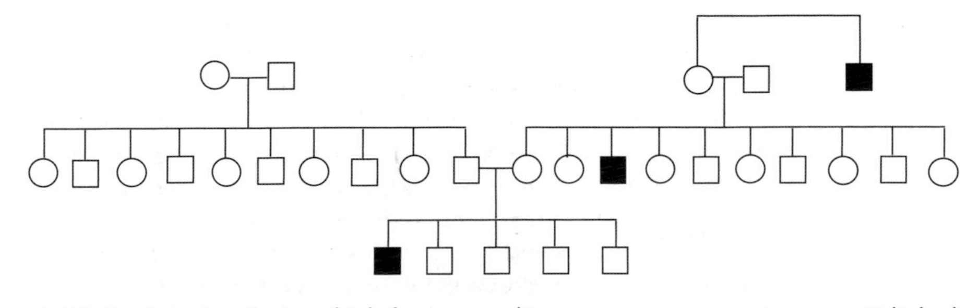

FIGURE 21.6 Family in which homosexuality appears to segregate as an X-linked recessive trait. (Redrawn from Hamer et al., *Science* 1993; 261:321–327.)

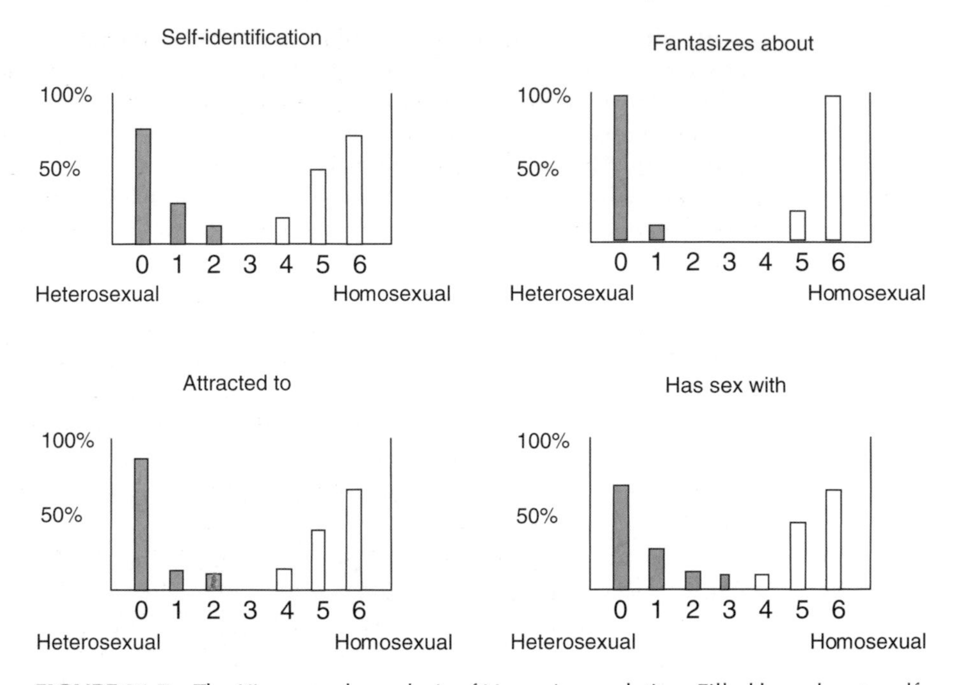

FIGURE 21.7 The Kinsey scale analysis of Hamer's population. Filled bars denote self-identified homosexual men, while open bars denote self-identifed heterosexual men. (Adapted from Hamer et al., *Science* 1993; 261:321–327.)

quote the Hamer paper, "All (69/69) of the relatives identified as definitely homosexual verified the initial assessment, as did most (27/30) of the relatives considered to be heterosexual; the only possible discrepancies were one individual who considered himself to be asexual and two subjects who declined to answer all the interview questions." Thus, again quoting Hamer, "describing individuals as either homosexual or heterosexual, while undoubtedly over-simplistic, appears to represent a reliable categorization of the population under study."

What Hamer is saying is that in this study population of these men and their relatives, homosexuality or heterosexuality can be considered as a discrete pair of traits, such that each individual can be reliably classified as one or the other. Hamer and co-workers are backed up in this assertion by their data from the use of the Kinsey scales mentioned. Using these scales, people rate themselves on four aspects of their sexuality: self-identification, attraction, fantasy, and behavior. The ratings range from 0 for exclusively heterosexual to 6 for exclusively homosexual. Thus a man who has never had even a fleeting attraction to another man would rate himself a 0 on the attraction scale, whereas a man only attracted to other men would give himself a rating of 6. As shown in Figure 21.7, the graphs are bimodal and basically not overlapping for each of these four characteristics.

The graphs in Figure 21.7 should worry you because they suggest a discrete bimodality in human sexual orientation that is not consistent with many peoples' experience. Do Hamer's data really suggest that bisexuals do not exist? No, Hamer's data say only that such people don't exist in his study group, which is a highly selected and precisely defined study group, so these graphs represent only his study group and not any other populations.

Hamer's analysis yielded some fascinating conclusions. They confirmed previous studies by noting that a brother of a gay man had a fourteen percent chance of being gay as compared with a one to five percent chance for males in the general population. Hamer and colleagues also noticed something even more interesting among more distant relatives of these gay men. Maternal uncles and sons of maternal aunts had a higher chance of being gay (seven to eight percent) than expected for the general population, but no such effect was observed for paternal uncles or sons of paternal aunts. This is highly suggestive of an X-linked determinant for sexual orientation. Remember, paternal uncles or cousins on the father's side cannot share an X with the homosexual male in question, but maternal uncles or cousins can. (See the schematic pedigree diagram in Figure 21.6.) Thus Hamer only saw a high frequency of concordance among relatives who could share an X chromosome, evidence again for sex linkage of a gene or genes.

To study this effect further, Hamer's group further refined their study group to thirty-eight families in which at least two sons were gay. They excluded any family in which the father was gay (ruling out any case of father-to-son transmission), or with more than one lesbian relative. It was hoped that, by excluding other causes of homosexuality, this population might be enriched for the X-linked form of homosexuality they were seeking to study. It sort of worked. Maternal uncles and sons of maternal aunts had a higher chance of being gay (ten to thirteen percent), and again no such effect was observed in paternal uncles or sons of paternal aunts. However, the ratios were still lower than those expected for a simple Mendelian trait (fifty percent for a maternal uncle and twenty-five percent for a son of a maternal aunt).

To sort this out, Hamer and colleagues fell back on this concept: *If a gene exists, you can map it to a precise position on a chromosome; and, if you can map a gene, then it exists.* They now focused on only forty pairs of brothers. Realize how important this was. No matter what other environmental conditions need to be met, or other genes need to be present to develop homosexuality, they must all be there in these males. If there really is an important gene on the X chromosome that determines homosexuality, and if the mother was heterozygous for that gene, then these brothers should share a specific region of one of the mother's X chromosomes, the region bearing the allele predisposing them to homosexuality.

Consider thinking about it this way: If a woman is heterozygous for the color-blindness allele (cb), and two of her sons are affected, it is because both inherited the cb allele from her. Because recombination is frequent on the human X, approximately five exchanges per bivalent on the long arm of the X alone, one doesn't expect the brothers to share the same alleles for all genes at other sites, but they should share the cb allele and other closely linked alleles as well.

Hamer and colleagues analyzed the inheritance of these pairs of brothers by studying a large number of genetic markers distributed at various points along the length of the X chromosome. By looking at markers that have two different alleles in the mother, it is then possible to ask whether both brothers received the same allele, or whether one brother received the first allele and the second brother received the other allele. If there were only one gene for male homosexuality on the X chromosome and it was completely penetrant, we would expect to find every pair of gay brothers to carry the same allele of a marker located next to the "gay" gene. Because of recombination events along the X chromosome, we expect that they will not be identical for all of the markers on the X. The closer a marker is to the "gay" gene, the more often the brothers will share it. Markers that are farther away will be shared less often, and markers that are a long way away will seem to be randomly assorted in the brothers. If there were not "gay" genes on the X chromosome, we would expect to find that each marker on the X would present the same allele about fifty percent of the time, and a different allele about fifty percent of the time.

So what did Hamer find? For most of the X chromosome, the brothers were as likely to have two different alleles as they were to both share one of the two given maternal alleles. However, for one region, Xq28, near the tip of the long arm of the X, the two homosexual brothers shared the same alleles in thirty-three out of forty cases. This finding is very highly significant and provides strong evidence for an important gene in this region. Although this is a very strong result, you do need to note that there were seven pairs of brothers who carried different alleles of the Xq28 region. Thus, even in this highly refined population, the Xq28 region cannot account for all cases of homosexuality. Nonetheless, the basic result is still indicative of some *correlation* between genotype at Xq28 and sexual orientation phenotype in a large fraction of these sibling pairs. In a more recent study, Hamer and colleagues have repeated this mapping and extended their studies to include heterosexual brothers of the two gay brothers initially studied. Not surprisingly, these heterosexual brothers carried the alleles in the Xq28 that were shared by their gay brothers much less often (twenty-two percent) than would be expected by chance (fifty percent).

We should point out that Hamer's data also suggest that whatever genes might be in Xq28 that affect sexual orientation in men, there is no evidence for that gene or any other genes affecting sexual orientation in women. Although the heritability of lesbianism is as high as it is for male homosexuality, very little is known about a genetic basis for lesbianism, or indeed if one exists at all.

THE FINDINGS ARE STILL CONSIDERED CONTROVERSIAL

So Hamer's data *suggest,* and only suggest, that there may be a gene, or genes, in region Xq28 of the X chromosome that affects sexual orientation. However, not everyone in the scientific community agrees with that suggestion. Many workers worry about the small sample size of the study group. Other workers are trying to repeat the results using different populations. The final verdict is anything but "in." Finally, even if Hamer and his group are correct and some region of the X chromosome is determining sexual orientation in their population, it is not at all clear how generalizable this result is to the general population. Even if Hamer is wrong about this particular gene, this seems like an approach that could eventually answer these questions.

WHAT IF THE RESULT MEANS SOMETHING ELSE?

If Hamer is wrong, does that mean there was something wrong with his study or that he has produced what we would call a statistical artifact? Is there some way his study could be right and yet not mean what he think it means? Consider this: they mapped something to Xq28 that is held in common between the brothers in the study. They were selected because they had their sexual orientation in common. What else might they have in common? It is interesting to note that another similar study conducted in Toronto, in a different gay population, did not find the same thing. One thing that comes to mind is that gay brothers who are very "out" about their orientation, who are very public and outspoken about their status, might not only share their sexual orientation but perhaps also other personality characteristics, such as assertiveness or self-confidence or rebelliousness. Recruitment among a group of gay men who are circumspect about their status, maintain a low profile, or even hide their status might identify a group of men who share sexual orientation with the first population but do not share certain other personality characteristics with them. This sort of thing is a big risk in studies of this kind—that you select a set of study subjects on the basis of sharing a particular trait while not realizing that they also share other things that might actually be the basis for your findings. We do not know that this happened in the Hamer study and if there are differences between the two study populations that are responsible for the difference in findings, we do not know that those differences have anything to do with factors we have suggested. There might be a "gay" gene on Xq28, and the failure of a second study to reproduce the finding might really represent some other methodological difference between the studies, or perhaps a difference in the genetic backgrounds of the two populations. There might not be a "gay" gene on Xq28. It remains to be seen.

Beyond those caveats, suppose Hamer and friends are right. Just what kind of things might such a sexual orientation gene specify? How might it work? In studies considered even *more* controversial than the genetics just described, Simon LeVay has presented evidence for a structural difference between a small region in the brains of gay and heterosexual men thought to be involved in controlling sexual orientation. Might genes play a role in the formation of such structural differences? As interesting as these results are, at this time the scientific community is far from fully persuaded on this matter.

And so we close both this section and our formal discussion of sexual differentiation in humans. There are clearly proven roles for specific genes in determination of gonadal and somatic sex. Much less is known about other aspects of sex, but evidence suggesting that genetic factors contribute to gender identity and sexual orientation also suggest that it may not all be genetic. As we go on to talk about other aspects of human genetics, we hope you will keep in mind some of the lessons of this chapter: One line of evidence, such as family studies, can help validate findings from other types of studies, such as twin studies or population-based studies. There can be both genetic and environmental effects on a trait, and some of those environmental traits can be societal rather than the usual environmental effects we think of, such as diet or exposure to toxic substances. You might get different answers depending on what population you look at or how you assay for the existence of the trait. A single trait may actually be made up of multiple independent traits, with different underlying causes of separate aspects of things that we think of as a being a single trait.

REPRISE

And so we return to the Parisian model (Figure 21.8). It has been hypothesized that she was a case of androgen insensitivity with undescended testes in

FIGURE 21.8 This photo shows a Parisian fashion model, who was hypothesized to have represented a case of androgen insensitivity because she had normal external female sexual anatomy but had undescended testes. More information about this fashion model and the ways in which the medical establishment handled her case can be found in Alice Domurat Dreger's *Hermaphrodites and the Medical Invention of Sex* (Cambridge, Mass: Harvard University Press, 1998).[2] (Photo courtesy of Alice Domurat Dreger.)

[2] Additional information on related topics can be obtained by viewing the video "Is It A Boy or Is It a Girl?" or by visiting the web site of the Intersex Society of North America (www.isna.org).

a body that is externally female but lacks the reproductive machinery present in most females (although we cannot know specifically that she had an AIS mutation, since no tissue samples were available to test). In 1909, when the doctors said, "you are male," medicine seemed ill equipped to cope with someone who did not fit neatly into one of the two sexual niches, male and female. As we gain further understanding of the underlying genetics and biology of sexual development, we expect this to be one of the topics that will push policy makers and society to arrive at reasonable reactions to situations that do not fit our preconceptions.

Consider how someone would feel about finding out that they had the karyotype and gonads of a male even though they seemed to be anatomically female. How would you feel if you found out that your karyotype or gonads were not those expected for the sex that you appear to be? If you discovered that the person you were married to had the same set of sex chromosomes you have, even though their anatomy is that of the opposite sex, what would you think of laws declaring your union homosexual and your marriage invalid under the law? Would it change your perceptions of yourself and how you fit into the world? What would you think of the Parisian model's situation if she had simply said, "that's ridiculous," and went ahead with marrying her fiancé? Alternatively, how would you feel if the model had compliantly responded to the doctors by taking up male attire and going in search of a woman to marry? Consider why you have the reactions you have and what your justifications are, and think about how you would answer the same questions if you knew you were considering an individual with a mutation in a gene controlling gender or sexual orientation, instead of somatic sex. We were quite interested in knowing how the Parisian model dealt with the dilemma that faced her, but Dr. Dreger tells us that historical records do not indicate whether the planned marriage took place.

Clearly, much of our society feels a strong urge to fit people into known classifications, to be able to react to them as a man or a woman and not as someone somewhere in between. Much of our society seems to want people to be congruent, to have different aspects of their sexual being all match according to conventional definitions of male and female, even though, as we now see, different aspects of the sexual phenotype are the very separable result of different genes and biological processes. However, wanting the world to be neat and tidy does not make it so. We have argued before that diversity is one of the greatest gifts ever granted humanity, and that applies not only to issues of race or culture but also to sex. Sexual diversity offers us lessons that can grant us increased understanding of ourselves and our sexuality if we can learn what that sexuality consists of and realize that some of the things we feel are carved in stone are actually variables with real biological underpinnings.

22 ANEUPLOIDY: WHEN TOO MUCH OR TOO LITTLE COUNTS

His name was Earl, and when Scott met him in high school, they were both freshmen in the same physical education class. Earl had Down syndrome, a disorder caused by an imprecise segregation of chromosomes into the egg from which he arose. Like many kids with Down syndrome, Earl was mentally retarded and had been that way since birth. His intellect had stopped somewhere around that of a five year old, but his body never got the message. Because of his limitations and because of the facial features that are characteristic of Down syndrome, Earl became the butt of an awful lot of high school humor. Kids couldn't resist making fun of the way he walked, ran, or talked. For four years, Scott spent one hour a day in class with Earl. For various reasons, they became friends. Earl never did figure out why people made fun of him, but he knew that they did. Once one of the high school sports heroes tripped him in the hallway during break. The humiliation he felt seemed to hurt worse than the bloody lip. During Scott's junior year of high school, the March of Dimes held a public lecture on the basis of birth defects. For reasons that have long faded into a mist of high school memories, Scott made his father drive him to that lecture at a nearby college. There he learned for the first time about genes and chromosomes, but mostly about Earl and about himself. Scott developed a passion for understanding how heredity works and how our genes make us what we are. This book, especially this chapter, is a child of that obsession.

As we have talked about meiosis, we have emphasized the importance of getting the right numbers of chromosomes into the sperm and eggs that will be used to produce the next generation of human beings. Each sperm and each egg must end up with twenty-three chromosomes (and they have to be the right twenty-three chromosomes) if the resulting child is to have exactly the right number of copies of each of the genes in the human genome. Unfortunately, the cell machinery does not always succeed in its goal of getting all of the right chromosomes to where they are supposed to be by the end of meiosis, and missing or extra copies of chromosomes can mean illness or even death to a zygote produced by a sperm or egg with the wrong number of chromosomes.

The failure of two homologous chromosomes to *segregate* (separate from each other into the daughter cells after the first meiotic division) properly is called *nondisjunction*, and it can result from defects in any of several different kinds of structures and functions in the cell (Figure 22.1). Nondisjunction can occur either because two homologues failed to pair and/or stay together as they move to the metaphase plate at meiosis I, or because of a failure of the cell to properly move the segregating chromosomes along the meiotic

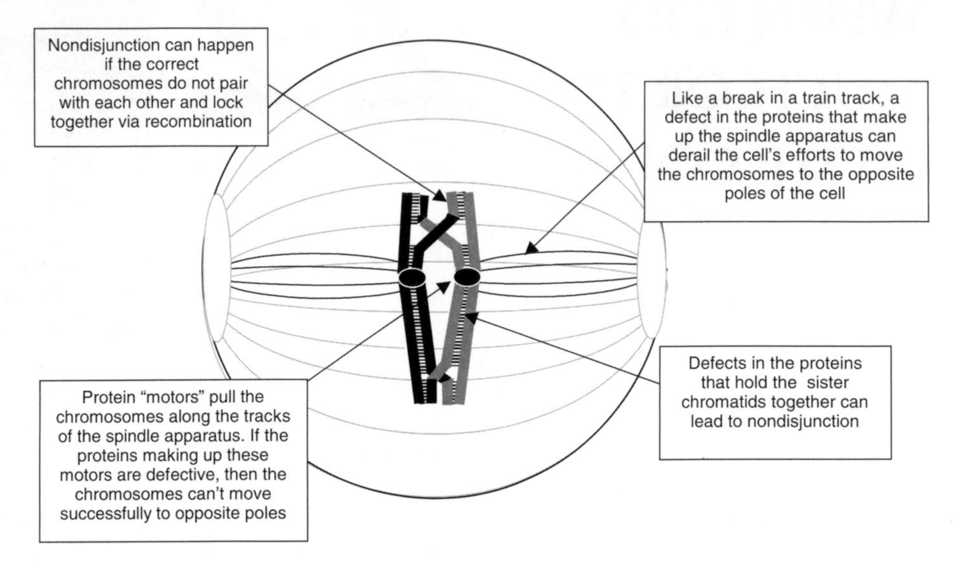

FIGURE 22.1 Nondisjunction can result when necessary steps in meiosis fail, which sometimes may be due to defects in proteins that carry out those functions or make up the structures used to carry out the functions.

spindle (those tracklike structures made of microtubules on which the chromosomes pull themselves to the poles) after the pairing and recombination steps. Indeed, considerable evidence exists that much of human nondisjunction may be due to failures of the processes that move chromosomes to opposite poles of the dividing cell at meiosis I. These failures or errors can include defects in the proteins that hold the sister chromatids tightly together during the first meiotic division, in the protein motors on the chromosomes that move them to opposite poles of the cell, or in the structural integrity of the spindle apparatus along which the chromosomes move.

Regardless of which structures and functions go wrong during meiosis, the result of nondisjunction is that the resulting sperm or eggs turn out to be *aneuploid* (i.e., having the wrong number of chromosomes). When an aneuploid sperm or egg is involved in a fertilization event, the resulting zygote is also aneuploid, and human biology is remarkably intolerant of aneuploidy. This is especially true for *monosomies*, zygotes with only a single copy of a given chromosome. With the exception of the X or Y chromosome (which we talk about more below), monosomy is simply not compatible with life and leads to early spontaneous miscarriage. We are not aware of a baby ever being delivered alive that carried one or three entire copies of large chromosomes (trisomes) such as chromosome 1, chromosome 2, or chromosome 3. It's not that such zygotes don't arise; they do and then are lost, sometimes so early that the mother may not even be sure of whether or not she was pregnant. With the exception of four cases described below, most human trisomies do not survive long after conception.

More dramatic cases of aneuploidy, such as full triploidy (three copies of every chromosome per cell), occur as well. Again, these are not compatible with early fetal development. Since meiosis in humans is really pretty sloppy, these errors are fairly common among human conceptions, and autosomal nondisjunction occurs at a reasonably high frequency (perhaps as many as forty percent of human conceptions are aneuploid). As such, aneuploidy has to be considered perhaps the most common cause of death in human beings.

The lethality of most types of aneuploid conceptions shows just how critical proper gene copy number is for correct development of a complex organism. While an organism might tolerate changing the dosage of genes encoding some kinds of enzymes, there may be serious deleterious effects from changing the dosage of genes whose products regulate the expression of other genes, carry out cell-to-cell communication, or serve as a structural component of a complex protein structure. Although even those changes might be tolerable in some cases if only a single gene were affected, realize that each human chromosome carries hundreds or thousands of genes. The additive effect of increasing the dosage of many genes is usually death.

However, a few types of trisomic zygotes are capable of survival, at least sometimes. These are *trisomy 21* (Down syndrome), *trisomy 18* (Edward syndrome), *trisomy 13* (Patau syndrome), and trisomy of the sex chromosomes including XXX, XXY and XYY. At least one factor that may make extra copies of chromosomes 21, 13, and 18 compatible with survival is that they carry a relatively smaller number of genes than do the larger chromosomes in the human complement. There are some very different factors that allow survival with an incorrect number of X chromosomes. We begin with a discussion of Down syndrome.

DOWN SYNDROME, OR TRISOMY FOR CHROMOSOME 21

Down syndrome, or trisomy 21, is the state of having three copies of chromosome 21 instead of two. It is perhaps the best known genetic defect, partly because it is the single most common cause of mental retardation among individuals outside of institutions, and partly because of the very distinctive characteristic appearance, including slanting, or epicanthic, eyes and small, frequently low-set noses (Box 22.1). Babies with Down syndrome grow slowly and have poor muscle tone. They have rather short fingers and short, broad hands. They have a wide skull that is somewhat flatter than usual at the back, and the irises of the eyes often have obvious spots. In many cases, the mouth appears to remain partially open due to a protruding tongue.

Perhaps the most commonly known aspect of Down syndrome is mental retardation. Intelligence quotients (IQs) normally ranges from 25 to 50 (compared to an average IQ of 100 in individuals who do not have Down syndrome); however, some children do show higher levels of mental function, with some individuals with Down syndrome having near-normal IQs and the ability to read and write at high school or college levels. There is serious controversy, and some increasing degree of optimism, regarding just how much children with Down syndrome can be expected to achieve. Clearly, some children with Down syndrome greatly exceed our expectations and grow up to be happy and reasonably self-reliant adults, but many are severely

BOX 22.1 PEOPLE WITH THE SAME DISORDER CAN BE QUITE DIFFERENT

Not all Down Syndrome individuals will have all of the characteristics listed in this chapter. Whether you are considering the information in this chapter or in the rest of the book, whatever descriptors we use for Klinefelter Syndrome, Turner Syndrome, Down Syndrome, or any other human disorder, not all features apply to *all* the people affected by that disorder. People are unique: these disorders can manifest themselves quite differently from one person to the next. Keep in mind that, even if a discussion of someone with Down syndrome focuses on chromosome 21, that person's overall characteristics are affected by differences on all of the other chromosomes, too. Some individuals with Down syndrome are born with heart defects, some are not. Many suffer from substantial cognitive deficits, sometimes substantial enough to warrant institutional care, but in contrast, we have heard about at least one specific young woman with Down syndrome who is attending college. When a disorder is discussed, we try to give you a general description of the common features of that disorder, things that are found much more commonly in the disorder than in the general population. We know we are making generalizations. We know there will be exceptions, but it is the best we can do. We hope that you will carry this caution about variability away with you along with whatever generalizations you encounter here.

limited. Growing evidence shows that certain types of early educational intervention, especially computer-assisted teaching, may be of real help to children with Down syndrome. Moreover, in these times, many adults with Down syndrome may be expected to live either semi-independently or independently and often are able to enter the work force. In some cases, such individuals seem to do better in so-called "sheltered workshops," but other individuals are able to find work in various aspects of the public and private sector.

Half of the children born with Down syndrome are born with severe heart malformations. These and other life-threatening conditions are so severe that some of these children die before age five. However, for those children who survive the fifth year of life, the average life expectancy is fifty years. Even so, these individuals are at high risk for leukemia and for a degenerative brain disorder similar to Alzheimer disease. Men with Down syndrome are usually sterile, but the women are fertile; from the few scattered reports available, it appears that half of their children are born with Down syndrome. On one hand, this result makes good sense—half of the eggs produced by such a woman should carry two copies of chromosome 21. However, given that some eighty percent of Down syndrome fetuses spontaneously miscarry, we have to wonder why the final result should be a 1:1 ratio. Although we can imagine models for how this might happen, at this time it is still one of many mysteries about this complex phenomenon.

MOST CASES OF DOWN SYNDROME ARE DUE TO NONDISJUNCTION IN THE MOTHER

Most often, a baby with Down syndrome is found to have three copies of chromosome 21 if chromosomes in their cells are examined via traditional karyotyping to classify the chromosomes present by size, centromere position, and banding pattern (Figure 22.2). Most cases of trisomy 21 are due to nondisjunction at the first meiotic division in the child's mother. We know this because we have developed several methods for determining which chromosome came from which parent, one of which makes use of subtle differences in banding patterns for chromosome 21 that can sometimes be seen with some staining techniques used in karyotyping.

Let's start with chromosomes marked with banding differences that let us track all four copies of chromosome 21 separately, two maternal copies and two paternal copies (Figure 22.3). If we can visibly distinguish the four chromosomes, we can track where and how the nondisjunction took place. In fact, we have other nonmicroscopic techniques in our genetic bag of tricks to distinguish maternal and paternal copies of chromosomes being passed along to the next generation, but this use of visible differences in the chromosomes is the technology that makes it easy for us to show you how conclusions can be drawn about where the duplicated chromosome came from.

Clearly, the consequences of nondisjunction at meiosis I in the mother are quite different than the consequences of nondisjunction at meiosis II. In the case of meiosis I nondisjunction, both maternal polymorphisms are present in the cells of the child with trisomy (Figure 22.4). If the problem had

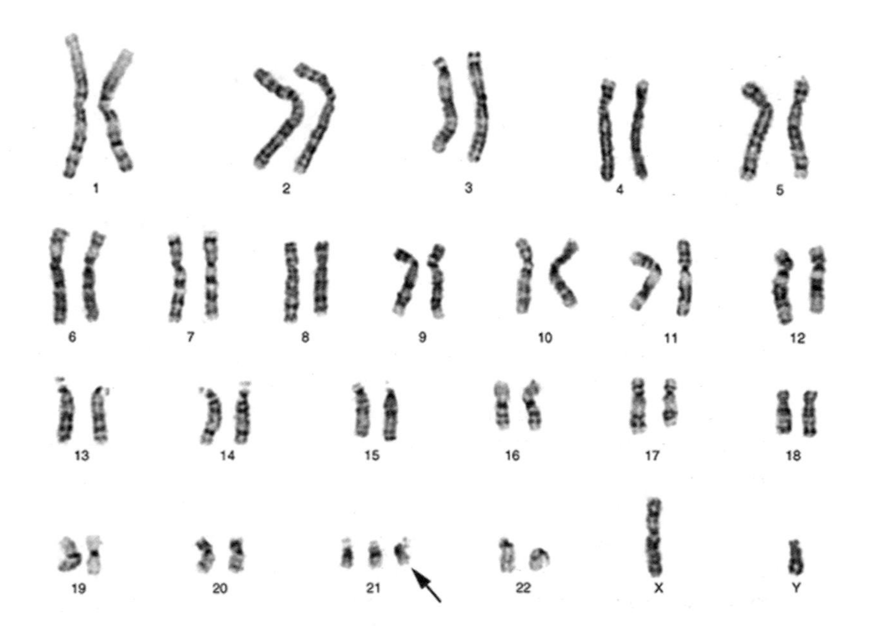

FIGURE 22.2 Karyotype of an individual with trisomy 21 shows the presence of an extra copy of one of the smallest chromosomes that shows the shape and banding characteristics of chromosome 21. (Courtesy of the Clinical Cytogenetics Laboratory, University of Michigan, Ann Arbor, Michigan, Diane Roulston, Director)

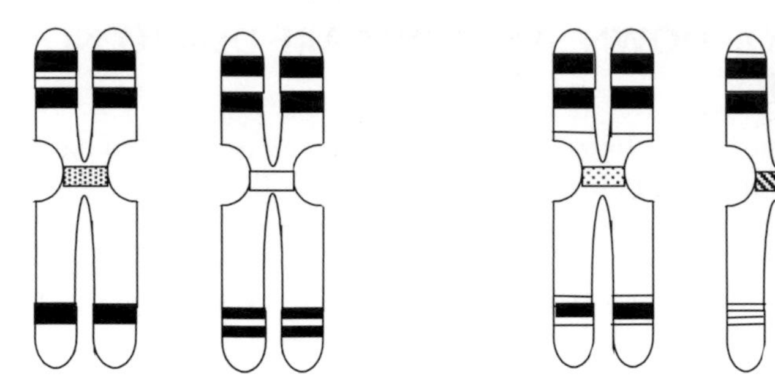

FIGURE 22.3 Cartoon of differences in banding pattern between individual copies of chromosome 21 that would let us tell whether the extra chromosome came from the mother or from the father. For purposes of illustration, we are showing banding patterns that are clear enough to allow distinction between all four parental chromosomes, but in real-life karyotyping situations, it won't always be possible to distinguish all four chromosomes.

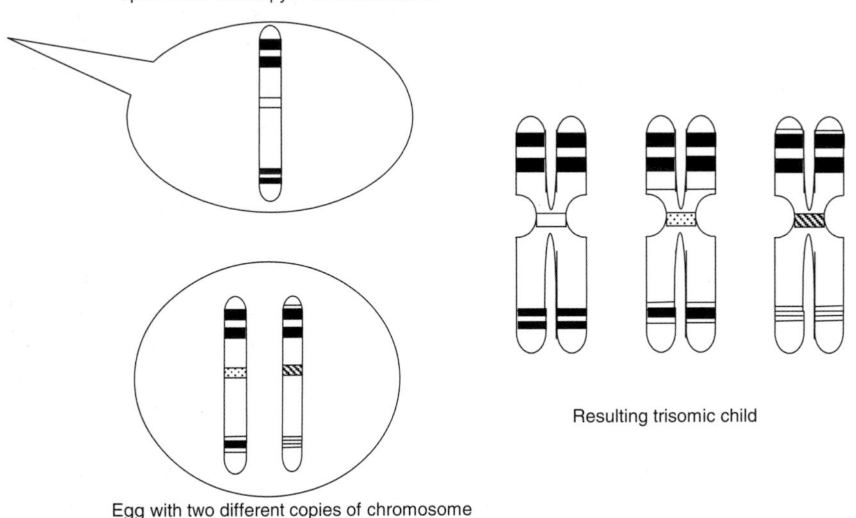

FIGURE 22.4 Cartoon of what we would see if nondisjunction of chromosome 21 in the mother happened at meiosis I. (Note that the egg has two copies of chromosome 21 instead of one, that both of the mother's banding polymorphisms are represented in the egg, and that three different banding polymorphisms are present in the child.) In cases in which banding polymorphisms can be distinguished, the most common outcome for trisomy 21 is the finding that both copies of chromosome 21 from the mother are present along with one copy from the father, meaning that nondisjunction happened at meiosis I in the mother.

resulted from nondisjunction at meiosis I in the father, we would still see three different banding patterns among the three copies of chromosome 21, but two of them would come from the father.

If nondisjunction happens during the second stage of meiosis, the child with trisomy will end up possessing two copies of the same maternal chromosomes, that is, both copies that come from the mother would look the

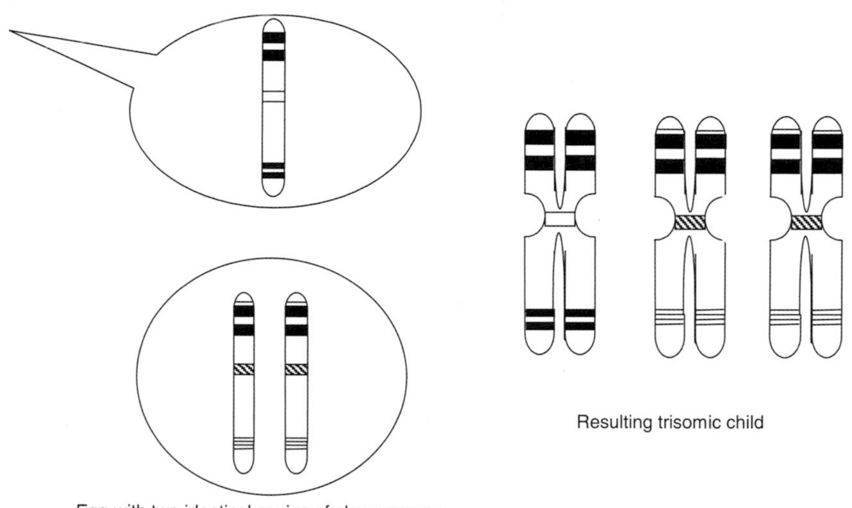

FIGURE 22.5 Consequences of nondisjunction of chromosome 21 in the mother at meiosis II. (Note that only one of the mother's banding polymorphisms is represented in the egg, but it is there in two copies. Also notice that the child has three copies of chromosome 21 but only two different banding polymorphisms.) This outcome is observed less often than the outcome in Figure 22.4.

same (Figure 22.5). Nondisjunction at meiosis II results when the two sister chromatids split and then go to the same poles as opposed to going to opposite poles as they would normally do. This results in the child having three copies of chromosome 21 that demonstrate only two different banding polymorphisms between them. If you look at the different banding polymorphisms, you can see that nondisjunction at meiosis II in the father would also have given only two banding polymorphisms among the three copies in the child, but it would have been one of the father's polymorphisms that would be present twice. Using these techniques, or other techniques that let us tell the chromosomes apart and track where the third chromosome came from, we can show that the extra copy of chromosome 21 almost always comes from the mother.

Why would the extra chromosome tend to come from the mother? The answer to that question is currently in rather hot dispute, but the best guess is that in male meiotic cells the failure of two autosomes to properly pair and segregate results in the cessation of meiosis and, indeed, in cell death. There appears to be a checkpoint in male meiosis that asks whether all of the chromosomes are properly paired and ready to segregate from their partners. If the answer to that question is "no," then the meiotic cell may be doomed and the potentially aneuploid sperm are never produced. Such checkpoints apparently do not exist in most female meiotic cells (oocytes). In oocytes the cell seems committed to completing meiosis despite whatever failures may occur. Thus, although the checkpoint system in sperm is not foolproof and some aneuploid sperm do get through, it does work efficiently enough to result in a substantially reduced frequency of this kind of nondisjunction in sperm as compared to eggs. So it is not that nondisjunction fails to happen in males, but rather that the sperm cells in which it has happened are unlikely to survive to fertilize an egg.

THE MATERNAL AGE EFFECT

Not only is Down syndrome normally a consequence of nondisjunction in the mother, but the frequency of Down syndrome births increases dramatically with advancing maternal age (Figure 22.6). According to the National Down Syndrome Society (www.ndss.org), Down syndrome children are found about once in every 800 to 1000 live births. However, if we look at what we know about the ages of the mothers giving birth to these children, we find that the risk that a baby about to be born will turn out to be a Down syndrome baby is lower, about 1/1500, for women under age twenty-five. On the other hand, by age thirty-five the risk increases to somewhere between 1/300 and 1/100. By forty it may be as high as 1/100 to 1/50. By the mid-forties the risk is in the range of 1/25, an increase of almost a hundredfold. If you are a college-aged woman reading this, you have probably started to do some math. What is the trade-off of time spent on more education, career advancement, and trial relationships versus the risks evident in Figure 22.6? The biological clock, after all, is not something invented by the stand-up comics who joke about it. We should note that age-dependent increases for the frequency of nondisjunction are not observed in men, perhaps partly for the reasons described in the previous section, although, as we see in the section on mutations, there are different age-associated risks for older men contemplating parenthood. Thus, there are actually important biological clocks ticking for both of the sexes.

Still, only about a quarter of kids with Down syndrome are born to mothers over age 25. Think about that for a minute. Yes, the risk that any one woman will have a child with Down syndrome is much higher for older moms,

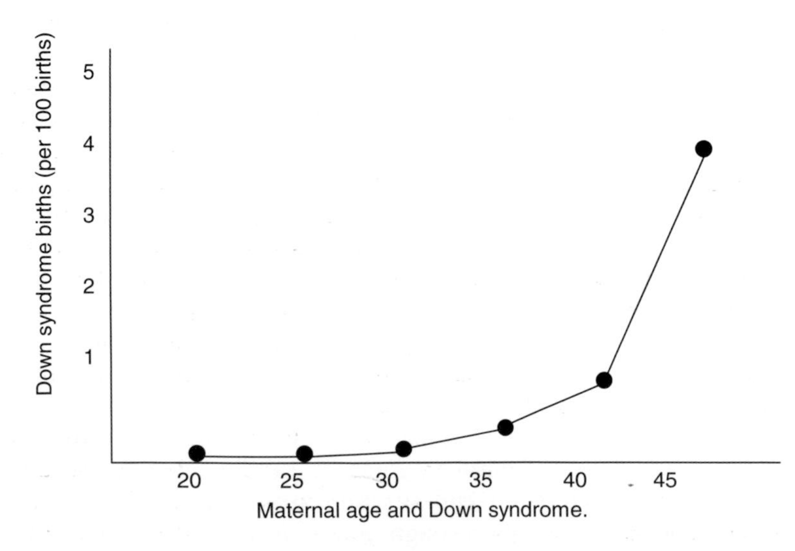

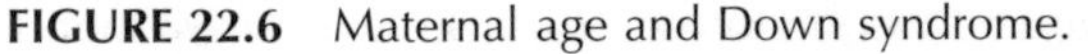

FIGURE 22.6 Maternal age and Down syndrome.

but then far more women are bearing their children in their twenties than in their forties.

The basis for the maternal effect is unknown, but it is likely to be a consequence of a long delay between prophase, when meiosis initiates, and the first meiotic division in human oocytes. Recall that human oocytes begin meiosis long before the female is born. Then the process stops in the late meiotic prophase, after pairing and recombination have taken place, but before the chromosomes have separated into the daughter cells. The chromosomes then stay that way for years. Beginning some months before birth, the chromosomes pair and then they just hang out, connected and poised to go their separate ways. They stay in this state of suspended animation until the girl hits puberty some ten to thirteen years later. Then several oocytes are given permission to restart meiosis each month, but usually only one of these is actually allowed to complete the first meiotic division! Realize then that the egg ovulated by a forty-five-year-old woman has been stalled part way through meiosis for forty-five years, with the chromosomes sitting there paired and recombined but not segregated into separate cells. One can imagine that quite a lot could go wrong in that period of time, and apparently often does. However, just what actually does go wrong remains a bit of a mystery. (Indeed, the absence of an age effect on nondisjunction in men may also reflect the fact that, unlike female meiosis, male meiosis is a continuous process, with no built-in pauses.) Some people argue that the egg's ability to build a normal spindle deteriorates over the years, whereas others propose defects in the mechanisms that hold sister chromatids together or in the resolution of some types of recombination events.

PARTIAL ANEUPLOIDY

We think of and talk about Down syndrome as being the result of trisomy 21, that is, we talk about Down syndrome in terms of there being three copies of chromosome 21 present. However, in about five percent of Down syndrome cases, a baby is found that does not have three full copies of chromosome 21 but instead carries an extra copy of just a piece of chromosome 21 (Box 22.2).

Why should triplication of a chromosomal region be bad? The answer lies in the complex interactions between various genes in our genomes and of the proteins they encode. Many genes produce proteins that act to regulate other genes, and thus the amount of protein they produce must be tightly regulated because either too much or too little can cause changes in levels of expression of many other genes. In other cases, many different proteins must combine together to form large, complex structures in the cell. In those cases, the exact amount of each protein component may well be critical to having the structures end up correctly formed. We still aren't sure exactly which triplicate genes are the real culprits in producing the various components of Down syndrome, but much progress is being made in the study of the roles of genes from this Down syndrome "critical region."

Could we cure the disease if we could answer the questions of which are the critical genes and why an extra copy is a problem? We just don't know. Our guess, and it is only a guess, is *maybe.* On one hand, even if we knew exactly what was wrong for any given component of the syndrome, we are unlikely to be able to fix the whole array of problems. We base this pessimistic

BOX 22.2 PARTIAL ANEUPLOIDY THROUGH TRANSLOCATION OF A CHROMOSOMAL SEGMENT

Sometimes a broken piece of a chromosome becomes attached to another chromosome through a process called *translocation*. The extra piece of chromosomal material gets carried along through meiosis when the chromosome it is attached to goes through normal pairing and segregation. What happens if a piece of chromosome 21 breaks off and sticks onto chromosome 13? Each of the germ cells produced by meiosis will have a normal copy of chromosome 21. One of the daughter cells will get the normal copy of chromosome 13 and the other will get the translocated copy of 13 that has some genes from chromosome 21. The result of meiosis will be some normal germ cells and some cells in which the translocated part of chromosome 21 is aneuploid. After fertilization, the resulting zygote will have three copies of the translocated part of chromosome 21 and two copies of the rest of chromosome 21. The resulting child will have some or all of the Down syndrome characteristics, depending on whether the translocated region includes some or all of the region thought to cause Down syndrome.

prediction on the evidence that some of the most profound problems arising from the extra copy of chromosome 21 arise during development before birth, in structures such as the lungs and heart. Once that damage is done, we may simply have to rely on traditional medical and surgical processes for help. However, some of the Down syndrome problems, such as leukemia and Alzheimer diseases, develop after birth, so there may be a chance for prevention or to improve medical intervention if enough is understood about the roles of the particular genes and gene products in these later developments of the disorder. On the other hand, if the most conspicuous component of this disorder, mental retardation, is truly due to the triplication of a single gene, we have to wonder whether it *may* be possible someday, with early enough prenatal diagnosis, to correct or at least ameliorate that problem through interventions before the baby is even born. It remains to be seen whether the study of triplicate chromosome 21 genes will give us any capability to intervene in postbirth developmental processes that might affect IQ and other capabilities.

PRENATAL DIAGNOSIS OF DOWN SYNDROME

Two things can suggest the possibility of Down syndrome: maternal age and a test for proteins present in the mother's serum, especially a fetal protein known as -fetoprotein or (AFP). Given that the risk that a woman over age thirty-five carries a child with Down syndrome well exceeds the risk of the various diagnostic procedures, most mothers in this age group are advised to seek testing. -fetoprotein is a fetal protein that can cross the placenta and is found in the mother's blood supply. High levels of this protein in the woman's blood can indicate that the nervous system of the fetus has failed to develop properly. However, a low level of AFP in the mother's blood may also indicate the presence of a fetus that is trisomic for chromosomes 21, 13, or 18. Because the number of false positives for this AFP test can be quite high, a better test has recently evolved. This test, called the *triple screen*, measures two other chemicals in the maternal blood stream, human chorionic gonadotropin (HCG) and estriol (E3), as well as AFP. The combined levels of these three components of the mother's blood predict the presence of a Down syndrome fetus in sixty to seventy percent of the cases and show a much lower rate of false positives. The triple screen is now in wide use as a screening tool for Down syndrome, as well as several other fetal anomalies, in women age thirty-five and over, and a more precise quad test is coming into use.

Two other risk factors are family history and chromosome anomalies in the parents. A previous aneuploid fetus or live-born child in a family increases the risk of a trisomic child in subsequent pregnancies. Similarly, if one parent is known to carry an altered form of chromosome 21 (e.g., a translocation in which pieces of the normal 21 have been rearranged to be on another chromosome), the risk of an aneuploid conception goes up greatly (Box 22.3).

Concern about possible aneuploidy in the fetus is also raised in cases in which a couple has had an unusually high number of miscarriages. The occurrence of those miscarriages raises the possibility that one of the parents might carry a genetic aberration or chromosome rearrangement that is causing their own health no problem but that increases the probability of producing a trisomic fetus. However, knowing whether aneuploidy is involved requires evaluation of the number and structure of the chromosomes in a person's cells, since there are many different things that can lead to miscarriages besides aneuploidy.

In those cases in which risk factors exist, or in which a positive result is obtained using a screening test such as the triple screen, there must still be follow up by other tests such as *amniocentesis*. Basically, a needle is inserted into the uterus around the thirteenth to sixteenth week of pregnancy, and a small amount of the fluid surrounding the fetus is removed. The withdrawn fluid contains a substantial number of fetal cells that can be used for chromosome or DNA analysis. A second less common test is called *chorionic villus* sampling (CVS). This test can be performed from eight to twelve weeks after conception and requires an actually biopsy of the tissue that will form the placenta. This test, as well as others like it, are described in more detail in Chapter 34.

As for so many diseases in which the underlying genetic cause has been identified, whether it happens through nondisjunction or mutation of a gene,

BOX 22.3 BALANCED TRANSLOCATIONS—TRISOMIC CHILDREN FROM HEALTHY PARENTS

Sometimes, translocation is a reciprocal process that causes what we call a balanced translocation; in these cases, it looks as if two chromosomes had traded pieces of DNA so that a piece of chromosome 21 ends up stuck to an incomplete copy of chromosome 13, and the material missing off of that copy of chromosome 13 turns up stuck to the broken copy of chromosome 21. The cell has the normal number of copies of each gene; they are just arranged differently. If one of the break points fell within a gene, that gene and the functions it controls may be damaged so that some aspect of the individual's health may be affected. In some cases, apparently healthy people might go through their whole lives without knowing their cells hold such a balanced translocation if the points at which the chromosomes broke did not disrupt any genes or if the defect is recessive and requires that both copies be charged to cause the trait; after all, they still have the right number of copies of every gene. A healthy individual might first find a doctor recommending that they be karyotyped when they have a child with certain kinds of birth defects or when they develop fertility problems that include spontaneous miscarriages. In looking at the children who could have been produced by one normal parent and one parent with a balanced translocation involving chromosomes 13 and 21, let's just follow the gray copies of chromosome 13 and black copies of chromosome 21 coming from these two healthy parents.

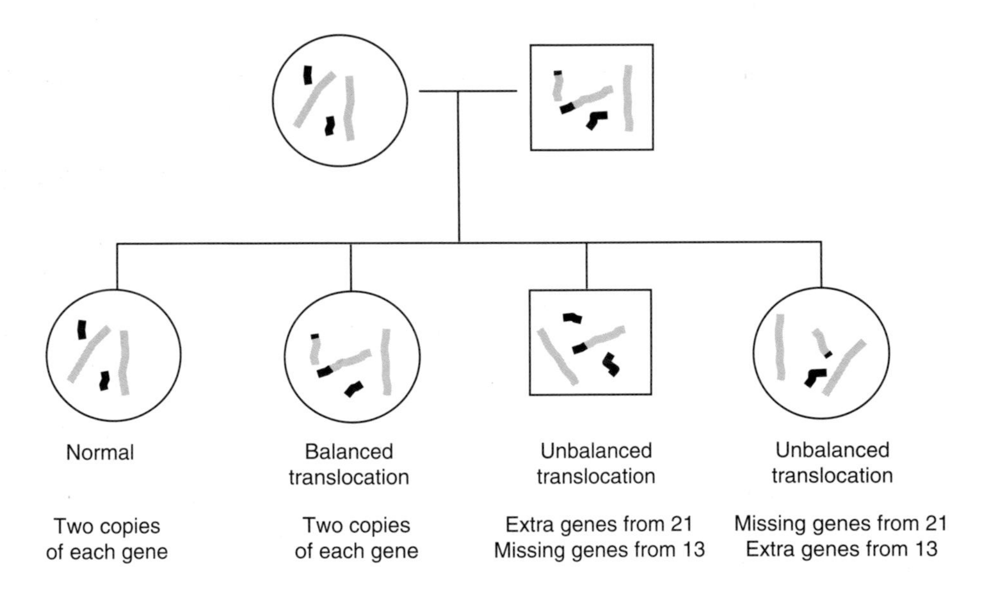

In this case, if no gene was damaged by the breaks that created the translocation, the first two children would both have been born healthy. The third child would have some features of Down syndrome due to some of the extra genes from chromosome 21, as well as additional major birth defects and health problems not typical of Down syndrome due to the missing genes. The fourth child would have had no characteristics of Down syndrome, different birth defects from those of her brother, and there was a strong chance that she would have died before she was six months old. If no genetic testing were done, no one would be able to tell whether either of the healthy girls carries a balanced translocation, but having their karyotype determined would tell them that the first girl carries no more risk of birth defects in her children than anyone else in the general population, but that the second girl carries a balanced translocation and, with every pregnancy, she will face about a fifty-percent risk that that child will face major problems comparable those of her younger brother and sister.

we are currently in a kind of limbo in which we can diagnose things that we cannot yet cure. In these cases, all that can be offered is information, and the consequence of that information is choice, choice in the current sense of the term—the choice to not become pregnant, the choice to terminate the pregnancy, or the choice to start making preparations for the birth of a child with very special needs. We offer no judgment here on the choices to be made but rather note that correct information, offered early, provides the soundest basis on which a woman or couple can make her or their own choices.

Clearly, the eventual goal is to be able to improve the alternatives that go with the word "choice." Can we go beyond the development of screening tests to the idea of being able to improve an older mother's odds with regard to nondisjunction? Only if we can learn enough about the mechanisms that cause nondisjunction, and the genes and proteins, the structures and processes, responsible for this complex process. Can we go beyond our original goals of an early detection system to think about a cure, to imagine a gene-based therapy or a new pharmaceutical product that could fix one or another of the problems caused by the genetic defect? Only if we learn enough about the genes and proteins that are causing the problems. For those of you who are concerned about this interim stage, in which the finding of genes offers only the traditional forms of choice, we want you to stop and look into the future with us. We want you to see that this current era in which the choices are too limited is a temporary state of affairs. We want you to join us in seeing that the long-term goal of these studies is to get beyond out current limitations to a point where the word choice includes the alternative of helping the baby. Frankly, though most of us in science have our immediate research anchored strongly in the near-future steps we are sure will work, it is possibilities such as this, farther from our grasps but kept firmly in the backs of our minds and discussed far into the night after the presentations are over at scientific conferences, that keep our colleagues and ourselves in our labs working until the wee small hours of the night most days of the week.

TRISOMIES FOR CHROMOSOMES 13 AND 18

Excluding rare exceptions, only two other human trisomies involving autosomes have been reported among live births. These are *trisomy 13* (sometimes called Edwards syndrome) and *trisomy 18* (sometimes called Patau syndrome). These usually have a characteristic set of congenital problems, including a high frequency of severe cardiac and neurological problems.[1] Trisomy 18 is observed at a very low frequency (about 1 out of 8000 live births). The incidence at conception is much higher, but most of these embryos are miscarried spontaneously. Trisomy 13 is even less frequent, occurring in about 1 in 25,000 live births. Like Down syndrome, the incidence of these trisomies increases dramatically with advancing maternal age.

Trisomies for other chromosomes are not viable, but, as noted above, they do indeed occur at the point of conception, and their frequencies increase with advancing maternal age. However, all of these other aneuploid conceptions lead to spontaneous miscarriages.

[1] SOFT (Support Organization for Trisomy 18, 13, and Related Disorders, found at www.trisomy.org indicates that less than ten percent of these babies live to see their first birthday).

CHANGES IN THE NUMBER OF SEX CHROMOSOMES (SEX CHROMOSOME ANEUPLOIDY)

The lethality of most autosomal aneuploidies stands in stark contrast to the viability of most sex chromosome aneuploidies. This may not be surprising given that mechanisms already exist for dealing with the dosage differences between human males and human females. In fact, the mechanisms that allow this difference in chromosomal dosage also allow the survival of individuals with a number of sex chromosome aneuploidies. Nonetheless, most of those with an aneuploid constitution do show some differences from people with a normal chromosome complement.

The most common examples of sex chromosome aneuploidies are Klinefelter syndrome (XXY males) and Turner Syndrome, denoted X0 because Turner syndrome has only one sex chromosome, which is an X.

According to Klinefelter Syndrome and Associates[2] Klinefelter males occur at a surprisingly high frequency of 1/500 to 1/1000 live births. Like XX females, XXY males undergo X inactivation during early embryonic development. Thus, half their somatic cells have inactivated one of their two Xs and the other half of the cells have inactivated the other. Klinefelter males often have small, nonfunctional testes, are sterile, and have some external feminization, such as breasts and hips. Because the testes have atrophied, not a lot of testosterone flows through their bodies. In many of these men, there is almost an even amount of estrogen and testosterone in the individual's system. It has been found that some Klinefelter boys and men have social problems and learning difficulties.

Women with Turner syndrome are normally sterile with ovaries that appear as a rudimentary streak. Because they only have one X, X inactivation does not occur in Turner females. They may also be shorter than average and show immature development of the breasts and genitals. According to the Turner Syndrome Society,[3] this is a very common genetic condition that affects 1/2000 to 1/2500 females. The actual frequency of conceptions with Turner syndrome is much higher, perhaps a few percent, but 99.9% of X0 conceptions are miscarried spontaneously.

Two other types of genotypic abnormalities of the sex chromosomes, which fail to have fancy names, are XYY males and XXX females. Although XYY men do not exhibit any characteristic set of abnormal phenotypes, they are often taller than average males. Many XYY boys have learning disabilities, and some fraction of both XYY boys and men may have behavioral problems. Although the great majority of XYY men lead normal lives, the frequency of XYY men is increased in various kinds of prison populations, especially among inmates greater than six feet in height. The frequency of XYY men is 1/1000 among newborn males but may be as high as five times that in general prison populations; it has even been reported to be as high as ten times that frequency in one juvenile prison population. More strikingly, if attention is restricted to male prisoners over six feet tall, the frequency of XYY males has been estimated to be as high as ten to twenty percent.

[2] www.genetic.org/ks/scvs/47xxy.htm

[3] www.turner-syndrome-us.org

However, XYY men are not always, or even usually, incarcerated for violent offenses. Rather, they are more often jailed for repetitive violations of probation agreements, possession of stolen property, writing bad checks, etc. The simple conclusion is that having an extra Y chromosome *does not* make a man more violent. It may predispose some men to get into trouble with the law, but it does not make them more violent. Indeed, it may be the case that the tendency of XYY males to end up incarcerated may be less reflective of an influence of an extra Y chromosome on criminal aggression, and more the result of learning difficulties created by the extra chromosome. *Indeed, in any instance in which we find that a particular genotype is more common in a particular population, we need to be very careful in determining what the precise phenotype is that actually correlates with the genotype in question.*

XXX females are not associated with any specific abnormal phenotype. Indeed, these females are most often detected only because they were karyotyped for some unrelated reason. The vast majority of them are fully phenotypically normal; however, there may be a decrease in fertility. An XXX female inactivates two X chromosomes. Thus, like a normal cell (XY or XX), the XXX cell has one functioning X chromosome. In contrast, a cell carrying three copies of chromosome 21 has no way to simply inactivate the extra copies!

SEX CHROMOSOME ANEUPLOIDY IN THE GERMLINE

The effects of both Klinefelter and Turner syndrome's on gonad development and fertility reflect the fact that inactivated X chromosomes of XXY males are reactivated in the germ cells of the testes create, which creates an excess of X-linked genes in the testes. Similarly, in normal XX females the inactivated X is reactivated in the germ cells and oogenesis requires two active Xs. Thus, X0 females will lack the necessary second X in their oocytes. Although one X does inactivate in the germ cell progenitors of early female embryos, it is eventually reactivated in oocytes before meiosis. This reactivation reflects a stringent requirement for two X chromosomes in oogenesis. The absence of the second X chromosome in Turner syndrome females causes rapid death (atresia) of oocytes during fetal development. The result is both sterility and small rudimentary ovaries. Similarly, Klinefelter men also reactivate the second (inactivated) X chromosome in the developing testis. The presence of the extra X in a male germ cell causes death of the male germ cells during early puberty and subsequent atresia of the testes. This testicular atrophy results in a great diminishment in the ability of many of these males to make testosterone. The resulting testosterone deficiency may explain many, if not most, of the characteristics of Klinefelter syndrome.

WHY IS TURNER SYNDROME SO OFTEN LETHAL IN EARLY EMBRYOS?

If the somatic cells only require one X, and all that the Y does is determine sex, why is Turner syndrome so often lethal to early embryos, and why are live-born females affected with any unusual phenotypic characteristics? As noted above, 99.9% of all X0 conceptions are miscarried spontaneously in utero. Thus possessing only one X chromosome, minus a Y as well, is almost always lethal to the zygote, yet males can survive with only one X chromosome; we also know that females inactivate one of their X chromosomes in all of their somatic cells. So why is that second X chromosome so important? One would think that if a normal female has only one X chromosome active anyway, what is the big deal with having only one X? Part of the answer may lie in the problems of gene dosage we talked about in Chapter 20: a small number of essential genes on the X escape inactivation in cells with two or more X chromosomes. The existence of such genes provides a straightforward explanation for Turner Syndrome. X0 females possess only one copy of such genes, whereas both XX females and XY males possess two copies.

The sterility of Turner females, like that of XY males missing the TDF gene, reflects the requirement for two functional X chromosomes in the female germline, but what about the other phenotypes? Where do they come from, and why are they so variable? Given the very high lethality of Turner syndrome, why do such individuals ever make it to birth, much less beyond? Perhaps the answer lies in other genetic variations in the genomes of these individuals that compensate for something missing from the X, or perhaps even differences in the copies of the X chromosome that turn up in X0 individuals that survive.

There also may be another explanation: perhaps the reason for the survival of the rare Turner female and the vast phenotypic variability among such live-born females may be attributed to what scientists call *mosaicism.* Some live-born cases of Turner syndrome may be due to the fact that these surviving girls are not composed solely of X0 cells: they are composed of both X0 and XX cells. The loss of a single X chromosome, during mitosis and not meiosis, in one cell out of several cells present very early in zygotic development may produce a combination of both XX and X0 cells. Thus the resulting individual possesses both XX and X0 cells. As long as an XX karyotype is present in those cells that *absolutely require* two X chromosomes, the individual will survive. Those cells that do not require two X's will be able to survive as either XX or X0 cells. The more X0 cells the individual possesses, the more severely affected the individual will be, whereas the more XX cells the individual possesses, the more normal the individual will be. If different Turner females each have a different fraction of X0 cells in their bodies, it makes sense that the phenotype would be so variable.

REPRISE

And so we have described the mechanistic origin of the story that was Earl's life, in which Scott was for some time a member of the chorus. His story started with an error of the meiotic process, an error whose seeds may have been laid before his mother was even born. We have to wonder, though, is it really fair to define Earl in terms of the meiotic error that produced him? Indeed, those of us who took the time and effort to know him, who have been touched by his life, know that he is much more than that, so much more.

BREAKING THE RULES

It is surprising how many things that look sporadic actually have underlying genetic components, and how many things that are not caused by a change in the DNA sequence can look as if they run in families. In previous chapters, we have shown you how a trait can suddenly pop up in a family that had never heard of that trait, especially in cases in which inheritance is autosomal recessive or X-linked recessive. Here we talk about the opposite situation, in which the evidence suggests that some gene, some locus, some change in the DNA sequence must be involved when really the explanation is quite different.

IMPRINTING

23

To our surprise, when Senator Bob Dole started doing commercials for Viagra and Pepsi, he turned out to be a very funny guy. However, in 2001, when he did a public service announcement for the Prader-Willi Syndrome Association,[1] no one was laughing. Senator Dole was reaching out to the public to tell them about Prader-Willi syndrome in an effort to educate the public and get proper medical care for the many undiagnosed children with this disorder. Prader-Willi syndrome first shows up in babies with poor muscle tone and feeding problems in infancy. As these children grow up, they turn out to be short with weak muscles, small hands and feet, distinctive facial features, a tendency towards morbid obesity, and a variety of other serious problems that vary from one child to the next. According to the Prader-Willi Syndrome Association, about one in every 12,000 to 15,000 babies born in the United States has Prader-Willi syndrome. In infancy, they may fail to thrive because weak muscle tone makes it hard for the baby to suck and gain nourishment, but once they are older, they have a very different problem—weight gain driven by the fact that they always feel as if they are hungry, even after they have eaten. Sometimes these children are in families with no history of Prader-Willi syndrome. Sometimes they turn up in families where siblings or other relatives are also affected. How can this set of features sometimes be familial and other times turn up in isolated cases? To understand, let's take a look at the unusual cause of Prader-Willi syndrome, which can sometimes happen in individuals with standard types of mutations that change the DNA sequence, but can also result from a different kind of change to the DNA that occurs in cases of a phenomenon known as imprinting.

In the previous chapters, we have said much about the rather enormous differences between meiosis in males and females and alluded to the very substantial differences in the processes by which eggs and sperm are made. After all, one of the primary objectives of sperm building is to condense the chromosomes into the smallest possible volume (to facilitate swimming!); no such constraint exists for eggs. Males accomplish this feat by stripping the chromosomes of the four meiotic products of virtually all of their usual DNA-associated proteins and replacing these proteins with a specific set of "DNA-packing" proteins that allow the genomes to be maximally compacted. However, once the race has been swum and fertilization achieved, the lucky winner of a sperm pronucleus must remove these "packing proteins" and rebuild the chromosomes in a fashion that can allow gene expression during interphase and chromosome movement during the ensuing embryonic

[1] http://www.pwsausa.org

mitoses. Realize that the changing of chromosomal proteins is one of the major players in controlling gene expression in eukaryotes. Perhaps, then, it might now be surprising if the two sets of haploid chromosomes, those from the mother and those from the father, begin embryonic development with rather different capacities for gene expression. In this case it, might matter a good deal whether an embryo heterozygous for a deleterious mutation in a gene whose function was required quite early received the normal copy from the mother or the father. If early expression was usually obtained (for example) from the maternal allele, a heterozygote that obtained the functional copy from mom might be far less impaired than a heterozygote who received the deleterious allele from mom.

One mechanism by which such *parental source effects* occur involves a phenomenon we refer to as *imprinting*, in which passage of a gene, chromosomal region, or even an entire chromosome through one germ line or the other "marks" that gene, region, or chromosome in a way that regulate's its expression throughout subsequent development of the organism. We will describe a number of examples of parental source effects and imprinting in this chapter. It is critical to remember that these effects are *epigenetic* (Box 23.1), that is to say, they change the degree to which a given gene or chromosome is expressed without changing the DNA sequence. They are "erased" in each new generation as each new germ line cell is formed. In that sense, these changes are fundamentally different from mutations. We will begin by discussing imprinting as a chromosomal phenomenon and then as a phenomenon affecting smaller genetic regions. As we go on, please remember that imprinting is rare. Fortunately for our efforts to understand many genetic phenomena, most genes do exactly what Gregor Mendel said they should do.

An example of imprinting that is easy to follow is the process of activation and inactivation of genes on the X chromosome in the kangaroo (Figure 23.1). In many ways, this process is strikingly similar to the events that occur in human females. The critical exception is that the inactive X in the cells of female kangaroos is always the X chromosome that they received from their father. When the X chromosome passes through the kangaroo male germ

BOX 23.1 GENETIC CHANGES AND EPIGENETIC CHANGES

Genetic changes take place through a process called *mutation* that permanently alters the DNA sequence so that subsequent generations receive faithful copies of the altered sequence without changing it back. Epigenetic changes leave the order of As, Cs, Gs, and Ts unaltered while making other temporary and reversible modifications to the DNA that have local effects on the ability of that sequence to be used by the cell to produce a phenotype. In the imprinting situations we are talking about, the effect is that of flipping an "on" switch to the "off" position, an "off" switch to an "on" position, or turning a rheostat up or down or changing when or where the gene is expressed. The lamp controlled by that switch is exactly the same lamp, but it will act differently depending on the position of the switch controlling it, and the position of that switch can be changed.

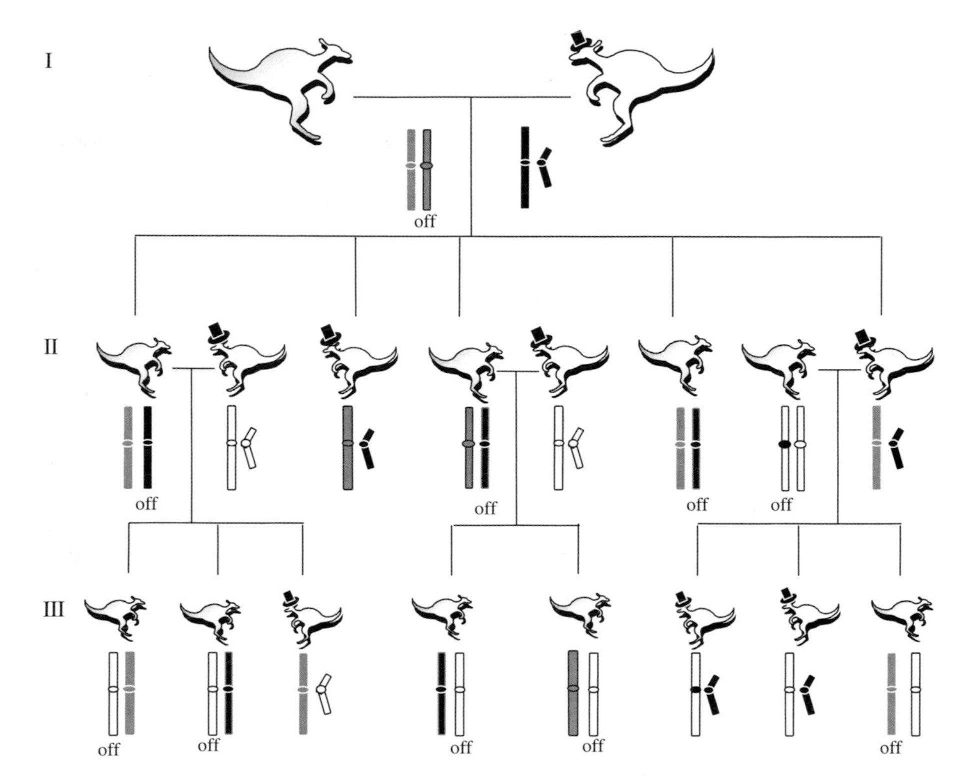

FIGURE 23.1 Imprinting made simple in a kangaroo family. Notice that the black copy of the X chromosome that came from Grandfather Kangaroo in generation I is inactivated in all of his daughters in generation II. However, when we get to generation III, the black X is not inactivated, since only Grandfather's daughters have that copy of the X to pass along to their children. Thus the black X chromosome that was inactivated as it passed through the male germ line has apparently become activated again after passing through the female germ lines of his daughters. Meanwhile, Grandmother Kangaroo started with one active X (from her mother) and one inactive X (from her father). Both of her X chromosomes are active in generation II in both sons and daughters, but the gray chromosome with white around the centromere, which was active in Grandmother Kangaroo and her children, is now inactive in her son's daughter now that it has passed through the male germ line. Since the X chromosome inherited from the father is always the one that is inactivated, and boys do not get an X from their father; only the girls have an inactivated X chromosome (marked "off").

line, it ends up inactivated in his progeny. This means that the genes on that copy of the X chromosome are not transcribed to make RNA and gene products from that chromosome are not present to affect the organism's characteristics. However, the switch gets reset as it passes through the germ line of the next generation, and genes on the X chromosome that were not used in one generation, because they came from a male, may be used in the next generation if they get passed along by a female. Thus, if you look at the X chromosome that the grandfather kangaroo contributes to the kangaroo family shown in Figure 23.1, it is easy to see that his copy of the X chromosome is turned off in all of his daughters, but is turned on in all of his grandchildren because they only inherit his X through his daughters. Whether that copy of the X chromosome is "on" or "off" in his great-grandchildren depends on whether it is being passed along by one of his male grandchildren or one of his female grandchildren.

The marking of the paternal X chromosome for inactivation is also observed in the extra-embryonic membranes of mice and human embryos. Thus, in all the cells of a kangaroo female and some of the cells of mouse and human embryos, the choice of which X is inactivated at the fifty-cell stage of embryogenesis can recognize some "mark" left on the X chromosome donated to the embryo by its father. The nature of this mark remains obscure. Perhaps some proteins loaded onto the newly rebuilt paternal chromosomes predispose the paternal X to inactivaton. Perhaps the DNA is modified by adding a small chemical group (such as a methyl residue) to the X chromosomal DNA during spermatogenesis. Other modifications, such as the binding of structural RNA molecules to the imprinted X chromosomes, have also been proposed. Although much remains to be discovered about the precise mechanism, we know that the changes do not affect the order of the bases in the DNA sequence.

There are numerous other cases of imprinting at the chromosomal level. In a more Amazonian example, there is an overgrown gopher called *Microtus oregoni*, in which all somatic female cells simply destroy and discard the father's X chromosome. The mealy bugs go a bit further: females inactivate the entire paternal genome! Notice that the word here is "*inactivate*," not "toss out". The paternal genome is still there, and still available to be passed on to subsequent generations, but inactivation remains during the daily life of the female mealy bug. Nonetheless, X inactivation is the only truly chromosome-wide example of this phenomenon that is well documented in mammals. There are, however, numerous cases in mice and humans that document the effect of parental source or imprinting in smaller genetic intervals.

IMPRINTING OF SPECIFIC CHROMOSOMAL REGIONS

Whether we are talking about a whole chromosome or some smaller region, the same basic concept of epigenetic activation and inactivation of an unaltered sequence of bases applies. The process of imprinting basically involves setting an on/off regulatory switch, that is, presetting an active or inactive status of a gene, chromosomal region, or entire chromosome as it passes through one of the two germ lines, male or female. Thus a given gene, or genetic version, might get turned off (made inactive) when it passes through the male germ line, but that same gene might get turned on (activated) when it passes through the female germ line. If the gene is inactivated, it will not be transcribed and the protein product will not be made. When active, the gene is transcribed and the gene product is available for the cells of the embryo to use.

Imprinting, the combination of on and off switch settings that control the use of specific genes, can have profound effects on developmental processes that take place during embryogenesis. Let's consider the following experiment. One can fertilize mouse embryos in a petri dish and manipulate the sperm and egg nuclei (called the pronucleus or pronuclei), which are present separately within the zygote just after fertilization and before their fusion. For example, the female pronucleus of one zygote can be destroyed and replaced by transplanting the female pronucleus from another zygote (Figure 23.2).

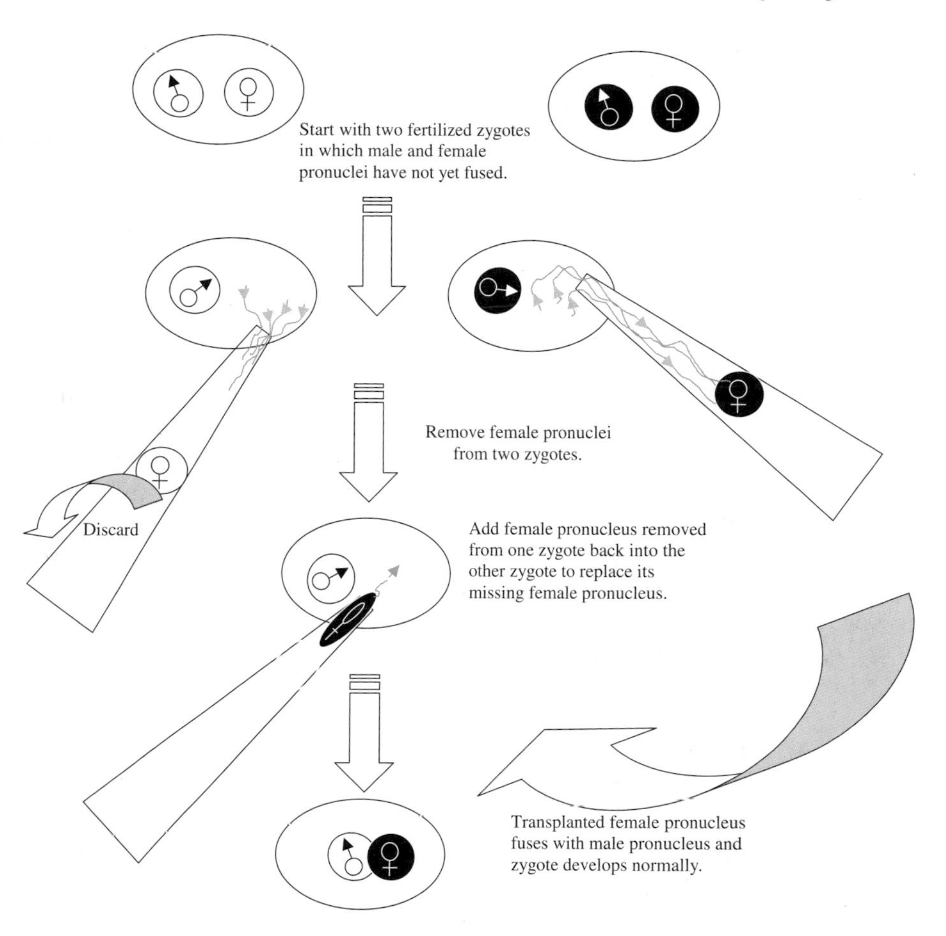

FIGURE 23.2 Pronuclear transplantation.

This results in a perfectly good, healthy embryo, which can produce a very expensive mouse when implanted into the uterus of a female mouse. It is also possible to replace the male pronucleus with some other male pronucleus. As long as the zygote has one pronucleus that came from a female and one that came from a male, embryonic development will go as it should. *What can't be done successfully is to fuse one male pronucleus with another male pronucleus or a female pronucleus with a female pronucleus* (Figure 23.3). If one attempts to create an embryo with two female pronuclei, the fetal tissues start out developing quite nicely, but the extra-embryonic tissues develop improperly. If the experiment is reversed, namely fusing two male pronuclei, the extra-embryonic tissues develop with no problems, but the embryo itself develops abnormally. *Apparently, both male and female pronuclei are needed if normal development is to proceed for both the fetus and the very essential placenta.*

The observation described above likely reflects the fact that some genes are differentially programmed or differentially inactivated during passage through gametogenesis in the two sexes. An example of this at the level of individual genes is provided by Prader-Willi and Angelman syndromes. These two disorders, although quite different in terms of their pathology, are due to mutations or deletions in the same small region of chromosome 15 that

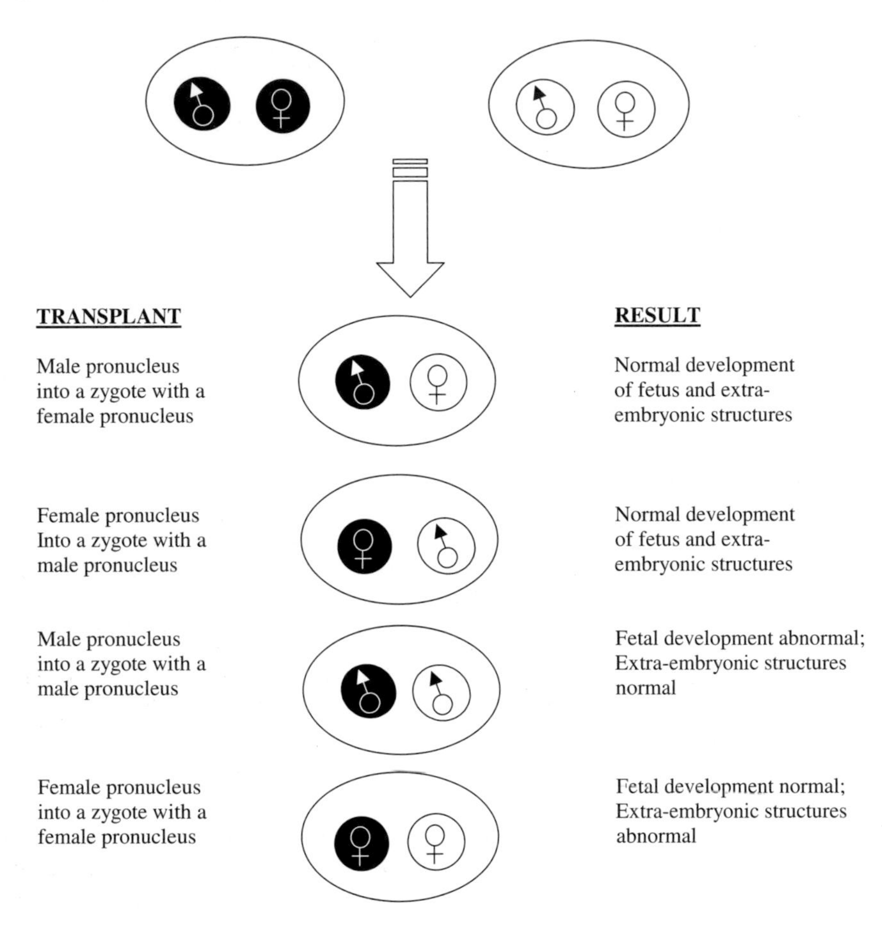

FIGURE 23.3 Mix-and-match nuclear transplantation. Using a pronucleus from one newly fertilized zygote to replace a pronucleus in another newly fertilized zygote can produce different combinations of pronuclei with very different consequences.

are affected by whether they were passed along by the mother or the father (Figure 23.4). Both Prader-Willi and Angelman syndrome are usually associated with newly arising deficiencies that include band q11 on chromosome 15. When that deficiency is inherited from the father, the offspring exhibits PraderWilli syndrome (obesity, small hands and feet, hypogonadism, and mental retardation). When the deficiency is inherited from the mother, the offspring exhibits Angelman ("happy puppet") syndrome.

Occasionally, one finds patients with Prader-Willi who do not display a deletion. Astoundingly, these individuals can be shown to possess two maternal copies of chromosome 15 and no paternal copies! Such individuals are examples of what we call *uniparental disomy*: the result of meiotic or mitotic nondisjunction events produce an individual bearing two identical copies of a chromosome derived from a single parent. As shown in Figure 23.5, one way this could happen would be through a process called *zygote rescue*, in which a trisomic zygote with three copies of chromosome 15 is restored to having the right number of copies through loss of one copy during mitosis after embryonic cell division begins, which causes a normal zygote if one of the two mater-

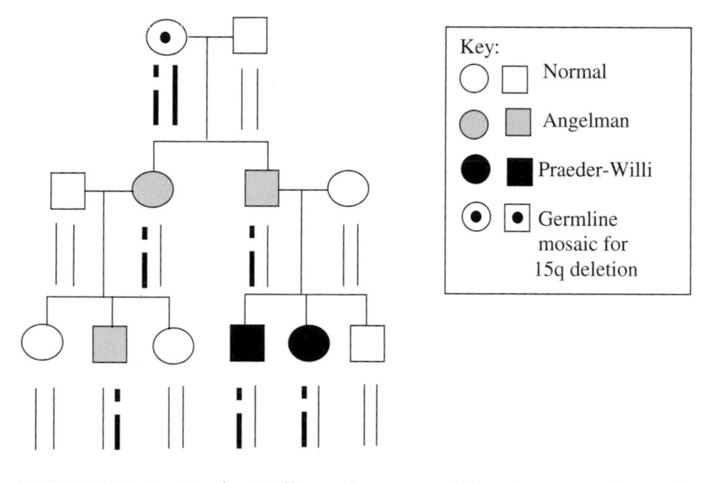

FIGURE 23.4 Prader-Willi syndrome and Angelman syndrome from the same deleted chromosome. If a three-generation family passed along a copy of chromosome 15 with a deletion of 15q11, we might see this mixture of phenotypes. All "affected" family members have the same deleted copy of chromosome 15 inherited from the grandmother, but they are not all affected with the same thing. The ones who got it from their mother have only a paternal copy of 15 and develop Angelman syndrome. The ones who got it from their father have only a maternal copy of 15 and thus develop Prader-Willi syndrome. The grandmother in this case is mosaic, with the deletion present only in some of her cells.

nal copies is lost and uniparental disomy if the only paternal copy is lost. Another way this could happen would be if an aneuploid egg with two copies of chromosome 15 is fertilized by a sperm that is also aneuploid in a complementary manner, that is, missing chromosome 15 (Figure 23.5). Thus it is not the lack of two copies of this region that causes Prader-Willi but the lack of a paternal copy. Clearly, the paternal copy is essential for some function that the maternal copy cannot provide.

Similarly, one also finds patients with Angelman syndrome who do not display any deletions in the critical region of chromosome 15. These individuals can be shown to possess two paternal copies of chromosome 15 and no maternal copies! Again, it is not the lack of two copies of this region that causes Angelman; it is the lack of a maternal copy. Clearly, the maternal copy is essential for some function that is quite different from that for which the paternal copy is required, a function that the paternal copy cannot provide. The bottom line here is that the same genetic region is programmed to perform two different and essential functions, depending on which of the two germ lines it has passed through. Obviously, daughters can erase and reprogram the genes they obtained from their fathers, and sons can reprogram what they received from their mothers.

IMPRINTING AND HUMAN BEHAVIOR (MAYBE)

Recently, psychological studies of women with Turner syndrome have raised the possibility that a region on the X chromosome that controls behavior and/or personality might also be subject to imprinting. According to work by

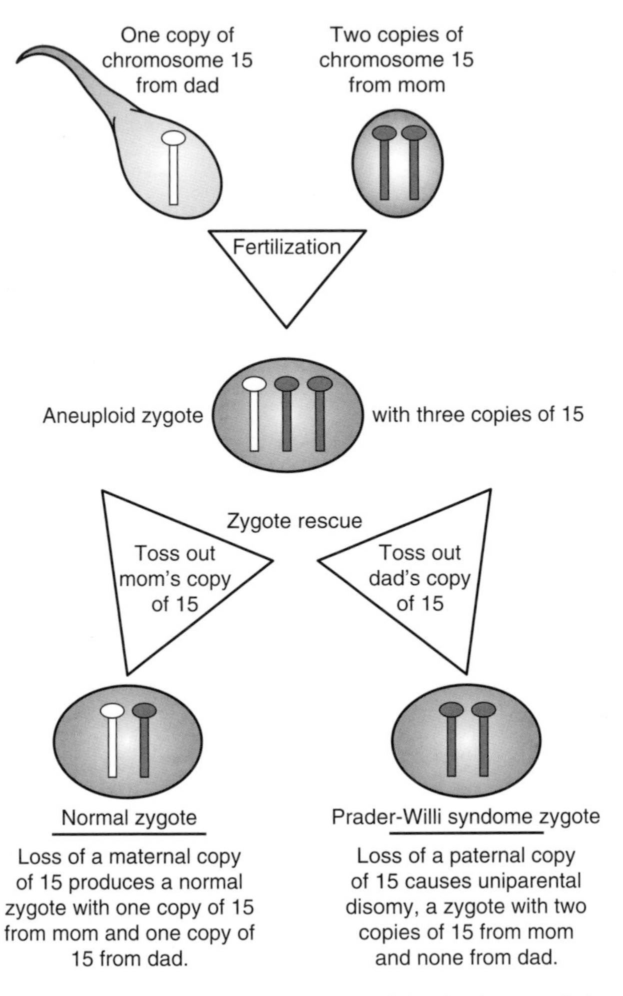

FIGURE 23.5 Zygote rescue. Loss of the third copy of chromosome 15 from a trisomy 15 zygote can produce a normal zygote or a zygote with imprinting effects. If both of mom's copies of 15 are kept, the child will go on to develop Prader-Willi syndrome, but if one of mom's copies is lost, a normal genetic complement is restored.

David Skuse and his collaborators, Turner syndrome women who received their single X chromosome from their mother had more problems in social adjustment than did Turner syndrome females who received their sole X from their father. According to these researchers, Turner syndrome females who inherited their X from their father displayed superior skills in those areas that mediate social interaction. These authors infer from these data that there is a region on the X chromosome that plays an important role in establishing the patterns of "social cognition," that is, imprinted in such a fashion that it is not expressed from the maternally derived X chromosome and thus would not be expressed in males.

There are problems with this interpretation, primarily because the distinction isn't really "all or nothing." It simply isn't the case that all Turner syndrome females with a maternally derived X chromosome behave one way and all Turner syndrome females with a paternally derived X chromosome behave

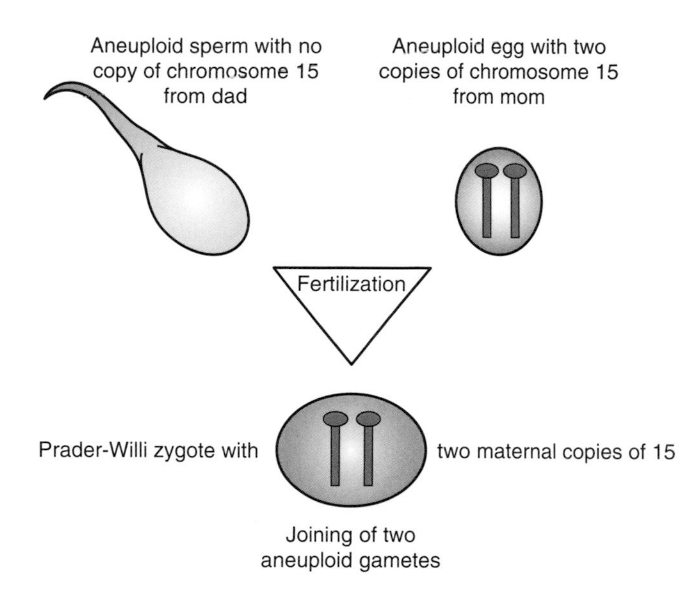

FIGURE 23.6 Joining of two aneuploid gametes. Sometimes an error in meiosis results in a sperm that lacks a copy of chromosome 15. Sometimes an error in meiosis produces an egg with two copies of chromosome 15. If these two aneuploid gametes come together, the resulting zygote will have restored a normal number of copies of chromosomes 15, but the resulting child will have Prader-Willi syndrome because both copies of chromosome 15 came from the mother.

another; rather, it is a matter of degree. (For example, seventy-two percent of the Turner syndrome girls in this study with a maternally derived X exhibited difficulties in social interaction compared to only twenty-nine percent of the paternally derived cases.) Nonetheless, it is an intriguing finding that raises some curious possibilities.

As you can imagine, this result quickly got the attention of the scientific and popular press. One possible interpretation of these data is that men, who get their X from their mother, might be more vulnerable to disorders of social condition (such as autism) than are women, who receive one X chromosome from each parent. The popular press took things even further in somewhat recklessly suggesting that the studies of Skuse and his colleagues provided a way of understanding why men and women sometimes seem to differ in their patterns of social interaction and aggression. However, it seems to us that extending results from Turner syndrome females to the general population might need to be done with some great degree of caution.

Regardless of the more global implications (or lack thereof) of this result, the work of Skuse provides a dramatic suggestion that imprinting may have quite significant implications for human development. It thus behooves us to figure out just how it might work.

WHATEVER WORKS

If the truth must be told, we're not at all surprised by imprinting. The different histories of a male and female pronucleus are reflected in the proteins bound to their DNA and in their degree of compaction. This difference must have provided a fertile substrate on which differential systems of gene regulation could be developed. Indeed, we would be shocked if this phenomenon didn't exist. The differences in the biology of the two types of germ lines and the existence of erasable DNA modification systems created an ideal substrate for the evolution of a curious set of ways of regulating genes. It may seem a bit messier than we would like, but as we sometimes say, *evolution is not Michelangelo and the Sistine chapel. It is a teenage kid with a broken car and no money. It just does whatever works!*

IMITATING HEREDITY: ONE TRAIT, MANY CAUSES

24

According to the World Health Organization, more than 100 million people worldwide are believed to have glaucoma, and more than 5 million of them are blind because of it. Glaucoma is the leading cause of blindness among African Americans and the second leading cause of blindness among North Americans of European ancestry. Among the millions of individuals in the United States who are affected with glaucoma, about half of them do not yet realize that they have the disease and many of them will not realize it until after irreversible damage has occurred. Although increasing age is one of the risk factors for glaucoma, and most cases of glaucoma are found among those who are over forty years of age, glaucoma can also be found less often in young adults and children, and one form of glaucoma even occurs in newborns.

Fundamentally, glaucoma is a disease in which the nerves that carry visual signals from the eye to the brain die in a characteristic pattern, usually slowly over a long period of time. This loss of nerves is accompanied by the loss of visual field (or area of vision) from the regions served by those nerves. Typically, visual field is first lost in local regions while vision remains excellent in surrounding regions. The local and arc-like regions of missing vision can gradually merge to form large regions of visual deficit if the disease remains untreated. Many people think that having glaucoma means having increased pressure inside the eye, because the most common form of glaucoma involves elevated intraocular pressure. However, elevated intraocular pressure is no longer part of the definition of the disease since the characteristic pattern of nerve cell death in glaucoma happens to people who have never shown any increase in the pressure inside their eyes.

There is a lot of evidence for genetic causes of glaucoma. As many as half of the patients being seen for glaucoma may have relatives who also have glaucoma. In some cases, autosomal dominant inheritance of glaucoma can be traced for three or four generations, with one family we know about showing transmission of glaucoma through eight generations in a row. With the mapping of more than a dozen different loci that can cause glaucoma and the identification of disease-causing mutations in more than a half dozen glaucoma genes, glaucoma now ends up being classified as a genetic disease.

There are many cases of glaucoma in which a clearly nongenetic cause of the disease can be seen. Not only can some kinds of injury cause glaucoma,

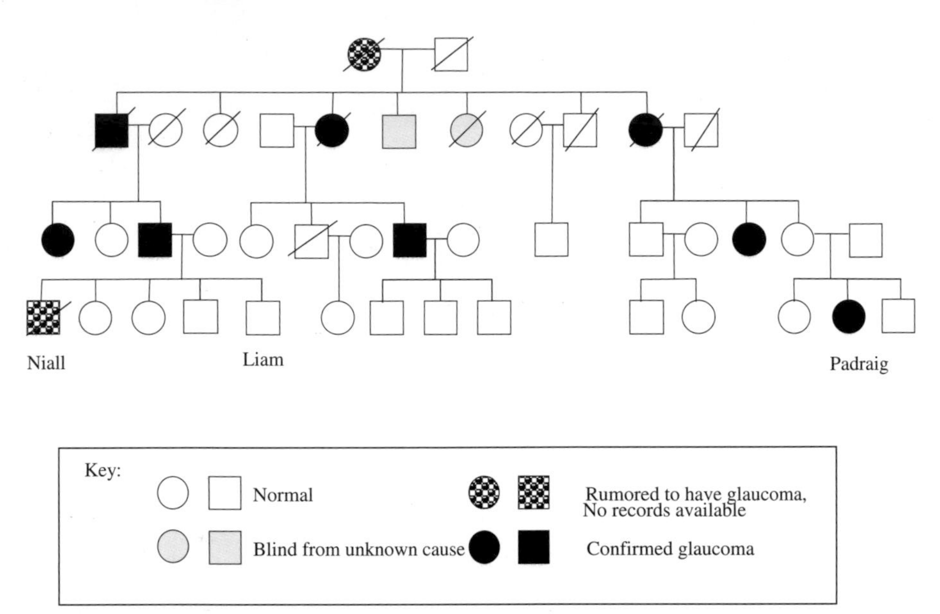

FIGURE 24.1 If Liam's family history looked like this, what might we conclude about his chance of developing glaucoma, which depends partly on whether or not his father was carrying the family's defective glaucoma gene? Liam's father ended up with glaucoma after years of using a prescribed corticosteroid that is known to cause a form of glaucoma called steroid glaucoma. Liam's brother Niall moved to Australia, where he was rumored to have developed glaucoma, but his death last year left the family with no access to information about whether his glaucoma might have resulted from injuries suffered when he was hit by a car.

but it can also be caused by some medications or inflammatory situations and may sometimes be an unwanted consequence of some kinds of eye surgery. Although we can easily see how genes play important roles in the processes that lead to glaucoma, in some cases it is equally clear that the actual event that started the disease was an external nongenetic event.

The term *identity by descent* refers to a situation in which the same underlying cause is producing the same trait among relatives because they have inherited the same genetic defect from a shared ancestor. The term *identity by state* refers to a situation in which two people have the same trait but we cannot tell whether the cause of that trait is the same. *Phenocopy* is the term for a situation in which someone has an environmentally-caused trait that looks like a genetic trait under consideration. In some families where people share a trait, we assume identity by descent as the simplest explanation for what is going on, but sometimes someone in the family is a phenocopy whose trait is the result of something environmental. As a further complication, a case of *nonpenetrance* shares the genetic defect with the affected relatives, but does not manifest the trait.

Phenocopies become a problem in glaucoma genetics studies in several ways. First, when we see glaucoma running through a family, and think that glaucoma in that family results from a genetic defect, we expect that there is a chance that some family member affected with glaucoma might actually not have the familial form of the disease but rather glaucoma due to some outside, unrelated cause. In some cases, as in Figure 24.1, there may be clues to help

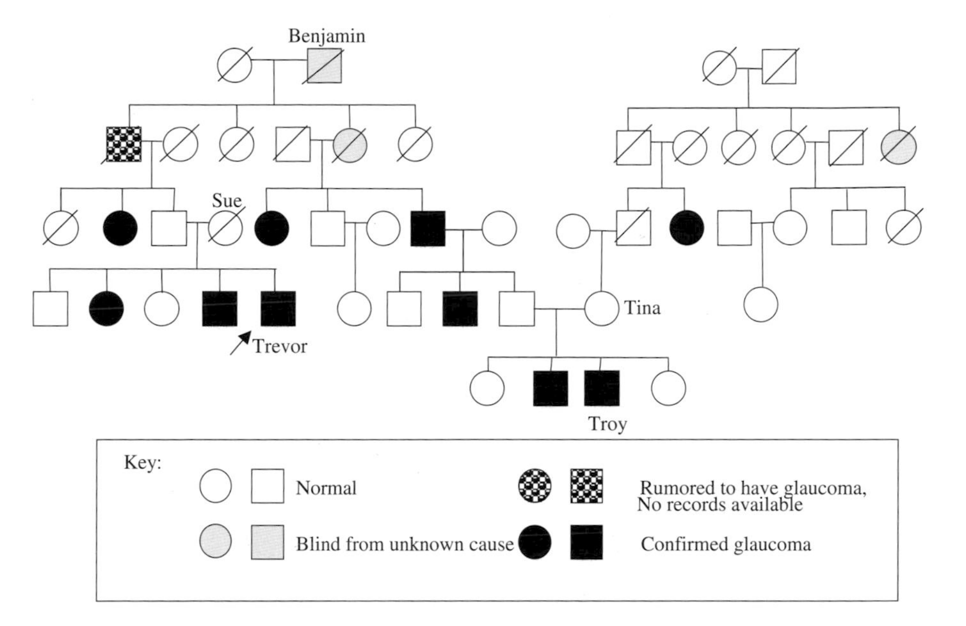

FIGURE 24.2 Let's consider what we would see if Benjamin has seven descendants with a form of glaucoma that begins in young adults and teenagers, one who was rumored to have had glaucoma, and two who were blind from unknown causes. If neither Troy's father nor Trevor's father have glaucoma, we might think this form of glaucoma has incomplete penetrance, leaving some people unaffected who carry the mutation. This would affect the predictions we can make if Trevor asks about the chance of passing glaucoma along to his children. However, before we can draw such conclusions, we need to further investigate the family histories of Trevor's and Troy's mothers. Sue, who had a grandmother with glaucoma, died young, as did Tina's father, so we do not know whether they would have developed glaucoma. Did Trevor's glaucoma come from his mother or from his father? Did Troy's glaucoma come from his mother or his father? Missing family history information makes it hard to answer these question, and the presence of possible additional glaucoma genes in this family lineage complicates everything ranging from efforts to map or clone the gene(s) responsible to use of information in genetic counseling.

us tell who the phenocopies might be, but in other cases, we may not have enough information to tell whether there are any phenocopies at all. Especially when families are large, and information about deceased relatives is coming from distant family members who might not have known them well, information can sometimes be incomplete without it even being obvious that anything is missing.

Identity by state becomes a confounding factor when someone with glaucoma in their family marries someone who also has glaucoma in their family. The possible input of more than one glaucoma genetic defect into the same family might seem an unlikely thing, but when we are dealing with such a common disease, we find that people with a family history of glaucoma often marry into a glaucoma family. Since most people do not develop glaucoma until late age, and some people who carry glaucoma genes pass away before their disease ever manifests itself, it is also possible for some people with a glaucoma defect to marry into a family without realizing that they have a family history or that they will later develop the disease themselves (Figure 24.2). In fact, some people do not realize their relatives have glaucoma even when the affected relatives are currently quite alive and under active care for their vision problems. So it is possible to have different genetic defects coming into the same family from different points in the family structure for several different reasons, and to have very similar medical courses even if the underlying causes include more than one gene and at least one unrecognized environmental factor.

Serious situations can arise when a genetic defect exists but is not recognized as being genetic in origin, or when something is deemed genetic that is not. In the case of osteogenesis imperfecta, some parents have been jailed for child abuse, accused of causing broken bones through physical abuse, when in fact the child had a genetic defect that causes incredibly fragile bones that could not tolerate even the normal stresses of daily life without breaking. In other cases, it can be a big problem if neurological problems in multiple children in a family are dismissed as genetic, if what was really needed, for example, was to identify the source of lead exposure in the home environment that needs to be remediated.

HOW DO WE TELL IF IT IS GENETIC?

One of the ways we tell whether a trait has genetic determinants is by studying the trait in families, looking for the kinds of patterns of inheritance we have talked about earlier (Box 24.1). We have to watch out for some pitfalls.

Among geneticists, a favorite example of nongenetic familial traits is the finding that in some families, attendance at Harvard Medical School runs in the family. Now, although we might hypothesize that a variety of genetic factors affect intelligence, specific talents, temperament, and other factors that contribute over the course of growing up to influence what someone will decide to do with their life, there are a great many other very obvious factors, such as geography, family wealth, family traditions, opportunities for relatives of alumni, and support or pressure from relatives that also contribute to decisions about attending college.

If we study a population of people with a trait, we can ask what fraction of those people have a family history of the trait. In the case of fully penetrant, early-onset, autosomal dominant disease, it will be very easy to identify the familial nature of the trait whether through study of individual families or through examination of family history information in a population. Some studies use information from pairs of sibling—or sib pairs, as they are known—or from twin studies (Box 24.2). For a recessive trait, it may be harder to sort out, especially if the trait is rare. However, occasional large families can provide answers that we cannot get from pooling information from many small families. One of the other clues that a trait is recessive and inherited may come from the study of *consanguineous* families, that is families, in which a married couple shares ancestry (Figure 24.3).

BOX 24.1 ONE OF SCOTT'S FAVORITE RULES

If a given trait is genetic you can map it AND if you can map it then it is genetic.

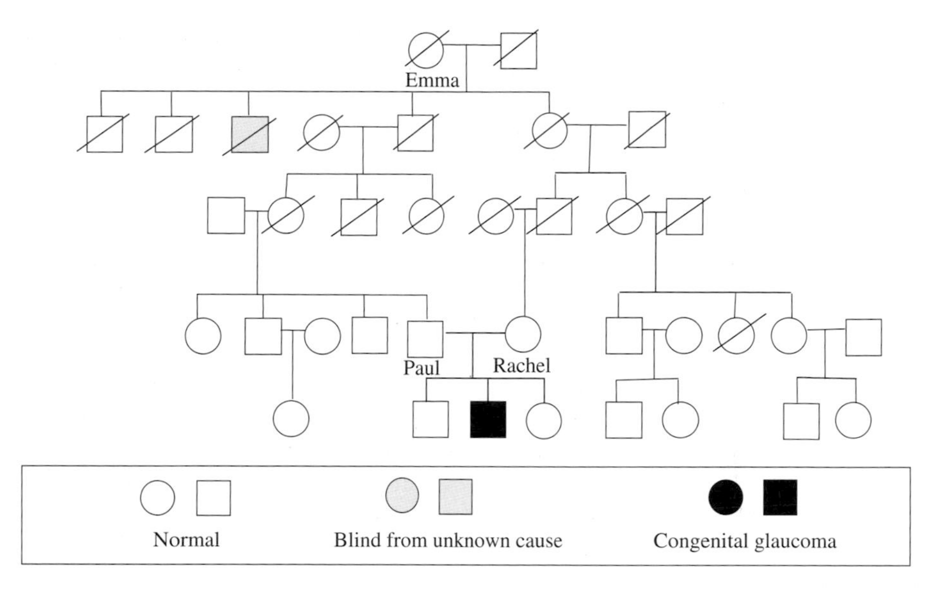

FIGURE 24.3 If Paul and Rachel had a family tree like this, they might know that they share a pair of great-grandparents, but it is unlikely that anything about the family medical history would have told them that they were each carrying one copy of a recessive congenital glaucoma gene inherited from their shared great grandmother Emma. Even if there had been clues available from Emma's generation that she had an uncle with congenital glaucoma, most families would not be sure of what had caused his blindness so long ago, if they had even kept any information about his health status for four generations. Increased frequency of recessive diseases such as congenital glaucoma in consanguineous families like this one helps contribute to our understanding that they are inherited. For some diseases, such as the premature aging syndrome Werner's syndrome, study of this kind of consanguineous family assisted in the discovery of the disease gene.

BOX 24.2 TONGUE-ROLLING TWINS

A trait that has previously been reported as running in families is the ability to roll the tongue so that the sides of the tongue come up towards the roof of the mouth while the whole center of the tongue stays down. Tongue-rolling was studied in fraternal twins and identical twins. It turns out that the fraternal twins are about as likely to share their tongue rolling ability (that is, that they would both be able to or both be unable to roll their tongues) as are identical twins. So pairs of people with very different percentages of gene sharing show the same frequency of trait sharing. This suggests that the primary differences between people who can roll their tongues and ones who cannot may not be genetic.

Identification of a biochemical defect underlying a disease process can also point towards a disease-causing role for genes in a particular biochemical pathway. Once a gene defect is identified, we can test many individuals with that disease to determine how many of them have that particular genetic defect. In some cases, such as cystic fibrosis, we find that everyone with the disease has a defect in the cystic fibrosis gene. In other cases, such as epilepsy, we may find that there are many different epilepsy genes and that only a tiny fraction of the population has a defect in any one of them.

Thus, although you may be able to look at your family and easily identify some simple autosomal dominant traits, for other traits you may need to consult a medical geneticist or a genetic counselor to find out if the item is hereditary, since any one family may not hold enough clues to provide the answer.

WHOSE GENOME SHOULD WE WORRY ABOUT?

Now stretch your brain around this concept: in some cases, your phenotype might not be caused by genetics, but it might be caused by someone else's genotype rather than your own. In the case of *Rh incompatability*, when the baby has an Rh blood type and the mother has an Rh- blood type, the baby can develop anemia after birth as a result of exposure to maternal antibodies directed against the Rh proteins in the baby's blood. However, this can only happen after the mother has borne a previous baby with the Rh blood type. How many other factors that affect a baby's development before birth are also affected by the maternal genotype? We can imagine that a variety of things, including hormone levels that affect whether the baby is delivered prematurely, and other factors affecting things such as nutrition and oxygenation could all affect the traits that will be observed in the new baby that do not depend on the new baby's genotype at all. So in this case the problem has underlying genetic and environmental factors but only some of those genetic factors are found in the baby since one of the critical gene-based elements is present in the mother.

A prime example of this is the case of a mother with *phenylketonuria* (*PKU*) who has two defective copies of the *phenylalanine hydroxylase* (*PAH*) gene that makes her unable to correctly metabolize phenylalanine (Figure 24.4). Even if she ends up healthy because she grew up eating a diet low in phenylalanine, her baby will be damaged by high phenylalanine levels if she does not keep her diet controlled during the pregnancy.

Logically, in addition to *maternal-fetal incompatibilities,* we can imagine *in utero* situations by which the genotype of one fraternal twin could lead to the production of a gene product that could affect the other twin who is not making that gene product. This would be especially important in the case of hormones or other exported proteins that could travel from one sib to the other by getting into the amniotic fluid. In studies of mice, it has been shown that male mice that undergo prenatal development surrounded by female siblings can come out with phenotypic differences from male mice that developed surrounded by other male mice. As with unusual genotypes leading a mother to expose her child to something the baby is not producing, so we can imagine human twins potentially influencing each other while still in the womb. This should only apply to certain kinds of proteins, such as hormones that can leave the first baby and get into the amniotic fluid that contacts the second baby in that shared environment.

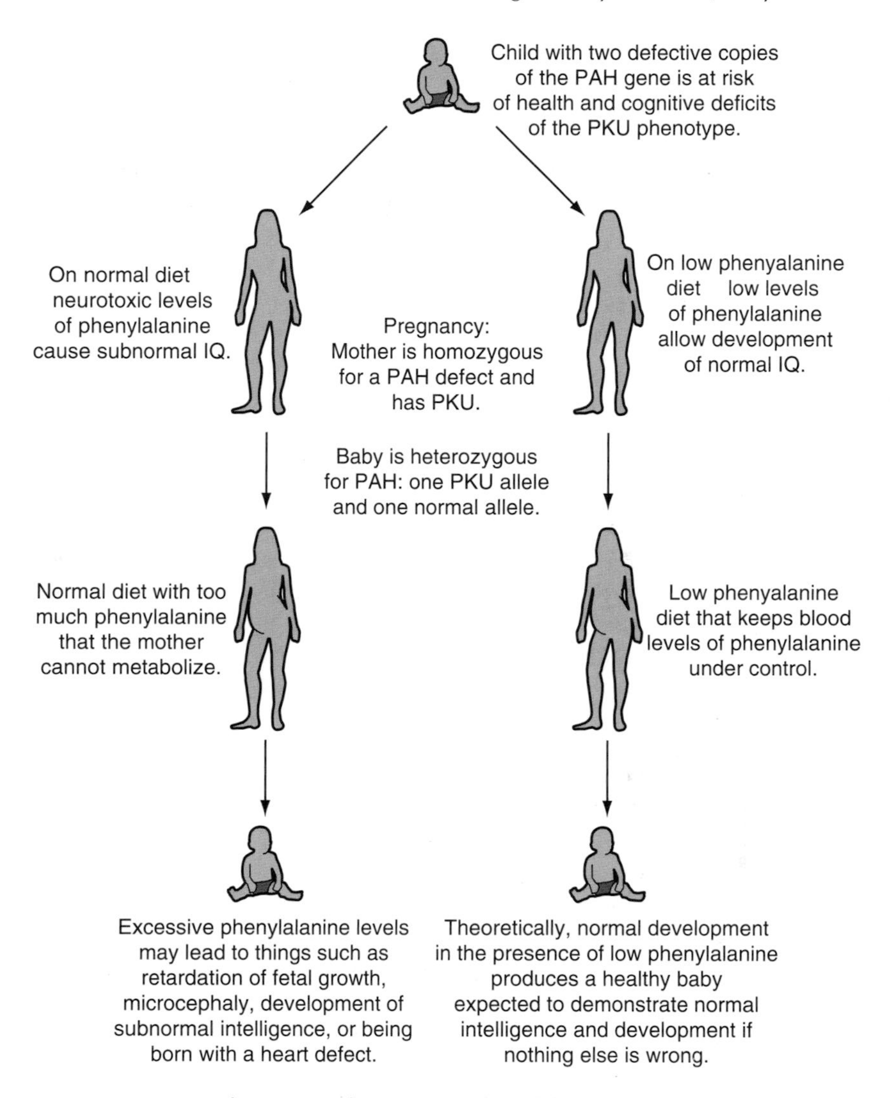

FIGURE 24.4 Deficiency in the enzyme phenylalanine hydroxylase can have devastating consequences not only on the person with the deficiency if they grow up eating a normal diet but also on their child if they do not use careful diet management during the pregnancy. This turns out to be a major problem because it is hard to stay on a diet that has just the right amount of phenylalanine, since damage can come from too little phenylalanine as well as too much.

CONDITIONAL TRAITS

One of the other factors that can confound our understanding of genetic causes of disease are the situations in which the phenotype only manifests if there is also some external eliciting event. In fact, we might think in terms of many people being phenocopies of "normal" for a particular trait not because they lack the genetic component of that trait but rather because they have not been exposed to the necessary stimulus that turns a genetic susceptibility into a disease in reality. When an environmental influence is needed to bring about the manifestation of a trait that is not always present in an individual, we call it a *conditional trait.*

In a disorder called *malignant hyperthermia,* individuals exposed to general anesthetic experience "explosive" elevation of body temperature, and many of them die. It has been suggested that there might be links to sudden infant death syndrome (SIDS) or heatstroke in some members of malignant hyperthermia families. Individuals who have never had general anesthesia do not know whether they carry a malignant hyperthermia defect. For most people, it might not seem like a great risk because the frequency of malignant hyperthermia has been reported to be about 1 in 50,000 cases of general anesthesia. However, the risk is substantially greater for those who have had a relative manifest malignant hyperthermia. This is an example of what is called a *pharmacogenetic disorder,* a disease state that occurs in response to a pharmaceutical agent (Box 24.3). For many pharmacogenetic traits, we expect that there will be multiple genes in the genome that will affect the combination of effectiveness and side effects that can result from use of any given drug. When dealing with a pharmacogenetic trait, it can be difficult to tell who is affected or at risk, since often there will be many family members who have never been exposed to the eliciting event.

Environmental components of dietary sensitivities and allergies can also complicate our efforts to tell whether a particular item is genetic because the

BOX 24.3 PHARMACOGENETICS

The emerging field of *pharmacogenomics* involves the study of human genetic differences that contribute to differences in reactions to drugs. In some cases, such as *malignant hyperthermia,* the variability may take the form of an *iatrogenic,* or drug-induced, illness. In other cases, there may be a difference in response to the drug, as has been seen in the cases of some asthma sufferers who do not respond to an asthma medication because they have a mutation that changes the receptor protein that binds the drug. One form of glaucoma results when individuals who are susceptible to the effects of corticosteroids are exposed to dexamethasone, which causes elevated intraocular pressure that will lead to glaucoma if left untreated (see Figure 24.1). In addition, some patients do not respond to one or more of the glaucoma medications they are given, possibly because of genetic differences in the body's ability to react to the drug. Often, the real array of variability in response to a drug is expected to be complex and involve differences in many different genes in the human genome. Thus, for example, we might expect the efficacy of a drug to be affected not only by the sequence of the proteins that the drug will interact with in the human body but also by differences that affect proteins that will transport the drug into the cell, degrade and eliminate the drug, pump the drug out of the cell, target the drug to a particular place in the body, or carry out an immune reaction to the drug. Eventually, it is hoped that it will be possible to carry out genetic tests that will tell us whether one drug will work better than another for a particular person, or what side effects or risks there would be for that person if they take the drug.

trait will only be manifested by people who have been exposed to the food or allergen. A trait called *favism* is characterized by a form of hemolytic anemia that happens after consumption of fava beans. Someone who has never consumed fava beans would not know whether they would be susceptible to favism or not. The genetics of lactase persistence/lactose intolerance, discussed in Chapter 17, is very hard to evaluate in countries where dairy products are not part of the diet. Sorting out the genetic components of allergies can be especially difficult because often people with allergies react to many different allergens, not just one, while being exposed to a vastly larger array of potential allergenic culprits. However, in some cases, such as penicillin allergy, the eliciting event (taking the antibiotic) and the unusual nature of the allergic reaction (a rash) make it easier to identify than some more generalized allergies to airborne allergens.

In a world in which many things start going wrong with people as they age, we need to think of aging as one of the eliciting events for conditional traits. *Age-dependent penetrance* is the term used to express the increased expression of a trait in older populations when compared with younger populations. Issues of aging and conditional expression of traits are genetically complex. This becomes complicated as we consider that genetic variation affects the rate at which we age, which in turn affects the rate at which aging causes us to express other traits caused by genetically variable susceptibilities.

INFECTIOUS DISEASE SUSCEPTIBILITIES

The human body offers many different *host defenses* against infectious diseases, and each of these defenses uses structures or processes involving proteins encoded by one or more genes in the human genome. In some cases, susceptibility to infection may be accounted for by some generalized reduction in effectiveness of the immune system, such as that seen in individuals with mutations in the adenosine deaminase gene who develop *severe combined immune deficiency* (SCID).

However, many factors affecting susceptibility to infectious disease involve particular proteins that carry out very specific interactions with the infectious agent. For instance, a naturally occurring mutation in a gene encoding the immune cell protein CCR5 results in a shortened protein that cannot be used by the human immunodeficiency virus (HIV) to enter the cells in which it replicates itself, resulting in rare individuals who are considered long-term nonprogressers who may go for decades without developing *acquired immunodeficiency syndrome* (AIDS). People with type O blood are more likely to have more serious cases of cholera if they become infected with *Vibrio cholerae.* In some populations, sequence differences in the *Nrapm1* gene are associated with predisposition to leprosy and tuberculosis.

When we look at the number of different genes whose variants affect susceptibility to malaria (such as genes that encode -globin, -globin, glucose-6-phosphate dehydrogenase, nitric oxide synthase 2, the Duffy blood group, and others) and the dramatic differences in frequencies of the key variants in populations living in areas where malaria is endemic, it would appear that lethal infectious diseases can be associated with substantial shifts in frequencies of mutations that confer a survival advantage. For many infectious agents, many different genes can be involved in determining susceptibility, and the result is often *mitigation* (reduction in the chance of becoming ill or reduction in severity of the resulting illness) rather than outright prevention of the disease. Thus even determining whether anyone became ill as a result of exposure to an infectious agent may not offer a clear answer to what that individual's genotype holds by way of resistance or susceptibility to that infectious agent.

MAD COWS AND CANNIBALS

One group of conditional traits results from exposure to something that was originally thought to be an infectious agent, such as a virus or bacterium. This agent later turned out to be something much more perplexing, a simple infectious brain protein called the *prion* protein. Prions are not alive and contain no genetic information, but they nonetheless seem to be able to get the host organism to produce more of the defective protein that causes damage to the brain. In fact, it has been argued that what prions really do is get the brain to turn the normal version of the protein into an altered disease-causing version of itself.

Kuru, a lethal human neurodegenerative disease called a *transmissible spongiform encephalopathy* (TSE), is transmitted between members of the Fore tribe in New Guinea via cannibalism. In cases of kuru, plaques and lesions in the brain look like those found in other diseases, such as scrapie in sheep, mad cow disease in cattle, and similar transmissible spongiform encaphalopathies found in other animals, including some kinds of wild game consumed by humans. The disease appears to originate from exposure to a misfolded form of the prion protein. Exposure to the externally supplied, misfolded prion protein coming from an infected person or animal apparently causes misfolding of endogenously produced prion protein so that the patient ends up with far more misfolded prion protein than he or she actually ingested. Of great concern is the fact that prion infections can jump between species, as shown in the 1990s when a small number of people in Great Britain were reported to have developed transmissible spongiform encephalopathy from eating beef containing the "mad cow" version of the prion protein. We are left with questions about what genetic susceptibilities contributed to the production of disease in the rare few who developed transmissible spongiform encephalopathy out of a huge number of beef eaters. Questions are being asked about whether some of the increase in cases of Alzheimer's disease might be phenocopies that are actually undiagnosed transmissible spongiform encephalopathies rather than Alzheimer's disease.

This conditional trait is a phenocopy of a genetic disorder because the pathology of the disease resembles that of a human genetic disease called *Creutzfeldt-Jacob disease (CJD)*. This hereditary neurodegenerative disorder presents with the same kinds of plaques and lesions found in the transmissible spongiform encephalopathies but is transmitted genetically within a family instead of via consumption of infected meat. There seems to be a certain logic to the finding that kuru and mad cow disease, which are transmitted by prions, would resemble CJD because the CJD genetic defect is a mutation in the gene that encodes the endogenous prion protein. It has been pointed out that four out of six individuals who contracted a transmitted (rather than inherited) form of CJD all had a rare variant of the prion protein in which a valine is found at position 129 of the protein sequence. This suggests that genetic variants in the prion protein gene may help determine whether someone is susceptible to the development of transmissible spongiform encephalopathy in people who are exposed to misfolded prion protein. This might also explain why the predicted mad cow epidemic has not materialized.

GENETIC CLASSIFICATION ASSISTS RISK ESTIMATION

So far, the complex issues of interaction between genotype and environment are understood for only a small fraction of genetic diseases. This can make it difficult to sort out the genetic components of many traits that involve phenocopies or that may be conditional upon environmental effects, especially if there are many different genes and environmental factors that can affect the trait. Through the studies of geneticists, epidemiologists, and others, the complex interplay of genes and environment is gradually coming into focus. Will this eventually let us pinpoint who will become alcoholics if they take up drinking, or who it is that will develop lung cancer if they take up smoking? The goal is to end up being able to offer estimates of altered risk in the presence of environmental effects. Those risk estimates will tell us who has susceptibilities that should lead to more medical monitoring, different treatments, or advice on how we should change our diets or other behavior patterns. For many diseases that involve multiple different genetic components to susceptibility in addition to a complex array of environmental factors, it will continue being difficult to make certain predictions about the medical fate of a specific individual even if we can make statistical statements of risk for groups of people.

Sorting out which effects are genetic and which effects are not will assist those who are hunting for the genes that affect susceptibility. It will also take us a long way towards a better understanding of many different risks in life.

THE HUMAN GENOME LANDSCAPE

In the 1980s and 1990s, a revolution in the biological sciences applied high-throughput production line approaches to biology. The result was the unveiling of a draft version of the human genome sequence in 2001, with a major update to the sequence in 2003. Experiments that once took weeks, months, or even years can now be done quickly, sometimes being completed in a matter of hours or days. Sometimes we can bypass the lab bench altogether and ask our question entirely within the computer. In this section, we present the history of this biological revolution and describe some of what was found when the human genome was sequenced.

THE HUMAN GENOME PROJECT

25

On July 20, 1969, many of us watched, spellbound, as Neil Armstrong stepped off the bottom rung of the ladder to the lunar lander and made the first human footprints in the powdery dust on the surface of the moon. "That's one small step for man, one giant leap for mankind," resounded around the world and down through the subsequent years. All of us who witnessed it felt the magic of that astonishing moment that transformed mankind from an earthbound species to one that had walked on the surface of another celestial body. We had watched for years as launch after launch headed us inexorably towards that moment when we all held our breaths while Armstrong stepped off the ladder. Armstrong's famous declaration of a leap for mankind told us that humanity had just crossed over an historic divide, separating all of prior earthbound history from all of subsequent history. Often, the real watershed moments in history can be harder to pinpoint, as we have seen in the course of the human genome project. (Photo courtesy of NASA)

Over the last thirty years, genetics research has been building towards a great historic divide, the moment when the genetic information at the heart of every human cell would be unveiled. Like the new and successive launches that carried the space program towards the moonwalk, major breakthroughs in molecular biology have paved the way for the anticipated release of the human genome sequence. Those of us in the field of genetics have watched the series of breakthroughs that have led towards this genetic "leap for mankind" with increasing anticipation. In the spring of 2000, completion of the draft version of the *human genome sequence* was announced at press conferences involving scientific and political leaders. With the announcement of the draft sequence, the field of genetics began a kind of countdown from that initial moment towards the true completion of the sequence, slated for April of 2003, a countdown not unlike the period of time during which we watch in anticipation as the ball falls in Times Square on New Years Eve.

What is the human genome sequence? Simply put, it is the order in which the letters of the genetic alphabet are arranged along the chromosomal DNA strands that are millions of letters in length. The whole length of the sequence, from one end to the other, measures more than three billion A's, C's, G's, and T's arranged one after another to spell out what we are made of and how our cells operate.

The human genome sequence did not simply spring into existence, nonexistent one day and completed the next. When Armstrong stepped on the moon, there was a discrete event to celebrate, but it followed upon years of technical development and vast amounts of work by large numbers of people who did not get to share the spotlight. Similarly, the unveiling of the sequence has been a similarly continuous process, developed over decades by scientists all over the world working towards the announcement that it is done. Yet, after several years of working with a draft version of the sequence, now that completion is upon us, we find ourselves wondering when and how anyone can arrive at a point of saying, "A moment ago it was incomplete, but now it is done" (Box 25.1).

BOX 25.1 WHEN IS A GENOME SEQUENCE CONSIDERED DONE?

The International Human Genome Consortium has set a standard for when and how the genome sequence of an organism can be considered done. Being done might seem a very simple matter. After all, it's done when it's done. In fact, as with any task a human tackles, we can always strive for a bit more perfection before considering a job truly done. It must be complete, with no remaining gaps in the sequence, and the sequence must be of high quality and accurate to better than 1 one base pair in every 10,000. In the case of the human genome sequence, efforts to meet the Consortium's definition of completeness are confounded by many features of the human genome that make some areas harder to read than others. There are some areas in the sequence that initially defeated efforts to clone and sequence them using the earliest technologies available. Scattered throughout the genome are small regions that were not clonable because they are lethal to the bacterial cloning system used to isolate them. Fortunately, as technologies have advanced, and as the regions of sequence have been completed surrounding the "unclonable" regions, it became possible to use polymerase chain reaction or alternative cloning systems to obtain and read DNA from many remaining problematic regions. Another problem for the sequencers has been some of the very large regions that are filled with repeated sequences, a pattern of sequence that occurs over and over, such as the regions around the centromeres. As with the many other problems that have arisen in the Human Genome Project, each new problem that arose led to new technical developments that not only solved the sequencing problem but in most cases also enabled researchers in other arenas, such as cancer biology or cardiac genetics, to take new approaches to the questions they were trying to ask. So, for now, the Consortium standard defines the sequence as being done because the majority of the sequence the rest of us want to access has been completed, but to truly meet the definition of complete, there are still i's to dot and t's to cross. Thus the "final" April, 2003, version of the sequence will be succeeded by other later versions that will include small changes here and there because, really, when we consider it done is determined by how perfect we want it to be, or need it to be, for it to help us address the other questions that need to be asked. That will change as our use of the sequence points us in the direction of new questions that we do not yet even know enough to be able to ask.

The first pieces of sequence from human genes and chromosomes started coming out in the 1970s. At that point, during the very first years of the human gene hunt, deciphering of a few hundreds or thousands of letters spelling out a tiny segment of the genome was regarded as a huge achievement. Julia learned early in her graduate school research project that having even a piece of a gene in the late 1970s was enough to get her research published in one of the top scientific journals in the world. During the early work on generating sequence from human genes, each of the labs around the world was working on DNA from different individuals. During the late stages of the human genome project, the experimental design called for the sequence to come from multiple different individuals (Box 25.2).

To look at the rate at which the sequence was unveiled, we can examine the amount of sequence present in the publicly accessible database called Genbank. This electronic information repository contains the human genome sequence in addition to sequence information on all of the other organisms being studied (BOX 25.3). The information contained in Genbank comes from not only our close relatives, such as primates and mice, but also from much more distantly related organisms, such as fruit flies, zebra fish, yeast, bacteria, and even viruses. In 1982, the first year for which information

BOX 25.2 WHOSE SEQUENCE IS IT?

One of the key decisions of the Human Genome Project is that the sequence being read in the sequencing experiments would not come from one single individual. This was an important decision, since there actually is no such thing as *the* human genome sequence. Although there are many letters in the human genome sequence that seem to be the same no matter whose sequence you look at, many places in the human genome sequence show a lot of variation from one person to the next. We still do not know all there is to know about the human genome sequence, since the known sequence represents what is present on the chromosomes of a small number of individuals. However, a lot of work remains to be done to understand what kinds of variation in the sequence are present in populations not represented in the initial sequencing set, or of greater importance, to sort out which of the differences matter. Also, researchers need to figure out which differences in the sequence are associated with different levels of medical risk and which differences actually make no difference whatsoever in the characteristics of the person bearing that particular sequence variant. One of the current challenges to the bioinformatics branch of the human genome community is to find ways to elucidate and display information on human genetic variation at the level of individual sequence differences in ways that will allow us to understand the roles those differences play in the growth, development, life, and death of each human being. One of the other challenges, explored by the Ethical, Legal, and Social Implications (ELSI) arm of the Human Genome Project, will be to ensure that the resulting information will be used in ways that will be beneficial to individuals and to humanity as a whole.

BOX 25.3 THE OTHER GENOME PROJECTS

The human organism is not the only organism for which we know the genome sequence. More than 100 different bacterial and parasite genomes have been completed, including the genomes of important pathogens, such as ones that cause cholera and meningitis. As more bacterial genomes are being completed, additional work is going into sequencing of different strains with important phenotypes, with important findings on the "pathosphere" emerging from comparisons of strains that cause different diseases or symptoms. While work progressed on bacteria and the even smaller genomes of some important viruses, the list of more complex organisms being sequenced has grown. Plant genome projects have been driven by agricultural needs to improve the nutritional composition of grains and the ability of plants to survive pests and disease. At the same time, advances in our ability to manipulate genomes have led to debates over genetically modified foods. Animal genome projects have been driven not only by agricultural interests but also by breeders of purebred animals and veterinary interests in curing diseases affecting peoples' pets. Some of the most important advances so far have come from the projects aimed at sequencing the genomes of a key set of research organisms, some of them now pronounced done, such as the fruit fly, some hovering in a near final state of completion, such as the mouse, and some in the early stages as we write this and yet likely to be done by the time this book is published, such as the zebra fish. Sequence similarities and clustering of genetic groupings in related organisms allow scientists to rapidly take findings from one species and do studies to confirm them in other species. The result is that a gene found in a fruit fly, studied cheaply in large numbers of lightning-fast experiments, can lead to an understanding of something important in the mouse that begins the development of pharmaceutical products destined eventually for human use. Thus, although many of the many ongoing plant and animal genome projects are justified on some levels by direct applications of the information to that particular organism, in most cases the findings for any one organism will also affect our understanding of things going on in other organisms. This is especially true for the research organisms, including the mouse, the fruit fly, and the zebra fish, which allow for very powerful genetic and biochemical studies to develop materials and information that can then be applied to studies of other critters, including we humans.

is available, there were 606 files of sequence data containing 680,338 base pairs of sequence. By 1992 there were more than 200 million base pairs of sequence. By 1998 there were more than 2 billion base pairs. By the year 2000 the number of files in Genbank passed 10 million, and the amount of sequence available from all organisms passed 11 billion base pairs! By 2002, this had grown to more than 22 billion base pairs of sequence, according to the National Center for Biotechnology Information. Over the course of twenty-five years, we had progressed from a stage at which massive efforts by a team of people working together for months or years produced a small piece of one gene a few hundred bases in length, to the point at which the sequence was being read so rapidly that new computer technologies and mathematical algorithms had to be developed to be able to handle the massive rate at which the sequence was expanding.

PROGRESSION TOWARDS THE SEQUENCE

At the beginning of the twentieth century, as the lost works of Mendel were being rediscovered and introduced to the scientific world, no one knew what Mendel's proposed genes were actually made of. Because the presence of only four letters in the alphabet seemed to lack the complexity needed to encode the amount of information in the genome, the scientific world originally dismissed DNA as the possible source of genetic information. There was rather universal surprise when experiments in the 1940s showed that DNA is in fact the repository of genetic information. By 1953, unveiling of the secrets of DNA replication and transcription via base pairing answered some of the questions about how genetic information could be copied and transmitted from cell to cell and from one generation to the next. By 1961 the unveiling of the triplet code solved the problem of how to spell out twenty amino acids using only four letters.

As Armstrong stood on the moon in 1969, no one had yet deciphered the order of A's, C's, G's, and T's in even one human gene. That awaited the invention of *DNA sequencing*, the technical process by which we read the order of genetic letters in a piece of DNA. Two different methods for determining DNA sequence emerged into the scientific world during the 1970s, along with *gene cloning*, the process by which an individual gene can be separated away from the rest of the genome and replicated in many identical copies. In the 1980s, polymerase chain reaction (PCR) gave us the ability to study DNA without first cloning it, and artificial chromosomes gave us the ability to clone enormous pieces of DNA thousands of times the size of the first things cloned in the 1970s. The 1990s were swept along by the concept of *sequence tagged sites* (*STSs*), pieces of sequence that could be entered into a publicly accessible database as a mechanism of transferring genetic information between labs without having to physically transport any actual samples of DNA. The new millennium saw the emergence of *microarrays* and *gene chips* as tools for being able to survey the expression of tens of thousands of genes in one experiment. In the course of fifty years, we had gone from not even being sure of the fundamental nature of genetic material to a point at which we know so much about our genetic blueprint that we can now do meaningful virtual genetics experiments inside of computers without ever touching a pipette, experiments involving millions or even billions of pieces of information. In the course of reaching this point, we saw not only the invention of new scientific techniques and equipment, we also saw some fundamental changes in how we approach science.

THE HUMAN GENOME PROJECT—THE RACE TO THE FINISH LINE

So just what is the Human Genome Project? To understand this, let's start by looking at the practice of genetic science before the Human Genome Project. In the early stages, molecular genetic research was carried out by chipping away at the human genome one gene at a time, with each research group picking some gene or disease or piece of the genome that they needed to be able to answer their particular questions, and working to identify and study that gene in the context of a variety of genetic, medical, biochemical, structural, physiological, and cellular issues that made the cloning and sequencing of the gene only a small part of the project. Thus each new gene became a known bit of the genome only as fast as research groups moved ahead on their set of questions to the point of needing to get the next gene involved in the problem they were studying.

The Human Genome Project started in the 1980s as an idea discussed at a series of scientific meetings by researchers who thought the existing approaches were inefficient. It all began with the idea that determining the sequence of the whole human genome would provide an incredibly powerful tool that would make possible many things that could not previously be done. It would be much more efficient for some researchers to just go after the sequence and then let that sequence become a powerful tool in many forms of research than the existing system of having many different labs go after one gene at a time while spending time and resources on other aspects of their research outside of just getting the sequence.

To accomplish this goal, a set of Genome Centers were established that could just focus on the aims of the Human Genome Project. With the establishment of the International Human Genome Project, the efforts on the part of Genome Centers spread around the United States were joined by researchers from all over the world. By the time the formal Human Genome Project came into being, the goals of the project had expanded beyond simply obtaining the sequence. Additional goals were added, including:

- Cloning, that is, isolating small parts of the genome in individual clones (getting one needle at a time out of the haystack)
- Development of *genetic and physical maps* showing the relationships of the different clones to each other and to pieces of sequence, such as STSs
- Identification of all of the genes, something that might sound easy if you know the whole sequence, but which is actually tricky
- Development of databases, new technologies, and new analytical approaches
- Establishment of a set of projects on the *ethical, legal, and social issues* (known as *ELSI*) surrounding the use of the information being generated

As government-funded projects flourished, industry joined in with an interesting mix of collaboration and competition with the publicly funded projects. As the initial groundwork of getting the clones and getting them mapped was finished, the real work of sequencing took off and turned out to be a run to the finish line between the International Human Genome Project effort on the public side of the race and Celera Genomics, Inc., a biotechnology company using a different technical approach to the problem. The result of the photo-finish dash across the line was a joint press conference between Francis Collins, Director of National Human Genome Research Insti-

FIGURE 25.1 Since their press conference announcing the existence a draft version of the human genome sequence, Craig Venter (left) and Francis Collins (right) have continued to play major leadership roles in the field of genomics as the completed sequence of the human genome was being assembled. (Photo by Alex Wong, *Nature*, 2/15/01.)

tute, and other major leaders in the International Human Genome Organization on the one side, and Craig Venter, the founder of Celera, on the other side (Figure 25.1).

As the side-by-side publications from the two scientific camps came out, with simultaneous publication of Celera's paper in the journal *Science* and the Human Genome Organization paper in the journal *Nature*, each side clearly felt that they had achieved some kind of victory. Venter seemed victorious because his use of a different strategy started later in the game but then caught up. Collins and the Human Genome Organization, on the other hand, seem to hold a certain moral victory because all of their research information had been available to (and used by) the industrial researchers in the course of their work, but the Human Genome Organization did not have access to the information being generated by the industrial researchers, who kept their information to themselves until the end and left the Human Genome groups to function without the benefit of their data.

In the course of this race, major questions were raised about public release vs. private ownership of the sequence locked up inside every cell in our bodies. During the 1990s, questions were raised concerning whether the sequence of a human gene can be patented. The very idea of this horrified many researchers, even as others were rushing out to file patents on things they had cloned and sequenced. Vigorous debate in both legal and scientific circles has resulted in a shift to the current view encountered by those of us stepping into the legal system to initiate a patent. Currently, we are being told that we cannot simply patent the sequence of a gene, which is a relief to those of us who feel as if we own our own genomes! However, we are being told that the gene can be patented if there is an idea regarding some use that could be made of that gene for a particular purpose, such as screening people to find out if they are at risk for a particular disease. Where the real line will be drawn regarding what can be patented will likely continue to change as patents are challenged in court and as those who seek to patent their findings identify new loopholes in the ways the laws are written.

Arguments on one side of the issue say that only with patents will there be financial protections needed for a pharmaceutical company to be able to invest the massive amounts of money needed to develop a product and bring it to market. On the other side of the issue are those concerned that, if use of a particular gene is locked up in a patent that does not get developed or that does not get licensed to others for use in development of other applications, all of us will suffer. Certainly, we have already seen a situation in which the presence of competing patents on a particular gene have inhibited development of some potential applications of that gene. When a gene is not available to be generally explored by anyone with a new idea for its development, we all potentially lose. The system of licensing patents is supposed to allow for others to develop derivative products, but if someone's bright new idea would compete with the primary patent holder, can we expect that the long-term welfare of the human race is going to be the deciding issue if market share and profits are at stake? We would like to think that it will sort itself out so that the necessary developments will take place, but the system currently walks an uncomfortable line between protections that allow product development and restrictions that block product development. As long as further exploration of patented genes within the nonprofit context of academic research is allowed, we can hope that the feared stifling of ideas and developments will not be the result of allowing a company to patent the use of part of your genetic information.

The Human Genome Project has earned a lot of interest, but it has not captivated the imagination of the whole world in quite the same way the moonwalk did. Those of us who played at being astronauts and walking on the moon do not usually find our children putting on white lab coats, picking up toy pipettes, and dreaming of finding a gene. This is in part because a single publicized moment that shows someone stepping out of a spacecraft is fundamentally more glamorous and easier to identify with than putting in long hours pipetting things in a lab. Still, the completion of the sequence has garnered its share of attention, at least among adults. The sheer wonder of deciphering the entire set of genetic information inside a human cell has caused a flood of headlines, news articles, TV specials, and press releases as new findings emerge. Even some heads of state got caught up in the excitement and joined in the televised announcements of the completed draft version of the human genome sequence announced in the spring of 2000. Those of us who see our day-to-day lives in science revolutionized by the existence of the completed sequence look back at the scientists who proposed the Human Genome Project and wonder if they realized what astonishing things would come of their daring dream.

People sometimes ask what we scientists are going to do with ourselves after "the end," now that the human genome sequence is finally known. We find ourselves with a very different reaction. Once the human genome sequence is done, once we have read the genetic information inside the human cell all the way from one end to the other, that is really just the beginning. With the sequence in hand, we can finally begin asking things we could not have dreamed of asking back in 1969, when Armstrong walked on the moon and no one knew the sequence of even one human gene.

THERE'S CLONING AND THEN THERE'S CLONING

26

"SCIENTIST REPORTS FIRST CLONING EVER OF ADULT MAMMAL"
Headline in the New York Times, Feb 23, 1997

The invention of molecular cloning was one of the biggest advances towards being able to accomplish the sequencing of the human genome. Earlier, we described how sequencing works, but the other thing that was needed along with the sequencing technology was a way for scientists to get their hands on pieces of DNA on which to carry out the sequencing reactions. It might seem like a simple thing. If you want DNA, just break open a human cell, and the DNA will flow out into the surrounding solution. In fact, it is not at all simple, since the DNA that comes out of a human cell contains the entire genome,

FIGURE 26.1 Dolly, the first cloned mammal, has a genome donated from a mammary cell nucleus from her only parent. In veterinary science, cloning offers the possibility of being able to transfer copies of rare, agriculturally valuable animals into general use by farmers. It also offers the possibility of being able to make transgenic animals that can produce specific proteins of interest, especially human proteins to be harvested for pharmaceutical use. Here Dolly is shown with her firstborn lamb Bonnie. (Photo courtesy of the Roslin Institute, Midlothian, Scotland, UK.)

and it is not possible to sequence individual genes or regions of chromosomes in the presence of a billion base pairs of other DNA that you are not trying to sequence.

Cloning first turned up in the 1970s, and in its aftermath during the 1980s and 1990s, genetics swept the headlines one gene at a time, one disease at a time, one organism at a time. Then in 1997, the *New York Times* announced that the first adult clone of a mammal had been produced. Scientists talk about plans to use therapeutic cloning to grow new organs from cells of an individual in need of a transplant so that there would be no transplant rejection issues to deal with. The term *cloning* seems to turn up in some situations that sound quite different from each other, so given that cloning seems to be at the heart of quite a lot of genetic activity of great importance, we are faced with the question: Just what is cloning?

Unfortunately, for those of us who spend our lives trying to explain such things, the term has more than one meaning in contemporary genetics. In fact, to someone concerned with human genetics, cloning applies to at least three distinctly different situations, cloning genes, cloning organisms, and something else sometimes called therapeutic cloning. The fundamental concept in all of the uses of the term remains the same: *a clone is a genetically identical copy of an organism or DNA segment produced through the use of biological technologies.*

THERE'S CLONING . . .

One of the big breakthroughs that began the era of molecular biology in the 1970s was gene cloning (Box 26.1). This allowed researchers to trick

BOX 26.1 THE UPROAR OVER CLONING

In the early 1970s, when early observations on some obscure aspects of bacteria led to the concept of cloning, there were grave concerns about things that could possibly go wrong if human DNA were placed into bacterial cells and allowed to replicate. These concerns ranged from simple technical issues to major fears that new pathogens would be created that could sweep through human or animal populations. Scientists involved in the early molecular cloning studies put a voluntary halt to the work and convened to discuss the ramifications of the new technology, both technical and ethical. This led to the development of a set of recombinant DNA guidelines that put very strict physical and technical limits on the kinds of work done in the early experiments. Although this slowed down the earliest progress on studies of the human genome, it allowed researchers to gain enough information about molecular clones to better determine how to safely proceed in the future. The result is that enough was learned to allow researchers to know which types of experiments continue to need a great deal of physical and biological isolation to be safely performed, and to know that some kinds of experiments can now be done safely with much lower levels of containment. Three decades of safe and successful science have proven the value of the steps taken when the scientific world paused to consider the possible consequences of cloning.

bacterial cells into making them lots and lots and lots of very pure copies of a piece of DNA of interest, no matter what organism the piece of DNA originated from. Over the last thirty years, molecular cloning has led to the identification of an increasing number of important genes and proteins that were not known before cloning, culminating recently with the completion of the human genome sequence and the identification of most of the human genes. This kind of cloning involves making copies of DNA. The basic idea is that a desired section of DNA, often containing one or several genes, is separated away from the rest of the human genome in a way that lets researchers make many copies of that piece of DNA in a pure form.

Before we consider how we clone a human gene or a piece of human DNA, let's consider why would we want to separate a piece of DNA away from the rest of the genome and make copies of it. If we try to read the sequence of a single gene while it is still present along with the rest of the genome, we find ourselves trying to detect a small amount of information that is buried in a large amount of very similar information. It is like looking for the proverbial needle in the haystack, when the signal you are looking for looks too much like vast amounts of extraneous information. However, if we can take the piece of DNA we want out of the cell and away from the billions of base pairs of sequence we are *not* trying to look at, and then make many, many copies of it, we find that the things we want to look at are easy to detect and give clear, clean results. Instead of trying to visualize a single needle buried in straw, we find ourselves looking at a pile of needles on a smooth, clear surface.

One of the keys to cloning a piece of DNA is the use of something called a *vector*, a piece of DNA that can replicate itself if it is put into a cellular "replicating factory," such as a bacterial or yeast cell. The DNA bases used in the human genome, the bacterial genome, and the vector genome are the same four DNA bases, A, C, G, and T. *Cloners*—people who isolate and make copies of DNA by combining them with a vector—use a kind of biological glue called ligase to splice together the human DNA and the vector DNA in a way that leaves no seam to indicate where the vector DNA stops and the foreign DNA ends. The product that results is one new intact piece of DNA that combines DNA from two different sources that were originally separate. The systems that copy the vector cannot tell that part of this new large piece of DNA is foreign, so they happily copy the foreign (in this case human) DNA right along with the vector itself. This process of recombining DNA from two different sources into one new structure is the basis for the term *recombinant DNA* (Figure 26.2).

Once human DNA had been cut up and combined with a vector, the resultant recombinant DNA was put into a host cell (such as a bacterial cell), where large numbers of copies of the recombinant DNA construct (or clone) could be made. Each insert-bearing vector is put into a bacterial cell, and the bacterial cell is placed on an agar plate, where the cell makes many copies of itself (and the insert-bearing vector it carries) all sitting at the same position on a plate in a little mound called a *colony*. If many such clones are grown on one plate, it is then possible to make a copy of the pattern of colonies by pressing a piece of filter paper onto the surface of the plate. Washing the filter paper in a solution that contains a tag that can recognize the clone with the desired sequence lets the researcher identify the colony housing the clone of interest. Because the colonies all sit apart from each other on a solid surface,

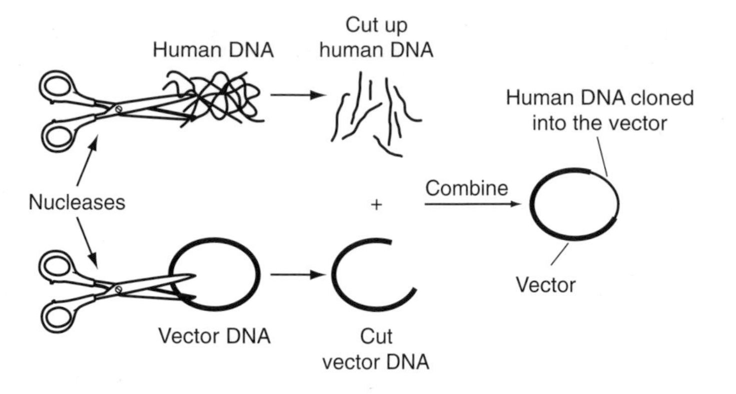

FIGURE 26.2 Recombinant DNA. Vector DNA is cut open and combined with cut-up human DNA. The cutting of the DNA is carried out by proteins called *restriction enzymes*, or *nucleases*, whose specific job in nature is cutting DNA. These nucleases act like scissors that can cut DNA at very precise positions within a piece of DNA. The two pieces of DNA are spliced together into one continuous piece of DNA that can be copied many times so that researchers can obtain large amounts of the clone in a very pure form. One of the key features of the vector is that it has an *origin of replication* used to initiate copying of the DNA. The hard part in cloning is not combining human DNA with a vector but getting the vector-human hybrid that contains the gene of interest separated away from the other hybrids containing the rest of the genes, a problem that was worked out in the 1970s.

once the right colony is identified, it can be lifted off of the agar plate and grown in large amounts away from the rest of the clones (Figure 26.3).

Such cloning techniques have been used to create clones that can then be used for a large number of different purposes. For instance, cloning of this kind can be used not only to isolate genes but also to obtain pieces of DNA from human chromosomes and mitochondria. Techniques like this have been used to make new versions of human genes, such as versions of human genes that contain particular mutations on which functional studies are needed, or versions of human genes from which key regions have been removed to test their functional significance. They have been used to make "expression" constructs in which the clone not only contains the DNA sequence of the desired gene but can even be used to produce the protein gene product of that gene. Molecular clones can be put back into human cells to determine whether addition of a particular gene can "fix" a particular metabolic defect present in a cell line. Molecular clones can be used to put a gene of interest into an experimental animal called a transgenic animal. Molecular clones were the main source from which sequence information was derived in the course of determining the human genome sequence, as we will discuss in Chapter 27.

AND THEN THERE'S CLONING . . .

Increasingly, headlines about cloning of human genes have had to compete with headlines about a very different kind of cloning. In 1997 the Roslin Institute announced that they had taken a mammary cell from a sheep and used its nucleus to create a cloned sheep named Dolly (Figure 26.4). A DNA

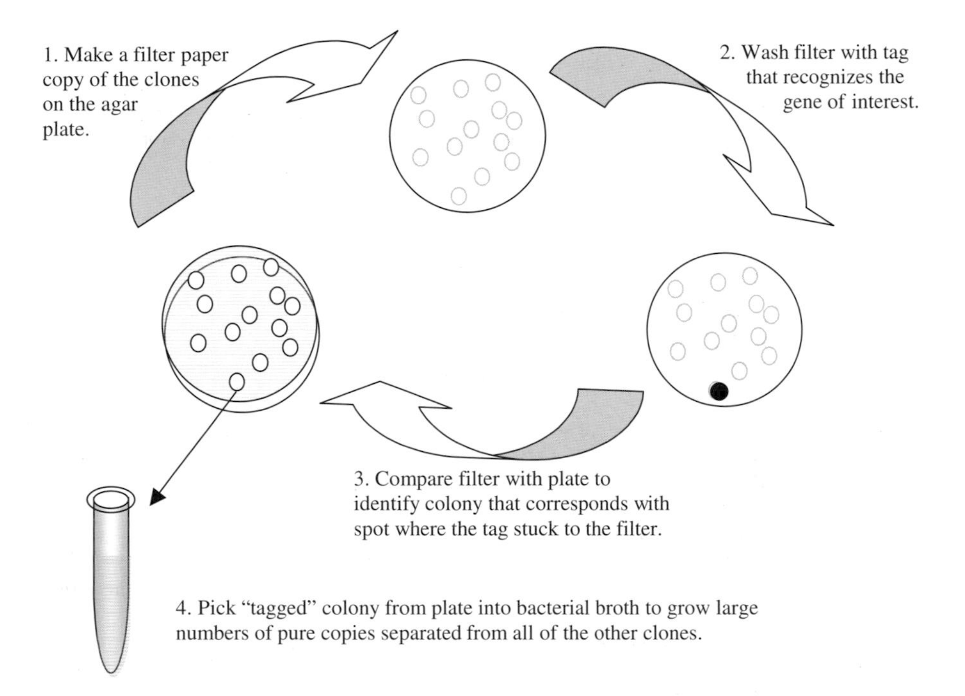

FIGURE 26.3 Finding the desired clone. After human DNA has been combined with a vector and introduced into the bacterial cell, a DNA tag is used to recognize the desired clone by base pairing with DNA in the clone that carries the gene we want to study. There are other cloning systems, such as some that use viruses as vectors, and others that replicate the cloned DNA in yeast cells instead of bacteria. In some more recently developed systems the assay might involve PCR instead of this kind of filter washing assay. The basic principle remains the same: combine DNA that contains the gene you are after with a vector system, introduce it into a host cell system, separate the clones from each other, use a tag that lets you identify the clone you want, and remove that clone away from the rest of the clones before beginning your studies of it. Often in an experiment with a purified clone, it is common to work with many billions of copies of the clone at a time.

fingerprint (an assay of repeated sequences at different points in the genome) confirmed that Dolly's DNA was derived from the genome of the donor nucleus and not from the genome of the animal that donated the egg into which the nucleus had been transferred. In 1998 the institute announced that Dolly had given birth and indicated that this confirmed that a cloned mammal is then capable of reproduction by natural methods. What is the difference between Dolly and her lambs? Dolly has her mother's complete genome, including both copies of every gene, but her offspring only have half of their mother's genome, since one copy of each of the lamb's genes came from the mother Dolly and the other half from the lamb's father.

How is cloning carried out? One approach (the one that produced Dolly) is to put a nucleus from some part of the body, such as the mammary gland, into an egg from which the nucleus has been removed (see Figure 26.4). This allows for generation of a clone of an existing adult animal, perhaps one with especially desirable qualities that one would like to reproduce. Another approach is to use embryo splitting, which essentially creates

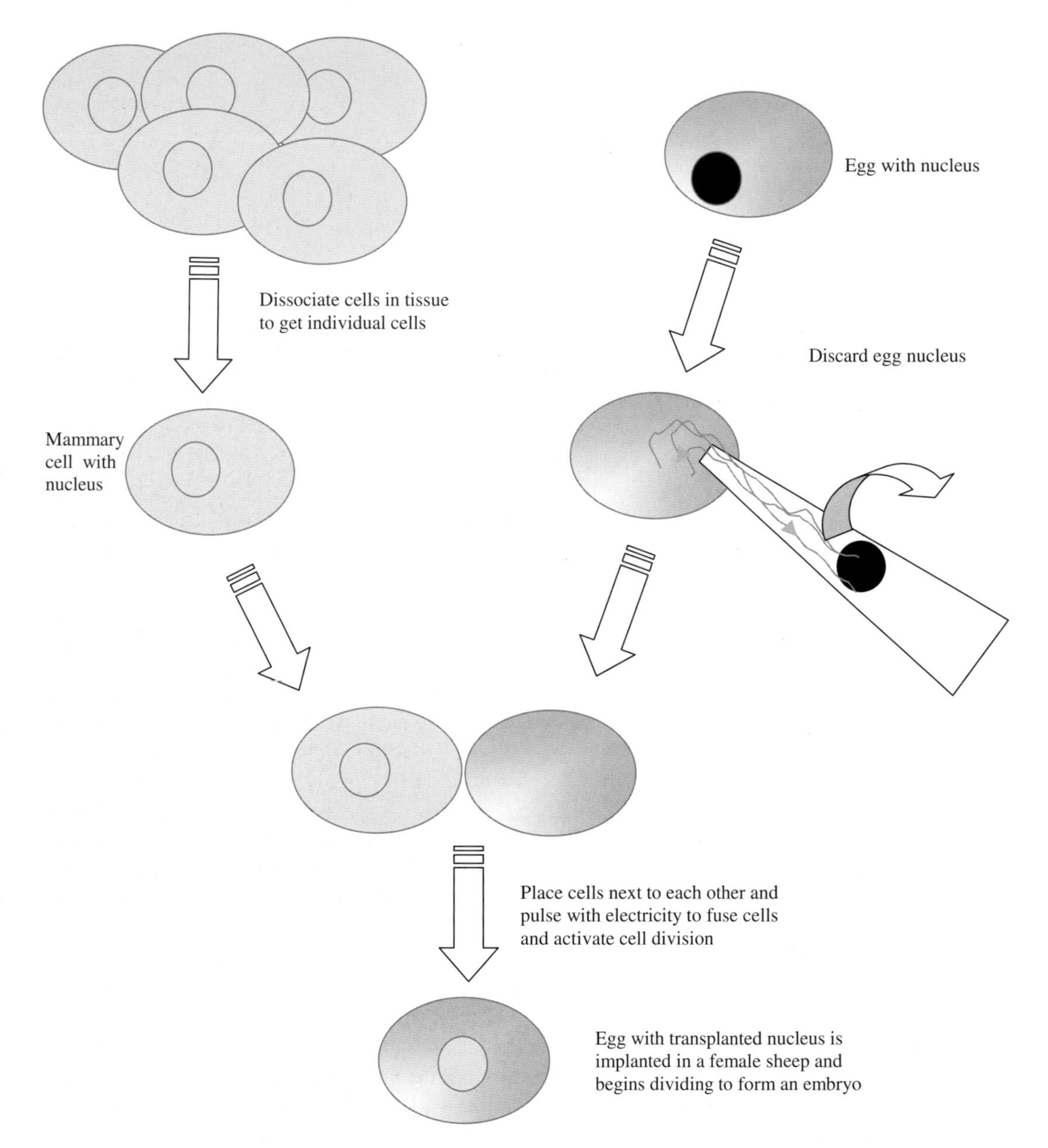

FIGURE 26.4 Cloning is carried out by transferring a somatic cell nucleus into an egg or embryonic stem cell from which the nucleus has been removed. One of the ways this can be done is by fusing a somatic cell with a nucleus to an egg from which the nucleus has been removed.

identical quadruplets or octuplets (Figure 26.5). A rhesus monkey named Tetra who was cloned this way is a sibling of the other monkeys that resulted from the embryo splitting process.

Why would it seem at all remarkable that this cloning would have been successful? Stop and recall our discussions of imprinting, the set of chemical and structural modifications to the DNA that control gene expression and

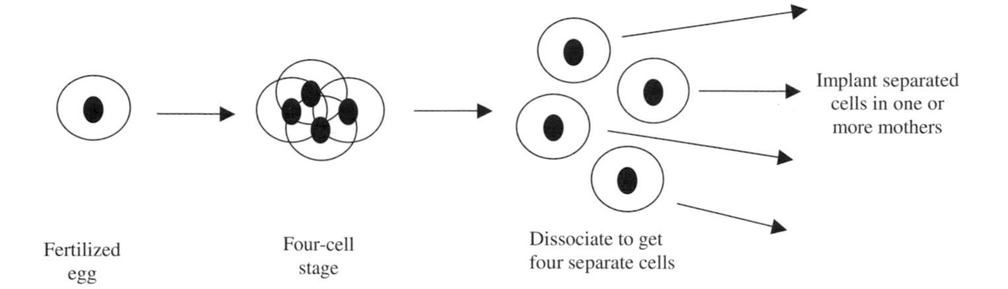

FIGURE 26.5 Embryo splitting at the four- or eight-cell stage can produce four or eight animals that are effectively identical quadruplets or identical octuplets if they all survive until birth. This has the disadvantage that it does not replicate an existing adult animal with a known set of properties. It has the advantage that, as with identical twins, the animals derive from a fertilized egg in which all of the appropriate imprinting signals (DNA methylation, chromatin structure) are present and (at least apparently) gene expression will occur normally, just as it would in the development of the single animal that would have derived from that egg if it had not been disrupted into multiple cells after cell division. Also, these animals will share the same mitochondrial genome and nuclear genome, just as identical twins or quadruplets do when produced naturally.

differ depending on whether the piece of DNA came from the mother or from the father. Since the DNA present in the first several cell divisions of the clone did not come from either an egg or sperm, we expect that the DNA from the mammary cell does not have the same pattern of imprinting signals that are normally present in a newly fertilized egg. Since the cloning succeeded, it appeared that a mammary cell may contain enough of the right signals to allow embryogenesis to proceed normally, even if we do not know whether other cell types could be used for the same purpose. As we subsequently learned from the follow-up on Dolly and other clones things in these cloned animals may not be entirely normal even if the initial product is a cute little lamb.

Is a clone really just an identical twin who happened to be born at a later date? Remember that, in addition to the chromosomes in the nucleus (the part that gets transferred during cloning), each cell also has mitochondria that have their own small circular chromosome. There are differences in sequence between mitochondria in different individuals as a result of mutation over the course of human history, so when a nucleus is transplanted from a mammary cell into an egg, the nuclear genome will be that of the donor, but the mitochondrial genome will be different unless the egg in question came from the donor or a very close female relative of the donor, such as his mother. So genetically, a clone who gets the same mitochondria as well as the same nuclear genome is in some senses just a twin born at a different time. However, we also have to consider environmental influences during pregnancy, since identical twins share not only their genome but also whatever effects there might be from maternal nutrition or other factors, such as stress during the pregnancy.

As we did for molecular cloning, let's stop to consider why we would want to clone an organism. Certainly it cannot be for the purpose of making large numbers of copies of the organism (in spite of the predictions of some science fiction stories), since it is a costly process that requires time and effort to produce even one animal. In fact, it is cheaper and faster to produce humans or animals the old-fashioned way, so any B-grade science fiction movie visions of someone filling the earth with armies of clones must first collapse under the practical realities that every single clone still takes the same old-fashioned nine months of gestation plus eighteen years of rearing to produce and costs vastly more to make in the first place. On the other hand, another movie plot in which people can have their deceased pets cloned, is looking as if it might become feasible if the scientists can get past the problems of imprinting, which causes differences in chemical modifications to egg and sperm chromosomes that may not be present in the cell used to donate the nucleus used in cloning. However, people are likely to be surprised at the outcome because at least some aspects of appearance, such as unusual color markings, may not reproduce in the cloned animal.

In the case of veterinary science, there are several arguments for the production of cloned animals. In some cases, cloners may want to reproduce multiple copies of special animals with unusual properties, such as high milk production or disease resistance, that have resulted from many years of hard work in a breeding program. Among the animals that have been cloned so far are mice, cats, sheep, pigs, goats, and mules.

In the pharmaceutical industry, the desire is to be able to make transgenic animals into which a particular gene of interest has been introduced. Production of human insulin, human growth hormone, or blood clotting factors would all offer not just industrial profits but also health benefits to patients who would no longer be at risk of hazards that accompany proteins isolated from human blood products. This process might or might not result in animals that are clones of each other, depending on how it is done. Certainly, the idea of being able to use transgenic technologies to make cows or sheep that lack the prion protein is terribly attractive, since they would then not be susceptible to mad cow disease or scrapie. Whether or not this is feasible remains to be seen.

Why would anyone would want to clone a human being? One can imagine a variety of reasons, ranging from vanity in someone seeking to clone himself to an effort to overcome grief through the cloning of a lost loved one or even an understandable variation on how to solve infertility problems. In the very first stages of efforts to clone humans, fame, publicity, financial gain, and egotism all suggest themselves. In the long run, once we are long past the headlines and attention that will go with any early cloning efforts, the reasons for electing cloning will likely be as complex as the reasons for many other reproductive decisions. The one thing that is clear, however, is that producing and raising a cloned human being will take just as much time, resources, and love as producing and raising any other child (and cost more in the beginning), so there will be few incentives for cloning to supplant traditional reproductive methods even if all of the technical and ethical bugs were worked out.

Why are so many saying that cloning of human beings should not be going on? First, there are a variety of religious, ethical, and legal issues that would need to be worked out before anyone should consider cloning a human being. Even if all of those issues were worked out to everyone's satisfaction (and that's a big if), to clone a human being would be unbelievably difficult

BOX 26.2 MAKING AN ORGANISM IS NOT EASY, EVEN WHEN DONE NATURALLY

"The concept of an embryo is a staggering one, and forming an embryo is the hardest thing you will ever do. To become an embryo, you had to build yourself from a single cell. You had to respire before you had lungs, digest before you had a gut, build bones when you were pulpy, and form orderly arrays of neurons before you knew how to think. One of the critical differences between you and a machine is that a machine is never required to function until after it is built. Every animal has to function to build itself."

Scott F. Gilbert[1]

(Box 26.2), and the consequences of something going wrong would be appalling.

Frankly, there are major technical problems with cloning organisms, at least given our current state of knowledge. After the initial optimism that came from the success in generating a cute little cloned lamb, concerns about a variety of issues, including imprinting, have arisen from the fact that Dolly turned out to not be so normal after all. As she aged, anomalies were observed. The lengths of the telomeres of her chromosomes reflected an age more in keeping with her mother's age than her own. She developed health problems her mother did not have and finally had to be put to sleep. It is unclear how many of Dolly's problems are the result of being a clone.

In other cloning experiments, imprinting effects appear to be responsible for fetal abnormalities. Experience with production of cloned animals indicates that currently less than two percent of efforts at cloning produce a live animal, and far too many of those that make it to live birth die or have major problems, such as lung abnormalities or failure of bone marrow to keep making blood cells. Even if the legal and ethical issues were all worked out, how can anyone set out to clone a human being if we expect imprinting to cause terrible health consequences for the baby? None of this is to say that human cloning will not some day be technically feasible, or that human cloning will be allowed even if the technical quirks are worked out of the system, but for now the reaction of much of the scientific community has been that, on a simple technical level, we do not know enough about how to make cloning work. Thus cloning of a human being is premature and should not be taking place at this time. However, this is not to say that permanent policies overseeing this technology should be based on our current state of technical capability. *As for a variety of situations in modern molecular genetics, ethical issues that exist because of current technical problems should be revisited in the future as our technical capabilities change.* Right now, we expect efforts to clone human beings to mostly not succeed, and if they did, to result in miscarriages and dead or damaged babies. We also expect that this could easily change as advances in veterinary cloning unveil the answers to the problems with imprinting.

[1] S. Gilbert, Developmental Biology, Sinauer Associates, 2005.

AND THEN THERE'S SOMETHING ELSE THAT IS NOT CLONING . . .

Unfortunately, a tendency to want to use short, easy slang terms instead of lengthy technical names for things can sometimes lead to the use of the same term for things that really have drastically different meanings. This results in imprecise communication, and in one case, it results in lumping some things together that are really substantially different. The use of the term *cloning* for molecular cloning and organismal cloning does not cause excessive confusion because the processes and objectives are so different that it is easy to tell them apart. However, when it comes to discussions of different types of cell-based technologies, application of the term cloning to situations that are actually quite different causes a problem, and it is clear that confusion has resulted.

There is a process called *somatic cell nuclear transfer* (SCNT) by which a patient's nucleus is transferred into an enucleated stem cell (Table 26.1). The cell can be used to grow an organ or needed cell type in cell culture. SCNT offers the promise of possible cures for some devastating diseases. Unfortunately, it came to be referred to as *therapeutic cloning*. The result is that when laws are passed to stop cloning of human beings, this nuclear transfer technology gets included even though the two processes are fundamentally different, are carried on for very different reasons, and result in profoundly different end products. Nuclear transfer technology, when used on cells in culture, does not create an embryo or a living being; it just creates cells in culture. The kinds of things that can result are designed for therapeutic use. If the nucleus from a patient is transferred into a legal and approved cell line growing in a petri dish and can be fed biochemical signals that will induce the cells to turn into insulin-producing cells that can be transferred into a child with diabetes without risk of transplant rejection, can we argue that this should not be done? Unfortunately, this strategy has been labeled therapeutic cloning, which normally would not be a problem but seems to contribute to legislative efforts to block this technology along with blocking efforts to clone human beings.

If we learn enough about how to control regulatory signals that tell a stem cell which developmental pathway to take, and if we use nuclear transfer tech-

TABLE 26.1 The Crucial Differences

	Nuclear Transplantation	Human Reproductive Cloning
End product	Cells growing in a petri dish	Human being
Purpose	To treat a specific disease of tissue degeneration	Replace or duplicate a human being
Time frame	A few weeks (growth in culture)	Nine months
Surrogate mother needed	No	Yes
Sentient human created	No	Yes
Ethical implications	Similar to all embryonic cell research	Highly complex issues
Medical implications	Similar to any cell-based therapy	Safety and long-term efficacy concerns

After B. Vogelstein, et al. Genetics. Please don't call it cloning! *Science* 2002; 295:1237.

BOX 26.3 TRANSFER OF A NUCLEUS INTO STEM CELLS

When a newly fertilized egg begins to divide, the very earliest cells produced have the capability to turn into any cell type in the body. As cells in the embryo divide, some of the cells begin to differentiate. Initially, some cells become committed to become the top of the organismm while others become committed to become the opposite end, but once a cell has committed to become part of the top of the organism, there remains a broad range of cell types that that cell might become, depending on what subsequent signals it receives. As cell division continues, cells become increasingly more restricted in their possible fates. Some cells eventually become terminally differentiated—permanently committed to be a particular cell type. Other cells stop at intermediate stages of differentiation, such as those that live in the bone marrow that can produce any type of blood cell but can no longer serve as precursors to the formation of a kidney or a neuron. Stem cells have now been found in a variety of tissues throughout the body, some with the capability of producing many cell types, some with much more restricted fates. Even in the brain, which was once thought to be unable to regenerate, stem cells have been found that can produce neurons and raise great hope for the treatment of a variety of neurodegenerative diseases. One type of stem cell, the embryonic stem cell, can be grown in culture and retains the ability to produce (eventually) all of the various cell types that can be produced during embryogenesis. A small number of embryonic stem cells exist that are approved for research use in the United States, but the complexity of transplantation antigens in human beings create problems with expecting to be able to treat any patient who comes along with one of a few existing cell lines. The use of nuclear transfer, to move the patient's own genome into these approved stem cell lines, offers the possibility that treatments could be developed that would meet currently existing ethical standards for use of stem cells while also meeting the patient's need to receive cells that his body will not reject. The other thing that is needed is to figure out which signals will turn a stem cell into the particular cell type or tissue needed in any given case. Initial successes in getting the relatively uncommitted stem cells to commit to a specific set of characteristics and functions suggest that important advances in health care could come about if the system is fully developed.

nology so that the cells responding to the regulatory signals have the patient's own transplantation antigens (Box 26.3), we hope to see a day when it will be possible to cure a baby who needs a heart transplant by growing a new heart from cells containing that baby's own genome so that the baby will not spend the rest of her life fighting transplant rejection problems. Such technologies would potentially allow researchers to create new skin cells to rescue a burn patient who is so badly burned that he cannot donate enough of his own skin to cover the damaged areas. It may some day allow doctors to cure victims of spinal chord injuries or people dying of Parkinson's disease or someone whose liver has ceased functioning.

The lifesaving potential for this technology is awesome. It does not involve growing human beings from whom organs would be harvested or any of the other nightmare scenarios that have been suggested. It does not require harvesting cells from embryos, since existing cell lines in culture can be used.

Instead, nuclear transfer technology involves cells growing in petri dishes, where they cannot possibly grow into a human being. Certainly, a lot will have to be learned to arrive at these terribly important applications of this technology, but that can only happen if development of the technology is allowed. Legislators can learn that nuclear transplantation technology aimed at transferring the patient's genome into the cells to be used in therapy in order to avoid transplant rejection problems. This does not involve cloning a human being and research into the development of nuclear transfer technologies should be actively pursued.

Organismal cloning offers the potential for advances in agriculture and solutions to some kinds of pharmaceutical problems. Nuclear transfer technology likely offers advances beyond anything we have seen from either molecular or organismal cloning, but it will not involve either molecular or organismal cloning and needs to not be lumped in with any of the various kinds of cloning when developing new policies.

GIVEN THAT GENES AND CHROMOSOMAL DNA CAN BE CLONED

Technological advances over the last several decades have brought us organismal cloning and molecular cloning. Each of these technologies has revolutionized not only our thinking about genes, cells, and organisms but have also given birth to a variety of related technologies of amazing brilliance, versatility, and practical use. As amazing as organismal cloning and nuclear transfer technology are, the main point of this chapter is really the molecular cloning technology that gave birth to much of modern molecular biology and genetics. The technology that made the Human Genome Project possible and that led to the sequencing of the human genome was molecular cloning. So join us in Chapter 27 as we tell you more about the Human Genome Project and what was found in the human genome sequence.

THE HUMAN GENOME SEQUENCE

27

"It's a history book—a narrative of the journey of our species through time. It's a shop manual, with an incredibly detailed blueprint for building every human cell. And it's a transformative textbook of medicine, with insights that will give health care providers immense new powers to treat, prevent, and cure disease."

—Francis S. Collins, Director, National Human Genome Research Institute

In June of 2000, great fanfare accompanied the announcement of the first draft of the human genome sequence. The following winter, the results of a massive set of analyses were published to the delight of a scientific community that had been all but holding their collective breaths in anticipation. In the journal *Genome Research*, Francis Collins reports:

"At the celebration accompanying the publications on February 12, 2001, my musical colleagues in "The Directors' Band" unveiled a few new songs about the genome. Most were rather tongue-in-cheek. But the chorus and final verse of the last song, sung to the tune of Woody Guthrie's most famous composition, sums up why we did all this and what some of our hopes are:

This draft is your draft, this draft is my draft,
And it's a free draft, no charge to see draft.
It's our instruction book, so come on have a look,
This draft was made for you and me.
We only do this once, it's our inheritance,
Joined by this common thread—black, yellow, white, or red,
It is our family bond, and now its day has dawned.
This draft was made for you and me."

In the spring of 2003, once many of the holes in the draft sequence had been filled and the level of accuracy of the sequence had improved, the completion of the sequence was announced. So what was it that everyone had awaited with such anticipation and greeted with such fanfare?

THE SEQUENCE

On the face of it, the result of the sequencing of the human genome is a list of As, Cs, Gs, and Ts that is more than 3,000,000,000 bases in length. It starts at the top of chromosome 1 and runs down each chromosomal arm in the order in which the chromosomes were originally numbered according to their apparent sizes.

Since it is supposed to be a completed sequence, you would think we would be quoting you some specific number like 3,095,784,273 but we cannot do that now and actually never will be able to. Why? Because the sequence is not exactly the same length in every single human being. Remember our previous discussions of simple repeat sequences and the fact that they are not always the same length in different human beings. There are tens of thousands of these simple sequence repeats, and even if only a fraction of those are different between any two people, and even if the ones that differ do so by only a few base pairs in length, the result is an overall sequence that cannot be tallied right down to the individual base pair and be considered *the* sequence length for *the* human genome. Even if you want to consider the genome of an individual human being, differences in lengths of the telomeres over time confound our efforts to say that your genome has an exact, countable number of base pairs.

Yet we do end up with a sequence in the databases to which we can point and say that the beginning of gene X starts at coordinate Y. How can we do that? Because the human genome project has assembled a reference sequence, assembled from the sequence of multiple different human beings, and in most cases, we do not actually need to know that a particular base is exactly 25,849,578 base pairs from the tip of chromosome 3; we just need to all agree that we are going to say that that is the coordinate of that particular base pair so that we know that we are talking about the same point in the sequence. Because small changes are still being made to the sequence, it is important that scientists who make use of the sequence report which version of the sequence they are talking about or indicate when they accessed the sequence to find out the information they are reporting. If they didn't, it would be possible for three different scientists to quote three completely different numbers and yet each be correct.

A completed sequence of a genome should be, by definition, truly complete, lacking gaps. Yet this "complete" sequence is not quite complete. The sequence released in 2003 is missing some of the highly repetitive sequences, such as the telomeres (repeated sequences at the ends of the chromosomes that protect the ends of chromosomes) and centromeres (repeated sequences in the region of the chromosome that attaches to the spindle apparatus of the dividing cell). There are also occasional very small bits of sequence missing where the sequencers encountered a region of the genome that simply could not be cloned or sequenced. The tally of base pairs in the genome also does not include the sequence of the "other genome" in the human genome, the mitochondrial chromosome (Box 27.1). Although a variety of technical approaches have gradually gotten the sequencers past one after another of these gaps, an occasional missing bit of a few thousand base pairs does not keep us from making good use of the overall sequence. In fact, the current tally shows the completed sequence to be about ninety-two percent of the estimated complete sequence; however, much is known about that other eight percent of the sequence at this point, even if it is not being listed as complete.

If there are pieces of the sequence missing at this point, why do we consider it a complete sequence? Partly because almost all of the important information is now available for use by the community of geneticists and molecular biologists whose work depends on the sequence. Even for the parts that are not listed in the completed sequence, such as the telomeres and centromeres, the basic structure and sequence of these regions is known even if they are

BOX 27.1 MITOCHONDRIA—THE OTHER GENOME IN THE HUMAN GENOME

Thousands of mitochondria, organelles of roughly the size and shape of a bacterial cell, exist inside of the human cell and serve to produce energy for the cell's use. Each of these organelles has its own mitochondrial chromosome that is tiny compared to the human genome or even the smallest of the human chromosomes. The thirty-seven genes of the 16.6-kilobase mitochondrial chromosome produce some of the proteins used by the mitochondrial energy factory, but there are other genes in the nucleus of the cell that also encode mitochondrial proteins that make up the rest of the energy production machinery of the mitochondria. The proteins produced by the nuclear mitochondrial genes and the proteins produced by the mitochondrial genes work together in complexes to create energy in the course of turning complex molecules such as sugars into simple substances such as carbon dioxide and water. The mitochondrial chromosome is a circle, like the circular structures of bacterial chromosomes and plasmids. One of the most interesting things about mitochondria is that their genetic code is similar to that of the rest of the genome, but it is not identical. This works because mitochondria have their own ribosomal machinery for translating the codons on the RNA into a protein sequence. Mitochondrial sequences get passed from mothers to their children but do not normally pass from fathers to their children. The result is that mitochondrial sequences present in a woman have been derived generation after generation straight down the line from her earliest female ancestors. By studying mutations in the mitochondrial genome found in different populations, it is possible to tell which populations are closely related (those whose mitochondrial sequences are very similar) and which ones are more distantly related (those with more differences in their mitochondrial sequences). Comparison of mitochondrial sequences has been used to trace the human lineage back into Africa and to tell some things about migration patterns over very long periods of time.

not known with the level of precision and completeness needed to include them in the final sequence. Enough of the important sequences are present to allow us to ask either local questions about a particular sequence or genome-wide questions of global proportions. The sequence was considered a draft as long as big chunks of gene-containing sequence remained unfinished, and it was considered a draft as long as most of the sequence was there but still in pieces that needed to be assembled like a puzzle. Now, the sequence is complete rather like a new house missing a cupboard doorknob and one light plate cover but otherwise possessing so many of the essentials that the new owners say, "It is finished," and move in.

A long string of genetic letters sounds pretty boring and uninformative, rather like a group of preschoolers endlessly singing the alphabet song with only four letters randomly arranged. So once all of that sequence was sitting in the computer, did they actually find anything interesting or do they really just have a pot of alphabet soup waiting for someone to unveil its mysteries?

The final outcome of the sequencing project could have been a boring combination of answers we already had in hand before the sequence was completed plus a lot of things we don't know enough about to recognize or understand. Instead, the result was a delicious series of surprises and revelations that we expect to continue for years as people continue mining information from the sequence.

HOW MANY GENES ARE THERE?

As we awaited the emergence of the sequence, one of the first questions on everyone's mind was, "Just how many genes are there?" This turns out to be a very difficult question to answer, and although we have approximate answers now, it is not clear when we will ever really know the final answer. So before we look at the numbers announced by the Human Genome Project, let's consider some of the issues that complicate telling what the real number is.

There is nothing in the sequence waving a red flag at us to say, "Look at this patch of sequence, it is a gene." However, there are a number of ways to tell where genes are located within a whole genome worth of sequence. In some cases, it is easy to recognize a gene in the completed sequence because the gene is already known and has been studied for many years by research groups investigating a particular disease, biochemical pathway, physiological process, or cellular structure. In such cases, we know the sequence of the gene and can tell whether it is in our new batch of sequence simply by comparing the two sequences to each other to find where the two sequences match.

For many genes that have not yet been studied, we still know something about the sequence of the RNA because of the expressed sequence tags (ESTs) transcribed sequences that have been stored electronically. ESTs allow for tracking and transfer of information about a gene in the context of a computer database instead of a test tube. ESTs contain bits of sequence from transcripts produced by each gene. Again, in this case a comparison between an EST sequence and the genome sequence can show us where in the genome that EST comes from.

In some cases, though if the gene is not a known gene and no one has obtained any sequence from the gene's transcript yet, it may still be possible to recognize it as being a gene. Computer programs have been developed that screen genomic DNA sequence for regions of sequence that have properties that are similar to the properties of known genes and ESTs. Another strategy is to use a computer to compare the human genome sequence to the sequences obtained from genomes of other organisms. Many genes are conserved across species, which means that the sequence of the human copy of a gene is similar to the copy of that gene in a monkey, mouse, or even in some cases a fruit fly. However, keep in mind that the programs that have been developed to search for genes are limited in their ability to detect genes and can't find everything that really is a gene.

The answer to the gene-number question provided one of the first big surprises. When the draft version of the sequence was announced in 2001, the two different genome groups that published papers on the sequence each came up with a different number, although the numbers were not terribly far apart. The Human Genome Project indicated that they estimate there to be about 31,000 genes, but it is clear from reading the description of their

TABLE 27.1 How Many Genes Are there?

Organism	Gene Tally at NCBI in 2003
Human	35,709
Fruit fly	13,821
Baker's yeast	6,304
Escherichia coli bacteria	4,288

NCBI address: http://www.ncbi.nlm.nih.gov

methods that they consider this an estimate and not a final true tally of the number of genes. The commercial venture at Celera Genomics gave out estimates ranging from 24,000 to 40,000 genes, depending on different methods used to identify genes. An interesting question lingers: If they each found thirty some thousand genes, did the two groups find the same thirty some thousand genes? No. In fact, since the release of the draft sequence, more genes have been found by various methods, including one group that said that they identified more than 10,000 additional genes by looking at sequences that are very similar (conserved) when the human genome sequence is compared to sequences of other organisms, and another group tallies more than 41,000 genes spread out across the different chromosomes. Another recent report dropped the number back to less than 25,000.

What was surprising about the announcement that humans have about thirty to forty thousand genes? To begin with, there appear to be fewer genes than were expected. For years, the debate over the number of genes had sent the number ranging from about 50,000 genes up to more than 100,000 genes and back again. When the commercial and academic versions of the sequence were released, each reported finding less than 40,000 genes. This was especially surprising to the egotistical humans of this world, since it is only about two to three times the number of genes found in a fruit fly! (See Table 27.1.)

The findings brought up other questions: Can thirty thousand genes really produce all of the elegant complexity of a human being? For those thinking in more concrete terms, can thirty thousand genes even produce as many different proteins as there appear to be in a human cell? In fact, the situation cannot be adequately portrayed by a simple tally of the number of genes. If you think back to Chapter 8, you may recall that one gene can make more than one final mRNA (and thus more than one final protein product) through alternative splicing. One of the revelations arising out of the human genome project is that there may be large amounts of alternative splicing going on, leading to a much, much larger number of functionally distinct proteins than there are genes. Thus this small number of genes can account for a large number of proteins produced by many cell types and developmental stages of a human being.

One of the next questions that researchers considered was: "Really? How do they know that is how many genes there are? How can they be sure they aren't missing something?" One of the curiosities that raised questions about whether we know how to recognize all of the different kinds of genes was the discovery of a phenomenon called a *gene desert.* This is a large region of sequence in which there are no known genes, in which no EST sequences are present and no genes are predicted by the search algorithms designed to predict genes. The existence of these deserts raises an interesting question: Does the genome really contain large regions devoid of genes? Or are there genes with properties that differ enough from the known genes that the computer programs don't know how to find them? Although the answer remains to be seen, we will offer this observation: there is a gene desert in the sequence of the fruit fly. It is possible to introduce a mutation into this gene desert and get a trait that can be passed from one generation to the next. In the strictest early sense of a Mendelian gene, a region in the middle of this gene desert acts like a gene.

We suspect that there are indeed genes in the human genome that the programs don't yet know how to identify, and genes in the human genome not yet represented in the collection of sequences taken from transcripts because they have not looked at transcripts from every possible cell type, stage of development, or environmental condition. Maybe the number of genes we can't detect yet is small, but we can't really know just what we will eventually learn about the gene deserts. We also can't yet know how large the number of genes will end up being by the time we look at every different cell type at each stage of development and in response to a variety of conditions that alter gene expression. Certainly, we do not expect the number to be vastly different from the current estimate, but we also expect that the genes have not all been found yet. In fact, the number of genes has continued evolving, with the number of genes going up a bit as the sequence became more complete and as researchers tried novel approaches to identification of genes. However, even as some additional genes are being found, other transcribed regions that might have been thought to be separate genes are being found to be part of the same transcript. One recent estimate indicated more than 41,000 human genes (Tables 27.1 and 27.2). Thus the number of human genes is currently a bit like the ocean with the tide lapping at the shore, with the bulk of the information now a solid constant, unlikely to change, surrounded by a fraction of changing information that comes and goes as additional experiments tell us more about things that we thought we already knew.

HOW BIG ARE THE CHROMOSOMES?

Another interesting item revealed by the human genome sequence is that those who named the chromosomes based on their size did not assign them all in the right order (Table 27.2). Yes, chromosome 1, with more than two hundred forty million base pairs of sequence, is still the largest chromosome. No, chromosome 22, with almost fifty million base pairs of sequence, is not the smallest chromosome. Instead, chromosome 21, at more than forty-six million base pairs of sequence, wins that honor. Given that the chromosomes were supposedly name in order according to size (except the X and Y), we also see that chromosomes 9, 10, and 11 are apparently named in the wrong order relative to the actual length of DNA present in the chromosome.

How could the chromosomes be numbered in the wrong order if they were numbered based on size? The original numbering was based on apparent size when viewed under a microscope. If we look back at Figure 10.3 we can see that 9, 10, and 11 are actually very similar in size and hard to distinguish if the banding pattern is not also available to assist the viewer.

TABLE 27.2 Gray Highlights Discrepancies between Chromosome Number and Sequence Length

Chromosome Number	Rank Order Size by DNA Length	Chromosome Length[a]	Amount of Sequence Completed on That Chromosome[a]	Number of Genes Reported by euGenes Database[b]	Official Tally of Number of Genes from Entrez at NCBI
1	1	245,203,898	218,712,898	3,926	3,232
2	2	243,315,028	237,043,673	3,485	2,653
3	3	199,411,731	193,607,218	2,519	1,906
4	4	191,610,523	186,580,523	2,146	1,631
5	5	180,967,295	177,524,972	2,447	1,772
6	6	170,740,541	166,880,540	2,616	1,935
7	7	158,431,299	154,546,299	2,393	1,891
8	8	145,908,738	141,694,337	1,991	1,470
9	11	134,505,819	115,187,714	1,780	1,468
10	9	135,480,874	130,710,865	2,118	1,463
11	10	134,978,784	130,709,420	2,319	2,027
12	12	133,464,434	129,328,332	1,995	1,673
13	13	114,151,656	95,511,656	1,167	820
14	14	105,311,216	87,191,216	1,423	1,212
15	15	100,114,055	81,117,055	1,461	1,206
16	16	89,995,999	79,890,791	1,667	1,327
17	17	81,691,216	77,480,855	1,889	1,693
18	18	77,753,510	74,534,531	1,009	670
19	19	63,790,860	55,780,860	1,993	1,761
20	20	63,644,868	59,424,990	1,179	956
21	22	46,976,537	33,924,742	561	440
22	21	49,476,972	34,352,051	978	844
X		152,634,166	147,686,664	1,835	1,438
Y		50,961,097	22,761,097	295	221
Other sequence not yet assigned to a chromosome		25,263,157	25,062,835		

[a] From www.ncbi.nlm.nih.gov/genome/seq/, accessed July, 2003.
[b] From iubio@bio.indiana.edu:8089/man/, accessed July, 2003.

Comparison of different kinds of information reveals a variety of interesting things about the human genome. For instance, the shortest chromosomal arms undergo recombination at about twice the rate found for the long chromosomal arms, something that probably assists in ensuring that at least some recombination takes place on these short chromosomal arms to help hold chromosomes together during meiosis. On the other hand, recombination takes place at a much lower rate around the centromeres. Although these ideas have been around and did not emerge with the completed sequence, having the whole sequence has allowed a much more complete overview of where the major differences in recombination rates fall. Placement of positions of human chromosomal rearrangements onto the map containing the positions of the genes that have been found has allowed for identification of disease genes not previously identified. Comparison of sequences within the Y chromosome has identified a novel mechanism by which the integrity of the Y chromosome is maintained (Box 27.2).

BOX 27.2 THE SEQUENCE SHOWS HOW Y CHROMOSOME SEQUENCE INTEGRITY IS MAINTAINED

One of the revelations that came out of the human genome sequence is that the Y chromosome maintains its sequence integrity by a different mechanism than the one used by chromosomes 1 through 22 or the X chromosome. Because there is a constant process of mutation, the sequences of independent chromosomes that never exchange information could drift substantially apart from each other. However, the process of pairing between chromosomes results in a continuous series of information exchanges between chromosomes. Since the Y chromosome has a large region that never pairs with another chromosome and thus lacks a partner from which to obtain information with which to potentially repair a damaged gene, we might expect that the Y chromosomes of different men would be very different from each other (which they are not). Analysis of the Y chromosome revealed a most interesting alternative mechanism for solving the problem of maintaining the sequence integrity over long periods of time. Researchers report the existence of backup copies of genes within the Y chromosome that are used to help keep the functional copies of the genes in good repair. This means that the process that will keep the Y chromosome intact and functioning in the far distant future will be a fundamentally different mechanism from that maintaining the other chromosomes, in that the backup information is contained on the same chromosome. Yet, in a way the mechanisms are quite similar and make use of related mechanisms for handling DNA repair, whether the cell is copying the information off of the same chromosome or a different one.

WE ARE ALL AMAZINGLY SIMILAR

One of the other findings to emerge from analysis of the sequence is just how similar human beings are to each other. The initial sequence was determined from more than a half-dozen different individuals from a variety of ethnic backgrounds, including African, European, and Asian ancestry. Since then, additional information on human variation has come from ongoing projects aimed at identifying vast numbers of sequence differences that can be used to ask a variety of medical and nonmedical questions about the human race. The result, we see, is that, if we compare the sequences of any two human beings, we find a difference in those two sequences about once in every thousand base pairs of sequence. That means that any one of us is on average more than ninety-nine percent identical to the next guy at the level of the DNA sequence. Many of these differences are detected not just when we compare DNA in people from different populations but also when we compare people within the same population.

More than a million such differences in the sequence, called *single nucleotide polymorphisms* (*SNPs*), will serve as important tools for a variety of studies. Although human blueprints are amazingly similar, there is more than enough diversity to allow SNPs to serve as very powerful tools for the investigation of many different questions. DNA chips with thousands of SNPs can be used to screen individuals with complex diseases to look for regions of the genome where small differences in the sequence turn out to be associated with the disease. As described in Chapter 33 SNPs can be used in the study of cancer. SNPs are used in anthropology and archaeology to look at the genetic relationships between ancient samples and current populations. SNPs are also used in tracing relationships of current populations to shared ancestral populations. SNPS are gaining importance in forensics in the identification of remains. SNPs will also become increasingly prominent in diagnostic situations and in the arena of pharmacogenomics, where the dream of every doctor is a test that will identify which drugs will work best for any particular patient.

Information about SNPs that sit near each other on the same chromosome can be combined to create a kind of local genetic fingerprint characteristic of a particular bit of chromosome in any one human being. One of the very interesting findings to emerge from analysis of the SNP fingerprints, or haplotypes, is that for any small region of a chromosome, most people in a population have one of only about a half-dozen different haplotypes that trace back through long tracks of history to shared ancestry in the far past. However, because recombination events have exchanged pieces of DNA between chromosomes during meiosis, we see that Person A may share the same haplotype with Person B for a spot at the end of chromosome 1 and have a different haplotype from Person B at a position thirty million base pairs further down the chromosome. Meanwhile, Person B shares the same haplotype in that region with Person C. By studying these blocks of SNPs that have traveled together through time, scientists are developing tools to assist in mapping of genes involved in complex disease, while also finding ways to look back through time at genetic events that may have happened thousands of years ago. One of the things we see from such studies is that African populations tend to have more different haplotypes in any given region than other populations, which is expected for a population that is older and has had more time for mutations to diversify the set of haplotypes (Figure 27.1). Fewer haplotypes in younger populations in Europe and Asia would be expected if small founder populations representing only a subset of the total available haplotypes settled a new region and then had a shorter time over which to have mutations diversify their set of haplotypes in any one chromosomal region.

WHERE ARE THE GENES LOCATED?

Some things that we can tell about the human genome sequence are not new to us, since they were things we could tell by looking at even a part of the sequence or the genetic map. One of these items that is interesting but not novel has to do with where the genes are located relative to each other. We see that some areas are rich in genes and others are relatively gene-poor. We see that some families of genes (genes with related sequences and/or functions) are located in clusters in the same region of a chromosome, but that other families of genes are scattered all over the genome and nowhere near each other. In some cases, we can see *paralogous* regions of the genome, which are regions that appear to share some ancestral sequence, as if a piece

African ancestry

% with each haplotype	More common (red) and less common (blue) variants at different points in the sequence
28%	
9%	
8%	
7%	
5%	
4%	
4%	
4%	
4%	

European ancestry

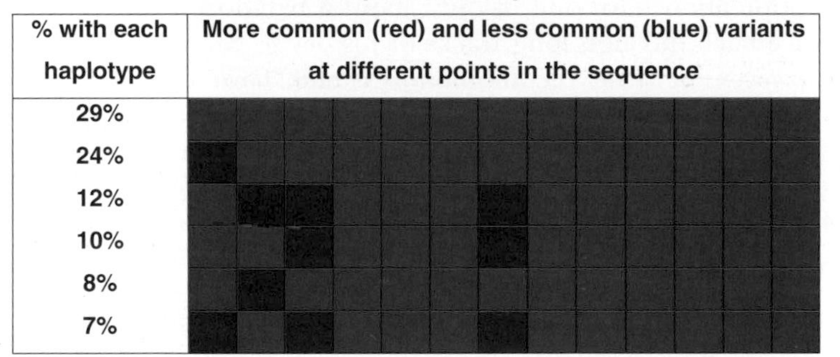

% with each haplotype	More common (red) and less common (blue) variants at different points in the sequence
29%	
24%	
12%	
10%	
8%	
7%	

FIGURE 27.1 Haplotypes in two different populations. Each line represents a different haplotype showing a genetic fingerprint in the region of one gene, the LMX1B gene, with the same haplotype being the most common in both populations. Percentages show what fraction of the population has that haplotype. A larger number of haplotypes is thought to indicate a population that was founded farther back in time than the population with fewer haplotypes. Blue squares mark one base of difference in the sequence, with the whole haplotype spanning thousands of based pairs of sequence. (Courtesy of Goncalo Abecasis, University of Michigan.)

of DNA had duplicated and the duplicated regions ended up sitting at different points in the genome gradually diverging from each other. We can also see that both strands of each chromosome end up having many different genes transcribed. For one gene we select, one strand gets used as the template strand for making RNA, but for a neighboring gene, the other strand might be used as the template. Because of the polarity of the two strands (they point in opposite directions), genes that get read off of one strand read in the opposite direction from genes read off of the other strand. Thus a chromosomal region containing six genes might show a pattern of transcription similar to that shown in Figure 27.2.

Sometimes the situation can be more complicated. Take the case of the very large *NF1* gene that is responsible for a disease called *neurofibromatosis*. The NF1 gene covers about 350,000 base pairs on chromosome 17. It has 59 exons that become part of the 13,000-base mRNA produced by the gene. In

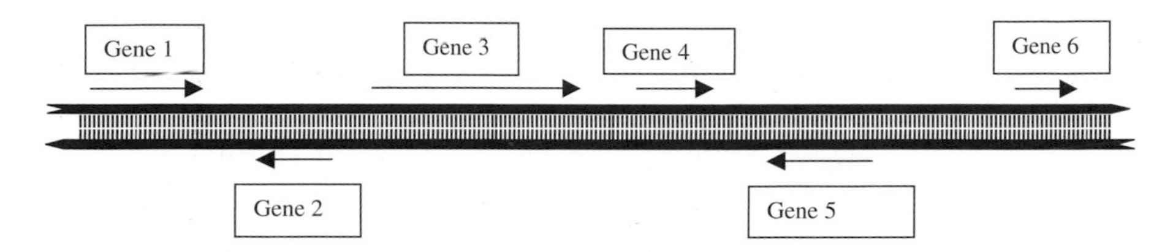

FIGURE 27.2 Transcription of several genes from one chromosomal region. Arrows mark genes, with genes 1, 3, 4, and 6 all being copied from one strand of the DNA going in one direction, and genes 2 and 5 being copied from the other strand of DNA going in the other direction. Notice that some genes are longer than others and that the amount of space between the genes is not always the same.

one of the introns, if we look at the opposite strand, we find three small genes OMGP, EVI2B, and EVI2A that produce proteins that are not involved in causing neurofibromatosis.

WHAT IS OUT THERE BESIDES THE GENES?

In previous chapters, we talked about genes, structures such as centromeres and telomeres, and repeat sequences. If we add up all of the genes and chromosomal structures we can account for, the amount of DNA they encompass falls far short of the total. Less than two percent of the DNA is used up on the sequences that make it into the final spliced transcripts. Only about a quarter of the DNA is taken up with sequences that get spliced out of transcripts in the course of making mRNA. What might the other approximately seventy-five percent of the DNA be for? Some people talk about "junk" DNA doing nothing or at best serving as filler or spacer sequences. In some cases, they might be right, since it appears that sometimes all that is needed is to keep two points on a chromosome a precise distance apart from each other without it much mattering what the sequence is that fills up that space. Some of the junk qualifies as evolutionary relics such as *pseudogenes* (nonfunctional copies of genes) and copies of viral genes left behind after a virus infection. However, there are many other functions we have not accounted for in our discussions—regional control of gene clusters, replicating DNA, interaction with scaffolding proteins, and more. For some of these functions, the DNA sequences and chromosomal locations of those sequences are known; for others, they are still being sought. Its not such an outrageous bet that some of the supposed "junk" will eventually account for functions that we don't yet realize exist and don't know enough to ask about.

WHAT COMES NEXT?

One of the concerns at this point is: What are we going to do with the flood of new information and ideas coming out of the Human Genome Project. The biggest point of concern is not whether any one person can keep up with enough of it while going about lives filled with other goals and necessities. The real concern is how the people who need to use this information, like our doctors, can keep up with the rapidly expanding wealth of information.

One of the ways this is being dealt with is through the development of new educational programs aimed at doctors who are beyond the end of their formal education. Medical policy requires that doctors who want to maintain their licenses must engage in a process of continuing medical education. The National Coalition for Health Professional Education in Genetics (www.nchpeg.org) works on the development of curricula that can be used to teach doctors about advances in the field. Other resources that can help doctors get in touch with new information come from a variety of government funded resources at the National Center for Biotechnology Information, including Online Mendelian Inheritance in Man, as well as outside resources such as GeneClinics. Computerization of the scientific literature has also begun helping people in genetics to keep up with a faster pace of information flow through computerized searches of the literature, mechanisms for downloading references directly from databases, and access to online versions of papers that previously would have required separate trips to the library to track down.

Fortunately, current funding policies include efforts at further development in education and ethics, as well as technical development, all of which are going to be needed to avoid pitfalls while taking advantage of the wonders contained in the compendium of genetic information that has been called the Book of Life. Now, having established the order of base pairs with which that book is written, the next step is to sort out what it all means. One of the most important next steps in discerning the message in the long string of genetic letters is that of pinning down which genes or regions of the sequence go with which traits. So join us in the Chapter 28 as we talk about the process of finding the spots in the genome that are responsible for human characteristics such as hereditary diseases.

FINDING GENES IN THE HUMAN GENOME

28

"Ready!" four voices shouted, and Angie opened the door to the room, where she paused and slowly passed her gaze across the contents of the room and the four giggling children clustered in the middle. After a moment's indecision, she moved to the left toward the windows. "Cold," said the other children. Realizing that she must be heading in the wrong direction, she redirected her course towards the fireplace. "Colder," they shouted. She shook her head at this news that she was still heading the wrong way, and continued into the room, moving towards the remaining wall of the room. "Freezing," they chorused, falling over each other in squeals of laughter. She stopped, confused, then turned around to head back towards the door. "You're getting warmer again," called out one of the boys. As she kept heading back towards the door, the volume of their voices rose with each word. "Warmer. Warmer. Warmer. REALLY HOT! BURNING UP!" they shouted as she reached out and touched the doorknob on the inside of the door, the object the children had selected as "it" before Angie came into the room. After laughing over their cleverness in picking an object back at the door as "it," they picked a different child to leave the room and started the game again. Surprisingly, this game of Hot and Cold is not unlike the process by which human genes are mapped. When searching for a gene, we use a process called a *genome scan* to test many different positions along the human chromosomes to determine whether that is where our gene of interest is located. However, we usually don't get a lot of information back from each test, since the main question we get to ask is, "Is this particular spot on the chromosome close to the gene we are looking for or not?" It is a yes-no question to which the answer is almost always "no" because our gene is located at only one place and thus almost all of the other positions in the genome are not "it." Once we get close, then we start getting more information about just how close we are, but if we are not close to the gene, the answer we get back basically just tells us that we are "cold" and looking in the wrong place. Because the genome is so large, we usually have to do many tests for whether we are hot or cold before we find a gene, but usually if we just test enough locations, at some point we turn from cold to warm and can begin following the signals from warm to hot to "burning up" at the spot where the gene we want is located.

The genomic haystack in which we search for genetic needles is enormous. We are going to start this chapter on gene hunting with a daring assertion: If we can find out where a gene is located, we can get our hands on copies of it and find out what it is. In fact, we can do that even if we do not know what it encodes, or what the gene product does. We can do this even if our models for disease pathology are incorrect. This approach—called *positional cloning*—led to some of the most inspiring breakthroughs in human genetics in the 1980s and 1990s, including the identification of the genes for cystic fibrosis, neurofibromatosis, and Huntington's disease. Since then, the approach has been modified into a *positional candidate* approach that also uses information about gene product function and gene expression that were often not available when the earliest positional cloning experiments were taking place.

This chapter will tell you a little bit about how we go about searching for human genes and what it means when we say that we have "found" a gene (Box 28.1). We will show you how the Human Genome Project has revolutionized this search for the underlying causes of our traits, the good ones and the deleterious ones alike.

RECOMBINATION AS A MEASURE OF GENETIC DISTANCE

Our search for a gene responsible for a human trait really involves two steps. First, we want to find out where the gene is located. Then we want to use that information to help us find the actual gene so we can find out what it normally does and how its altered function results in the trait. Occasionally, we can find out where a gene is through physical processes, such as identifying

BOX 28.1 MORE THAN ONE WAY TO "FIND" A GENE

Sometimes, when headlines announce that a gene has been found, the news media means that the location of that gene has been found. Because the location of a gene is so often a critical piece of information for actually getting our hands on the gene itself, the mapping of its location is important and often newsworthy, especially if the disease is especially common or severe. Other times, when the newspapers announce that a gene has been found, they mean that what has been found is the actual gene itself, complete with the order of As, Cs, Gs, and Ts that make up its sequence. Often, some of the most important findings, such as what the gene normally does or how it causes disease, may sometimes take years to arrive at after the gene has been found. Somehow, the same level of front-page fanfare is often missing from these terribly critical later steps. Perhaps it is hard to identify a point in time when we can say, "Before, we did not know what the gene does, and now we know." So even though some of the later steps may be at least as important, if not more so, there are perhaps several reasons that they do not get the same fanfare. The functions of genes are often pieced together gradually through a long, slow series of increments of information. When a gene is found, there is an identifiable, discrete moment, a "eureka" moment, when each new person who hears the news experiences that sudden transition from wondering where the gene might be to knowing its exact position in the human genome. When a gene is first found, it has the feel of a breakthrough because it is an initiating event rather than another in a long string of events. So in some ways, the most critical questions we want to answer all follow upon the finding of the gene, but it is that moment of finding it that seems to warrant a special kind of notice.

a chromosomal deletion that has removed the gene or a translocation that has broken the gene. In most cases in which there are no such landmarks visible under a microscope to help us, we need to localize the gene genetically. For this, there needs to be a kind of genetic geography called a *genetic map* so that we can relate the gene's location to the position of other genetic landmarks whose locations we know.

The simplest example of mapping occurs when we determine that a given trait shows sex-linked inheritance. This tells us that the gene is on the X chromosome. This gives us our first level of geographic localization: assignment to a chromosome. The mapping of genes to other chromosomes is more difficult but also possible. Once we map the gene to a chromosome, we need to know exactly where on that chromosome it maps. To get refined localization of the gene, we have to build up a map of landmarks that we can use as a point of reference for any new item we want to "find."

The fundamental principle behind the processes by which we build maps of gene locations is the concept that things that are close together on the same chromosome have a tendency to stay together as the chromosome is passed through generations of meiotic recombination and segregation. Things that are farther apart on the same chromosome are more likely to be separated from each other by recombination, and things on different chromosomes will be separated by segregation. Thus, if two mutations are present together on the same chromosome, the closer together they are, the more likely that people in subsequent generations will inherit both mutations at once. We recognize linkage between two genes when a particular combination of alleles of those two genes turns up in the offspring more often than expected by chance. Genes that are close to each other will recombine rarely. Genes that are physically far apart will recombine frequently.

If we look at two traits present in the *founder*—the person with both traits from whom the rest of the family is descended—we can ask about cosegregation, that is, how often the two traits stay together or separate from each other as they are passed down through multiple generations of a family. This information can be used as a measure of how close together the two genes are. The earliest genetic experiments looked at traits that are transmitted together from one generation to the next. We rarely find a single family in which two different traits that we want to study are being passed along in the same family.

Let's consider two dominant traits, a hypothetical taster trait (that causes people to detect the bitter taste of a particular substance) and an eyelash trait that involves a double row of eyelashes. If the founder with the taster trait had the extra eyelashes and married someone who is a nontaster and has normal eyelashes, we could start trying to trace the transmission of these taster and eyelash traits through multiple generations of the family to see whether people with the taster trait tend to have extra eyelashes. If the two genes are right next to each other on the same chromosome and the two dominant alleles are physically located on the same copy of that chromosome, we might see what is shown in Figure 28.1, where almost all of the tasters have an extra row of eyelashes and all of the nontasters have normal eyelashes.

What would we see if the two genes were far apart? Let's reconsider the same fictitious traits and look at the kind of outcome that would have resulted

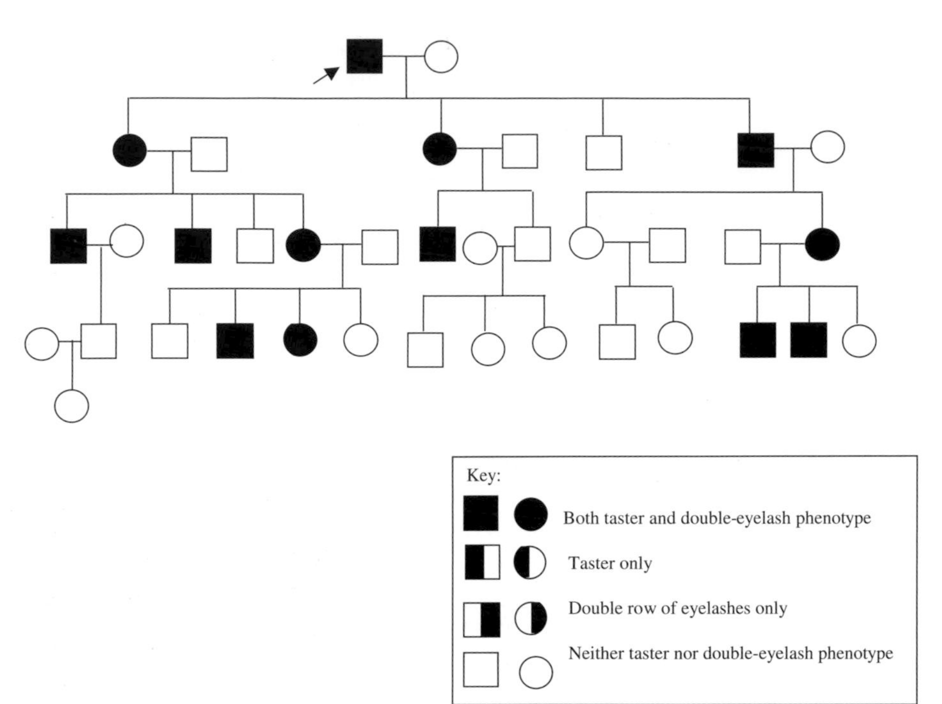

FIGURE 28.1 The hypothetical taster and double-eyelash traits are usually transmitted together in this family. In this case, out of twenty-three individuals who were at risk of inheriting the taster trait, twelve of them have it. Out of twenty-three who were at risk of inheriting the double-eyelash trait, twelve of them have it. What is most noticeable is that every one of the tasters also has extra eyelashes and every one of the non-tasters has normal eyelashes. This suggests that the two genes are very close to each other on the same chromosome, and even that there is a chance that the two traits are caused by the same genetic defect. If these two traits are normally each seen without the other, we would favor the hypothesis that there are two genes located very close together on the same chromosome.

if the two genes were located on different chromosomes (remember how the chromosomes get passed along independently of each other to the next generation) or when recombination events fell between the two genes frequently (Figure 28.2). In Figure 28.1, when the two genes were very close together, most people who inherited one trait also inherited the other. In Figure 28.2, when we hypothesize that the two traits are far apart (on different chromosomes), we see that the chance of any one at-risk individual inheriting one of the traits is the same as what we saw in Figure 28.1, but it is much more rare for one individual to inherit both traits.

What happens when the two dominant alleles are located on the same copy of the same chromosome but there is enough distance between the two genes so that sometimes a recombination even can fall between the genes? Then we see what is shown in Figure 28.3, a family in which the two traits are transmitted together from one generation to the next in most cases, but every once in a while someone turns up with only one or the other of the two traits. The rate at which children have one trait but not the other gives some indication of the distance between the two genes. Since recombination is more

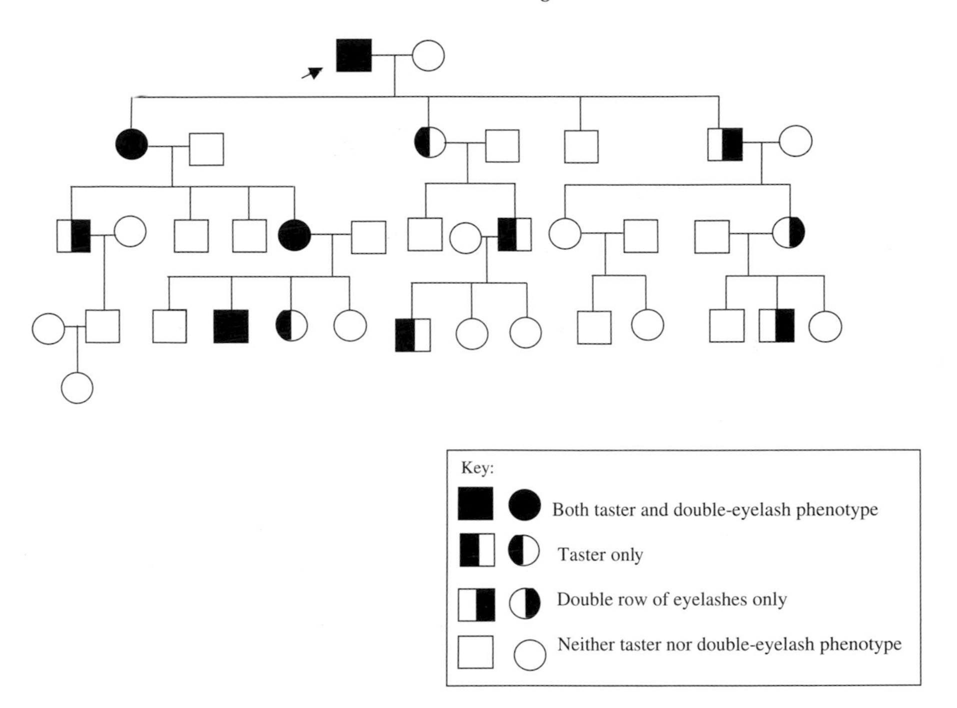

FIGURE 28.2 In this case, in which the two hypothetical genes are actually located on different chromosomes, not different copies of the same chromosome, we see that out of sixteen people at risk for the taster trait, eight have it, and out of sixteen at risk for having double row of eyelashes, eight have them. In many cases the individual received one trait or the other but not both. Notice that, occasionally, both traits get passed to the same individual when simple chance results in passing a chromosome carrying the taster allele and a different chromosome carrying the eyelash allele along to the same person.

frequent in some regions than in others, the recombination frequency can only be used as an approximation of the physical distance.

By looking at the rate of co-segregation of two traits (the rate at which they are transmitted together from one generation to the next), we can arrive at an estimate of the genetic distance between the genes responsible for the two traits. When one recombination event (separation of the two traits from each other as they pass from one generation to the next) is seen out of one hundred offspring who are at risk for the traits, this is considered to be a genetic distance of 1 centimorgan, which can also be written as 1 cM. Thus two traits that are 10 cM apart are farther apart than two traits that are separated by a genetic distance of 1 cM. 50 cM, or fifty percent recombination, is the largest genetic distance that we can measure, so things that are actually 50 cM apart along the same chromosome and things that are 200 cM apart along the same chromosome and things that are on different chromosomes will all show the same result (approximately fifty percent recombination), and we will not be able to distinguish those cases from each other. (For a review of recombination, see Chapter 13.)

Any time we look at co-segregation of two items in the genome, we end up with two different numbers of importance. One number is the recombination fraction, which is our best estimate of what the distance is between the

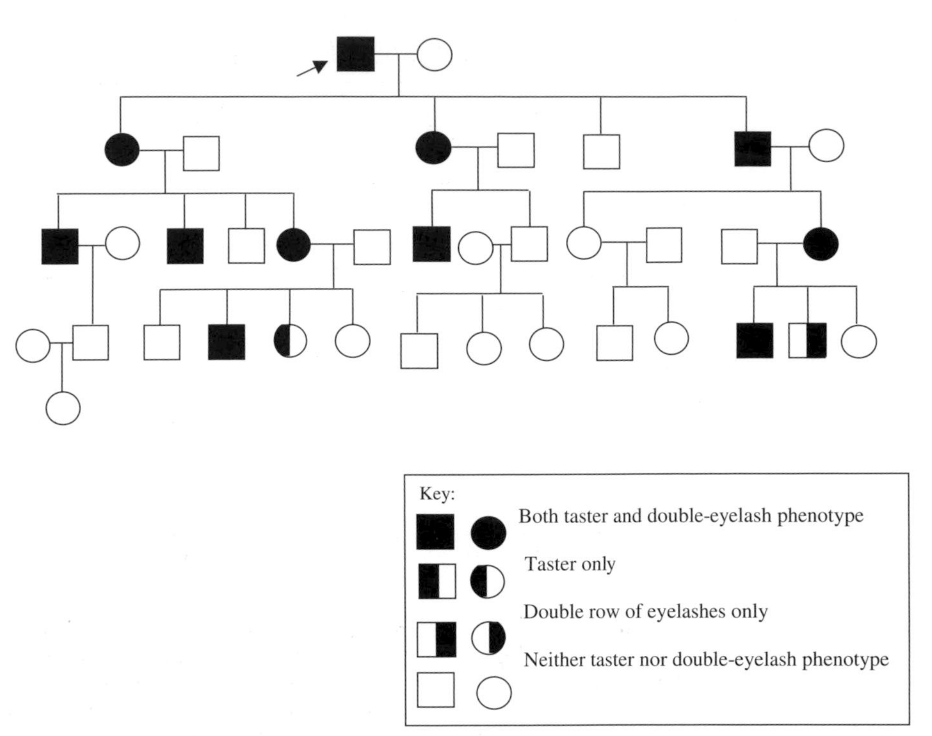

FIGURE 28.3 If the hypothetical taster and eyelash genes are close together on the same chromosome but far enough away to allow recombination events to fall between them sometimes, we will see the effect shown here. In most cases, affected individuals will be affected with both traits, but occasionally, someone will inherit only one of the two traits. Out of twenty individuals at risk of inheriting these defects, two of them received only one of the defects, suggesting a rate of recombination of about ten percent.

two items. The other number is called the *log of the odds score, LOD score,* and estimates our confidence in the accuracy of the recombination measurement. A LOD score of 1.0 indicates that we think the odds are 10 to 1 in favor of the recombination fraction identified being right. A LOD score of 2.0 gives us 100 to 1 odds in favor of being right, and a LOD score of 3.0 gives us 1000 to 1 odds in favor of being right. So if we find a recombination fraction of 1% associated with a LOD score of 5.0, we are highly confident that the two items are fairly close together, but if we find that same recombination fraction of 1% associated with a LOD score of 0.5, we have little confidence in the accuracy of that 1% assessment.

For some markers, we will not only fail to see evidence that the two items are next to each other, but in some cases we will actually see evidence that they are probably not anywhere near each other. If we find a LOD score of 2.0, the odds are 100 to 1 against the gene being at the proposed location. A LOD score of 3.0 gives us odds of 1000 to 1 against the gene being at the tested location. Whenever we see the amount of recombination approaching 50% recombination, we also doubt that the gene is near the tested location.

The standard in the field is that a LOD score of 3.0 or greater (for a proposed distance from the tested marker) is considered to be highly significant

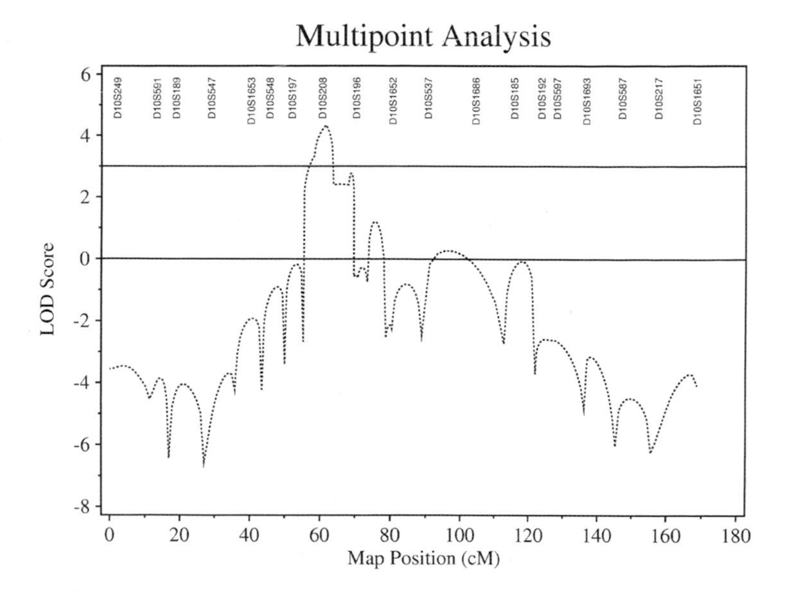

FIGURE 28.4 A graph of data from a gene mapping project called a genome scan. The highest peak on this graph shows the position in the human genome most likely to contain a newly discovered disease locus. The distance along the chromosome is marked across the bottom, and our confidence that the locus is located at that position on the chromosome is compared to the LOD score scale along the side. LOD scores above 3.0 indicate greater than 1000 to 1 odds of the gene being at that location. The peak marking the location of the disease gene on this graph rises above the level of 4.0 on the LOD score scale, so we expect that the chances of the gene being at about the 60 cM mark on chromosome 10 exceed ten thousand to one odds in favor of this being the correct location. (Courtesy of Edward H. Trager.)

evidence in favor of linkage—that is, of the two things being close together on the same chromosome. A LOD score of 2.0 or lower is considered to be highly significant evidence against linkage—that is, to say, about 100 to 1 odds that the gene is not at that location. Figure 28.4 shows a peak that reaches above the LOD score of 4.0, meaning that we think the odds are greater than 10,000 to 1 in favor of that being the location of the gene. At other points along the chromosome, we see LOD scores reaching as low as 6.0, suggesting odds of more than 1 million to 1 against that being the location of the gene.

MAP BUILDING

If we look at pair-wise combinations of many different traits, we can arrange things based on the measure of recombination, so that things that rarely recombine are placed close together and things that recombine often are placed far apart. By doing this, we can begin building a map of relative spacing between the genes that cause different traits.

We are often faced with the problem that two traits we want to map may not both be present in the same family, so we cannot ask how often recombination events fall between the two traits. What we need is some outside point of reference to serve as a landmark so that each trait can be tested for recombination between the trait and the landmark. The landmarks we use are called genetic markers.

Technically, a genetic marker can be anything that differs between two copies of a chromosome in a way that lets us use that item to separately track the transmission of two different copies of the chromosome from one generation to the next (Box 28.2). Thus, if we wanted to know whether two things

BOX 28.2 GENETIC MARKERS

A *genetic marker* is any inherited difference between individuals that can be tracked from one generation to the next in a way that effectively tracks the two copies of a gene present on the two copies of a chromosome that is being passed through meiosis. Any kind of inherited difference among different individuals can potentially be a genetic marker: a blood type, an eye color, even the shape of an earlobe. Genetic markers can also be proteins in which a genetically determined difference can be detected, such as a blood type or tissue transplantation marker. Most genetic markers these days are differences in the DNA that can be detected by molecular biology technologies, such as sequencing or PCR. Early DNA-based, markers were often *restriction fragment length polymorphisms (RFLPs)* detectable by differences in the sizes of DNA fragments produced by the cutting of enzymes at specific points in the sequence. The use of RFLPs was initially a major breakthrough in genetics, but RFLPs have gradually been replaced with more efficient technologies. Next came *microsatellite repeat markers*, differences in the lengths of simple repeats that can be monitored by simple PCR reactions. Microsatellite markers have a number of advantages: there are a lot of them, so you are likely to have choices of many markers in any region you want to study, and they are *highly informative* because many of them are heterozygous in more than eighty percent of human samples tested. However, microsatellite repeat markers are more often between genes rather than in them, so most tests that use microsatellite repeat markers can only hope to get near to the gene at best, with the possibility remaining that recombination could happen between the marker and the gene of interest. A more recent form of marker, the *single nucleotide polymorphism (SNP)*, has the advantage that we can select markers that are actually located in genes, but has the disadvantage that a binary system (the base is either the original base or it is the mutated base) cannot achieve the high levels of heterozygosity and informativeness that we see in the microsatellite repeat markers. For all of these markers, their key feature is that we can study someone who has one allele of the marker on one copy of a chromosome, and a different allele of that same marker on the other copy of the chromosome, and we can tell in each case which allele (and thus which copy of the chromosome) they passed along to each of their children. If everyone in a family who inherits a particular trait also gets the same allele of a particular marker, we suspect that the trait and the marker might be close together on the same chromosome *if* the marker was informative in that family.

were close to each other, we did not have to comb the earth in search of some incredibly rare family with those two specific traits. Instead, we could pick a genetic marker and ask how far it is from that marker to each of the two traits we want to study, each of which is present in only one family.

During the 1990s, useful *framework maps* of the human genome began to emerge, maps that had placed genetic markers at known positions all along each of the chromosomes. A map of usable genetic markers made possible the process called a genome scan. A genome scan evaluates each of many markers from throughout the genome, and for each marker asks the question: How far is this marker from the gene we are trying to map? If we want to screen markers that are separated from each other by a rate of ten percent recombination, we would consider this a 10-cM genome scan, and we would expect to have to screen more than 300 markers spread across the genome with a distance of about 10 cM between each pair of markers. DNA from members of a family with a trait would be screened to determine which allele sizes are present for each marker, and for each marker a statistical test would be performed to evaluate the amount of recombination between the marker and the gene encoding the trait.

PUTTING MULTIPLE GENES OR MARKERS IN ORDER ALONG A CHROMOSOME

Knowing the distance between a marker and the gene that determines the trait of interest is only the first step in placing a gene on a map. Let's consider a hypothetical gene that causes the presence of teeth at birth in one large family. When DNA samples from family members at risk of having the baby tooth trait were tested with marker M4, recombination events were seen to fall between the gene and the M4 allele only four percent of the time. If we have a map that shows the location of a set of genetic markers that includes marker M4, we can see that knowing there is a four percent recombination rate is not enough to let us place the gene on the map, since the gene could be located 4 map units to the left of M4 or 4 map units to the right of M4 (Figure 28.5).

If we test other markers in the region and find that the baby tooth gene seems to be only 6 map units from M3 but is 18 map units away from M5, we can then place the gene on the map (Figure 28.6). In fact, it often takes testing of multiple markers on the map to get a precise placement, but this theoretical experiment shows how the map placement process works.

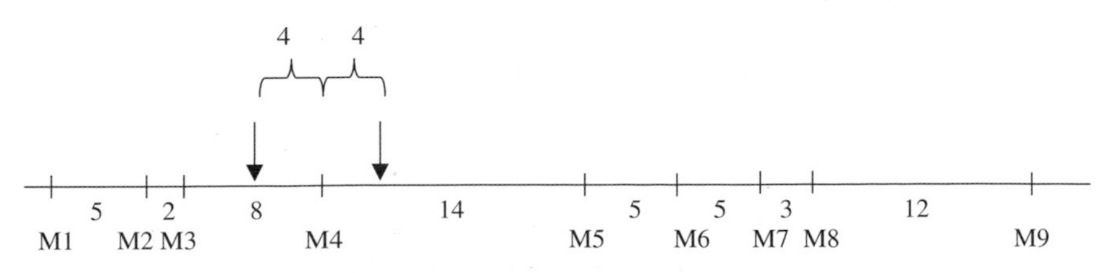

FIGURE 28.5 Finding out the distance from only one marker is ambiguous. In this case, we know that the gene is located 4 map units from M4, but we still do not know where the gene is because the gene could be located either to the left or right of that marker on the map.

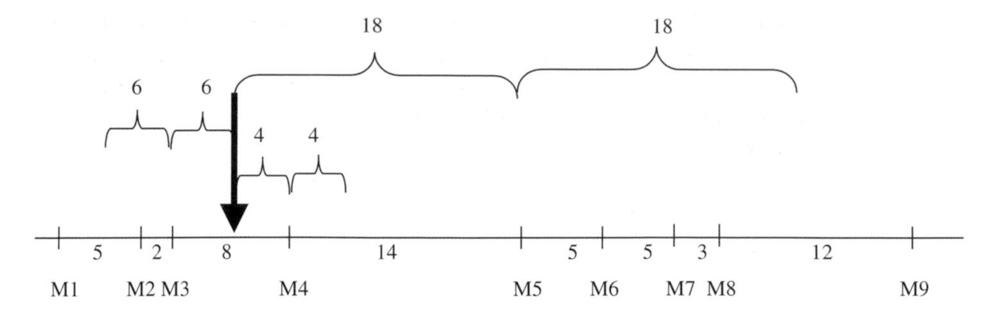

FIGURE 28.6 Adding in information from more than one marker gives us an unambiguous position on the map. Above each bracket is the distance measured from the gene to the marker. For each marker, use of information for that marker only results in two possible locations for the gene as indicated by the pairs of brackets, but when we combine information on the rate of recombination between the gene and each of the markers, we find that there is one place on the map that is consistent with all of the different distances measured. On this figure, that unambiguous position is marked with an arrow at a position that is 4 units from M4, 6 units from M3, and 18 units from M5. In a real experiment, the numbers would not be so tidy. We might expect to see 3.8% recombination with M4, 6.4% recombination with M3 and 16% recombination with M5.

So we now have two different things we want to measure in the course of building a map, or in the course of placing something new onto an existing map: the distance from the trait gene to the markers on the map (the *recombination fraction*), and the measure of how likely it is that the data are right (the LOD score mentioned above). If we combine information from multiple different markers, the process is often more complex than what was shown in Figure 28.6 and may require some sophisticated biostatistical genetic calculations. Fortunately, the results can be displayed in ways that make it easy to see where we think the gene of interest turned out to be on the map (Figure 28.5).

Now that the human genome sequence is known, you might think that such genetic maps would be unnecessary and that we would simply look at the sequence to see where a particular genetic marker is located. In the 1980s and 1990s, large amounts of work went into getting a map of markers constructed. The human genome sequence lets us tell where things are in the sequence even if they have never been studied in genetic mapping studies. However, for many genes located on those maps, we do not yet know what they do or what traits are associated with them. For many genes now placed on the sequence, we will not know what traits they cause until the genetic defects causing the traits are genetically placed onto the map.

There are limits to what we can tell from just looking at the sequence because, although the order of the genetic and physical maps is the same, there are some critical differences. Just looking at the map might give us the correct order of genes and markers along a chromosome, but it does not give us enough information because the rate at which recombination events take place are different for different parts of the genome. On average, we might see that a one percent recombination rate that we call a 1 cM distance on the genetic map would equal about 1 million base pairs of sequence. In some areas of the genome, such as near the centromeres, a one percent recombination rate corresponds with a much larger physical stretch of DNA, and there are other regions of the genome that contain hot spots for recombination where a one percent recombination rate may be seen between things that are separated by distances much, much smaller than a million base pairs. If we look at a particular point on a chromosome, we find that the rate of recombination in that region may be quite different in male meiosis (producing sperm) and female meiosis (producing eggs).

AFTER THE MAP: WHAT COMES NEXT?

In the early days in positional cloning, finding out where a gene was located on the map was the first step in a long series of processes once described as "laborious slogging" by Francis Collins, the leader of the United States Human Genome Project efforts. Once the general location on the map was identified, many additional experiments in the lab would lead to refinement of the location to a much smaller region than that initially identified. The next step required that the DNA from the chromosome region containing the gene be obtained and studied, and that the genes contained in that region be found and identified. Once a gene in the region was identified, sequencing of copies of that gene from individuals with the trait would show whether the gene was mutated in the affected individuals. Once a mutation was found, others with the trait were also tested for mutations. If a gene was found that showed mutations in affected individuals and it was shown that the mutation co-segregated with the trait in families, it was considered likely to be the cause of the disease and additional studies were initiated. This process of moving from a map position to the identification of the actual gene and mutations usually took years (Box 28.3).

Now that the human genome sequence is available and the locations of many genes have been marked on the sequence, the strategies that come after the mapping stage are very different. Instead of going to the lab bench to use biochemical tools to fish around in the chromosomal region for physical pieces of DNA to be searched for evidence of the target gene, most often the next step is to go to the computer, use the Internet to visit one of several sites with extensively annotated versions of the sequence, and begin using a combination of database technologies and search algorithms to sort through the genes in the region electronically. In 2003, we find that in most regions of interest there are:

- Some already known genes of known function
- Some genes not previously known for which we can make intelligent guesses about function based on their resemblance to known genes in other organisms or their similarity to members of a family of genes of related function
- Some genes that we think are real because ESTs give us is evidence that they are transcribed
- Some genes that are being predicted on a strictly theoretical basis for which there is no concrete experimental evidence yet

BOX 28.3 TWO STEPS TO THE POSITIONAL CLONING OF THE HUNTINGTON DISEASE GENE

When a research team at Harvard University first set out in the 1980s to find the gene responsible for Huntington disease, they expected that with the new strategy that they had proposed, it would take many years to carry out the search for the location of the gene. To their surprise, they almost immediately bagged linkage—by finding that a genetic marker they tested was located right next to the gene they sought. However, the hunt for the actual gene, carried on by the Harvard group and an international consortium of collaborators, took almost ten years to get from that initial localization to the identification of the gene. Later on, there were jokes about whether they had used up all of their luck right at the beginning. Throughout the gene hunt, luck was not the issue. The fact that they found the gene was a matter of talent and hard work, not luck. Also, the fact that it took so long had nothing to do with luck. Rather, it was the result of a variety of obstacles that arose directly from unusual things about the region of the chromosome they were working in, the size of the gene itself, the nature of the causative mutation, the complexities of diagnosing the disease, and the difficulties of carrying on some of their work in the field thousands of miles from their home base. In fact, they faced a variety of very real biological problems that did not plague many other projects that sailed past them to find their target genes in less time. Many genes responsible for human traits have gone through this two-step process of first finding the location of a gene, and then using that location as part of the basis for identifying the gene itself. This approach, variously called positional cloning or positional candidate cloning, depending on some of the details of how it was carried out, has become one of the major approaches to identification of human disease genes. Back in the 1980s, before the location of the Huntington disease locus had been pinned down to a small region of chromosome 4, the idea of being able to clone a gene simply based on knowing where it is located was a daring and brilliant idea that helped reverse our view of molecular genetics. The first genes identified were genes that were already well known indirectly from years of study of the gene products. In some cases, their location in the genome was known, but if the location was not already known, getting a copy of the gene was a sure way to be able to find out where the gene was located. What could be done with such information? At first, just catalog it. Soon, however, the localization of a gene went from being an interesting intellectual exercise to being the critical first step in finding a new gene responsible for a trait of interest.

After the mapping step, the next steps in the gene hunt are to:

- Limit the size of the region we have to search to as small a region as possible
- Identify the known and theoretical genes in the region
- Prioritize the genes based on what we know about their function

- Prioritize the genes based on what we know about where they are expressed
- Prioritize the genes based on when they are expressed during the organism's lifetime
- Screen for mutations in the most likely genes by sequencing those genes in individuals with the trait
- Show that an identified mutation is not also present in individuals who lack the trait (or, in the case of recessive disorders, show that individuals without the trait have at most one copy of the gene defect)
- Show that the mutation co-segregates with the disease in families

Gene hunts that used to take as long as a decade can now proceed in months or a few years. The production of the human genome sequence has revolutionized how we do genetics and made it possible for researchers studying a particular trait to spend their time and resources on studying the trait instead of spending their energies trying to generate the tools with which to carry out their studies. We are a long way away from eliminating the need for the molecular biology and biochemistry used to identify mutations in disease genes, and the initial mapping steps still take huge amounts of work that can sometimes generate tens or even hundreds of thousands of data points to get the answer, but the in-between step of finding the genes so we can evaluate whether one of them is "it" has been reduced from years of laborious slogging to a matter of hours or days online. This allows scientists to spend more of their time asking questions of great functional importance, and it allows their operations to cover more territory in less time, with all of the accompanying implications for advances in understanding human health issues of great importance.

Analysis of the human genome sequence revealed more than 30,000 known or predicted genes. A query at Online Mendelian Inheritance in Man shows more than 10,000 entries, a large number of which are loci for which the genes still need to be identified. In the summer of 2003, a search of the LocusLink resource at the National Center for Biotechnology Information shows 1521 disease loci for which the gene sequence has been identified. Thus, as the completion of the chromosomal sequencing portion of the Human Genome Project nears a close, we find ourselves not at the end, but at the beginning of the most important work that will grow out of the Human Genome Project and the human genome sequence: the determination of which genes are responsible for which traits.

BOX 28.4 THE HUMAN GENOME SEQUENCE: A STARTING POINT FOR GENE DISCOVERY

"This is the beginning of genomics, not the end. Critical understanding of gene expression, the connection between sequence variations and phenotype, large-scale protein protein interactions, and a host of other global analyses of human biology can now get seriously underway."

Francis S. Collins, Director, National Human Genome Research Institute

Section 8

COMPLEX AND HETEROGENEOUS TRAITS

In the space of this book, we cannot begin to cover the broad array of traits that are complex (more than one causative factor in the same person) and/or heterogeneous (different causative factors in different people). Here we offer a sampler of topics that illustrate points we especially want to make about the relationship between the genotype and the resulting characteristics in individuals and populations, and we talk about some of the complications that affect our ability to tell what is going on.

GENOTYPE PHENOTYPE CORRELATIONS 29

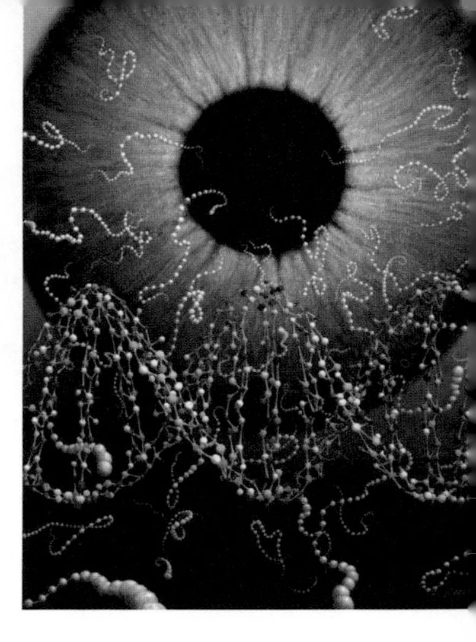

Dana hates going to movies because she cannot see well enough in the dark to navigate her way out of the darkened theater to get to the bathroom in the faint runway lighting used when the theater lights are turned down. She has no problems watching videos at home on her TV, where she can keep the lights turned up while she watches. Dana has a trait called congenital stationary night blindness. It is called congenital because she has always had it. It is called stationary because it does not change as she ages, which is good news if it means that her condition will not get any worse than it is now. It is called night blindness because it affects her ability to see things at night like the stars or to see in dark situations like in the theater. Her eyes work just fine for anything she wants them to do during the day or in a lighted room. Dana has a mutation in the gene that makes the rhodopsin protein. Dana's vision problems actually make sense when we consider rhodopsin's normal function. The thing that might be harder to predict is that most people with rhodopsin mutations do not have the trait she has. Most people with rhodopsin mutations have much greater levels of visual disability up to and including blindness. What are some of the factors that affect the phenotype that results from any particular mutation, and are there theoretical considerations that let us look at a newly discovered mutation and predict what trait will result from the mutation?

One of the surprising revelations that emerged as molecular geneticists began identifying human genes was the finding that sometimes apparently different traits can be caused by mutations in the same gene. If you think back to our analogies regarding absent essentials and monkey wrenches, it makes sense that you might get a different effect from eliminating a gene product (and its corresponding function) than you would get by adding a new function for the same gene. Is it really that simple and straightforward, that knocking out a gene causes one trait, partially kocking it out causes another, and adding a new function to it causes the other trait? What kinds of variation can there be, and how much can we predict just from knowing which gene it is and which kind of mutation? In the long run, sorting out such issues will be important to be able to make the most use out of genetic testing information.

ONE GENE: MULTIPLE TRAITS

One of the most dramatic examples of the one gene–multiple phenotypes phenomenon is the androgen receptor (AR) gene. However, the story is much more complicated than the one we presented in Chapter 21. If we look at the known types of mutations in the AR gene, we see not two but three phenotypes that are dramatically different:

- If the AR gene is knocked out, that is, if a mutation eliminates the gene product or prevents the AR protein from performing its function, the result is the AIS phenotype described in Chapter 21—external female anatomy, no uterus, and presence of undescended testes in an individual with both X and Y chromosomes who is apparently female but infertile.
- If a trinucleotide repeat sequence located within the AR gene is expanded from its normal copy number of about twenty-one copies up to forty or fifty copies, in the same way the Huntington disease gene repeats are expanded, then the result is spinal bulbar atrophy. In this case, there are questions about whether the phenotype results from anything about AR function or loss thereof, or whether the interjection of the amino acid repeat is adding a new function that actively causes a set of events unrelated to the protein's normal function, as we see in Huntington disease.
- In some cases, an AR missense mutation turns up in cases of prostate cancer, with the altered receptor chronically signalling the cells to proliferate even when testosterone is not present. Since binding of testosterone to the androgen receptor signals proliferation of prostate cells (and prostrate cancer cells), some treatments for prostate cancer inhibit testosterone production to stop tumor growth. However, the "always on" mutations in the AR gene bypass this prohibition, allowing continued tumor growth in the absense of testosterone.

This finding of such disparate phenotypes associated with the same gene seems surprising but makes sense when we look at the underlying mechanisms in the three cases. We expect loss of the gene product to prevent hormone signaling, and we now know that loss of that hormone signal changes the pathway by which secondary sexual characteristics develop. Based on studies of other diseases, we might expect that a repeat expansion could cause a gain of function (such as making the protein sticky) that might bring about effects completely unrelated to whether or not the hormone is successfully transmitting its signal to the receptor. A missense mutation might be expected to alter how the protein folds and/or to change the binding properties of the protein, perhaps causing changes in the interaction of hormone and receptor but not eliminating the signaling process. Far more work is needed to understand how these different mutation types actually lead to the phenotypes in question, but at least we have a framework for the formation of hypotheses to be used in designing new experiments.

A TRAIT WITH BOTH GENOTYPIC AND PHENOTYPIC HETEROGENEITY

More than 120 different loci have been mapped that can cause forms of retinal degeneration that involve the death of photoreceptor cells in the eye, and the genes corresponding to more than 80 of those loci have been identified. One of the genes in which mutations were found that can cause retinitis pigmentosa (Box 29.1), one of the main forms of retinal degeneration, is

BOX 29.1 GENETIC AND PHENOTYPIC HETEROGENEITY

Retinitis pigmentosa is a form of retinal degeneration characterized by death of rod photoreceptors in a pattern that begins with loss of rods in the periphery with slowly progressive loss of photoceptors moving inwards towards the macula, the central region used for activities such as reading. Retinitis pigmentosa is considered *genetically heterogeneous* because it can be caused by mutations in many different genes, including rhodopsin. Rhodopsin phenotypes are considered *phenotypically heterogeneous* because mutations in rhodopsin can cause multiples phenotypes, including retinitis pigmentosa and congenital stationary night blindness. Modes of inheritance of retinitis pigmentosa include autosomal dominant, autosomal recessive, and sex-linked recessive inheritance, and different rhodopsin mutations can cause autosomal dominant disease in some people and autosomal recessive disease in others. There is currently no cure for retinitis pigmentosa.[1]

the gene that encodes *rhodopsin*, the protein that detects faint light at night. Mutations in rhodopsin are responsible for about a third of the retinal degeneration that is caused by autosomal dominant retinitis pigmentosa.

When we consider rhodopsin, we cannot simply look at the type of mutation and predict what the phenotype will be like or, frankly, even what the mode of inheritance will be. Missense mutations have been found that are located throughout the gene affecting many different positions in the protein (Figure 29.1). Most of those missense mutations cause retinitis pigmentosa, which results in progressive death of the photoreceptor cells that make rhodopsin. However, several different missense mutations in rhodopsin instead result in a fairly simple form of night blindness, with any retinal degeneration being very minor and occurring very late in life, if at all.

The rhodopsin molecule had already been studied extensively before it was identified as a disease gene. It's a fascinating molecule that sits in the rod photoreceptors and catches photons of light from faint sources in the dark, such as starlight.

Because the processes by which rhodopsin "catches" light and sends the signal along to the brain had been so well studied, it was possible to make some intelligent predictions regarding some specific amino acids that would be expected to cause disease if changed. For instance, it was not surprising to find that one of the disease-causing mutations changes amino acid 187, which had been shown to contribute to the correct folding and shape of the protein by forming a bond that holds two specific points in the molecule together (see Figure 29.1). Another mutation was found that changes amino acid 296, the amino acid that was already known to bind to the *vitamin A derivative* that is essential to rhodopsin function.

However, for many of the amino acids, their importance and their role in disease pathology only became evident once a mutation at that position was

[1] More information about retinitis pigmentosa and other forms of retinal degeneration can be obtained from the Foundation Fighting Blindness at www.blindness.org.

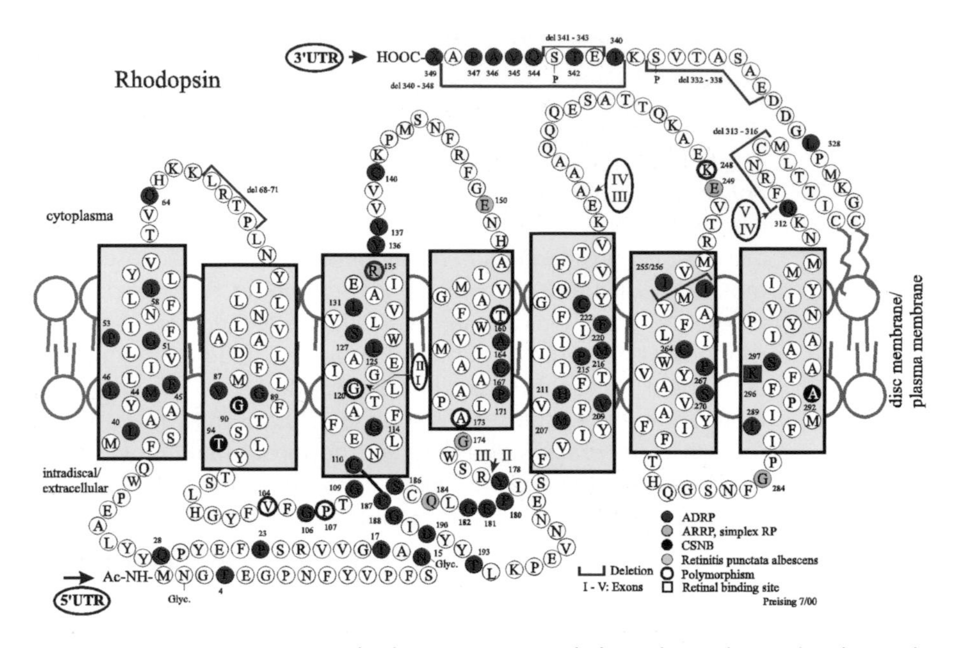

FIGURE 29.1 Mutations in rhodopsin are spread throughout the molecule. Both missense and nonsense mutations can cause either dominant or recessive retinitis pigmentosa, and some missense mutations that are not located anywhere near each other can cause a much milder disorder that involves night blindness without retinal degeneration or daytime blindness.[2] (Courtesy of Markus Preising and the Retina International Scientific News letter.)

identified and the phenotype examined. For instance, one missense mutation at position 90 (Gly90Asp) had no particular known role until it was found to cause night blindness, at which point extensive studies revealed its role in the process of detecting light. When the rhodopsin molecule is folded up into its three-dimensional structure (Figure 29.2), it contains a binding pocket in which the vitamin A derivative sits. The glycine at position 90 is one of the amino acids that lines the surface of this binding pocket. Glycine is a neutral (uncharged) amino acid and asparagine has a negative charge, so a Gly90Asp mutation introduces an extra negative charge into the binding pocket in the vicinity of the vitamin A derivative. Rods normally function only in dark environments and shut down in daylight, perhaps a mechanism that protects them against high light levels that they are not designed to use. Adding a new negative charge into the binding pocket seems to be resulting in a chronic low level of signaling from rhodopsin that makes the rods think they are in a high light level environment, so they shut down even at night just as they would at noon on a sunny day.

Sometimes when a new disease gene is identified, it turns out to encode a known protein such as rhodopsin, one of the globins, an immunoglobulin, or other well-characterized proteins that had been studied for years before

[2] More information on genes that cause retinitis pigmentosa can be found at Retnet at www.sph.uth.tmc.edu/Retnet/, and information on rhodopsin mutations can be found at the Retina International Scientific Newletter Database at www.retina-international.com/sci-news/rhomut.htm.

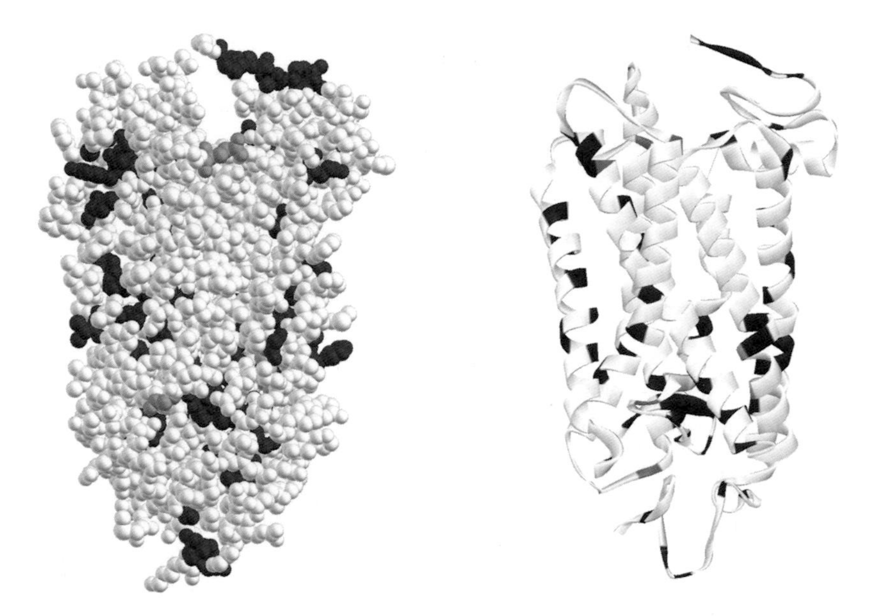

FIGURE 29.2 Three-dimensional view of rhodopsin showing locations of different mutations that cause autosomal dominant retinitis pigmentosa (black) or other forms of disease, including autosomal recessive retinitis pigmentosa and congenital stationary nightblindness (gray). On the left is a space-filling model showing the positions of the different atoms within the molecule. On the right is a ribbon model that gives a view of how the protein is folded into helical, pleated, and other structures to achieve the three-dimensional structure. These images were generated with Accelrys' Discovery Studio Viewer Pro software from the results of submission to SWISS-MODEL the Automated Protein Modeling Server. The resulting model was based upon the coordinates of PDB entries 1HZX, 1L9H, and 1F88, all of which are x-ray diffraction crystal structures of bovine rhodopsin. Positions of human mutations of the rhodopsin gene were taken from Retina International's mutation database. (Courtesy D.M. Reed.)

cloning and sequencing came along. However, when many genes are first found to be disease genes, we do not yet know what role they play in normal cellular processes or disease pathology. This makes it even harder to figure out how a mutation is bringing about the trait being studied.

MUTATION TYPE AND DISEASE SEVERITY

Sometimes the phenotypic differences between mutations in a given gene do not cause entirely different diseases but rather cause differences in some aspects of a single disease such as severity, rate of progression, and age at diagnosis. This phenomenon—different mutations causing differences in disease severity—can be seen for a variety of diseases, although there is no simple generalization that a particular type of mutation will automatically cause a more severe form of disease. Nonsense mutations and deletions or insertions that cause frameshifts can all eliminate the DMD gene product and cause the early severe disorder Duchenne's muscular dystrophy. Missense mutations and in-frame deletions that alter but do not eliminate the gene product cause the much less severe, later-onset Becker's muscular dystrophy (see Chapter 18).

However, for many genes, such as the MYOC glaucoma gene, we cannot make simple generalizations. Although many different MYOC missense mutations cause a very severe autosomal dominant form of glaucoma in children or young adults, some other missense mutations in MYOC do not cause glaucoma at all, even though they also change the amino acid and in some cases are located in regions where other missense mutations have been found to cause disease. A nonsense mutation at codon 368 in one copy of the MYOC glaucoma gene causes disease that is milder and starts later than the disease caused by many of the missense mutations, and another nonsense mutation at codon 46 was found in someone older who does not have glaucoma at all.

We also have trouble predicting what will happen when someone is homozygous for a dominant mutation. For one MYOC missense mutation that causes disease in heterozygotes, homozygosity for the disease allele results in a normal phenotype! With a different missense mutation, homozygosity causes disease that is even more severe than what we see in the heterozygotes.

Thus for some genes, such as the DMD gene, we seem to be able to generalize from the mutation type to the phenotype. Absence of gene product results in the severe phenotype; reduced function due to missense mutation in DMD results in a later-onset, more mild phenotype. However, for MYOC, we clearly cannot make simple predictions about phenotype based on mutation type. These complications mean that every new MYOC mutation that comes along requires additional analysis to determine whether it is causing disease or not. That analysis may include screening populations to find out whether that particular mutation is present in many unaffected people or looking at co-segregation in a family to find out whether the presence of the mutation correlates with the presence of the disease. Once they are developed, biochemical assays to determine whether the gene product still carries out its normal function would greatly assist in interpreting the implications of a new missense mutation.

VARIABLE EXPRESSIVITY

So far, even if the situation looks complex, the underlying concepts make sense. Different mutations in a gene may produce a different phenotype because the two mutation types are having very different functional effects on the resulting protein. A gene in which an absent essential is associated with a trait is not thereafter limited to mutations that remove its activity or eliminate its gene product. It is still free to mutate in a fashion that creates a few monkey wrenches or to take on some new function for which we might not be able to predict the trait. So it makes sense that different mutations in the same gene might do very different things. Different mutation types in the same gene may produce very different traits because the two mutation types have very different functional effects on the resulting gene product, with one type of mutation causing a gain of function and another type of mutation in that same gene knocking out the protein's ability to carry out its function.

What is harder to understand is the level of variation in phenotype within a given trait that we can see between individuals whose disease is caused by exactly the same mutation. We use the phrase *variable expressivity* to refer to the situation in which different individuals with the same disease-causing mutation show quantitative or qualitative differences in the severity of the trait. Let's take the example of a family in which everyone affected with glaucoma has a Val426Phe MYOC mutation. A total of twenty-two members of this family have glaucoma caused by the Val426Phe mutation. The average age at which glaucoma was diagnosed in this family is twenty-six years of age, which is decades earlier than the age at which the common forms of glaucoma usually turn up. One of the most obvious signs of variable expressivity in this family is the great variation in the age at which the disease first manifests itself. The earliest age at which anyone was diagnosed was age sixteen, and the latest diagnosis was at age forty-six. One individual with the Val426Phe mutation still had not developed the disease by the time she was sixty, although she was starting to show faint signs that she might eventually become affected.

When we look at other MYOC mutations, we see a similar pattern: the average age at which the disease starts is young, but there is a big difference in the age at diagnosis of different individuals with the same mutation. Table 29.1 shows information on when glaucoma was first diagnosed in six families with six different MYOC mutations. When we look at the last mutation in the table, we see the amazing range of four to eighty years of age. If we look further at what we know about each of these families, we discover that the family with the Ile477Asn mutation is an enormous kindred with almost a thousand known members, including seventy-four affected individuals spread across eight generations. So perhaps the phenotypic effects of this mutation are really even more variable than the others, but we have to wonder if we were to look at a comparably large number of individuals with each of the other mutations whether we would find a similarly large range of ages. We also have to wonder whether we are looking at identity by descent, or whether there could be one or more additional glaucoma genes playing a role in a family this large.

Some information makes us think that there are real differences in the phenotypes associated with these different mutations. One missense mutation that replaces the proline at position 370 with leucine causes a very early onset form of the disease. If we compare the Pro370Leu family from Table 29.1 with other known Pro370Leu families around the world, we can confirm that this is on average the earliest of the known MYOC mutations. Also, if we compare

TABLE 29.1 MYOC Mutations Show Great Variation in the Age at which Glaucoma is First Observed in six Families

MYOC Mutation	Average Age at Diagnosis (years)	Earliest Age at Diagnosis (years)	Latest Age at Diagnosis (years)
Glu323Lys	19	9	43
Gln368Stop	36	28	49
Pro370Leu	12	5	27
Thr377Met	38	34	44
Val426Phe	26	16	46
Ile477Asn	26	4	80

the range in age at diagnosis for the Pro370Leu data (5–27 years) with data for one of the other mutations in the table, Gln368Stop (28–49 years), we see that the ranges for the two families are so different that they apparently do not even overlap. Comparing information between individuals and between families helps to convince us that the different mutations in this gene show some real differences in how early the disease starts, on average, but that each mutation shows a lot of variability in the age at which the disease is diagnosed in different individuals with that mutation.

MODULATORY FACTORS

How can the same gene end up causing more severe disease in one person than in another? In some cases we have to ask, how can the same mutation leave one person affected and another person with that mutation apparently unaffected? In some cases, this happens because an environmental factor is needed in addition to the genetic predisposition. In the case of rhodopsin, questions have been raised about whether individuals with the Pro23His mutation who have mild cases might have experienced especially low light exposure over their lifetime (wearing sunglasses, living in Seattle) while their more severely affected relatives might have been exposed to especially high light levels (jobs with high light exposure such as welding, living in regions of the country with a lot of sunlight). The role of vitamin A in rhodopsin function has also raised questions about whether there could be a nutritional component to retinitis pigmentosa, although studies are still ongoing to determine which derivatives of vitamin A might help for which types of retinitis pigmentosa.

We also expect that there are differences in what we call the genetic background that will account for some or all of the variability in any given trait. If we consider some of the ways in which mutations have their effects, we can even predict which kinds of genes could alter the effects of a mutation. We call these additional genes and gene products that can alter the phenotype *modulatory factors.* Two different individuals with the same mutation may show great differences in disease severity because of environmental and genetic modulatory factors. Let's take a look at examples of some kinds of potential modulatory factors.

Sequence variants in genes that normally protect us against various forms of damage have the potential to either increase or decrease our ability to compensate for or protect against such damage. If part of the disease pathology involves oxidative stress, a normal biochemical process that goes on at some level in all of us, there are quite a number of genes involved in generating and protecting against oxidative by-products that might affect a large number of traits influenced by oxidative stress. If the disease pathology leads the cell to die through the process called programmed cell death, we can point at a variety of genes that could play a differential role in the programmed cell death pathway if mutated. These modulatory factors are acting on fundamental processes that could well apply to many traits and disease processes, so something that modulates one trait might very well be able to modulate many others that involve the same common damage mechanism.

Another important class of modulatory gene encodes a type of protein called a *chaperone*. These proteins were originally discovered as part of a category of proteins called heat shock proteins that are produced in response to the stress of high temperature inside the cell. In some cases, we see chaperones helping to "escort" proteins to where they need to go, an important role in the intracellular trafficking process. However, in some cases, what we see chaperones doing is helping newly synthesized proteins fold correctly. This is especially important in the case of mutant proteins that are having their effect if they are misfolded. We can imagine that someone whose cells make more of a particular chaperone, or make a genetic variant of that chaperone that is more efficient at helping proteins fold, might end up with less misfolded copies of the mutant protein clogging up the cell. Although this might not completely prevent problems, it might very well slow down the initial development of the disease and the rate of subsequent progression. For some mutations and some proteins, the amino acid substitution might cause a major change in the local chemical properties of the protein that is causing the primary problem. The resulting misfolding problem might be so severe that it can't be overcome by a chaperone. For other combinations of protein and mutation, the new amino acid might be only a bit different from the amino acid normally used at that position, so it would be much easier for a chaperone to refold it. In the latter situation, we would expect a slightly more efficient chaperone or a slightly higher level of the chaperone to have a beneficial effect on the disease pathology. Of course, it is easy to see that a mutation that makes a chaperone less effective at its job might make the disease pathology even worse. Because each chaperone interacts with many different proteins, we might expect that many different traits could potentially be modulated by sequence variants in the same chaperone. Thus the mutation causing the primary disease pathology might be something very rare, but the mutations in modulatory genes could potentially be much more common.

GENOTYPE PHENOTYPE COMPLEXITY

Even with the human genome sequenced, it is still necessary to carry out *genotype-phenotype studies* for each new gene identified as causing a particular human trait because we are still very limited in our ability to make predictions about what effect any particular mutation will have on an individual's characteristics. Some of that limitation arises from not being able to simply predict that a missense change will affect function, or that a missense mutation will or will not be more severe than a nonsense mutation in a particular gene. In some cases, we may understand that loss of function in a particular gene will lead to one trait while misfolding of the protein gives some very different phenotype, and yet not be able to tell whether a given newly discovered missense mutation will eliminate function or cause misfolding. Of course, even if we do genetic testing for mutations in a particular disease gene, there are still the complications of environmental and genetic modulatory factors. Hopefully, as we study more genes and more mutations, we will make advances in our ability to predict phenotypes. However, it is more likely that we will first accumulate a large enough database of known mutations that we will have a set of known information to work with when making our predictions.

Some of the problem with arriving at a sensible understanding of the relationship of genotype to phenotype arises when we consider situations involving *genetic complexity*: traits that will only happen if mutations are present in more than one gene in the same individual, as we discuss in Chapter 30.

HOW COMPLEX CAN IT GET? 30

Amidst the various concepts originated by Mendel, we find that the "one gene–one trait" concept was an important one for helping simplify and organize the genetic models that were being developed. Although this was a very useful model that helped Mendel and others sort out some of the initial rules of heredity, it is an oversimplification. The more we know about the roles of individual genes in any given trait, the more complexity we find; and yet, looking at it from the molecular viewpoint, Mendel was fundamentally right. Any given protein in fact derives from one gene, or rather from the two copies of that one gene that reside in our cells. Whatever trait may be caused by loss or alteration of that protein corresponds with that one gene. This really shifts the problem backwards a step. Wouldn't we then say: "one protein–one trait?" The answer, we discover, is no. In Chapter 29, when we looked at the many diseases that can be caused by mutations in rhodopsin, and when we talked about mutations in modulatory genes, we were just looking at the tip of the complex genetic iceberg.

THE SIMPLE TRAIT IN THE HETEROGENOUS POPULATION: MANY GENES FOR ONE TRAIT

What do we mean by complex genetics, and how complex can it get if it is all based on this "one gene–one protein" variant of Mendel's original "one gene–one trait" concept? We use the same term—complex genetics—to cover two rather different situations, those in which multiple genes cause the trait or those in which a combination of genetic and nongenetic factors lead to the trait.

A trait is heterogenous when multiple people with the trait each have a mutation in a different gene. Let's start out by looking at the situation we started to explore when we talked about the many different rhodopsin mutations that can cause disease: the simple trait (only one thing is causing his trait) in a heterogenous population (different things cause the trait in different members of the population).

THE BIOCHEMICAL ASSEMBLY LINE

A model for genes and traits that are simple in the individual but complex in the population is the *assembly line model* in which you can get the same effect—a car that you can't drive out of the factory—from messing up any of a large number of different steps in the assembly of the car (Figure 30.1).

In a *biochemical pathway*, multiple steps may be needed to reach the desired end point, with each step carried out by a different gene product. Whether a *metabolic defect* results in the failure to make something essential or from the accumulation of toxic levels of some material that the pathway normally eliminates, there may be multiple different ways to get that effect (Figure 30.2).

So in some cases, breaking a step in the pathway may have two consequences—failure to make things beyond the break point is one consequence, and accumulation of intermediates before the break point can be another consequence. Sometimes only one of those two items causes a problem.

FIGURE 30.1 An early twentieth century Ford assembly line demonstrates the sequential nature of the events that lead to the completed assembly of a series of cars that have gone through the same assembly stages. The auto assembly line is a model for some complex genetic traits. A defect created at any of many steps along the assembly line can result in the same trait—a car whose engine won't run. In theory, a car that fails to turn on and makes no noise when the key is turned could be distinguished from a car in which the engine turns over and then stops. The idea that we can tell the difference between two cars that won't start presumes that we have the ability to distinguish subtleties beyond the fact that this car can't drive away from the factory because the engine doesn't run. Similarly, hits at any of a number of points in a biochemical pathway may cause related traits with the same main characteristics. Such traits might have additional features that can help distinguish the two situations, but if we do not understand enough about the causes of the disease, we may not be able to detect the subtle differences that would let us tell whether we are studying one disease or several different diseases with very different causes but the same final outcome. (Courtesy Detroit News.)

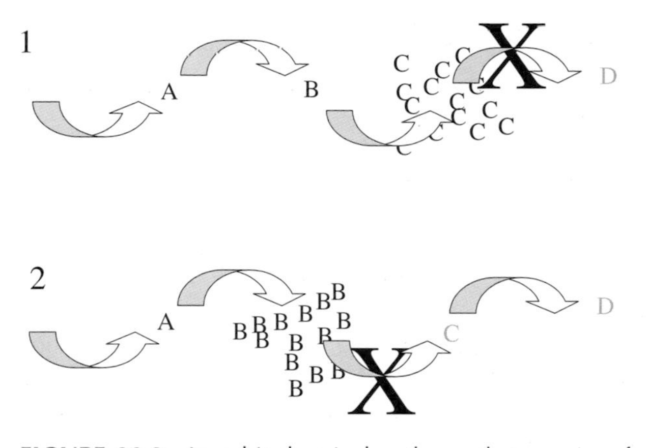

FIGURE 30.2 In a biochemical pathway that requires four steps to make the item labeled D, knocking out one of the steps in the pathway by mutating the gene that makes one of the enzymes in the pathway can cause problems one of two different ways: by preventing the cell from making D, or by accumulating too much of an earlier intermediate metabolite in the pathway. In the first example, end product D is eliminated by knocking out the last enzyme in the pathway (by mutating the gene that makes it), the enzyme that converts biochemical C into biochemical D. If the only way the cell can get rid of C is by turning it into D, a cell with this defect will accumulate an excess of C. In the second example, if the second gene in the pathway were knocked out, excess B cannot be converted to C, so excess B would accumulate, and both C and D would be missing. If the only essential item in the pathway is D, if the body has a way to get rid of excesses of A, B, and C, and if the trait is caused by the absence of D, mutations at any point in this pathway would produce the same trait. However, if the trait is caused by accumulation of excess C, we might see very different effects from knocking out the fourth enzyme and accumulating C than we would get from knocking out the third enzyme and accumulating excess B. In real-life situations, we might see loss of D causing the some aspects of a trait, with accumulation of B or C causing some of the differences between the two traits. If loss of D causes the primary characteristic for which the trait is known, hits on any of the four steps in this pathway would cause the trait.

For instance, if the item that is not being made is something that you can also obtain from your diet, there may be no major consequence to your body's inability to make it as long as you are consuming it. In biochemical pathways that convert something like a protein or fat or sugar into some other biochemical form needed by the body, accumulation of an unwanted intermediate may not always be a big problem. In some cases, the intermediate may be something readily excreted from the body. In other cases, the intermediate that is created may also participate in some other biochemical pathway that is capable of eliminating it (Figure 30.3). Thus, even though both consequences can theoretically be there, often the problem may involve only one of them and it is not always obvious which one it will turn out to be.

A good example of the phenomenon shown in assembly line/pathway models shown in Figures 30.2 and 30.3 is the *retinoid cycle* that is responsible for processing and delivering the vitamin A derivative used by rhodopsin when it detects light (Figure 30.4). A variety of different diseases result that share the major feature of retinal degeneration but that can be distinguished by

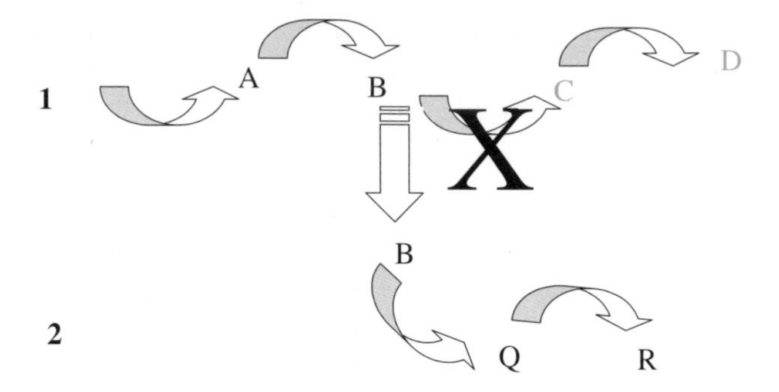

FIGURE 30.3 In a biochemical pathway that blocks the conversion of B to C, we may not see accumulation of B if there is a second pathway that uses excess B to make Q and R. For some pathways, this will solve the problem, and for others it will simply create a different problem if excess Q or R is toxic.

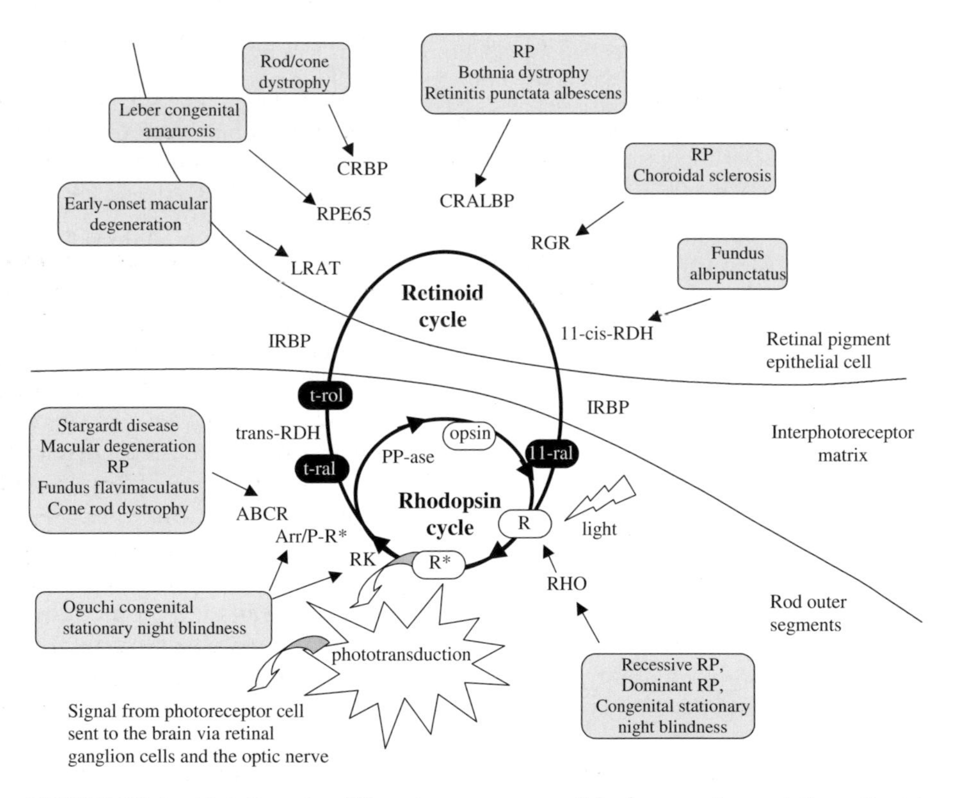

FIGURE 30.4 Mutations in different genes responsible for carrying out the retinoid cycle can cause more than a half-dozen different but related traits. Arrows point from the names of different retinal degenerative diseases (against gray background) to the names of genes that can cause those diseases, and that also carry out the different steps in the retinoid cycle (the process by which the different vitamin A derivates are formed that are needed to complete the cycle) and the rhodopsin cycle (the process by which a complex of rhodopsin plus a vitamin A derivative detect a photon of light and send the information that light hit the complex down the line via a process called *phototransduction* that leads to the transmission of a signal to the brain). Vitamin A derivatives are shown in white letters on black background, and different forms of rhodopsin are framed in white ovals. You don't have to understand any of the chemistry of the retinoid cycle to understand that mutations in the many different genes of the retinoid cycle can cause a related set of diseases by affecting different steps in the same pathway. CSNB stands for congenital stationary night blindness (Modified from McBee and colleagues.)

differences such as the age at which the disease begins, the appearance of the retina as viewed through the pupil of the eye, the rate of progression of the disease, the extent of vision loss that can result, which subtypes of photoreceptors are dying, and the types of electrical responses made by the retinal cells when responding to a light signal. So if someone has a mutation in certain genes of the retinoid cycle, they may not be able to convert vitamin A consumed in the diet all the way through to the final chemical form of it that the body uses. This has inspired researchers to begin working on ways to feed some intermediate form of vitamin A (or other forms of vitamin A that are normally not part of the cycle at all) to bypass the block in the cycle. This general principle may eventually provide answers for a number of metabolic disorders, if the details can be figured out: if the step that creates intermediate four is blocked, then deliver intermediate four and it won't matter that you can't manufacture it yourself. For a variety of reasons, this will not always work, but in some cases, it might someday transform the lives of people with some retinal dystrophies that involve in at least some parts of the retinoid cycle.

Another level of complication is that some genes have close relatives, genes that can carry out the same or similar functions. Thus some traits may require alteration of more than one gene to get the effect. In these cases, the trait that could potentially result may virtually never be seen or might be incredibly rare because people who have a mutation in one of the genes would be protected by the fact that other members of that gene family or a related pathway are still functioning. In other cases, the other members of the gene family may carry out a similar function but not one that can substitute for the missing function. Sometimes, the genes may be functionally similar enough but may not be expressed in the same tissues or at the same stage in development. If you need to replace the activity of gene A in the kidney but its potential substitute gene B is only expressed in the brain, it cannot come to the rescue. This raises questions about whether some therapies will eventually result from getting a gene to change where or when it is being expressed so that it can cover for a defect in a different gene. We already can see an example of this in treatments that compensate for the HbS sickle cell anemia defect by turning back on some expression of a fetal hemoglobin that can help compensate for the defect. Of course, there are many genes whose loss cannot be compensated by another gene simply because nothing else performs that function, even in a different cell type or at a different point in development.

Thus, if we try looking for THE gene that causes the primary pathology—cell death, toxic metabolites, absent essentials, inflammatory reactions, susceptibility to infection, etc.—we would discover that we are actually looking for many different genes that could each lead to that main functional defect if they were mutated, but we may still be looking for only one gene causing the trait in any given individual. Sometimes, in science, we ask a simple question and get back a complex answer. If we did an especially good job of designing the experiment, or at least if we are flexible in our thinking, we may even recognize the complex answer as telling us something important and not just dismiss it because it is not the answer we expected.

MULTIFACTORIAL TRAITS: THE COMPLEX INDIVIDUAL

So now we come to some examples of *genetic complexity*—a trait that arises from multiple factors that come together in one individual. In some cases, we will find that everyone with that trait has all or most of the same set of multiple factors involved in the trait. In other cases, the trait may be complex in the individual and heterogeneous in the population.

Let us begin with the simplest complex case, a form of retinitis pigmentosa that is called *digenic* because it is caused by the simultaneous presence of mutations in two different genes. The discovery of digenic inheritance was the product of a large project on inherited eye diseases at Harvard University. Samples from patients with retinitis pigmentosa were being screened for mutations in genes that were known to cause this disease, or at least suspected of being able to cause it. A person was found that had a mutation in only one copy of the RDS gene, which is normally known to cause autosomal recessive retinitis pigmentosa if both copies are knocked out. That same person turned out to have a mutation in one copy only of ROM1, a gene suspected of involvement in the disease. Further study has shown that a digenic form of retinitis pigmentosa can result when each of these two genes has one defective copy. This actually makes some sense when we consider that the proteins made by these two genes interact in a key structure in the rod cell. We have to wonder how many other situations like this will be found, in which a gene can play a simple role in recessive disease and yet also play a more complex role in the disease when only one copy is defective.

In another trait, called Bardet-Biedl syndrome, a model has been proposed for a tri-allelic cause for the disease. This is a situation in which three genetic defects must come together, two of them in one gene plus an additional defect in a different gene.

We expect that we will eventually find a whole series of increasingly complex situations, some involving one gene, some involving two, three, four, or five genes. In some cases, we expect that we will see that hitting any point in a particular pathway will cause a trait. In other cases, we expect to find that it will require multiple hits on the pathway, each causing a slight change in the efficiency of the pathway without actually shutting down the pathway, with some traits being more dependent on the number of hits on the pathway than on exactly which of the genes have defects.

THE COMPLEXITY OF CARDIOVASCULAR DISEASE

Some of the most complex traits, those involving the conjunction of many different factors in one individual, are not yet fully understood. Some of the most common of the complex traits may well turn out to be the result of the combined effects of many different genes, as well as environmental effects. If we just consider cardiovascular diseases, it is easy to identify many categories of genes that we expect will turn out to be involved, including genes that control:

- Cholesterol levels
- Strength and flexibility of blood vessel walls

- Formation of plaque
- Adherence of plaque to vessel walls
- Inflammatory processes
- Heart muscle strength
- Heart valves
- Valves in blood vessels
- Heart rhythms
- Blood clotting

This is just to name a few. You do not necessarily have to have things wrong with all of those to have a problem, and every once in a while someone will have a truly major defect in just one of them that will lead to a substantial chance of developing a health problem. For many people, simply having several things on that list, each operating suboptimally because of a minor change in how well they work, may be enough to add up to major health consequences over the course of a lifetime. If we then consider that each category on the list is going to have a variety of complex structures and biochemical pathways affecting how well that part of the cardiovascular system works, it is easy to see how we could come to see cardiovascular disease, or even any one particular form of cardiovascular disease, as being truly *polygenic* with involvement of many different genetic risk factors. If we then add in environmental effects such as diet, exercise, and exposure to smoke, we can see how any dozen people arriving in the emergency room with symptoms of a heart attack may effectively have a dozen different diseases, each with a slightly different combination of the many factors that potentially lead to such events.

Does this mean that every heart attack that happens is its own customized genetic event, distinct from what has happened to anyone else? No; we expect that there will turn out to be a limited number of combinations of a finite (although possibly large) number of causes, so that there may be many people who share the same main combination of causative factors. If we can identify these different genetic risk factors, there will be several important benefits. It will become possible to screen people for genotypes associated with the highest risks (something we currently try to estimate from a combination of family history and current health status) so that we can intervene before a problem develops. Identification of the genes involved may offer us new insights into the underlying mechanisms, allowing for development of new approaches to treatment or prevention. Identification of a subset of patients who all share the same primary causes of disease will finally let us do clinical studies in which we pool information on people who actually have the same disease, not just a related set of symptoms resulting from very different causes. This should greatly increase our ability to learn things about how to help people with that particular genotypic combination, and should help researchers' efforts to identify the particular environmental factors that pose the greatest risks to individuals with that at combination of genetic risk factors.

This brings us back to our assembly line model and the problem of telling why we can't drive the car out of the factory. If we could really just look at the cars with defective starter motors and know that they had a different problem than the cars with uncharged batteries, we might have a better chance of fixing the problem or maybe even keeping it from happening.

So we come back around to the title of this chapter: How complex can it get? The answer is very, very complex indeed. The existence of quantitative traits, to be talked about in Chapter 31, adds even more complications.

QUANTITATIVE TRAITS

31

Ira listened in frustration as his doctor explained that his blood pressure was borderline. Today it had registered 142 over 90. On the last visit it had been 120 over 73, and the visit before that it had been 135 over 85. "So, what does that mean?" he asked. "Do I have high blood pressure or not?" His doctor smiled ruefully and shook her head, saying, "It's not really that simple. Although I sometimes treat people at this blood pressure, I don't always. It is absolutely clear that we want to treat someone at 190 over 110, and we are unconcerned about someone at 110 over 60. In the past, I would have been unlikely to treat someone with these pressures, but as we are learning more about the effects of supposedly borderline blood pressures on long-term health, it has become clear that the pressures that should alarm us are lower than those that alarmed us in the past. You are in a range that we consider borderline, an area that is hard to interpret because the range of values for blood pressure is so continuous that we cannot say that there is some dramatic step from 130/75 to 131/76 that suddenly moves you from a healthy category into a disease category. So I am going to start by having you do a blood pressure diary, recording blood pressure each day to see how it is varying over time so we can get a better idea of whether you are actually spending most of your time at pressures that are lower or higher than what we are measuring here in the office, when you have just rushed in to see me in the middle of a stress-filled day. On your next visit, we can talk about whether or not to do a trial run with blood pressure medication, depending on what happens to your blood pressure over the next few weeks." Ira walked out of the doctor's office shaking his head, wondering why the answer could not be some clear-cut yes or no. As we will discuss here, many human traits are quantitative traits that differ not by being present or absent, but rather by differing continuously over a range of values.

Many of the traits we have talked about so far have been *binary traits*, that is, traits with two states—present or absent. We have talked about secondary aspects of variation in the trait, such as how severe it turns out to be in any given person, but these have been traits for which most people can usually be divided up into those who have the trait and those who do not. However, for many human traits, such as height, weight, blood pressure, or IQ, the variation is not binary but rather continuous across a broad continuum of values. Although some of those at extreme ends of the weight and height continuums may sometimes be suffering from pathological conditions, it is difficult to draw a simple line that lets us say that everyone above the line has the trait and everyone below the line does not. Moreover, many of these traits, although strongly genetically determined, may also be affected by the environment. A real understanding of these traits requires a more sophisticated view of the mechanism by which a given genotype produces a given phenotype.

MANY TRAITS ARE SPECIFIED BY ADDITIVE EFFECTS OF MORE THAN ONE GENE

Unfortunately for most geneticists who are trying to find genes responsible for human traits, the traits in which we are most interested, those posing the greatest public health risks to the most people, are often not simple yes/no traits but rather complex *quantitative traits* in which there may be no obvious place to draw a line that separates those at risk of harm from those who are not at risk. Many human characteristics are continuous traits that show some fairly continuous pattern of distribution. These quantitative, polygenic traits are the result of many separate effects that are added (and subtracted) together to arrive at the resulting phenotype.

Thus, for a trait such as height (Figure 31.1), a very large number of gene products (as well as additional environmental factors) go into determining

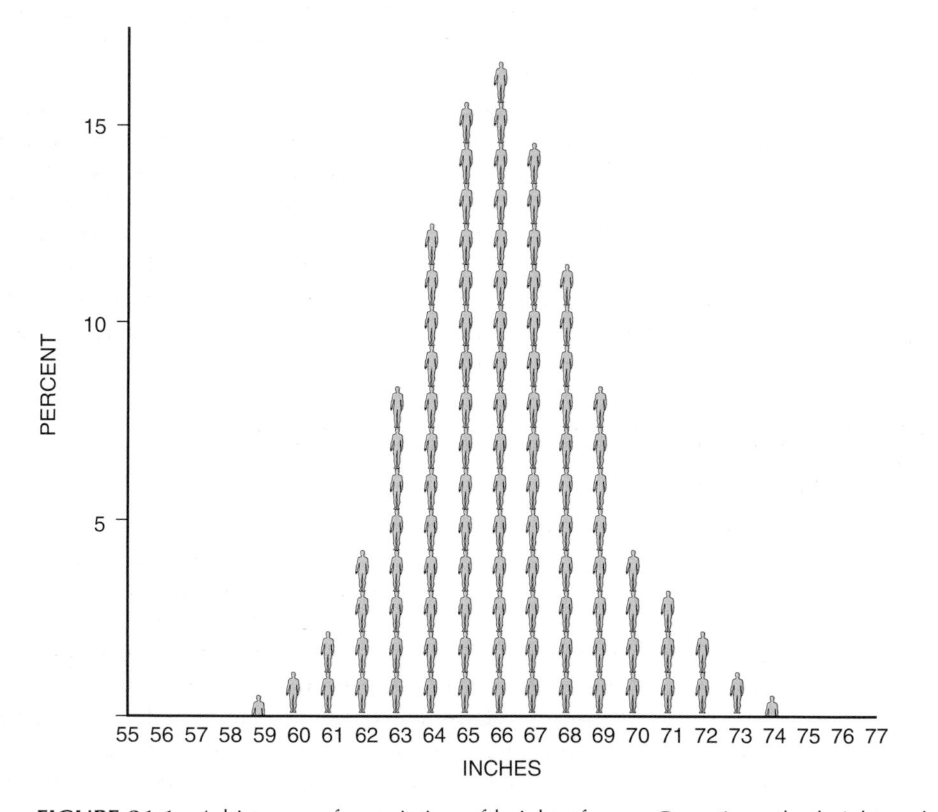

FIGURE 31.1 A histogram for variation of height of men. Over time, the height and weight charts used decades ago have gradually become outdated, raising questions about whether the different current distribution of height may represent some combination of altered nutrition in the population and changes in the overall racial/ethnic composition of the population in this country, since we would see very different histograms if we plotted this information separately for Caucasian, Hispanic, African-American, and Asian men. (Adapted from Harrison et al., *Human Biology*, 2nd edition, Oxford, UK, 1977, Oxford University Press.)

one's height. Although we can sometimes see simple single-gene effects on the height of an individual, such as the defect in human growth hormone that causes a form of dwarfism, most human height is the additive result of contributions by genes such as growth factors, transcription factors, growth hormones, and their receptors, some of which may be more important than others but each of which makes some contribution. Thus the final height of an individual will, in part, reflect the sum of the activity of many genes. This point can be made explicitly with the following bit of whimsy.

Let us imagine a hypothetical trait, the football-quarterback trait, which confers the ability to accurately throw an oblong ball to a receiver who is fifty yards away and moving fast, while dodging a whole bunch of very strong guys who want the ball for their very own. This trait reflects the additive effects of genes encoding skeletal structure, musculature, hand-eye coordination, agility, speed, vision, reflexes, and intelligence. Further suppose that for each of these 500 or so genes there are "athlete" alleles, "normal person" alleles, and, "couch-potato" alleles. One could then imagine that an individual exhibiting the quarterback trait would possess athlete alleles at most or even all of these genes. Conversely, couch potatoes might have received few of the athlete alleles and have many of the couch-potato alleles. The rest of us, with some combination of each kind of these various alleles, might well fall on a continuum between those two ends of the spectrum.

However, even an individual possessing the right combination of alleles required for the quarterback trait will never make it to the Super Bowl without proper nutrition, health care, training, and encouragement during growth and development. It seems less likely that training, coaching, and nurturing could create the quarterback phenotype in an individual who lacked the requisite genetic makeup. However, every once in a while an athlete like figure skater Scott Hamilton, who went on to great athletic prowess and fame in the wake of substantial health problems, makes us reconsider just how complex all of the contributing factors really are.

SOME TRAITS MAY REQUIRE A THRESHOLD NUMBER OF DELETERIOUS ALLELES

Imagine that there was a trait that was quite variable in expression but not truly continuous. *Cleft palate* is a good example. Although the severity of this trait varies between affected individuals, it is not a continuous trait within the population. Babies are either born with a cleft lip or palate or are normal. Yet cleft palate does seem to "run in families" and follows the rules for multifactorial inheritance that are listed in the next section.

We think of cleft palate and other similar disorders such as *anencephaly* and *spina bifida* as being *threshold traits*. These are traits in which individuals with the trait are thought to carry more than a certain threshold number of "deleterious" or "advantageous" alleles that are required to create a phenotype, as shown in Figure 31.2. Forming the top of the palate during embryogenesis requires that two masses of tissue in the head of the developing fetus arch up over the tongue and the mouth cavity and fuse properly to form the upper palate and the two sides of the upper lip. Those tissue movements surely require the products of many genes acting in concert. These movements also must occur in a very short window of fetal development.

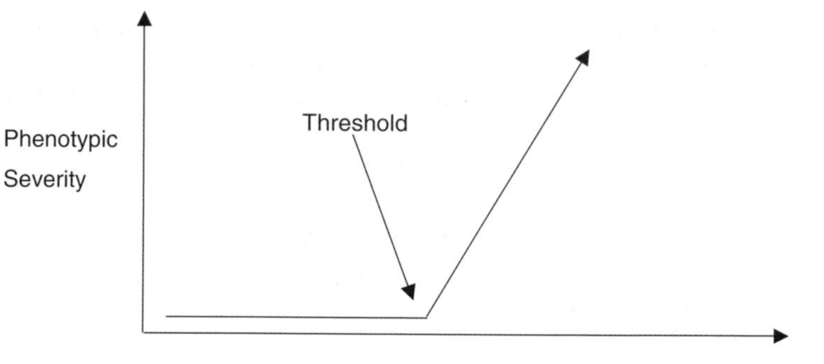

FIGURE 31.2 A schematic view of the threshold model, in which the key issue is how many total genes are affected. Thus different individuals with the trait may not all have defects in the same genes, but out of a set of many genes that can affect the trait, the ones who will be affected will be those with more "hits" on the set of relevant genes, and individuals who have some mutations in such genes will not be affected if the number of "hits" they carry is below the threshold level.

Clearly, some alleles might produce defective proteins that retard this process. As long as they don't retard it too much and the arch of the mouth is built before the window of time is closed, things will be fine. However, if more deleterious alleles are added to the genome of this embryo, progressively more impairment of movement is observed. Finally, as shown in Figure 31.2, in true "the one straw or allele that broke the camel's back" form, one too many proteins is impaired, and the arch fails to be completed before the temporal window closes.

One of the more interesting consequences of proposing such thresholds is that it becomes easy to suggest that thresholds might be altered by environment, such that two identical genotypes might display different phenotypes in different environments. For spina bifida, this idea is supported by the observation that differences in prenatal nutrition affect the chance that a baby will be born with spina bifida.

If all of this seems less solid or more confusing than simple Mendelian inheritance, it is. The understanding of polygenic traits and of genotype-environment interactions is becoming an increasingly interesting and profitable avenue of inquiry in human genetics. Multifactorial inheritance presumably underlies some of the more clinically important human traits, such as susceptibility to several major diseases or illnesses (e.g., heart disease, stroke, diabetes, schizophrenia). Some data also support the notion that alcoholism may involve multifactorial inheritance. It is also possible, as noted earlier, that this type of inheritance might play a role in establishing some crucial aspects of personality, as well. Thus it is worth considering just how one would recognize multifactorial inheritance in a pedigree.

THE RULES FOR MULTIFACTORIAL INHERITANCE

As is the case for autosomal recessive and autosomal dominant traits, there are some rules that let you determine that a given trait is best explained by multifactorial inheritance.

1. Although the trait obviously runs in families, there is no distinctive pattern of inheritance (autosomal dominant, autosomal recessive, or sex-linked) within a single family. In other words, when nothing else makes any sense, start thinking about multifactorial inheritance. Nonetheless, a few rules are helpful in identifying a trait whose expression reflects multifactorial inheritance. These are:
   - The risk to immediate family members of an affected individual is higher than for the general population.
   - The risk is much lower for second-degree relatives (aunts, uncles, grandchildren, etc.), but it declines less rapidly for more remote relatives. This latter point is a hallmark of multifactorial inheritance and distinguishes it from autosomal recessive inheritance.

2. The risk is higher when more than one family member is affected. Traits reflecting multifactorial inheritance are controlled by the number of deleterious or advantageous alleles segregating in the family in question. If there are several affected individuals within a family, the odds go up that a large number of deleterious or advantageous alleles are segregating within the family. Obviously, in such a family the risk to subsequent children is increased when the parents are consanguineous.

3. The more severe the expression of the trait, the greater the risk of recurrence of the trait in relatives. The model for this is that the phenotype is presumed to be proportional to the number of deleterious alleles distributed over a large number of genes carried by that individual, with any one gene having a very small effect on the phenotype. An individual with many such alleles will, on average, pass on half of those alleles to their children and share half of them with their siblings. Imagine that some trait, such as cleft lip or cleft palate, was governed by alleles at 400 genes and that "normal" and "bad" alleles exist at all of those genes. Assume that any 50 such "bad" alleles distributed among 400 genes are sufficient to produce a phenotype and that the severity of the phenotype gets worse as the number of "bad" alleles increases. Thus an individual carrying 60 such "bad" alleles will be mildly to moderately affected. On average, he will pass only 30 "bad" alleles onto his children, which is usually too few to cause a problem. However, a more severely affected individual with, say, 150 "bad" alleles might be expected to produce children with 75 such alleles. Such offspring are very likely to be affected. Note that these are 400 hypothetical genes since many of the genetic factors affecting cleft lip or palate remain to be determined.

4. Because of the influence of the environment on such traits, identical (or monozygotic) twins need not always be concordant with respect to the trait. Indeed, the concordance can fall anywhere between one hundred percent and the concordance observed between siblings. This may be especially true in cases where the number of alleles is just over the threshold at which characteristics of the trait can occur. Keep in mind that if the trait starts late in life, lack of concordance among younger twins could indicate variability in onset rather than genetic complexity or environmental influences.

Obviously, most of these rules work best with traits that are to some degree discontinuous, and they may not all apply in every case. However, there are methods for applying these rules, with some degree of difficulty, to fully continuous traits, such as height. Let us close with a discussion of one of the most controversial of all multifactorial traits, intelligence.

THE "GENETICS" OF IQ

Human beings can't fly and don't swim very well by the standards of other more aquatic mammals, such as seals or dolphins. Even the best of us cannot outrun a gazelle or outclimb a bear. Our hides are thin, we have lousy claws, and the acuity of our senses of sight and smell don't match those of many other animals, especially at night. We are the dominant critter on terra firma because of our intelligence. We developed language and a method to pass what we learn and know onto both our young and our whole species. We create and build tools. We work, hunt, and kill well in groups. All of this requires a quality, or large set of qualities, called intelligence, something that turns out to be very difficult to define. A question often raised is that of just how much of intelligence is attributable to genetic factors.

The current best guess is that tens of thousands of our genes are expressed in our brains, and for many of them the nervous system is the primary place where they are expressed. It would be foolish to imagine that allelic variations in those genes don't exist or that such variation is without phenotypic consequences (i.e., that this variation doesn't explain some of the differences we see among us for various aspects of that elusive thing we so crudely call intelligence). Moreover, studies in lower organisms, such as fruit flies and even mice, support the idea that genes play a role in such processes as learning, memory, aggressive behavior, and various other aspects of cognitive functioning.

In human beings, we see a number of genetic conditions, such as Down syndrome and fragile X syndrome, that are associated with mental retardation. We have also noted the behavioral problems or anomalies sometimes associated with Klinefelter syndrome. Clearly, these are genes whose proteins are required for the proper functioning of our nervous systems. Those genes, like all genes, are expected to be mutable, and there is every reason to believe that in people, as in animals, these allelic differences will have phenotypic effects. However, from what we know about mutations and the genetics of other traits, we expect that mutations, whether deleterious or advantageous, should be well spread out throughout the human population such that no one group is likely to have a monopoly on mutations in these "intelligence" genes. However, the interplay of effects on so many different genes combined with a truly complex set of environmental factors affecting child development leaves us not at all surprised that there are problems getting people to even agree upon a definition of intelligence.

Often when people ask about whether intelligence is inherited, what they are really asking about is heritability of IQ scores. They see reports in the news talking about heritability of IQ and come away concluding that heritability is high (for some populations, it is) and that there are reported differences between average IQ scores for the various racial groups in the American population. Some people incorrectly argue that, if IQ is heritable, and if IQ values differ between two populations, the difference must be genetic. Some take such arguments as an indication that people of one race or another are "biologically smarter" than other people. One can imagine the horrid social consequences of accepting such a view. Speaking solely on scientific grounds, there are at least two major things wrong with this argument.

First, *heritability cannot be compared between two populations.* This is because the proportion of the variance in IQ scores that is due to environmental factors in the two populations may be very different, even if the genetic structure of the populations is similar or identical. This central tenet of quantitative genetics tells you that you cannot use heritability to compare the genetic structure of two populations. Nothing in available data argues for a genetic difference with respect to IQ between human populations. That differences in average IQ scores between races really exist or are biologically meaningful even if they do is arguable. Even if such differences do exist, there is *no* reason to believe that they are reflective of differences in genotype between races. This caveat is especially important given many studies that suggest a strong environmental component to performance on IQ tests. (All of which avoids a whole additional problem involving questions about possible scientific error or even fraud in some early studies of IQ and heritability).

Second, although IQ scores may well measure some aspects of intelligence, exactly which aspects of intelligence they measure is an area still open to dispute and investigation. More importantly, there are also types of intelligence, such as creativity, intuition, or some types of abstract reasoning, that are not addressed by IQ tests. Clearly, IQ tests are around and may be useful in identifying and helping people whose scores fall toward the low end of the curve. However, exactly what the tests measure in individuals toward the middle and upper ends of the curve is unclear. Moreover, the variation for IQ within any population group assayed is quite wide and various populations overlap significantly, despite differences in the mean.

Before we can meaningfully discuss the relationship of IQ to genotype and of genotype to intelligence, we need a better understanding of just what an IQ test measures and a better set of definitions of intelligence. We will also need to free this issue from the trappings of racism that have shrouded it for the last two decades. As stated earlier, we have no doubt that there may be genetic differences relating to various cognitive processes, but getting at meaningful answers is going to be difficult and may require advances in multiple fields, including genetics.

However complex the issues are surrounding genetics and intelligence, far more complicated and frankly vexing are efforts to study behavioral and psychological issues. Chapter 32 addresses a frankly uncomfortable issue, that of the genetics of violent aggression in humans. Many of the issues considered here will be relevant and should stay in your minds as we continue with the story of MAOA and questions about whether it plays a role in human criminality.

THE MAOA GENE: IS THERE A GENETIC BASIS FOR CRIMINALITY?

32

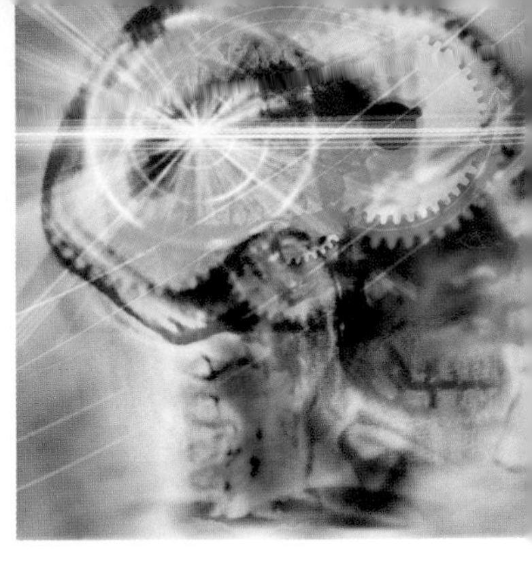

In 1997 in California a man was tried and convicted for a terrible crime, the murder of a young girl. It was reported that he displayed a total lack of remorse or respect for our society and its laws. During the coverage of the trial a local television station aired a report in which someone who had known this man for quite some time claimed, "he was just wrong from the beginning, just a bad seed." More recently, another man on trial for murder is actually trying to use genetic inheritance of criminality in his family as a legal defense that implies he was predestined to do bad things. If there were such a thing as a "natural born killer," a genetically predestined "bad boy" or "bad girl" from birth, would such a person be missing some crucial gene product essential for building the parts of the human psyche that proscribe most of us from such behavior, genes that make empathy, caring, and guilt possible? Can the situation possibly be so simple? Do such genes even exist, or are they just part of a prejudice ingrained in our culture? What follows in this chapter is an attempt, however unsatisfying, to gain some insight into this complex issue.

GENETICS OF VIOLENT AGGRESSION IN A DUTCH FAMILY

Figure 32.1 displays a pedigree for a family that was studied in the Netherlands. Indicated males in this family were often subject to seemingly unprovoked and uncontrolled violent outbursts. These aggressive episodes ran the gamut from ranting and shouting to exhibitionism and serious crimes, such as rape, arson, and assault with deadly weapons. (One of these men forced his sisters to undress at knife point. Another man raped his sister and then later, while incarcerated, attacked the warden with a pitchfork. A third member of this family attempted to run over his employer with a forklift.) All of the affected males were mildly retarded, with an average IQ of only 85 (100 is considered "normal"). Although there were no affected women, sisters of affected males frequently gave birth to affected sons.

All of this evidence pointed strongly not only to a genetic basis for this trait but indeed to a sex-linked mutation in a gene on the X chromosome underlying this behavior pattern. To test this possibility, researchers set out to determine if they could find the gene responsible and determine what mutation caused this set of behaviors. The gene they found, called MAOA, encodes a protein known as monoamine oxidase A that is required to break down molecules known as neurotransmitters in the brain.

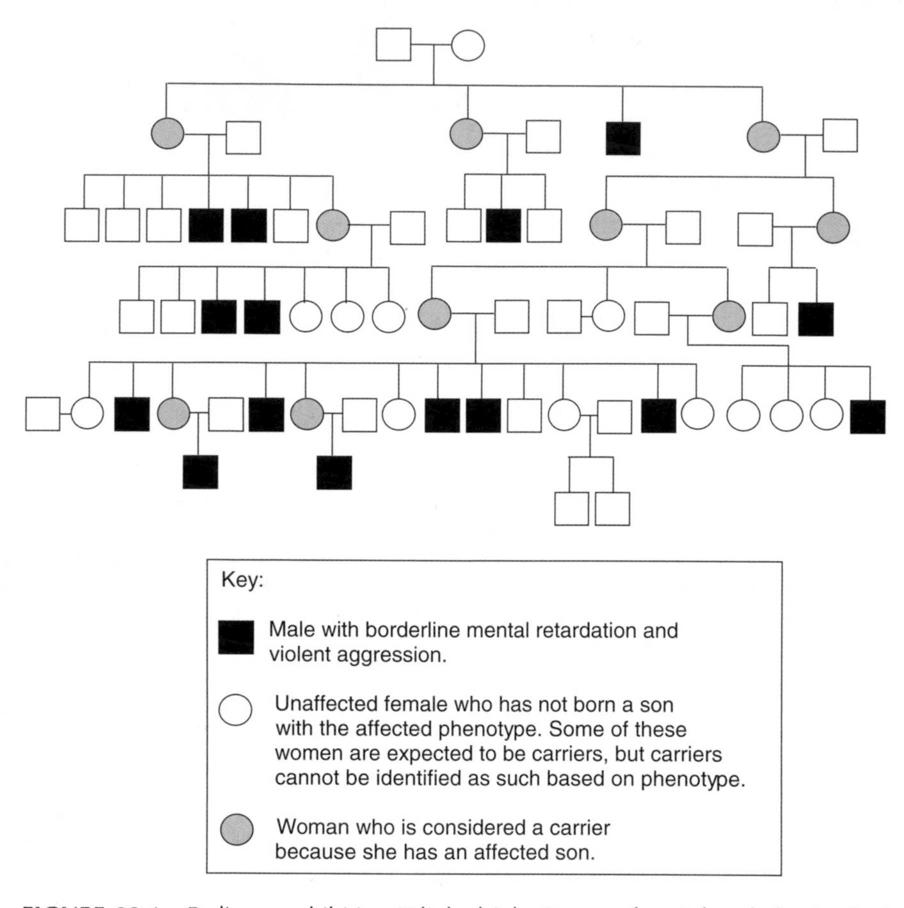

FIGURE 32.1 Pedigree exhibiting X-linked inheritance of a violent behavioral phenotype. (Adapted from Brunner *et al.*, *American Journal of Human Genetics* 1993; 52:1032–1039.)

Neurotransmitters are small molecules that facilitate communication between cells, known as neurons, that comprise the nervous system (Figure 32.2). Obviously, the presence of neurotransmitters at the synapse needs to be very tightly controlled if the nerve cells are to function properly. They must be released rapidly by the stimulating cells and absorbed and/or degraded by the responding neuron. One of the enzymes required to break down some neurotransmitters is MAOA. A variety of biochemical studies on urine samples taken from the aggressive males in the Dutch family indicated markedly abnormal metabolism of neurotransmitters that are normally broken down by MAOA, including dopamine, epinephrine, and serotonin. Very high levels of these compounds were found in the urine of these males, consistent with an inability to break down these compounds.

As diagramed in Figure 32.3, in a subsequent paper, affected males in this kindred were shown to carry a point mutation in the eighth exon of the MAOA gene that changes a glutamine codon to a stop or termination codon. This nonsense mutation was not found in the MAOA genes carried by unaffected brothers of affected males; obligate female carriers were also found to carry one normal and one mutant allele, although they were phenotypically

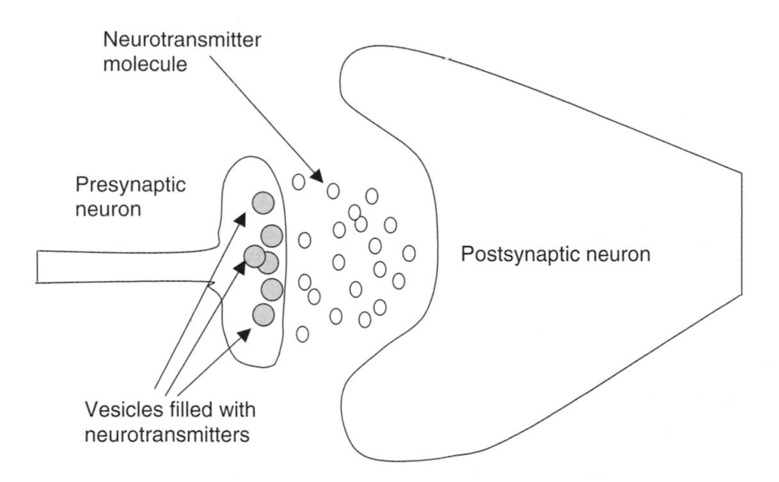

FIGURE 32.2 Neurons and neurotransmitters in synapses. Although signals within a neuron are carried electrically from one end of the cell to the other, a given neuron communicates with the next neuron in the sensory or motor pathway by releasing neurotransmitters into the small space between these cells, known as a synapse. The second neuron absorbs the neurotransmitters, triggering it to fire an electrical signal along its length, and so on.

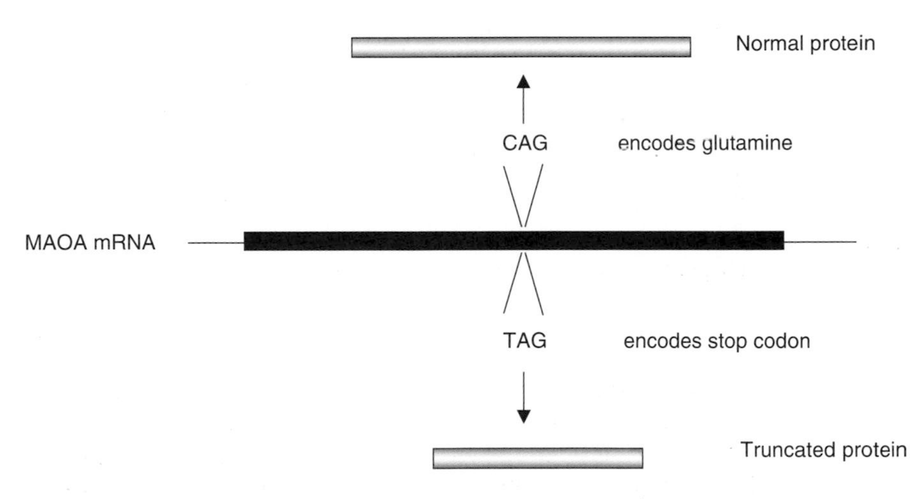

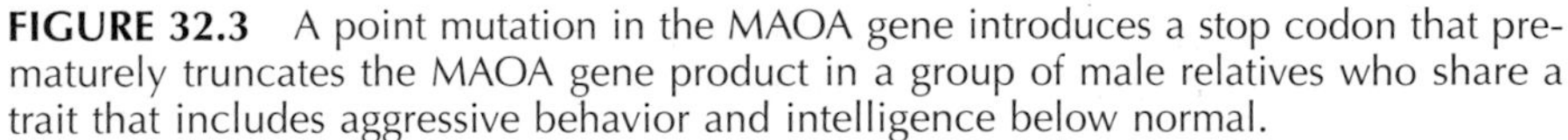

FIGURE 32.3 A point mutation in the MAOA gene introduces a stop codon that prematurely truncates the MAOA gene product in a group of male relatives who share a trait that includes aggressive behavior and intelligence below normal.

normal. Thus there was a precise correlation between whether males carried this nonsense mutation and whether they expressed violent aggressive behavior.

Similar observations were made in a strain of mice that were genetically engineered in such a fashion as to delete the MAOA gene. Mice homozygous for this mutation displayed highly increased levels of aggression and greatly increased levels of neurotransmitters. More critically, adding back a functional copy of the MAOA gene to the mouse genome restored both the ability to break down the neurotransmitter molecules and normal levels of aggression. More recently, studies in Macaque monkeys also confirm these findings.

How is the defect in MAOA correlated with the violent outbursts exhibited by these men? Realize that, among their many roles, neurotransmitters function as part of the body's "fight or flight" response to threats or danger. In most of us, as the levels of these neurotransmitters increase in our brain in response to various stresses, they in turn are broken down by MAOA. Thus most minor stimuli produce only transient increases in neurotransmitter levels. One can then imagine that the degradation of some neurotransmitters is greatly impaired in males lacking the MAOA enzyme, thus allowing levels of neurotransmitters to rise far in excess of a normal level. Indeed, in several cases, crimes committed by these males closely followed traumatic family events, such as the loss of a loved one. It is then possible that stressful events might overstress these males to the point that violent outbursts become more likely.

How important is the genotype for MAOA (and the environment) in terms of violent aggression in the general population? Most workers now seem to accept the conclusion that this null mutation at the MAOA gene largely explains the violent phenotype seen in this family. However, this type of mutation at MAOA is extremely rare in the human population. The differences that do exist among most human beings result in altering the level of MAOA expression, not its abolition. Attempts to correlate the polymorphism that was observed with violent aggressive behavior in the general population remained inconclusive for several years after the report of of this family from the Netherlands.

A recent study in Australia and New Zealand by Caspi and colleagues reveals that variations in the level of MAOA expression do not simply create a predestined behavioral phenotype; rather, the MAOA genotype appears to mediate the effects of mistreatment of young boys, at least in terms of whether or not those children develop into adult men who exhibit antisocial problems, specifically: a disposition towards violent antisocial behavior, an antisocial personality disorder, or eventual conviction for a violent offense. They found that mistreated or abused male children with high levels of MAOA expression were much less likely to develop into adult men that exhibit these phenotypes than were male children with low MAOA levels (Table 32.1). For example, among severely maltreated boys with low MAOA expression, greater than eighty percent of these children developed into adults with behavioral difficulties, while the fraction of mistreated children with high levels of MAOA expression that developed similar difficulties fell just above forty percent. Because it was not asked in this study, we cannot know what other kinds of environmental difficulties might similarly confound a "low-MAOA" genotype. Sadly, similar differences were observed when the metric was the percentage of individuals of a given expression level who had been incarcerated for violent offenses by age twenty-six.

TABLE 32.1 Different Outcomes for Abused Children with High and Low MAOA Genotypes

Genotype	Violent Antisocial
High MAOA	40%
Low MAOA	80%

Derived from Caspi *et al. Science* 2002; 297:851–854.

The critical finding in this study was that the genotype for MAOA, by itself, did not predict violent or aggressive behavior unless something environmental was factored in. Differences were observed only when childhood abuse or mistreatment was superimposed on these differences. Thus the final phenotype (violent aggression or antisocial personality disorder) seems to result from the combination of two separate components, genotype (low MAOA levels) and environment (childhood abuse). The biochemistry and pharmacology of all of this makes good sense. It turns out that in both animals and humans there are data to suggest that early maltreatment or stress can alter neurotransmitter levels in a fashion that can last well into adult life. Perhaps higher levels of MAOA in a child may make that person more resilient, or more resistant to the stresses of abuse and maltreatment. Similarly, lower threshold levels of MAOA might sensitize a child to the same effects. The critical finding was that, although the MAOA genotype contributed to the behavioral phenotype, the environment in which the child was raised also played an important role in whether antisocial, violent behavior turned up in the men with the low MAOA genotype. Development of violent, antisocial behavior was not a foregone conclusion for these individuals but rather highly dependent on environment.

Given the well-documented correlations that have come out of two different studies in two different parts of the world, you might be wondering why we don't just start sequencing the MAOA gene from every serious criminal who will stand still long enough to let us draw blood. There are several reasons why such studies are both ethically difficult and scientifically incomplete.

- First, there is a serious problem of "informed consent" when doing this sort of research with inmates. The first principle of informed consent is that the individual participating in a study must be participating freely and of their own will, without being under pressure to participate. Can people in prison truly give free informed consent (imagine what type of pressures they might feel to agree even if they don't want to)?
- Second, what would be the legal status of this kind of genotype information? Suppose we did find inmates bearing such a mutation. What effect would this have? Would we change anything about how we handle that individual's case? Would his legal status change? Before trial, would the finding of such an MAOA mutation constitute a legal defense? Might the governor be more or less likely to pardon a condemned man if he thought the crimes were driven by his genes, and would that inclination be valid or not?
- Third, do we know enough to understand what the information is telling us? As with XYY males discussed Chapter 21, would finding a higher fraction of inmates with such a mutation really prove "cause and effect"? Here we have to be very concerned because a finding of co-occurrence does not always tell us what we think it is telling us, and our ability to interpret the answers we get is very much limited by our ability to frame the right question in the first place.
- Fourth, even if it were all true, if there really were a cause-and-effect relationship between MAOA genotype and behavior, do we know enough to know what should be done with the information? Do we know enough

> about how to design an environment in which to raise a low MAOA child so as to avoid his development of unwanted behavioral characteristics? If someone has already grown up in an abusive environment, do we know whether changes in his adult environment can help undo any of what developed during his childhood? If someone has a low MAOA genotype, are there other genetic factors modulating that effect? To put it differently: How do we account for the twenty percent of low MAOA cases raised in abusive environments who did not develop the predicted phenotype?

Still, the question in most of our minds is: Just what do we do if all, or even some, of this pans out? Would we, as a society, screen male babies or fetuses for this mutation? If we did such screening, what action would we take? Would bearing this mutation be a cause for termination of a pregnancy? Twenty students in a senior seminar class were once asked if they would choose to terminate a pregnancy if they knew that the male fetus carried the MAOA mutation that had been found in the family in the Netherlands. The answer was an overwhelming "yes." Stop and think what your answer might be, then ask yourself why. Then ask yourself whether we know enough to be basing such decisions on a genotype, and if not, what else do we need to know to make such decisions valid and fully informed?

So the question arises: How many other genetic influences might there be? Unfortunately, we cannot answer this question. Even more unfortunately, the issue has led to some rather careless speculation. Scott once listened in horror while a professor told a class in medical genetics that he thought that virtually all criminal behavior was genetic: that the likelihood of dying in a hail of bullets over a bad drug deal was as genetically influenced as other human traits, such as blood clotting and color vision. There is currently no hard data to support such an assertion, and in fact the Australia/New Zealand MAOA study raises very serious questions about this proposed genetic model of criminality. Moreover, as was said in the discussion of the genetics of IQ, or lack of genetics thereof, in Chapter 31, at least some of these discussions seem to be more about social politics than social science.

Still, this is an issue that is not going to go away. To lay our prejudices on the line, we suspect that, while much criminal or violent behavior may have roots in environmental causes, such as child abuse, hunger, drug addiction, and seemingly hopeless poverty, there will be more cases like the MAOA mutation that will render some people more susceptible to a poor outcome from being raised in such circumstances. Our society is going to have to find ways to cope with the crime and punishment of such individuals, and to be sure that decisions about how to handle such cases are based on real knowledge about the cause-and-effect relationships, and real knowledge of the level of complexity involved. The existence of genotypes that render an individual susceptible to the development of violent or antisocial tendencies in response to environmental influences challenges the basic and cherished concepts of free will and individual responsibility. If some mutations turned out to apparently make crime essentially inevitable, how would we arrive at a truly just views of the punishment of such crimes? Perhaps in those cases we will refocus our interests as a society from punishment to treatment. Perhaps.

THE MULTIPLE-HIT HYPOTHESIS: THE GENETICS OF CANCER

33

For more than eight years, Scott's father fought a long, hard battle with prostate cancer. During the course of his treatment, the doctors considered and/or used each of the three primary modes of cancer treatment: *surgery* (to remove the dividing cells), *irradiation* (which preferentially kills dividing cells by fragmenting the chromosomes), and *chemotherapeutic drugs* (which act to block cell division, often by inhibiting DNA replication or mitotic spindle function). Because the cancer in question was prostate cancer, hormone-based treatments were also used to try to remove the testosterone signal that was giving a "go ahead and divide" signal to the prostate cells. Over time, each treatment stopped working, a reflection of the mutations ongoing in the multi-hit process by which initially benign tumors become malignant and increasingly aggressive, and tumor cells become more resistant to treatments. In addition, some of the side effects were seriously affecting Scott's father's quality of life and, at times, seemed worse than anything the cancer cells were doing. Why would cancer treatments come with so many bad side effects when other medications we take, such as antibiotics, have so little effect beyond curing the malady at hand? The answer is that cancer treatments are almost the only therapies we experience that are aimed at killing actively dividing human cells, and we walk a fine line between killing the cancer cells and killing other cells in our bodies that we need to keep around. So in the vicinity of the tumor, the cancer treatment may do an excellent job of distinguishing between the actively dividing tumor cells and the nondividing cells in the surrounding tissue, but in other parts of the body these same treatments have deleterious consequences when they kill the normal cells that are also dividing. For example, blocking the normal active division of white blood cell precursors in the bone marrow leads to a compromised immune system and susceptibility to infections, whereas impairing the normal ongoing division of gastrointestinal cells can produce debilitating nausea and other complications. If we just had a pharmaceutical "bullet" that would target the cancer cells, the other normally dividing cells in the body could be left alone. Building such a truly specific cancer killer requires that we can answer the question, "What makes the cells in a tumor different from all the other normal cells in the body—especially the actively dividing cells?" To answer that question, we need to understand how tumors arise and how they proliferate. As we will show you, many of those answers are found in the genes that play a role in cancer, and those answers are not the same for all kinds of cancer.

Cancer is a disease that results when a single cell in your body overrides the normal controls of cell division. Most of the cells in your body are not supposed to divide anymore, or if they do divide, they are supposed to do so slowly and rarely under very tightly regulated control. There are some notable exceptions that continue dividing frequently, such as the cells that make up the inside of your intestines, the cells that comprise your bone barrow, the cells in the lowest layer of your skin, and your germ cells. The basic pathway

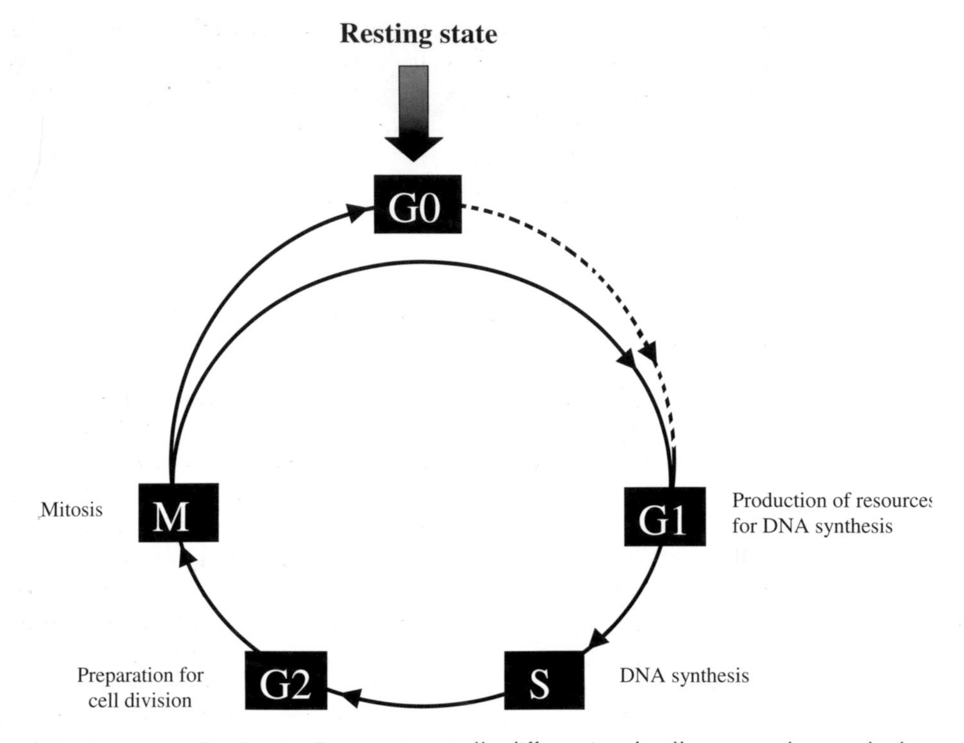

FIGURE 33.1 Resting in G zero (G0). Fully differentiated cells may end up parked in the G0 resting state, in which specialized metabolic activities may be taking place but cell division has stopped. Some specialized cell types continue dividing throughout your lifetime, such as cells that line the gut, but many other cell types have either given up on dividing or will only divide in specialized circumstances, such as during repair of injury. This simplified version of the cell cycle figure that we showed in Figure 11.3 emphasizes the G0 pause that is the resting state for cells not trying to divide.

by which each cell must divide is diagrammed in Figure 33.1. As this diagram of the *cell cycle* shows, cell division begins in a resting or preparatory state called G1. Once it commits to cell division, it then enters the S phase, where replication occurs, moves to another preparatory phase called G2, and then enters the mitotic division (M). For the most part, your cells are permanently parked in a sidetrack of the cell cycle called G0 (pronounced "gee zero"). Sitting quietly in this stage, your cells have permanently foregone the possibility of division in favor of a stable commitment to execute their particular function.

Sometimes, however, something inside a single cell overrides that inhibition. The cell loses that commitment to just "being" and begins to divide. This is the start of a tumor, the hallmark of the most awful word in our language: cancer. Most of the time, the daughter cells descended from one miscreant cell will divide slowly, staying together in a dense, well-defined, and tightly bordered mass that can usually be removed by a surgeon. Such tumors are referred to as *benign*.

However, sometimes the daughters of those cells change further and lose their inhibitions regarding the invasion of normal neighboring tissue and begin to spread throughout the organ or tissue in which they arose. Such tumors are described as *invasive.* In some cases these tumor cells, now committed to rapid unbridled cell division, find their way via the lymph nodes or bloodstream into other sites or tissues, establishing new sites or nodes of tumor formation. This movement to new sites is called *metastasis.* Invasive or metastatic tumors are referred to as *malignant.* Unfortunately, malignant tumors have the potential to kill people if their growth and their spread to new locations can't be stopped.

ACCUMULATION OF MUTATIONS IN SOMATIC CELLS CAN CAUSE TUMORS

Tumors begin from single cells within our bodies. In the case of prostate cancer, a single cell in the prostate gland leaves the "G0 parking lot" that we described above and begins dividing repeatedly—something that none of the cells in the prostrate gland of an adult male is supposed to do! There are actually two counterbalanced processes going on during the cell cycle: signals that tell the cell to move forward in the cell cycle (to grow and divide) and signals that tell the cell to stop (to pause before moving on to the next step in the cycle or to stop and wait indefinitely). So the simple little diagram of the cell cycle in Figure 33.1 actually represents a very complex combination of start signals and stop lights that advance the cell from one step to the next until the entire cycle has been completed, or until the cell is left paused in G0. There are two ways that a cell can escape the normal controls of the cell cycle: by failing to stop when it should or by getting a go-ahead signal that it should not be getting. Normally, this series of stop and go-ahead signals take place in a very tightly regulated series of events in which many of the signals are "on" only part of the time.

The stop signals and the go-ahead signals of the cell cycle are each given by proteins encoded by genes in the human genome, and it is through changes in these genes that regulate the cell cycle that cancer comes into being. The ability of the rogue cells to restart cell division after many decades in a nondividing state develops as a consequence of mutations in three types of genes:

- First, there are *tumor supressor genes* (or stop lights) whose protein products protect the cell against unwanted cell division. Mutations that inactivate both copies of a tumor suppressor gene can allow the cell to slip past a point in the cell cycle at which the cell should have stopped. As a consequence, the ability of the cell to stay in the resting state is compromised. In some senses, we can think of the two copies of a tumor suppressor gene as being a pair of guards protecting a step in the cell cycle and preventing it from advancing in an uncontrolled manner, with the presence of even just one of the two copies of the gene being sufficient to keep that step protected and correctly regulated.
- Second, there are *tumor promoter genes* (or go-ahead signals) known as *proto-oncogenes.* Although these genes are normally silent during adult life, they play critical roles in promoting cell division during the early stages in development. Sometimes genetic events occur (described below) that activated these genes in nondividing cells. These activated cell division genes, known

as *oncogenes* or tumor promoter genes turn on cell division by supplying a go-ahead signal on a continuous basis. This go ahead and divide signal, which should have been there only transiently as a tightly controlled event at a specific point in the cell cycle, forces the cell into continuous division. Protooncogenes, whose normal cellular functions involve supplying a go-ahead signal in the cell cycle on an occasional, highly regulated basis, can become oncogenes if mutation or translocation turns them on continuously.

- Third, *DNA repair genes* act as cancer genes by affecting the rate at which the other two categories of genes (especially the tumor suppressor genes) are mutated and begin causing unwanted cell proliferation. Mutations that impair certain DNA repair systems impair the ability of cells to properly replicate their DNA without making errors and/or diminish the ability of the cell to repair the DNA damage done by the environment. The loss of these repair proteins allows damage to the DNA to escape repair or to be repaired improperly. Thus there ends up being a drastically increased probability of a cell acquiring a cancer-causing mutation in a tumor suppressor gene or a tumor promoter gene. DNA repair genes can cause cancer by greatly increasing the rate of mutations in the other two classes of cancer genes.

HEREDITARY RETINOBLASTOMA: A MODEL FOR UNDERSTANDING THE GENETICS OF TUMOR FORMATION

Insights into tumor suppressor genes have come from the study of families in which a strong predisposition to develop a specific type of cancer is passed along in a family through multiple generations. In such families, half of the children are at risk because they are born carrying one mutant allele in a gene whose normal function is to prevent improper cell division. Although every cell in the body has this mutation in one of the two alleles of the gene, those cells all go through cell division normally, under tightly regulated control. However, during the life of that child, at some point the body's normal low level of mutation knocks out the other normal allele in a cell that needs that particular suppressor gene to regulate its cell cycle. This leaves that cell and its mitotic descendents defenseless. They have lost both copies of a gene that makes a critical guard protein, and they are now prone to inappropriate division in the cell that is missing both copies of the tumor suppressor gene.

An example of this type of disease and the genetic defect that underlies it is found in the inherited form of the ocular cancer retinoblastoma. *Retinoblastoma* is a cancer of the retinal cells of the eye that is most commonly diagnosed in young children. In certain families, retinoblastoma appears to be caused by a simple autosomal dominant mutation. These cases of inherited retinoblastoma make up about forty percent of the total cases of retinoblastoma identified each year. Children with this inherited form of the disease usually end up with multiple tumors in both eyes. Other cases of retinoblastoma are sporadic (turn up in people with no family history of retinoblastoma). Inherited retinoblastoma is often manifested earlier than the sporadic form, and the sporadic form may turn up in only one eye, not both. This inherited form is the result of a mutation in the retinoblastoma gene that maps to the long arm of chromosome 13. When a child inherits the mutant retinoblastoma gene, or RB gene, from only one parent, he or she is likely to develop tumors in both eyes at an early age. In this case, we can point our finger at one mutation in just one gene and say that mutation causes the cell proliferation, and we would be right.

It is important to realize that the normal product of the RB gene plays a crucial role in the nondividing cells in the retina. This RB protein is required to block cells from entering S phase and starting the mitotic cell cycle. If it is missing, cell proliferation begins and the cell starts on the path to tumorigenesis.

THE TWO-HIT HYPOTHESIS

When we consider inherited forms of cancer, one of the first and most important questions to ask is: Why do only some retina cells in patients carrying a inherited RB defect form tumors? If every cell of this child carries the one copy of the RB mutation, why doesn't every cell initiate tumor formation? The normal copy of this allele, obtained from the unaffected parent, is sufficient to control cell division and prevent tumor formation in every cell that keeps an intact copy of the normal allele.

Then why are any tumors formed? Don't all the cells have the normal copy of RB that they inherited from the other parent? The retinal cells of heterozygotes only become competent to form tumors when they lose that normal allele of the RB gene, that is, they must carry two loss-of-function alleles of the RB gene to form a tumor. In these children the first hit was inherited and is present in every cell. The second hit happened in a single retinal cell at some later point in the child's development and is only present in cells that are descended from the cell in which that second hit took place (Figure 32.2).

This is the basis of the *two-hit* model of tumor formation. To form an eye tumor, cells have to knock out both good copies of the RB gene. Cells throughout the eyes and bodies of these RB heterozygous children already have one bad copy, but tumors will only occur in those cells that lose the other copy, as well. The inherited RB mutation is thus dominant in terms of pedigree analysis: most offspring that receive this mutation will develop tumors. However, at the cellular level we may think of it as being recessive because both copies have to be missing to manifest the trait.

How can cells lose the normal allele of the RB gene? Mutations in any gene, including the normal allele of RB, do occur at a very low frequency (approximately 1 per 100,000 or 1,000,000 cells per cell generation) during the process of DNA synthesis in each cell cycle. Many cycles of cell division are required to produce millions of cells in the retina, called retinoblasts. At each of those divisions there was a very low risk of mutating the second normal ( ) allele of the RB gene. So even though the somatic mutation frequency is low, if we look at enough cells, we will see more than one independent case in which a cell loses the second allele. As a consequence, several cells in each retina will endure such mutations and thus be left without a functional RB gene.

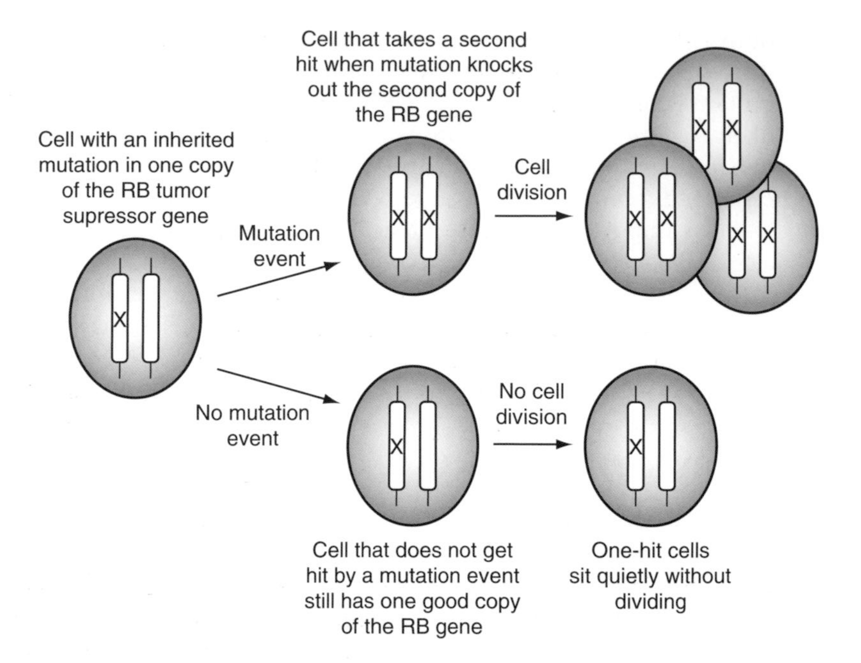

FIGURE 33.2 The two-hit model for tumor formation. The first hit, an inherited defect in a tumor suppressor gene in every cell in an individual's body, does not itself result in cancer. It is only when an additional mutation event (the second hit) takes out the second copy of the tumor suppressor gene in a single cell that loss of regulation of cell division occurs and the cell begins to divide in an uncontrolled fashion. It is unlikely but possible for defects to develop in both copies of a tumor suppressor gene in the same cell in an individual who did not start out in possession of a defective copy, but such events are very rare. In an individual who has inherited an RB copy with the first hit already present, most cells will retain their protective good copy and only rare cells will receive a second hit and become cancer cells. Second hits like this can happen independently more than once in the same individual bearing an inherited tumor suppressor defect. The two-hit hypothesis for tumor formation was proposed by Dr. Alfred Knudson.

One mechanism by which the second copy can be lost is through deletion of one copy of the RB gene. When this happens the cell is left with only one copy of RB and that copy is detective. This one cell is no longer heterozygous.

This process of losing the remaining normal allele is referred to as *loss of heterozygosity (LOH)*. It is a common event at many sites in the genome in human tumors. It is important to realize that the normal product of the RB gene plays a crucial role in the nondividing cells in the retina. This protein, the RB protein, is required to block cells from starting the mitotic cell cycle. In its absence, cell proliferation begins and the cell starts on the path to tumorigenesis.

If mutations are going on throughout our lifetimes, can't a cell sometimes have separate mutations hit both alleles of the same gene even though the person does not carry a mutated allele in all of their cells? In both inherited and sporadic retinoblastoma both copies of the gene are defective. They differ only in whether it takes one mutational hit, or two hits both falling on the same cell, to bring about a functional defect. The sporadic form is quite rare (about 1/40,000) because it is so unlikely that both copies of a gene will be knocked out in the same cell, and two hits on different cells will not cause tumors because each of those two cells will still retain one good copy. Because it is so unlikely that this double hit will happen even once to a particular person, the chances become vanishingly small that someone with two normal copies of the RB gene will have a double hit happen independently in two different cells at different times. Thus sporadic cases of RB characteristically present with a tumor in only one eye. The inherited form is found in only those who have inherited a first hit, but it is likely that they will experience a second hit more than once and end up with multiple tumors originating from different cells at different points in time. Thus individuals with inherited forms of RB are more likely to end up with tumors in both eyes.

CELL TYPE SPECIFICITY OF CANCER GENE DEFECTS

Some children suffering from the inherited form of retinoblastoma are also at risk for some other kinds of cancers later in life, especially those children with retinoblastoma who are treated with radiation therapy who may sometimes develop tumors of the eyelids or elsewhere. However, these people are not at risk for all types of cancers. That is to say that, even though many of the proteins protect and direct the cell cycle are expressed in many cell types, defects in a particular tumor suppressor gene will often be limited to causing cancer in a small number of cell types.

Other types of inherited cancers show similar relatively specific effects. The *Wilm's tumor* gene causes hereditary development of kidney tumors if a second hit causes the loss of the one normal allele that was passed along from the unaffected parent. Two *breast cancer* genes, BRCA1 and BRCA2, lose their ability to block uncontrolled cell division in individuals who lose both copies of either gene. Although BRCA1 and BRCA2 are primarily known for causing breast cancer, functional loss of either one can also predispose to other cancers, especially *ovarian* cancer, in some individuals, with the risk of ovarian cancer being lower for BRCA2 mutations than for BRCA1 mutations.

However, the mutational inactivation of some tumor suppressor genes leads to a broad array of different kinds of cancers. One such gene is p53, which produces one of the most important tumor suppressor proteins in our cells. This protein plays a crucial role both in preventing unwanted cell division and in regulating the response of the cell to DNA damage. More than fifty types of cancer have been shown to carry new mutations in the p53 gene; indeed, more that seventy percent of all colorectal cancers carry mutations in this gene. The same is true for many other kinds of tumors. So what happens if one inherits one defective copy of the p53 gene? The result is a hereditary disorder known as *Li-Fraumeni syndrome*, in which heterozygotes are at risk to develop a wide variety of different tumors in different tissues of the body, including tumors of the ovary and sarcomas; the type of tumor depends on which cell types experience a mutation in the second, normal copy of the gene.

THE MULTIPLE-HIT HYPOTHESIS

Like the tumors just described, *adenomatous polyposis of the colon (APC)* is inherited as a simple autosomal dominant disorder. Affected individuals in these families already have the first hit on one copy of the APC gene on the long arm of chromosome 5. The first step in the formation of the tumor(s) is the loss of the normal (second) allele of the APC gene. Again, the protein product of the normal APC gene appears to be required for the control of cell division. However, in this case, the result of the second hit taking out the second good copy of the APC gene is not cancer but rather a precancerous polyp (Figure 33.3). The further progression of the small polyp into a malignant tumor can be divided into clear stages that can be distinguished by a pathologist. A study of these various stages reveals that the development of an invasive and metastatic tumor requires multiple new mutations at other places in the genome. Thus, although the formation of the early polyp appears to require only one mutation, additional mutations in other genes (most notably the gene encoding a protein called p53 discussed above) are required for that early tumor to become a dangerous malignancy (see Figure 33.3). Other kinds of cancers can also be found in individuals who inherit an APC gene defect, such as stomach cancer, and other nonmalignant features such as pigmented scarring of the retina can sometimes help in identifying individuals who are carrying an APC mutation.

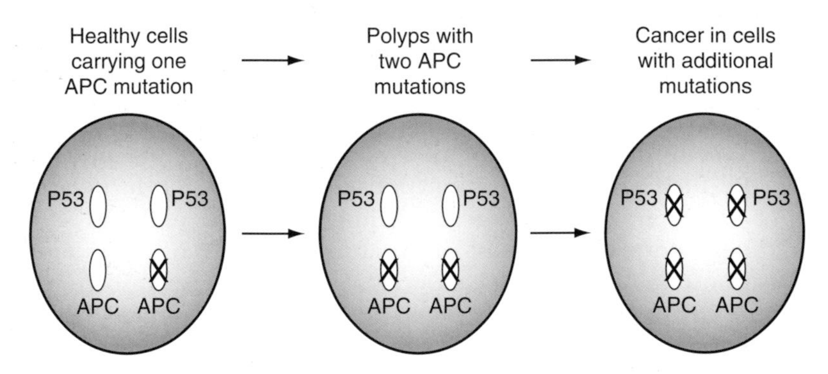

FIGURE 33.3 The more-than-two-hit hypothesis. The two-hit hypothesis deals with the initiating event that leads to the initial loss of control of cell division. In many cases, as an initial tumor cell divides and the tumor grows, additional mutation events affecting other genes in the genome can enhance cell growth and help confer other properties, such as invasiveness, that contribute to metastasis. In the case of the APC gene, the second hit results in growth of a polyp, but if the polyp is removed, those cells are not available to be hit with further mutations in other genes that could turn that polyp into a metastatic form of cancer. Since the probability of mutation is proportional to the number of cells, if there are less cells, there is less overall chance of a mutation. Thus a strategy used in APC is to screen for and remove polyps before they have a chance to acquire the additional mutations that will convert a polyp into metastatic cancer. X marks a copy of a gene that has been inactivated by a mutation.

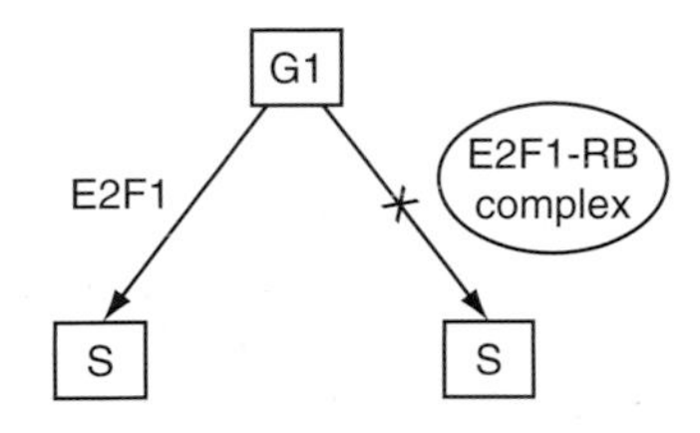

FIGURE 33.4 E2F1 is a critical key that unlocks the ability of the cell to enter S phase and proceed with cell division. RB is a tumor suppressor protein that blocks control of entry into S by helping to sequester E2F1 in a complex that contains a number of other proteins important to the cell cycle. The cell can release E2F1 from this complex when it needs it to act, and sequester it again when it needs to keep from proceeding into S phase. If there is even one good copy of the RB gene in the cell, there is enough RB protein being made to keep E2F1 under control. In a normal cell, E2F1 carries out its role in initiating S phase under very carefully controlled conditions, so that cell division only happens when it is supposed to. If there is no RB in the cell, E2F1 can escape suppression and activate entry into S phase and the subsequent events leading to cell division.

HOW DO THE PRODUCTS OF TUMOR SUPPRESSOR GENES ACT TO PREVENT TUMOR FORMATION?

Some tumor suppressor proteins, such as the RB protein, act by shutting down the function of other proteins that activate steps in the cell cycle. The RB protein is a critical member of a suppressor complex that keeps a number of cell cycle proteins locked up in a multiprotein complex, and specifically regulated events are normally required to free a protein from this complex so that it can act. One of the key gene products that the RB "guard" protein locks up in this complex is a protein called E2F1, which is the protein that signals the cell to move from G1 to S in the cell cycle (Figure 33.4).

Other tumor suppressor genes, such as BRCA1, appear to encode proteins that regulate the transcription of other genes, most notably genes that control the cell cycle. Another such example of this phenomena is found in *Von Hippell syndrome*, a hereditary predisposition to brain tumors. In this case the mutated gene is a known component of the enzyme complex that transcribes genes into mRNA.

DEFECTS IN DNA REPAIR

Tumors may often begin with mutations in genes that impair the ability of cells to repair their DNA and thus to fix either errors that occur during the replication process or DNA damage that occurs as a result of exposure to environmental mutagens (Box 33.1). The relationship between defective DNA repair and carcinogenesis is well illustrated by a hereditary form of colon cancer called *hereditary nonpolyposis colon cancer (HNPCC)*. Researchers have found two sets of families segregating for HNPCC. One set of families allowed them to map a gene that predisposed individuals to HNPCC to chromosome 2, and the second set of families allowed them to map a second cancer-causing mutation to chromosome 5. Analysis of tumors from both families revealed an unusual genetic instability in the tumor cells. This instability was most easily

BOX 33.1 MISMATCH REPAIR

Mismatch repair is one of several processes that repair the damage done to DNA both by exogenous agents (such as sunlight, radiation, or carcinogens in food, water, or smoke from tobacco) and by everyday life in the cell, such as errors resulting from mistakes that happen in the normal course of copying DNA. Because such errors during replication are not infrequent, our cells have a fairly efficient mismatch repair system for dealing with them. In the absence of a mismatch repair system, those errors that do occur during replication are not repaired but rather end up being incorporated into one of the two daughter strands following the next round of replication. The resulting daughter strand will then carry a new mutation at the site of the replication error. Cells that cannot repair errors through mismatch correction will have greatly increased mutation rates. Please note that errors occur throughout the genome, not just at cancer genes or at microsatellite repeats. However, errors in mismatch repair are detectable through monitoring of changes in lengths of microsatellite repeats because errors seem to be more frequent where DNA polymerase has been trying to copy a length of repeated simple sequence in the absence of mismatch repair.

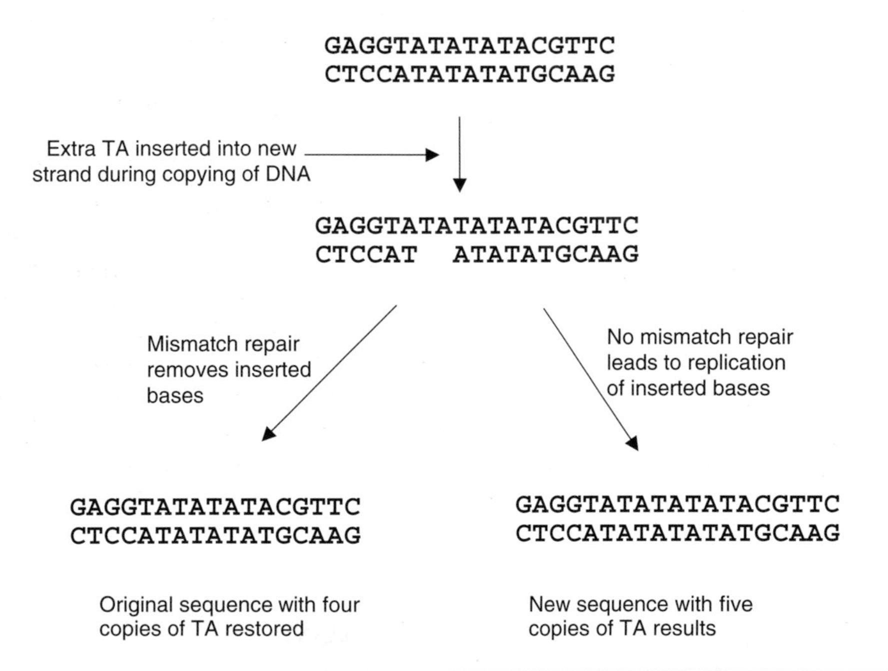

manifested as the expansion or contraction of sequences in the human genome called *microsatellite repeats,* the short runs of repeated simple sequences such as di nucleotides and tri nucleotides (e.g., CACACACACA-CACA) that are scattered throughout the genome (see Chapter 19 for a review). Although the number of repeats at each site was constant in normal cells of these individuals, cells in the tumor showed huge expansions of the repeat (e.g., CACACACACACACACACACACA CACACACACACACACA) or contractions of the repeat (such as CACACACACA). Indeed, investigators quickly found that HNPCC resulted from repair-deficient mutations in genes (MSH2 and MUTL) that encoded enzymes involved in a DNA repair process called mismatch repair.

Indeed, the colon cancer–causing mutations on chromosomes 2 and 5 were, in fact, mutations in one copy of the MSH2 gene or the MUTL gene. In the tumors themselves, the normal allele of these genes often appear to be deleted. This is another example of the "two-hit" model. Individuals inheriting a cancer-causing allele of one of these genes are at high risk for HNPCC. The "second hit" is the mutational "knockout" of the normal copy of the MSH2 or MUTL gene in one or more of these cells, which creates a cell with no functional copies of these genes, a cell that can no longer accomplish mismatch repair. That cell is now going to experience a high frequency of new mutations every time it replicates its DNA.

Why should a defect in DNA repair cause cancer? If you have an error-prone system for replicating DNA, one that cannot repair the occasional errors made during replication or spontaneous damage to DNA, every round of replication is a potentially mutagenic event. Every round of replication gives you a chance to lose another tumor suppressor gene by mutation. Every round of replication makes the loss of control of cell division more inevitable. If DNA repair defects really make a cell susceptible to cancer, shouldn't defects in other repair systems also make cells more cancer prone? Yes! Indeed, there are a number of human diseases in which individuals who are demonstrably repair defective are highly cancer prone, including *xeroderma pigmentosa, Bloom syndrome, ataxia telangiectasia,* and *Fanconi anemia.* Each of these disorders results from homozygosity (or compound heterozygosity) for inherited recessive mutations in genes required for various aspects of DNA repair. For example, patients with xeroderma pigmentosa are defective in various aspects of a process called excision repair, whereas children with ataxia telangiectasia are deficient in a process that forces cells with unrepaired DNA damage to stop dividing until they can repair their DNA. In each of the disorders, mutations in genes whose protein products are required for DNA repair predispose their bearers to develop tumors. How? The best guess is that these repair-deficient disorders effectively raise the mutation rate and, in doing so, increase the probability of mutating the tumor-suppressor genes.

THE END RUN: DOMINANT TUMOR-PROMOTING MUTATIONS PUSH THE CELL INTO THE DIVISION CYCLE

Chromosome rearrangement and instability are hallmarks of tumor cells. Indeed, such rearrangement may play a crucial role in the initiation of some tumors. One of the best examples is *chronic myeloid leukemia (CML).* Greater than ninety percent of individuals with CML carry a specific translocation, referred to as the *Philadelphia chromosome,* involving chromosomes 9 and 22. This translocation is not present in normal cells of these patients. In this case a normally inactive gene (called *ABL*) on chromosome 9, which acts to promote cell division, is moved by translocation to fuse it to a gene called

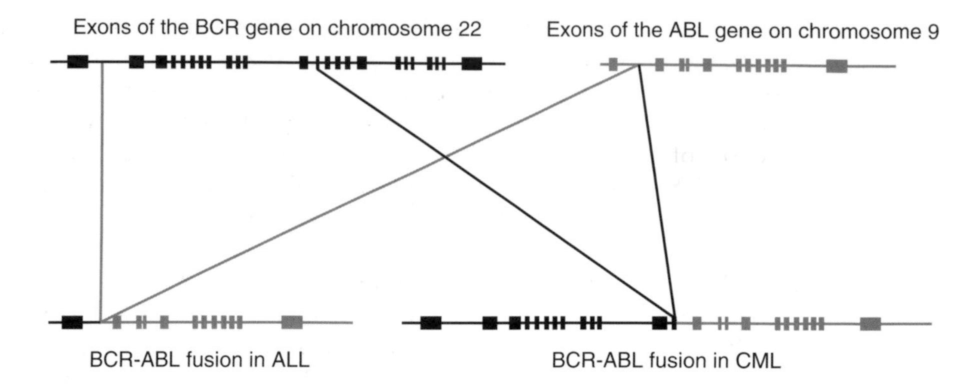

FIGURE 33.5 Translocations that connect the BCR gene on chromosome 22 with the ABL gene on chromosome 9 can cause either acute lymphoblastic leukemia (ALL) or chronic myelogenous leukemia (CML), depending on where the breakpoint occurs. (Adapted from Meltzer *et al.* in *The Metabolic Bases of Inherited Disease* 2001; p. 558.)

BCR from chromosome 22 (Figure 33.5). After translocation, the activity of the BCR-*ABL* fusion protein is now carried on continuously instead of happening transiently in response to signals from the cell. The activity of the BCR-ABL fusion protein triggers the activation of a regulatory cascade that promotes cell division. As a result, these white blood cells begin to constitutively enter the division cycle, repeating the cycle over and over instead of moving into G0 and parking until cell division is needed again.

In the case of acute *lymphoblastic leukemia (ALL)*, a translocation chromosome that looks like the Philadelphia chromosome when viewed under the microscope has actually created a different gene structure that includes a smaller portion of the BCR protein than occurs in the CML translocation. *Burkitt's lymphoma*, another form of cancer, results from a similar translocation mechanism, in this case involving chromosomes 8 and 14, that perpetually turns on another tumor promoter gene called *MYC*. The Metabolic and Molecular Bases of Inherited Disease lists dozens of different combinations of chromosomes that cause various forms of leukemia as the result of a translocation in a single cell bringing pieces of genes together in a way that turns on the activity of a cell cycle regulator, which normally should be expressed only under tightly controlled circumstances.

There is a crucial difference between such rearrangements and the inherited tumor promoting mutations discussed earlier. These rearrangements are found only in tumor cells. They are *not* present as a "first hit" throughout the body. Thus they are dominant on the cellular level (in the sense that hitting only one of the two copies can cause the problem even when a second "good" copy is present) but are not transmitted as a dominant disorder within a family because the mutation is not present in any cells of the body except the tumor cells.

BUILDING MAGIC BULLETS—THE GLEEVEC STORY

This chapter has cancer as a genetic disease, in which mutations in a variety of genes comprise the numerous systems that prevent cell division. Some mutations, such as those in DNA repair genes, act by increasing the mutation rate, thus making mutants in the critical tumor suppressor genes more likely, but other mutations directly inactivate the tumor suppressor genes themselves or activate tumor promoter genes. The question then becomes, What if we could restore the function of the lost tumor suppressor gene or suppress the action of the product of a tumor promoter gene: could we then cure the cancer? Such is the type of molecular biology-based therapeutic that most of us describe as the *magic bullet*, that drug that specifically targets a critical defect in the tumor cells themselves.

One magic bullet that already exists destroys a type of cancer cell that we have just described, chronic myelogenous leukemia. As we noted, this cancer is due to the inappropriate activation of a division-promoting gene called ABL. The issue here is that, if one could inactivate the function of the BCR-ABL protein in CML cells, one might be able to stop the tumor growth. Work by numerous investigators had shown that ABL is a *tyrosine kinase*, a protein that chemically modifies other proteins in the cell. It turns on and off at different times depending on signals being received by the cell. The BCR-ABL protein also acts as a tyrosine kinase, but it has lost the ability to respond to the signalling systems that would allow it turn off when that is what the cell needs. There are many types of tyrosine kinases in our cells, but the BCR-ABL held a special position in the cellular signalling cascade. Suppose one could build a small drug that specifically and uniquely inactivated the ABL protein, leaving the functions of the other tyrosine kinases intact? Could that cure the cancer? A scientist named Brian Druker identified a small compound, initially named STI-571, that can inhibit the BCR-ABL protein.

Working with the pharmaceutical company Novartis, Druker was able to show that this drug works astonishingly well against CML. In the first clinical trials reported in 1998, remissions were observed in one hundred percent of the first thirty-one patients tested! After dramatic subsequent trials, STI-571 (trade name Gleevec) sailed through the government's drug approval process in record time. Although there are few side effects and most patients get long, substantial remissions from Gleevec, the drug loses effectiveness and eventually fails in patients with end-stage disease. Sadly, even some patients treated early may become resistant to Gleevec as a consequence of new somatic mutations in the tumor cells that survive the treatment. Gleevec also works well against one other type of tumor, known as *gastrointestinal solid tumors (GISTs)*. However, from the specificity of its design, so far it does not appear to have a broad therapeutic range against other types of cancers.

WIELDING A MOLECULAR LANCE

Maybe each type of tumor will require its own magic bullet, but the success with Gleevec is informative. Back in Chapter 1, we described the tragic loss of one of Scott's former graduate students, Brenda Knowles, to another type of leukemia, known as *acute myelogenous leukemia (AML)*. Unlike CML, AML cannot be attributed to any one single mutational lesion (such as the expression of the BCR-ABL protein). Worse yet, it seems likely that two separate mutations are required to trigger the disease. However, mutations in a gene that encodes another tyrosine kinase called *FLT3* is present in the tumors of about thirty percent of patients with AML, and most often in those patients with the worst prognosis. Once again, these mutations serve to activate, in tumor cells, a kinase that should only be expressed in the rapidly dividing *stem cells* of the marrow. In the mutated state, this activated kinase keeps the leukemic cells in a state of rapid cell division. Professor Gary Gilland and his research team at Harvard University are working on drugs that specifically target the FLT3 kinase. These drugs are now in clinical trials. Others like them will follow. It is simply one of the real sadnesses of life that these drugs come a decade too late for Brenda and the others who have died of this disease.

One can imagine other such drugs that would target a molecule that is having specific effects of importance in a particular tumor cell type. In the case of prostate cancer, a drug called flutamide has been developed that can inactivate the ability of the testosterone receptor to respond to testosterone, which offers a tool with which to impair the progress of testosterone-sensitive prostate cancer cells. Already, women with certain types of breast cancers, those that overexpress the *HER2* protein, are treated with a protein called *Herceptin*. Herceptin is an antibody that binds to cells expressing HER2 and slows their growth. Herceptin treatment doesn't work for every tumor, and even when it does work, often it only delays the progress of the disease. Nonetheless, for many patients, it does extend life. These are clearly the first steps in building a diverse molecular pharmacy for the treatment of cancer. Even these limited successes make it clear that if we ever want to treat the vast majority of tumors, we are going to have to be able to understand the genetic lesions that underlie each of those tumors. In the last and final section of this chapter, we will discuss the development of modern tools for the study of gene expression in tumors that let us assess the activity, inactivity, or hyperactivity of every gene in that tumor.

BETTER GENETIC DIAGNOSIS, BETTER TREATMENT: THE STORY OF MICROARRAYS

In Chapters 7, 8, and 9 we discussed the basic processes by which a cell produces a transcript. A new technique called *DNA microarrays*, measures the activity of virtually all the genes in the genome simultaneously. Basically, it is possible to build small chips containing many thousand of spots, each of which carries DNA molecules corresponding to a given gene. It is possible to label mRNA from a cell with a fluorescent dye, and combine the labelled RNA with one of these chips and have each mRNA stick to the spot on the chip that corresponds to the gene it came from. The more RNA molecules bound to a given spot, the more copies of the fluorescent tag are bound and the stronger the signal at that spot on the chip. The result is an array of spots whose brightness varies from gene to gene, based on the transcriptional activity of that gene and the rate at which the cell gets rid of the transcripts. One could imagine then comparing the mRNA in tumor cells to mRNA from the cells surrounding the tumor and looking to see which genes show a change in level of expression (as manifested by a change in the brightness of the spots corresponding to the gene you are interested in). As chips are being made that have a large fraction of the human genes represented on the chip, it then becomes possible to do what are effectively whole-genome experiments looking for genes whose expression correlates with the presence of a tumor, with a progression from benign to malignant, or with a progression from malignant to highly invasive. Indeed, exactly such strategies have created a new generation of molecular diagnostics for cancer (Figure 33.6).

Tumor types can be more specifically identified, and differences in gene expression will be used to detect tumors that already demonstrate gene expression changes associated with increased risk of metastasizing. *Gene expression profiling* will also eventually let a doctor make predictions about optimal treatments. For instance, tumor cell expression of high levels of a gene whose product acts as a pump that pumps chemotherapeutic agents out of the cell is not good news for a patient being treated with chemotherapy. Knowing about this high level of pump activity would be important for a patient whose doctor has alternative treatments to offer. As more research proceeds, this gene expression profiling approach to pharmacogenomics will become an important part of the doctor's clinical testing repertoire, but far more must be learned to understand enough about the clinical implications of such test results.

Further innovations are coming, as well. New techniques for FISH (fluorescent *in situ* hybridization) allow researchers to assess the expression of important genes in single cells, thus providing far better tools for diagnosing tumor type in biopsies, detecting metastases, and measuring the response of a tumor to treatment. Also, *chromosome painting* (Figure 33.7) using a large number of colors allows the identification of chromosomes that are damaged or present in altered numbers in tumor cells, which can assist in interpreting how advanced the cancer is and identifying regions of the genome involved in progression of the cancer to later stages.

A CONCLUDING THOUGHT

In our best dreams, twenty years from now, your doctor will have available a pharmacy full of drugs that specifically impede the growth of (or, better yet, kill) the tumor cells without impacting their healthy neighbors. This not only has the potential to greatly reduce many of the terrible side effects of current cancer treatments by avoiding attacks on healthy cells but also offers the possibility of getting more complete elimination of cancer cells very early in treatment. Although such drugs may well only work in the early stages of the disease, we anticipate that a better understanding of molecular events in later stages of cancer will lead to breakthroughs in other medicines specific to the properties of later stage cells. Because there are so many ways to turn on cell division and because not all cases are diagnosed early enough, we don't imagine that these drugs will ever cure every cancer or even that every tumor will respond to any one treatment. However, that is not so different from the status of many "curable" things in current medicine, where someone can still die of an infection in a world with plentiful antibiotics. The difference is that we will progress from the current state of affairs, in which a diagnosis of cancer automatically summons up images of death, to a state of affairs in which survival will be expected and a poor outcome will be the rare event.

We see another plus to very specific pharmaceuticals that target the primary molecular defect in a tumor cell: too many of today's treatments, such as radiation and chemotherapy, affect the immune system along with the

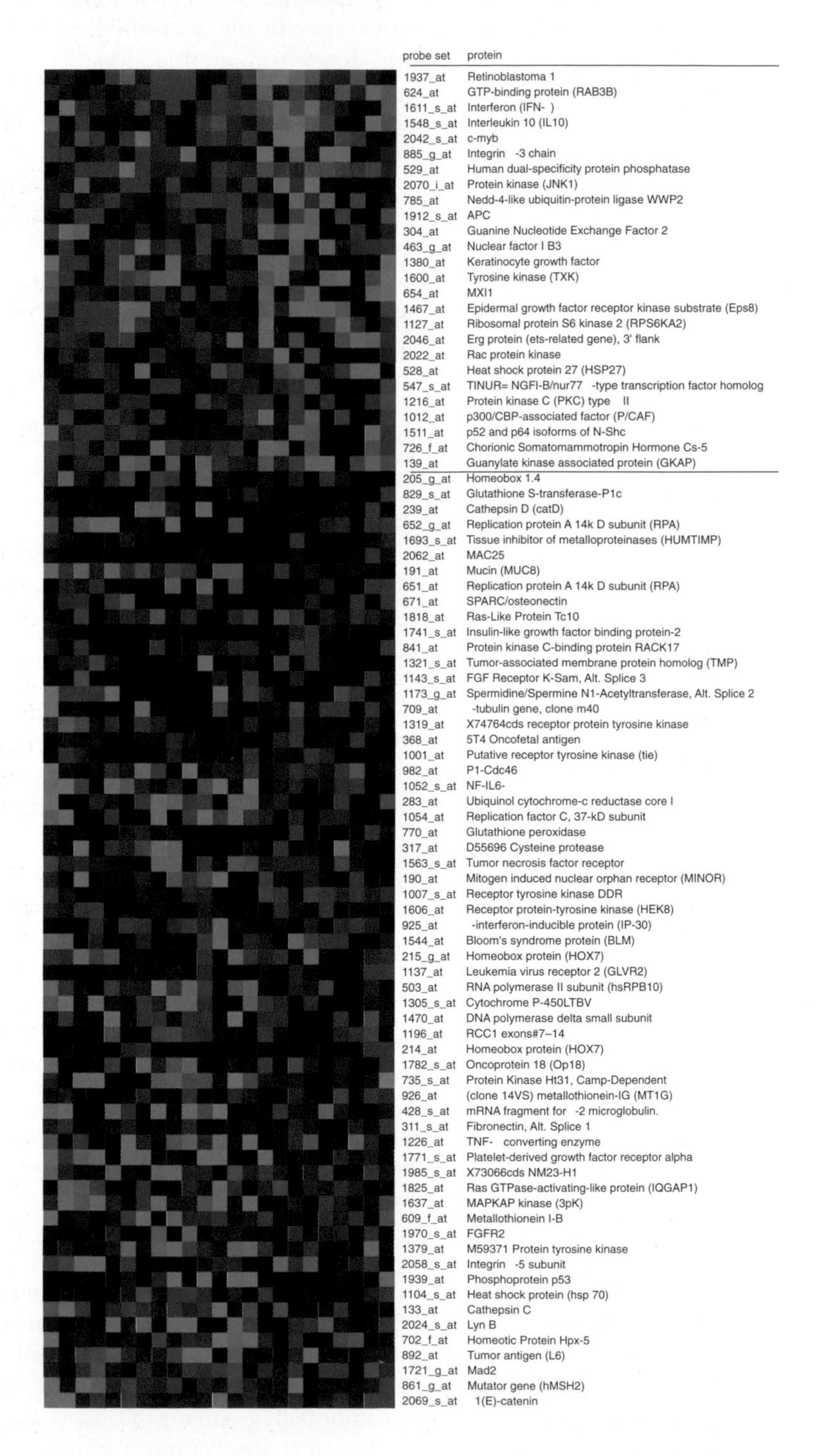
RNA from non-metastatic tumors
RNA from metastatic tumors
Gene names
probe set protein
1937_at Retinoblastoma 1
624_at GTP-binding protein (RAB3B)
1611_s_at Interferon (IFN-)
1548_s_at Interleukin 10 (IL10)
2042_s_at c-myb
885_g_at Integrin -3 chain
529_at Human dual-specificity protein phosphatase
2070_i_at Protein kinase (JNK1)
785_at Nedd-4-like ubiquitin-protein ligase WWP2
1912_s_at APC
304_at Guanine Nucleotide Exchange Factor 2
463_g_at Nuclear factor I B3
1380_at Keratinocyte growth factor
1600_at Tyrosine kinase (TXK)
654_at MXI1
1467_at Epidermal growth factor receptor kinase substrate (Eps8)
1127_at Ribosomal protein S6 kinase 2 (RPS6KA2)
2046_at Erg protein (ets-related gene), 3' flank
2022_at Rac protein kinase
528_at Heat shock protein 27 (HSP27)
547_s_at TINUR= NGFI-B/nur77 -type transcription factor homolog
1216_at Protein kinase C (PKC) type II
1012_at p300/CBP-associated factor (P/CAF)
1511_at p52 and p64 isoforms of N-Shc
726_f_at Chorionic Somatomammotropin Hormone Cs-5
139_at Guanylate kinase associated protein (GKAP)
205_g_at Homeobox 1.4
829_s_at Glutathione S-transferase-P1c
239_at Cathepsin D (catD)
652_g_at Replication protein A 14k D subunit (RPA)
1693_s_at Tissue inhibitor of metalloproteinases (HUMTIMP)
2062_at MAC25
191_at Mucin (MUC8)
651_at Replication protein A 14k D subunit (RPA)
671_at SPARC/osteonectin
1818_at Ras-Like Protein Tc10
1741_s_at Insulin-like growth factor binding protein-2
841_at Protein kinase C-binding protein RACK17
1321_s_at Tumor-associated membrane protein homolog (TMP)
1143_s_at FGF Receptor K-Sam, Alt. Splice 3
1173_g_at Spermidine/Spermine N1-Acetyltransferase, Alt. Splice 2
709_at -tubulin gene, clone m40
1319_at X74764cds receptor protein tyrosine kinase
368_at 5T4 Oncofetal antigen
1001_at Putative receptor tyrosine kinase (tie)
982_at P1-Cdc46
1052_s_at NF-IL6-
283_at Ubiquinol cytochrome-c reductase core I
1054_at Replication factor C, 37-kD subunit
770_at Glutathione peroxidase
317_at D55696 Cysteine protease
1563_s_at Tumor necrosis factor receptor
190_at Mitogen induced nuclear orphan receptor (MINOR)
1007_s_at Receptor tyrosine kinase DDR
1606_at Receptor protein-tyrosine kinase (HEK8)
925_at -interferon-inducible protein (IP-30)
1544_at Bloom's syndrome protein (BLM)
215_g_at Homeobox protein (HOX7)
1137_at Leukemia virus receptor 2 (GLVR2)
503_at RNA polymerase II subunit (hsRPB10)
1305_s_at Cytochrome P-450LTBV
1470_at DNA polymerase delta small subunit
1196_at RCC1 exons#7–14
214_at Homeobox protein (HOX7)
1782_s_at Oncoprotein 18 (Op18)
735_s_at Protein Kinase Ht31, Camp-Dependent
926_at (clone 14VS) metallothionein-IG (MT1G)
428_s_at mRNA fragment for -2 microglobulin.
311_s_at Fibronectin, Alt. Splice 1
1226_at TNF- converting enzyme
1771_s_at Platelet-derived growth factor receptor alpha
1985_s_at X73066cds NM23-H1
1825_at Ras GTPase-activating-like protein (IQGAP1)
1637_at MAPKAP kinase (3pK)
609_f_at Metallothionein I-B
1970_s_at FGFR2
1379_at M59371 Protein tyrosine kinase
2058_s_at Integrin -5 subunit
1939_at Phosphoprotein p53
1104_s_at Heat shock protein (hsp 70)
133_at Cathepsin C
2024_s_at Lyn B
702_f_at Homeotic Protein Hpx-5
892_at Tumor antigen (L6)
1721_g_at Mad2
861_g_at Mutator gene (hMSH2)
2069_s_at 1(E)-catenin

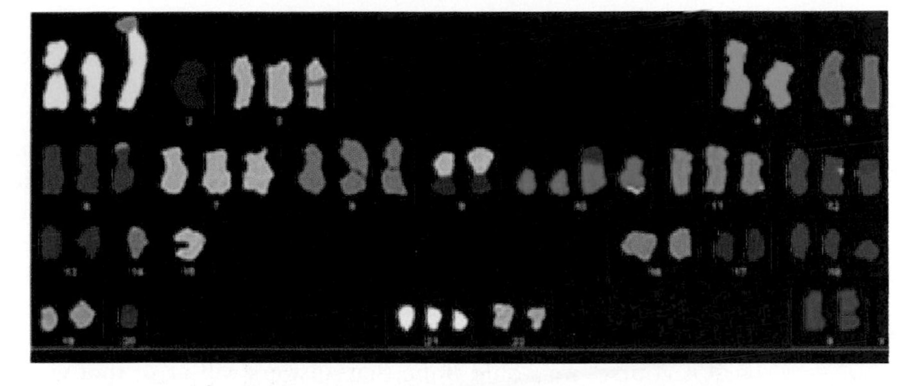

FIGURE 33.7 Twenty-four-color chromosome painting identifies multiple translocations and other signs of aneuploidy in a breast cancer tumor cell line. (Courtesy of Joanne Davidson.)

tumor cells. If specificity of treatment spared the cells of the immune system, might we also find ourselves able to fine-tune the immune system response to tumor cells to help eliminate any cells missed by the drugs?

We also see great gains to be made in diagnostics. If we can arrive at knowing not just what kind of cancer it is, but also which biochemical pathways have become involved in the tumor's progression, it may streamline targeting just the right treatment not just for that cancer but for that stage of that cancer. Of great importance, we anticipate better diagnostics arising from molecular genetics that will reduce the problem of people being diagnosed "too late." Imagine home monitoring systems akin to home pregnancy tests or the blood-sugar monitoring systems used by a diabetic patient, allowing those with especially high risk levels to watch for recurrence of a banished tumor, or appearance of an expected tumor that has not yet manifested itself.

Will the magic bullets yet to come provide a simple outright cure for cancer the way we cure an infection with an antibiotic? Perhaps. For sporadic cases who did not inherit a cancer gene, we expect that cures will be achievable and that curing their cancer once will mostly take care of the problem. Perhaps for individuals who inherited that first hit, a cure will really involve three different stages. First, as for the nonhereditary cases, there will be the problem of how to shoot just the right magic bullets early enough to end the

---

FIGURE 33.6 Use of an Affymetrix Cancer Array gene chip to examine gene expression in cancer. This image shows that certain genes show higher levels of mRNA (in red) in metastatic tumors than in tumors that have not metastasized. Other genes show the opposite pattern. Different kinds of cancers will show a different profile of gene expression levels that can point to where critical biochemical events in the cell are taking place, and eventually assays of changes of this kind will be important in distinguishing different stages of cancer. In this case the RNA samples came from medulloblastoma tumors, and analysis of images like this one allowed the authors to identify genes that play a role in medulloblastoma metastasis. Some gene chips and microarrays can display thousands or tens of thousands of genes, but use of specialized chips that do not contain the full array of human genes lets researchers focus in on where the critical differences are. (Modified from T.J. MacDonald et al, *Nature Genetics* 29:143–152, 2001. Used by permission.)

initial cancer. The second stage will involve ongoing diagnostics to detect any new tumors at the earliest possible stage. Third, there will be ongoing medical management to suppress development of subsequent tumors. However, if we identify key proteins produced by tumor cells, key changes in gene expression events, can we in fact also concoct a surveillance system that can knock out any cell that turns on the particular cancer pathways characteristic of that individual's hereditary tumors before the tumor ever gets large enough for anyone to know it is there? We expect that eventually harnessing key features of our own immune systems along with the magic bullets from the pharmacy may allow for many cancers to become a chronic disease, treated over a long lifetime through your local pharmacy. How much of an advance would that be? Of course, we would hope that answers will arise that will offer simple, clean cures, but twenty years from now, if cancer has moved out of the acute life-threatening category into a chronic management category, we will feel as if a dream has come true.

GENETIC TESTING AND THERAPY

Although the study of genetics has taught us many profound things about underlying mechanisms of processes in the human body, for many people the real question is one of how to apply this information to helping people whose problems are genetic in origin. In this section, we look at the two sides of one coin—genetic testing and gene therapy—that constitute the translational aspects of genetics. Not surprisingly, we find ourselves once again facing ethical issues along with the technical and medical issues.

GENETIC TESTING AND SCREENING

34

When he came into the world, he was greeted with all of the hope, love, and eager nervousness that greet so many newborns who go on to their parents' lives with chaos and delight. However, immediately following his birth, they knew there was a problem. In place of lusty cries at birth, instead of kicking feet or wiggling arms, he presented a picture of complete stillness, laying there silent and unmoving as a crisis erupted around him. They were not surprised that he was small, since prenatal ultrasounds had indicated short femur length that was waved away at the time as something consistent with the modest height of some of his relatives. They had not expected the low-set ears, the high palate, the muscle weakness, the heart defect, and the unresponsive stillness. Amidst a flurry of medical tests, anxious discussions, and fearful waiting, his parents were told that he needed heart surgery if he were to survive. Without a diagnosis, the doctors could not tell his parents whether the surgery would be enough to save his life or whether he would die anyway of other problems besides his heart defect. A decision had to be made, and it turned out that the rush of medical events forced his parents to make that decision before they could obtain the karyotype results that might have provided facts that could have informed their choice. They needed to know: did their son have an extra copy of a chromosome or piece of chromosome, and if so, would that information tell them that he had greatly reduced hopes of survival even if he had the operation? In agony, his parents agreed to surgery in the hope that there was hope to be had beyond the surgery. While he recovered, while his pain was reflected in his parents' anguish, the genetic testing results came in too late to prevent the surgery. Yes, he had an extra copy of part of chromosome 18, and he would most likely not live to see his first birthday, but no one could say for sure how long he would be with them. With a sense of guilt over having subjected him to a painful surgery that his parents would not have agreed to if they had realized it could not save him, his parents took him home to watch him around the clock. They hovered over him, willing him to breathe each next breath, watching as formula moved through a tiny tube down his nose to his stomach because he could not coordinate the movements needed to drink. So he continued for five months, until his body could do no more and he stopped breathing for the last time. The kind of genetic testing used to diagnose trisomy 18 takes time, and in the first days of his life, when the heart defect became a crisis that could not wait, decisions had to be made at a faster rate than the genetic testing could be completed. His parents have never gotten over some of their regrets. They wish they had insisted on a follow-up on the report of short leg bones. They wish they had had genetic testing done before he was born so that they would have been fully informed when they had to decide about the surgery. That he was born so ill, they could not prevent. That they were uninformed when they had to make decisions, that is something they wish they could have done differently. Although this is not the usual course of events, it shows us that many of the cases in which we end up wishing we had more information available are ones we don't anticipate. It also shows us that information gained from prenatal genetic testing can have important uses other than making decisions about whether to continue a pregnancy.

The field of medical genetics is practiced by doctors with subspecialty training in medical genetics (Box 34.1) and genetic counselors trained in counseling people about genetics (Box 34.2). Medical genetics offers help for every stage of genetics in our lives. Prenatal testing and medical diagnosis assist in cases in which a trait seems to run in a family or resembles a known genetic trait. The medical genetics practitioner can also end up playing Sherlock Holmes, sifting through clues to arrive at an answer to a medical mystery, or even providing information about a trait to someone who is unaffected but concerned because of family history or for other reasons. Although many problems in this field come to the doctor's attention at birth or during childhood, medical genetics also plays an important role in diagnosis of traits in

BOX 34.1 MEDICAL GENETICS SPECIALIST

Medical genetics specialists have an MD degree plus specialty training in an area such as pediatrics or internal medicine, *plus* fellowship subspecialty training in the diagnosis and treatment of genetic disorders, birth defects, and other types of malformations. As more and more is learned about the genetic causes of human health problems and effects of teratogenic agents, the role of the medical geneticist is becoming increasingly important and specialized. Your family physician can give you some information about genetics, but there is a limit to how much training can take place on any given specialty topic during the education of a family practitioner, a pediatrician, or an internal medicine specialist. If you have questions about a complex, severe, or rare medical condition in yourself, your child, or some other family member and are trying to make major decisions about having additional children after a child with a birth defect and/or genetic disease has been born or decisions about continuing a pregnancy, having a genetic test done, or undertaking surgery or other major intervention because of a genetic disease, you may find yourself wanting to see a medical geneticist. A medical geneticist's extra training includes not only information on genetics and birth defects but also training in techniques and resources for sorting out some very complex health puzzles with underlying genetic and environmental causes. You might go to a medical geneticist because you are concerned about whether you have been exposed to teratogens that can cause birth defects, to find out more about a birth defect even if that kind of birth defect doesn't seem to run in your family, or because you have had many miscarriages and want to find out why. The problem that leads you to a medical geneticist need not be life threatening nor does it need to concern a baby or a pregnancy. Sometimes medical genetics specialists solve medical mysteries that don't get brought to them until the patient is an adult. As you can see, medical geneticists really deal with prenatal, pediatric, and adult situations, and they deal not only with genetic disorders but also with birth defects and other situations in which the genetic origins may not be obvious to you. You don't necessarily have to have a referral from another doctor, but you should consult your health plan on this one because they might require that referral.

BOX 34.2 GENETIC COUNSELOR

The other half of the medical genetics team is the genetic counselor. These health care specialists have a master's degree in genetic counseling. Their education has trained them to be able to assist you with the medical, genetic psychological, and social repercussions of whatever it is that took you to the medical geneticist in the first place. The genetic counselor works with the medical geneticist to help determine the origins of a trait that runs in your family, the causes of a birth defect, or the probability that something present in your relatives could turn up in your children. Part of the genetic counselor's job is to help assess risks to individuals who are not yet born or not yet known to be affected. They will also educate you and be sure that you understand any tests that are offered. They play a later role in helping interpret the outcomes that result from genetic testing and screening. Once results are obtained, the genetic counselor explains the results and deals with questions and concerns you may have about what you have found out. The purpose of the genetic counselor is not to tell you whether you should have a genetic test done or whether you should have a child or continue a pregnancy. Their role is to be sure that you are armed with all of the information that you need so that you can make the best, most informed decision possible for you and your particular circumstances. As the amount of genetic information available increases and the complexity of the choices for testing or dealing with test results increases, the role of the medical genetics team becomes increasingly important.

adults with later-onset conditions or whose correct diagnosis was missed during childhood (Box 34.3).

SCREENING VS. TESTING

In this chapter, we will talk about both screening and testing. What is the difference? A screen is population based and is administered to most members of a group, such as newborn babies or pregnant women. A test is an assay requested by a doctor on an individual basis with the patient's consent, usually in response to some medical information or risk factors indicating the need for the test. Thus we talk in terms of newborn screening, which looks for infants in which a metabolic defect of genetic origin can be detected without actually testing the genetic material itself.

A genetic test may look at the patient's DNA or may be a biochemical test for an enzymatic function. Tests of genetic material mostly either test for mutations or look at chromosomes under a microscope. A biochemical test will evaluate whether a gene product is correctly performing its function, either by a direct test of the gene product (such as a test of enzymatic activity) or through assaying for levels of metabolites, such as sugars or amino acids, to determine whether the body is maintaining the correct levels of molecules handled by the biochemical pathway. Why do a biochemical screen

BOX 34.3 OTHER USES OF GENETIC TESTING

There are a variety of uses of genetic testing long after the baby is born. In some cases, it may be used to distinguish between several diagnoses so that doctors can know which disease they are dealing with when they develop treatment strategies. Although this will often deal with things that could have been diagnosed through prenatal testing, the majority of pregnancies in this world are completed without the assistance of genetic testing. Other nonmedical uses seem to be proliferating. DNA-based testing has resulted in freeing some men who had been wrongfully convicted of rape, and DNA evidence in other cases has helped secure convictions. In Russia, DNA screening allowed the determination that bodies found in a grave were those of that last Russian tsar and his family, including the finding that the Princess Anastasia, who had been rumored to have survived, was indeed among those who died. Immigration programs in some countries have started using DNA testing to evaluate whether people being brought into the country on the basis of being a relative are, in fact, related as purported. Paternity testing has nailed some fathers with their financial responsibilities and sent others on their way out of the child's life. In one recent story, DNA-based testing dashed the hopes of a family who were just sure that a recently found child was their long-missing child. In anthropology, screening of mitochondrial DNA sequences from around the world have created a view of ancient migration patterns that support the idea that humanity traces back to a small number of women in southern Africa, with much talk of tracing us all back to Eve.

instead of a genetic test? Time and cost are both factors. It only takes one rapid biochemical screen to detect a functional deficit that might be caused by fifteen different mutations in fifteen different affected individuals. Even if the mutation is something unusual and hard to find, such as a mutation in a promoter region or something out in an intron that affects splicing, the biochemical screen will find any mutation that has an effect on the biochemical pathway being assayed.

So why not always just do biochemical tests? Sometimes we don't do a biochemical test because the locations of the cells expressing the genetic defect could only be assayed through something invasive, difficult, and expensive, such as a liver biopsy. In other cases, there may not be a simple biochemical test that is adequately specific and sensitive to tell us what we need to know. Also, we often want to identify the problem before the development of functional deficits so that we can intervene before damage occurs, which may mean that we want to do the testing before the biochemical imbalance becomes large enough to measure through biochemical testing.

Why do screening on a population basis instead of testing individuals? One reason is that the disorders being screened for are rare recessive diseases that can easily turn up in individuals with no known risk factors. Except where there is a clue such as a known family history, many babies have about an equal risk of the disease. Testing on an individual basis can take place once the infant has developed symptoms of the disease, but for the kinds of things being assayed in the newborn, screening programs that may be too late. It is very important to find out that the infant is affected and begin intervention before the symptoms develop in order to minimize risk of permanent damage or death.

One of the most dramatically successful newborn screening programs involves screening for phenylketonuria (PKU), which can cause profound neurological and cognitive damage. The successes result from a combination of factors. The test can be done rapidly and in a cost-effective way, and once a PKU infant is identified, dietary intervention can make a major difference in the child's health prospects. Current newborn screening programs do not screen for all of the known disorders that are detectable and "fixable" through medical or dietary intervention. However, some technological advances allow for screening for multiple different biochemical defects in one test, so we are hopeful that the number of disorders covered in these newborn screening programs will improve.

In some parts of the world, programs to screen babies at birth for things such as PKU have made a dramatic difference for babies lucky enough to be born where such programs are in place. However, there are many places in the world none of these tests are done, and in many places where some kinds of newborn screening programs exist, there are still known tests that are not being done for many such metabolic deficiencies that can kill, cripple, or lead to mental disabilities, such as OTC deficiency. For some disorders, we do not screen because there is no way to remediate the defect even if we can identify it. For some problems that can create a crisis in the life of an infant, newborn screening programs are all that is needed to dramatically alter that child's future for the better.

PRENATAL TESTING AND SCREENING

There are two main tools available for prenatal diagnosis: noninvasive or minimally invasive tests, such as *ultrasound* and *maternal blood screening*, and invasive tests, such as *amniocentesis* and *chorionic villus sampling* (CVS). For any of these procedures, it is important to keep in mind why we do them (Box 34.4).

BOX 34.4 SAMPLE INDICATIONS THAT MIGHT LEAD SOMEONE TO SEEK ADVICE FROM PROFESSIONAL GENETICISTS

Advanced maternal age
Previously affected child
Presence of a chromosomal anomaly in a parent
Parents are possible carriers of a genetic trait
Family history of a neural tube defect
Some kinds of environmental exposures

Since most children are, in fact, born healthy and without major genetic anomalies such as extra copies of chromosomes, perhaps the primary value of such testing for many people turns out to be simply to provide peace of mind to couples who fall into one or another so-called "risk groups," such as older mothers or parents who have already had one or more children with a genetic anomaly. In such cases, the test can often reassure the parents that the fetus developing inside the mother is healthy, at least as far as we can assay it. However, since couples who elect such testing often fall into known risk groups, there obviously will be cases in which the answer will not be a happy one (i.e., the fetus does have an extra copy of a chromosome or bears some other significant hereditary defect). In such cases, this information is crucial in giving parents the choice to either prepare for a child who may have very special needs or perhaps to terminate the pregnancy. Neither of these options give parents what they want, and we eagerly await the day when new technological advances will offer a third, better option: to fix what is wrong, either in utero or shortly after birth.

Some efforts to evaluate risk factors are carried out as large scale screens of entire populations rather than being done on an individual basis by individual choice. Some of the simpler less, invasive screens, such as a blood test for *α-fetoprotein* (AFP), are quick, minimally invasive, and inexpensive enough that they are routinely done on most pregnant women in many technologically advanced countries. In the case of AFP, which is produced by the fetus, a low concentration of this protein in the mother's blood is a possible indicator for Down syndrome, whereas a very high value might be indicative of a neural tube defect such as spina bifida. This test alone is not sufficient to diagnose either disorder because there is a lot of overlap in the values that can result for each phenotype. There are too many "false positives," but a high or low value would justify additional testing that is less easily done but much more accurate (Figure 34.1). As methods for identification of fetal proteins that make it into the mother's bloodstream get faster and better in parallel with improvements in our assay techniques, we expect that we will see development of large-scale screenings for many additional disorders.

In other situations, testing is done simply because parents want the testing even if they are not in a high-risk group. Perhaps they had a friend who gave birth to a child with a genetic problem, or perhaps they read a newspaper or magazine article that raised concerns that could be addressed through such testing. In such situations, as in most other potential genetic testing situations, a genetic counselor can educate the couple about the risks and benefits of the test, as well as the risks of whatever they are concerned about testing for, so that the couple can make an educated decision.

SEX SELECTION

In some cases, genetic testing can use information on a *haplotype* (a kind of genetic marker fingerprint for a region of a chromosome) or even a known mutation to distinguish between affected and unaffected fetuses, which allows the parents to make decisions on a fully informed basis for each pregnancy. Sometimes, if the gene has not yet been identified or information on an affected haplotype cannot be obtained from other family members, the sex

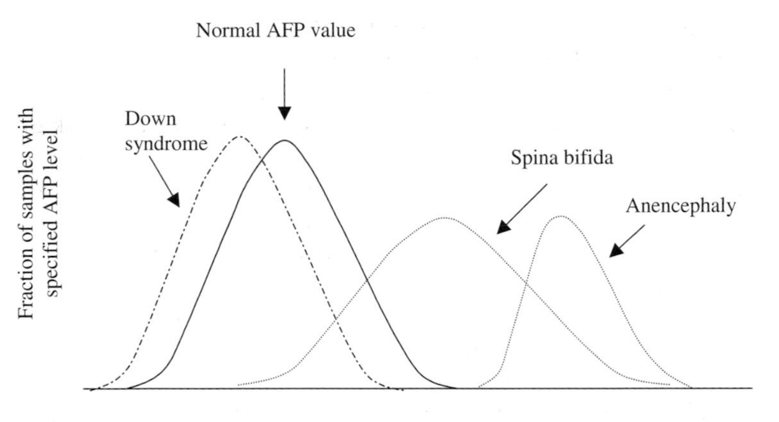

FIGURE 34.1 Distribution of levels of AFP in the mother's blood can suggest the need for additional testing. There is considerable overlap between normal values and the lower values observed on average for Down syndrome. There is some overlap between normal values and the elevated values seen in cases that later result in the birth of a baby with spina bifida. As you can see here, there is a region of overlap representing values observed in a small fraction of all three categories of infants—Down syndrome, normal, and spina bifida. Since these values change over the course of a pregnancy, interpretation of the values can be thrown off if there is an error in the estimation of how advanced the pregnancy is. (From Wald and Cuckle, *Am J Med Genet* 31:197–209, 1988; and Brock, Rodeck, and Ferguson-Smith, *Prenatal Diagnosis and Screening*, p. 161, London, 1992, Churchill Livingstone.)

of the fetus may be the only information available to parents trying to decide what to do. Testing for sex is sometimes done in cases of severe sex-linked disorders for which no specific test is yet available. The concept is that, in cases in which the mother is a known or obligate carrier of an X-linked disorder, the parents may decide that if they only have daughters, they will not have to face the fifty percent chance of having a child with the trait in question.

Can you imagine a situation in which a family might be faced with deciding whether to simply have only girls, even though half of the boys would have been unaffected? Duchenne muscular dystrophy (DMD), with its terrible consequences for the child and the relatives who love him, can be a problem to diagnose even now that the gene has been cloned. The gene is enormous, and available genetic tests only detect some kinds of mutations. If a couple has a son with DMD and genetic testing fails to detect a mutation that we know must be there in the mother or the affected son, the couple has several different choices: they might decide to have no more children and focus their attention on the needs of the child they already have, they might decide to have more children and hope that the next child is healthy, or they might decide to have only daughters because they know that their daughters are not expected to have DMD. As more and more disease genes on the X chromosome are identified and testing methodologies improve, we will move gradually away from having anyone face a decision to select for the sex of a child simply because they cannot test for the mutation they actually want to know about.

A much more problematic situation arises in some situations in which couples seek prenatal testing for the sole purposes of ensuring that the child they bear will be of the "right" sex. To some, it might seem absurd that anyone would have an abortion simply because the child's sex is not the sex they were hoping for. To others, the necessity of producing a child of the "right" sex may be a matter of grave importance. In some cases of cultural pressures to have sons, doctors find themselves faced with parents who want the doctor to carry out testing for the sex of the fetus so that the couple can use this information in deciding whether or not to continue the pregnancy. In other cases, a family that has had five sons may be found putting pressure on the doctor to help them assure that the next child will be a girl.

Although ultrasound offers the possibility that the sex might be determined without invasive genetic testing, thus limiting some risks, there are major ethical problems with choosing to terminate a healthy pregnancy should the fetus be of the "unwanted" sex. Currently, sex selection simply because a couple wants a child of a particular sex is not supported in the medical genetics community. Moreover the American Medical Association has advised physicians that sex selection of this kind in the absence of any accompanying health problems is not something that physicians should do. The issue of sex selection raises the larger overall issue of what constitutes an allowable basis for anyone to elect discontinuation of a pregnancy. Clearly, different doctors and different prospective patients hold different views on these subjects. There are certain levels at which the right of individual autonomy in health decisions leaves each individual to decide where they draw the line. But when the decision to select against a child is based on sex alone or on things regarded as cosmetic or falling within the normal range of human variability, little if any support will be found for such choices.

At some point we are going to need to ask ourselves, "Just what traits are we going to test for and how will we deal with the outcomes of those tests?" In doing so we are going to need to balance the "value" of the test (as defined in a host of ways) against whatever "risk" that test (and its results) may pose to the mother or to the fetus. And frankly it becomes important to consider where to draw the line. Terminating a pregnancy in which the child is unlikely to survive and the pregnancy threatens the mother's life sits at the far opposite end of an ethical continuum from terminating a pregnancy because the child is the "wrong" sex or does not have some cosmetic feature the parents desire. The field of medical ethics struggles with issues of where to draw the line along that continuum, but in most cases the final burden rests on the parents themselves, each of whom likely have different perspectives on what they can and cannot live with.

Finally, in any test that has even minimal possible risks, we face the problem of balancing the risks to the fetus or mother in the absence of testing against the risks that are incurred if testing takes place. So before we can continue this discussion, we need to discuss the tests themselves.

MATERNAL BLOOD SCREENING

The most common form of maternal blood screening is an assay for the concentration of a fetal protein called -fetoprotein (AFP) in the mother's blood. This is a major blood protein in the fetus and a small amount of it leaks out into the mothers blood supply. The AFP level alone does not provide a definitive diagnosis, just an indication that further testing such as ultrasound examination (sonar imaging of the fetus in the mother's uterus) is needed. The situation is complicated by the fact that levels of AFP that correspond with a normal pregnancy overlap with the levels of AFP found in situations where the fetus has Down syndrome (which will produce somewhat lower levels of AFP) or spina bifida or other disorders called neural tube defects (where the levels are somewhat higher). In fact, as we can see in Figure 34.1, there is a set of AFP values that can be found in any of these three situations- normal, Down syndrome, or spina bifida.

The requirement for further, more definitive tests, such as ultrasound, is underscored by the fact that there are pieces of information that could help reassign some of the "false positives" into the negative category. For instance, the level of AFP varies over the course of the pregnancy, so if the age of the fetus had been underestimated, an observed value (which changes with the age of the fetus) might be a normal value for the correct age but appears abnormal simply due to the age error. Similarly, if the value is high but there is a "multiple" pregnancy, the interpretation of the value is altered by that knowledge; a higher AFP value is expected if the product of two or three babies rather than one is measured.

Because other very rare genetic anomalies can also lead to elevated values, knowledge that such a genetic anomaly runs in a family would assist in the correct interpretation of the test values. For instance, certain kinds of kidney defects could lead to altered AFP levels because AFP passes through the fetal kidneys to get into the amniotic fluid.

There is also a worrisome possibility of false negatives: those cases in which the maternal blood AFP level for a fetus with a neural tube defect falls within the normal range. Indeed, the best estimate is that twenty percent of such cases are missed because of AFP levels that overlap with those of normal fetuses. However, as discussed later, when AFP levels are measured in amniotic fluid taken from the mother's uterus and the examination is coupled with ultrasound studies, most cases can be detected.

As shown in Figure 34.1, the maternal blood AFP concentration is often reduced in fetuses with Down syndrome and in those fetuses with some other autosomal trisomies, as well. However, note that the distributions for normal and Down syndrome fetuses overlap far more substantially than for neural tube defects. Thus a low AFP level, like maternal age, can only be considered a risk factor used to determine whether more accurate, but also more invasive, testing should be pursued.

Because the number of false positives obtained by assaying AFP alone can be quite high, a better screen has recently evolved. This test, called the *triple screen*, measures two other chemicals in the maternal bloodstream, human chorionic gonadotropin (HCG) and estriol (E3), as well as AFP. The combined levels of these three components of the mother's blood predict the presence of a Down syndrome fetus in sixty to seventy percent of the cases and show a much lower rate of false positives. The triple screen is now in wide use as a screening tool for Down syndrome, as well as for several other fetal anomalies, in women age thirty-five and over, and we are watching for the emergence of a quadruple screen that will further improve the accuracy of the screening.

ULTRASOUND

Ultrasound studies are becoming more and more common in the medical management of pregnancies and represent an especially attractive screening mechanism because the process is not invasive. Numerous studies clearly show that this test poses no risk to fetus or to mother and yet can provide extremely valuable information with respect to a variety of tests for neural tube defects, limb deformities, or some types of heart disease, as well as some disorders of the kidney and gut. High-resolution sonograms done later in pregnancy can also detect disorders such as cleft lip and cleft palate. Because of their safety and recent advances in technology, such tests are growing in their acceptance and usage. Moreover, prospective parents love the opportunity to see the baby on the imaging monitor months before delivery.

MORE INVASIVE TESTS

Think about this situation: Mom is thirty-seven years old, and the AFP level is on the low end of normal. There is clearly a risk that this woman is carrying a fetus bearing an autosomal trisomy. To be certain, we need access to a reasonable number of fetal cells. Currently, there are two standard methods for getting these cells: amniocentesis and CVS. Both techniques can be done after the tenth to twelfth week of pregnancy, and both provide the necessary cells for a variety of genetic tests, most notably karyotyping.

AMNIOCENTESIS

Amniocentesis is diagrammed in Figure 34.2. Basically, ultrasound is used as a guide while the doctor inserts a needle through the abdomen into the uterus and removes a small amount of fluid (known as amniotic fluid) that surrounds and cushions the developing fetus. Although many texts still say that amniocentesis should be done in the fifteenth to sixteenth week of pregnancy, improvements in the technique and in ultrasonography now permit the test to be done much earlier, often at ten to twelve weeks. A needle is put through the abdomen into the uterus. With the use of ultrasound, the doctor can see the tip of the needle: it looks like a very bright star on the ultrasound image. Because the doctor is also able to see where the fetus is on the ultrasound screen, it is possible to aim so as to miss the fetus, and the chances of damage to the fetus are minimal.

For these reasons, this procedure is now considered quite safe; the risk of miscarriage is estimated to be less than 0.25% at centers routinely performing amniocentesis, and knowledge of miscarriage rates in individuals who do not have this test suggest that at least some of those miscarriages would have happened anyway. The reason physicians now routinely encourage a pregnant woman over age thirty-five to undergo amniocentesis or CVS is that her age-dependent risk of an autosomal trisomy begins to exceed the risk of miscarriage from the procedure alone. The disparity of the two risks increases as the mother gets older.

The amniotic fluid withdrawn by this method provides a rich source of both fetal cells and fetal proteins. The cells can be cultured using newer

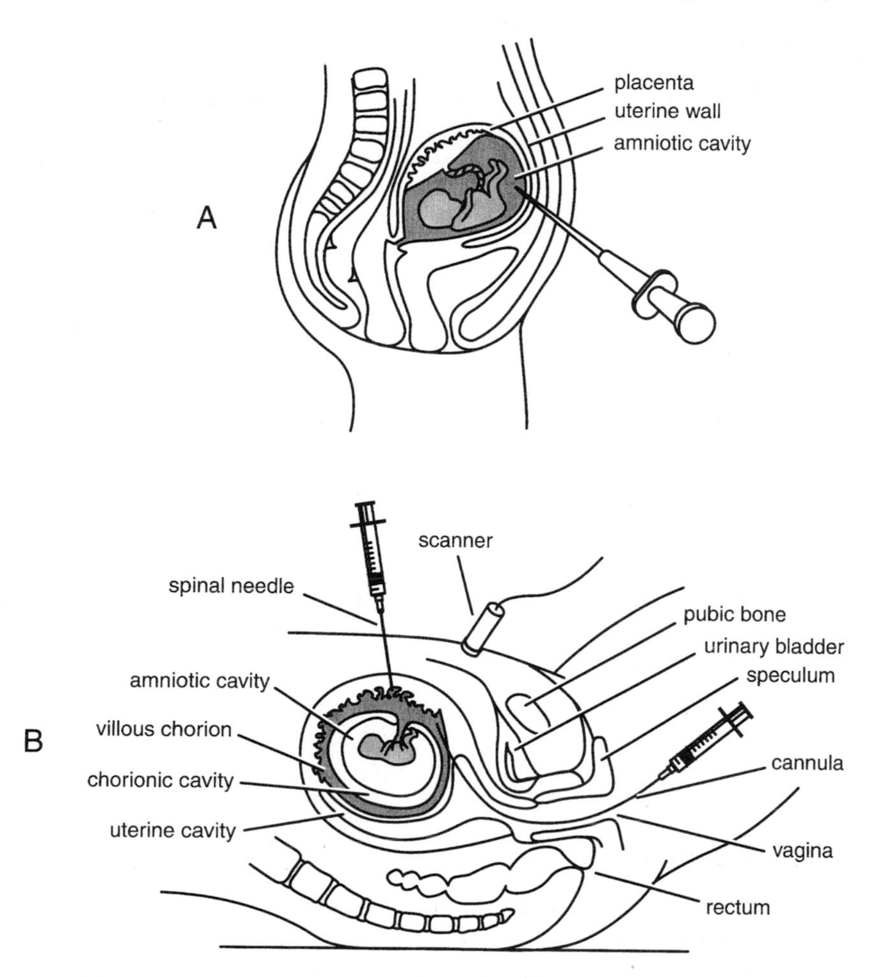

FIGURE 34.2 **A,** Amniocentesis, performed at fifteen to sixteen weeks of pregnancy. **B,** Chorionic villus sampling, performed at ten to twelve weeks of pregnancy, may be carried out either through a catheter via the vagina or though a transabdominal needle. In both cases, the noninvasive ultrasound imaging allows visualization of the fetus throughout the procedure. In addition, monitoring of both maternal and fetal status afterwards can help provide peace of mind that the procedure was completed safely. CVS samples extraembryonic tissues, and since extraembronic tissues can suffer from elevated levels of chromosomal anomalies compared to the fetus, CVS samples can sometimes carry chromosomal anomalies not present in the fetus. Amniocentesis samples skin cells shed by the fetus into the amniotic fluid and thus directly assays the fetus's genotype. The subsequent testing on the sample that is retrieved can take time, but the term limit for terminating a pregnancy is twenty-four weeks, which is after test results are available.

"microdrop" methods, and metaphase spreads needed for karyotyping can be obtained in a few days to a week. Cells can be even more quickly analyzed by FISH assays, although such techniques provide information only about the specific chromosomes for which probes were available or used. Sufficient cells are available for both DNA analysis and biochemical testing. Discussion of these sorts of analyses will be deferred while we consider another means for getting fetal cells, chorionic villus sampling.

CHORIONIC VILLUS SAMPLING

CVS, which can be performed as early as the ninth week of pregnancy, is diagramed in Figure 34.2. Basically, the doctor inserts a flexible needle, known as a cannula, through the center of the cervix and into the uterus. The doctor then removes a small amount of tissue from a fetal tissue known as the chorionic villi, a tissue that will go on to form the placenta. This tissue divides mitotically very actively, and thus metaphase cells for karyotyping can be obtained quickly. The tissue can also be subjected to other tests, such as mutation screening.

There appear to be regional differences in use of CVS vs. amniocentesis, with some areas preferring one procedure and other areas preferring the other. A decade ago the major advantage to the use of CVS was that it could be performed earlier in pregnancy than amniocentesis, but technical advances have allowed amniocentesis to be done at an earlier stage than it could be done a decade ago. Until some years ago, amniocentesis could only be done at sixteen weeks, but it has been improved to the point that it can now be performed before the twelfth week. Thus CVS can be done about three weeks sooner than the current earliest time point for amniocentesis, offering only a minimal advantage in terms of early detection. Second, some workers report a lower success rate of correct karyotyping with CVS, apparently because it samples extraembryonic tissues that have an elevated rate of chromosomal anomalies. Although there have been troubling reports that CVS might induce an elevated frequency of limb anomalies, this appears to have been disproven.

Both methods provide the same three things: metaphase cells for karyotyping, fetal DNA for DNA analysis, and cells and enzymes for biochemical studies. The second half of this chapter will consider each of these types of assays.

ANALYSIS OF FETAL CELLS

Karyotyping

A picture of a normal human karyotype is presented in Chapter 10. Such figures are obtained by taking dividing cells from the fetal sample and lysing them (breaking them open) on glass slides so that the individual mitotic chromosomes spread out in a loose field. After the slides are stained with dyes, the resulting clusters of chromosomes from each cell are photographed and examined. Skilled cytogenetic technicians begin with large photographs of each metaphase spread. Then they carefully cut the picture of each chromosome out of the photograph and match each pair of chromosomes side by side on a piece of mounting paper. The folks who do karyotyping in the cytogenetics labs are truly gifted at pattern recognition.

Some disorders, such as autosomal trisomies, and sex chromosome anomalies, such as Turner (XO) and Klinefelter (XXY) syndromes, are easily picked up by this method. However, a good karyotype can also recognize more subtle aberrations, such as deficiencies or duplications for small regions of the genome, translocations of material between chromosomes, inversions of the material on a given chromosome, and the fusion of two ends of a given chromosome to form a ring. Although such anomalies are not common in humans, they do occur and they often have phenotypic consequences. For example, being heterozygous for a deletion of material on the short arm of chromosome 5, which is to say possessing only one copy of that material, results in a disorder known as *cri-du-chat* (or "cry of the cat") syndrome. This syndrome, recognizable because such infants mew like kittens when they cry, results in severe mental retardation and craniofacial anomalies. We have also noted previously the role of deletions in causing Prader-Willi and Angelman syndromes. Note that if the right culture conditions are used, it is possible to recognize the presence of "fragile sites," such as those regions at which chromosomal breakage is observed in *fragile X syndrome* patients.

One might also note that karyotyping sometimes reveals abnormalities, such as an XYY karyotype, whose effects are not clearly understood or defined. Other abnormalities may be found, such as balanced or reciprocal translocations that may not affect the health of the child but might well affect his or her ability to produce children. Some care is required in explaining such outcomes to the prospective parents. This point also applies when karyotyping anyone. A student who accidentally discovered that she carries a balanced translocation as the result of using her own cells for a routine lesson in how to do karyotypes found that even though she was phenotypically normal, the translocation offered a potential hazard to her fertility and the health of any children that she would have. Although it was possible to tell that there are potential health hazards for some of her offspring, simply looking at the rearranged chromosome structures under the microscope cannot indicate what form those health problems might take.

Karyotyping will also, by default, tell you the sex of the fetus. Curiously, prospective parents differ in whether they want to be told what sex was revealed by the karyotype. Parents often say that they prefer to be surprised in the delivery room regarding the sex of the baby and thus ask that the report of the results of a normal karyotype be limited to "everything looks fine." However, some parents want advance information on this point. One even finds the occasional case of a split decision between parents.

Once the fetus is identified as having a complex chromosomal rearrangement, an important question is: Exactly which chromosomal bits are involved in the rearrangement? Much is known about the consequences of extra copies of even small regions of a particular chromosome, so if we can go beyond telling that there is a rearragement to telling exactly which chromosomal regions are going to have too many or too few copies, we may be able to make some predictions about what the consequences will be. In some cases a simple karyotype is enough to identify the additional piece of DNA, but FISH (the use of a specific probe from one specific chromosomal region) or chromosome painting increases the chance that we will be able to get a specific answer to our questions about the baby's karyotype. There are some specific regions of the genome associated with well-characterized syndromes, so in some cases, important questions can be asked by FISH-ing with probes specific to the region in which deletions are known to cause traits such as velocardiofacial syndrome or diGeorge syndrome.

DNA Analysis

In a real sense, this section is simply a summary of the last thirty-three chapters. Suppose a couple's first child had Duchenne muscular dystrophy

(DMD) and the couple came to you asking about the DMD status of their second, unborn child. Karyotyping might provide some reassurance because a female fetus is very unlikely to be affected. If the fetus is male, the answer is less clear-cut because the odds of the child being affected are fifty percent if the mother is a known carrier. However, if you already know the nature of the DMD mutation borne by the first child and have a sample of fetal cells available from the second child, it is straightforward to determine whether this child carries the mutation. Now you can provide truly useful information to the parents. Very similar things can be done in the cases of quite a number of other diseases, such as cystic fibrosis, when it is known that the fetus is at risk for that disease. Tests are being developed rapidly for a host of other disorders for which tests were not previously available. Even if the gene is not yet cloned, a closely linked DNA marker can sometimes be used to diagnose the genetic state of the fetus if DNA from other family members is also tested. There is nothing special about fetal DNA, at least in terms of its chemical properties; any of the tests described so far can be used to assay the genetic state of a fetus.

It usually will make a big difference if the primary genetic defect has already been determined for affected family members and carriers before tackling prenatal diagnosis. For many, the test for a known mutation can be fast. However, as stated earlier, many genes are quite large. An open-ended search for an unknown mutation somewhere within one of these genes can take some time, precious time that you do not want to spend during the period in which prenatal testing takes place. For some genes, such as CF, tests have been developed that can identify a large number of known mutations but cannot detect rare mutations or new mutations not previously observed. For other genes, however, the development of testing has not advanced as far, and it could take too much time to determine the primary mutation on the time scale of the prenatal test.

Thus, if you are concerned about a genetic defect and want to include mutation screening as part of your family planning, it will work out much better if you start asking questions before you are pregnant. You may be told that a standard test is in place that can do everything you need done at the time of the prenatal test, or you may be told that you qualify for some type of "preimplantation" testing. However, depending on what the gene is and what the genetic defect is, beginning your inquiries ahead of time might give you important choices that might not be available if you wait until week twelve of the pregnancy. Of course, there are many cases (new mutations or recessive diseases that you don't realize are lurking in the genomes of both you and your spouse) that you don't even know you should ask about until the first child with a problem is born into a family. Even then, asking your relatives questions about the family medical history can sometimes offer a warning. If you have a cousin with cystic fibrosis and your spouse had a great uncle with cystic fibrosis, you should be asking yourself whether you want to talk to a member of a medical genetics team before the first pregnancy. Sometimes the test result will be the happiest one of all—that you are not carriers. However, if that is not the answer, being informed can save you from later saying, "If only I had asked."

This ability to do genetic testing is not a panacea. Indeed, the facility of such tests becomes most worrisome in terms of the material we have presented on the inheritance of complex traits. One worries about people in the near future attempting to test fetuses for DNA markers associated with traits such as mental illness, obesity, and sexual orientation. It is perhaps a rather personal set of prejudices, but we draw a distinction here between testing for traits an individual will express and which will greatly diminish their quality of and length of life, such as DMD or fragile X, and those traits that they *might* express and whose effects on their quality of life are hard to assess.

Aren't there diseases for which the responsible genes are neither mapped nor cloned? Yes, there are, and some of these disorders can be diagnosed by biochemical or enzymatic assays performed on fetal tissues. Such tests are briefly described in the next section.

BE SURE THAT YOUR INFORMATION IS CORRECT

Our ability to assess the genetic health of a fetus is impressive, and our capabilities expand daily. To the extent that truth is good, knowledge is power, and informed choice is better than uninformed choice, advances in the technologies of prenatal diagnosis can greatly improve our quality of life by providing us with better information from which to make better choices. As more and more information emerges from the Human Genome Project and a large number of research projects involving specific traits, it becomes more and more important to understand enough about genetics to be able to interpret the information being provided by the press. It also becomes important to be good at evaluating the information and the sources providing it. There are, after all, many things on the World Wide Web that are completely false, in addition to many important, valid sources of good solid information. Although your doctor, medical geneticist, or genetic counselor will be the most reliable source of information when trying to make decisions about genetic issues in your own family, you will often find yourself getting information from other sources, also. There are a variety of information resources out there that can give a good overview of many traits or in some cases individual traits that are the focus of that organization (Box 34.5). Other good sources of information can be found at websites for organizations such as the National Institutes of Health, the Department of Energy, the Centers for Disease Control and Prevention, and the Food and Drug Administration, all of which include some information for the public.

Some Uncertainties of Genetic Testing

Many of the tests requested by your family practitioner or the medical geneticist will be done by clinical laboratories in your own hospital or nearby (Box 34.6). These tests are done under clinically certified conditions specially designed for use in the context of a medical practice. However, some very specialized testing, such as screening for mutations in a particular gene, may be done at only a few places in the country. If information on the gene responsible for your genetic disease is recent, the testing might not be available from any of the clinically certified labs and might only be available on a research basis, conducted by the labs that are doing research on that gene.

BOX 34.5 ORGANIZATIONS

There are many other places that you can turn to for information. An Internet search finds organizations such as the Alliance of Genetic Support Groups, the March of Dimes, and the National Organization for Rare Disorders. Some individuals who are trying to take an active role in communication about disorders in their families have established Web pages that present information or reach out to others with similar problems. There are several ways to locate an organization that provides information or support relative to a particular disease. For many different diseases, organizations raise funds for research, provide support groups, and provide information about the disease. One example is The Foundation Fighting Blindness, which supports research, carries on educational programs, has local chapters throughout the country, and holds national meetings attended by patients, family members, caregivers, and educators who want to understand more about forms of retinal degeneration, such as retinitis pigmentosa, macular degeneration, or retinoschisis. Often your doctor's office will have information about such organizations. If they don't, try checking with a medical geneticist or other specialist who sees many cases of the trait in question. Even just looking under Social Service Organizations in the yellow pages of the phone book for a large city can connect you with a variety of organizations that can help you get information or support. Looking in the phone book for a town with a population of 100,000 yielded organizations that deal with cancer, lung diseases, multiple sclerosis, kidney disease, sickle cell anemia, epilepsy, and birth defects (the March of Dimes).

In the early stages of development of a genetic test, the test may not be available from a clinically certified lab. There can be a lot of reasons for this. If some aspect of the technology still needs to be worked out, the test may be feasible but not yet meet clinical standards. In some cases, the test protocol meets clinical standards, but not enough is yet known about what can be predicted based on the kinds of results that are produced.

For instance, if a mutation is associated with the trait but does not seem to be an outright cause of the trait, researchers may elect to continue more investigation of the gene before making testing available for general clinical use so that they do not end up putting uninterpretable results into the hands of doctors and patients who will then not know what to do with information that is ambiguous. For instance, there is a sequence variant in the optineurin gene that is found more frequently in people with glaucoma than in people who do not have glaucoma. However, there are many people with glaucoma who do not have it, and many people who have it who do not have the disease. For now, it is considered a risk factor—something that tells you that you have a higher chance of the disease than the general population without indicating that you are certain to develop the disease. Until more is understood about what role this sequence variant is playing, further research is needed and results are not clinically very helpful.

In other cases, a test may involve a trait that is rare for which there has been no incentive to work out the test under the more involved conditions

BOX 34.6 CLINICAL LABORATORIES

Once a gene has been identified, researchers may spend years developing screening of that gene before it becomes available through a commercial clinical lab. Once optimal screening processes are worked out in the research labs, once it has been determined just exactly what can and cannot be predicted based on the test, and once the patent lawyers are done squabbling (and perhaps even before), the test will become available commercially. However, even at that point there are no guarantees that you will find out what you want to know, since finding out that you have the mutation can still leave you with many unanswered questions. For some genes, mutation screening results can provide clean, simple answers, but for many genes, sequence changes fall not into the two expected categories, sequence variants that cause disease and sequence variants that don't, but rather into three categories, those that cause disease, those that don't cause disease, and those for which we cannot tell whether they will cause disease. Even when we know that a mutation can cause disease, we may not know whether having the mutation assures you of ending up affected or whether it just puts you at higher risk. Hopefully, long-term studies of mutation screening results in many people will increase our ability to make predictions based on the results of mutation screening. Currently, however, for many genes and mutations, having the results explained by someone with clinical genetics training will greatly assist in being sure that you know as much as possible about what your test results mean. The critical things to figure out in trying to arrange for a test include determining whether the test is available from a clinically certified laboratory and determining if there is only one source of the test or if you will need to make a choice between several labs. Other issues will of course, include things such as cost and whether the test is covered by insurance, and what kind of sample you will need to provide. There are other, more complex issues, such as what level of predictions can be made from the test that might influence your decision about whether to be tested or not. A clinically certified lab will carry out the test according to a rigorously defined set of standards and protocols designed for use in this type of clinical screening. Normally, the doctor involved in carrying out the sampling procedure will also determine which lab will do the testing.

required for clinically certified testing. In such cases, a clinical test would actually be possible and informative but would require that someone make an investment in converting the test from research to clinical status. In any of these cases, even though there is no clinical test available, it may be possible to be tested on what is called a "research basis" (Box 34.7).

Gaining access to research-based testing will usually mean that you have to agree to participate in the research project doing the testing; often, however, the project only involves mutation screening, so you would only have to provide a blood sample or cheek swab from which testing could be done. Normally, research labs will not provide prenatal testing, but adult testing information from a research program can sometimes give you information that will tell you whether there is any reason to get prenatal testing done. For

BOX 34.7 RESEARCH LABORATORIES

All of the genetic tests that become available to patients are developed by researchers working in laboratories, first to identify the gene and then to learn about how much can be predicted about the phenotype that will be caused by any particular gene or mutation. The research lab may be at a university, in a hospital, at a pharmaceutical company, or at a research institute, such as the Stowers Institute. Such labs are headed by MDs or PhDs, who design and direct the research conducted by a team of researchers. Medical genetics research includes a broad array of topics, such as searches for genes, studies of how mutations are caused, investigation of how chromosomes are duplicated and distributed to the correct daughter cells, research on animal models of human diseases, and testing of improved methods for diagnosing genetic anomalies. Because the testing techniques first emerge out of research done in these labs, they are the only source of such testing available before commercial development of a test, and because they are research operations and not commercial labs, they are not usually set up to handle things in the same way a commercial lab might. The result is limited availability of testing. People who participate in medical research help contribute to the body of knowledge that has to accumulate before a test can move from the research environment into certified clinical testing. Research labs will normally not be found doing prenatal testing for a variety of legal reasons.

instance, if there are only two known genes causing your trait of interest and it is clear that most genes for your trait have not been found yet, it will not make a difference to know that you do not carry mutations in either of the two known genes. Thus prenatal testing would not have made any difference to your reproductive plans, even if you had gone to some great lengths, such as going to Europe or Asia, to try to get the test done. On the other hand, if the mutation in your family was already known, that information could affect the level of efforts you would make to get prenatal testing when it is not yet readily available.

THE CONDITIONAL PREGNANCY

Years ago, women attempting to bear children waited to seek a pregnancy test until their second missed period. They then went in for the so-called "rabbit test," and a positive result was often sufficient to warrant announcing the "happy news." These days, the home tests available at most supermarkets or drug stores are accurate on the first day of the first missed period. However, earlier knowledge, coupled with increases in prenatal diagnostic techniques, has not always resulted in earlier announcements of impending births. Rather, women are increasingly aware that, on average, one sixth of human pregnancies will result in miscarriage before the end of the twelfth week. Perhaps not surprisingly, some couples are then waiting to announce the pregnancy either until after the end of the first trimester or until they have seen a healthy fetus developing on a sonogram.

Couples are also becoming increasingly more guarded and concerned about genetic disorders, as well. Some of these couples who are concerned that a negative result will lead them to terminate the pregnancy prefer to keep the news of a pregnancy private until they are sure that they will go through with the pregnancy. Some women say that they also consciously try to avoid accepting or "bonding" with the pregnancy for fear of becoming attached only to have to lose the pregnancy through miscarriage or following the adverse result of one or another test. Clearly, as our technology gets better, the number of disorders that can be analyzed will increase dramatically. One cannot help but wonder just how "conditional" pregnancy can become, and what kinds of psychological effects this concept of conditional pregnancy will have on those who wait for the moment when they can decide that they can finally believe that they are going to have a baby.

GENETIC MATCHMAKING AND OTHER STRATEGIES

What can we do with genetic testing information to inform our reproductive decisions? Certainly there is the obvious: decide whether or not to continue the pregnancy, a choice that we hope will start to gain some alternatives in the future, such as perhaps curing the condition. We hope that there will come a day when the prenatal test will primarily serve to find out whether stem cell development of a replacement organ needs to be started before birth, or treatment in utero needs to be started to prevent a developmental mistake from causing damage, or something about the embryo's environment needs to be manipulated through changing the mother's diet. In some cases, testing can even be done before implantation, although this approach is costly and results in many nonviable pregnancies. However, there are other uses, such as in the case of the parents of the little boy at the beginning of this chapter who just wanted to be able to operate from full and correct information when making decisions about their child's life and quality of life. For others, testing offers no information that the parents would use, and it is important that people who do not want testing have the option to decline testing (Box 34.8).

Let's back up a step and consider this: what impact might there be on courtship and marriage if you could look into a genetic crystal ball before picking a mate? What if you could look at your latest heartthrob and see that they are also your perfect genetic complement—having a different set of genetic flaws than your own, or at least genetic flaws that you would not mind passing along to your children? Sounds great? What if you looked at that same "love of your life" and saw instead that the two of you share some terrible recessive genetic flaws right along with your passions and your values and your hobbies?

Use of this kind of parental genotyping information has led to substantial reductions in the rate with which children are born with *Tay-Sachs disease* in some communities, not through abortion or even people refraining from having children, but rather through selective arrangement of who will have children with whom among those at high risk of bearing such children. However, if we are going to put such genetic information into the equation when picking a mate, is the implication necessarily that we might have to walk

BOX 34.8 THE OPTION TO NOT TEST

Even a brief conversation with Jill shows her to be expressive and intelligent, with her striking prettiness often lit up by warmth and sympathy and humor that have to have been assets in her nursing career. However, when Jill was growing up, people who did not know her well sometimes decided that she was a bit stuck-up or standoffish. If you were to meet her, you would wonder how anyone could think her manner anything but friendly. That is, you might wonder, unless you looked down and saw her constant canine companion, a beautiful yellow lab wearing a sign saying, "Please do not pet me, I am working." Jill was born deaf and as a teenager began losing her sight. She was "mainstreamed" in the public school system, learning to speak words she could not hear and to "hear" spoken language through a combination of lip-reading and other cues. She was adept at carrying on a conversation with someone sitting across from her at the lunch table or standing and talking to her in the hall. However, if someone who did not realize she was deaf asked, "Hi, how are you doing?" as they passed her in the hall, they would get no reaction because she had no idea she had been addressed. As her visual field gradually shrank, she might not see if someone who was not right in front of her waved and then looked confused when she did not turn to wave back. Jill has Usher syndrome, a combination of traits that starts out with deafness and later adds progressive vision loss from retinitis pigmentosa. She and her brother participated in a study of Usher syndrome that identified the genetic defect causing their hearing and vision loss. Researchers identified the gene and mutation responsible for their combination of visual and aural deficits. Jill has never gone back to ask the details of what they found out. She says that the information that came out of that study is of little use to her. It will not tell her anything about herself that she does not already know, and if she were ever to have a child, she would not have prenatal testing done because she would go ahead and have the child whether or not there were hearing or vision problems. This option to not carry on genetic testing is an important aspect of the rights of the individual to make their own reproductive decisions, so it is important that the system for educating and counseling people about genetic information include the ability to meet the needs of those who would continue the pregnancy no matter what the outcome of the test. In such cases, the main question becomes whether there is anything that such testing can do that can meet some other aspect of the family's needs, such as helping them prepare ahead of time to be able to meet some special needs present right at birth. There will be many cases like Jill's, where testing is not going to tell her anything that she needs to know sooner than she would be able to find out the old-fashioned way. The system needs to continue accommodating the desires of people like Jill to not do prenatal testing right along with the needs of those who want such testing done.

away from our soul mate based on a printout from a genetic testing company? Perhaps not. As genetic and reproductive technologies improve, giving up on having children, walking away from the love of your life, or having an abortion will be increasingly pushed aside by alternatives that let us fix the problem instead. Other technologies will allow for testing in the context of in vitro fertilization to preselect embryos free of the defect in question to be implanted in the mother's womb. Parental genotyping before reproduction will improve the odds that the children who are born will be healthy, or at least free of the identifiable defects for which their parents are carriers, and will improve the likelihood that neonatal health management will be improved in situations involving serious genetic illness that impacts the first days of life.

THE "MAYBE" TEST RESULT

As we learn more about complex diseases, issues in genetic testing become more complex. Clearly, testing for a mutation that causes some terrible disorder that involves pain and death for an infant falls at some opposite end of a spectrum from testing for cosmetic traits, such as eye color or a cute nose. However, even in trying to talk about the relative ethical dilemmas involved in considering such a spectrum of trait severity, the way we talk about it implies something else very important. It implies that the test we do will provide us with some absolute answer: if the child has the mutation, they will have the trait, and if they do not have the mutation, they will not have the trait. Test results don't always give such simple answers.

As we begin looking at complex traits, we find that, in addition to the ethical dimension to the problem, there is a practical dimension. What do we do with "maybe" results? What do we do with a test result that says, "This child will have a fifty percent increase in risk of heart disease over the general population"? What do we do with a test result that says, "There is an eighty percent chance that this child's IQ will fall on the low side of normal"? What do we do with a test result that says, "This child will be at increased risk of a life-threatening illness, but only if exposed to identifiable environmental items that will be difficult but not impossible to avoid"? What do we do with a test result that says, "This child absolutely will develop the disease in question, but the disease severity can range from lethal to barely even annoying, and we cannot tell you where on the severity range this child's clinical course will fall"? What do we do with a test result that says, "This child has a ninety percent chance of developing Alzheimer's disease seventy years from now at a time when the unknown future of biomedical advances might (or might not) turn this frightening, uncurable, fear-inducing fate into something treatable over the counter"? The fields of medical genetics and bioethics struggle with issues of whether there are identifiable places where clear lines can be drawn, but we expect that many of these issues will remain very fuzzy for a long time to come as we struggle with the implications of information that tells us, "maybe, but maybe not," and, "eventually, in a future so far away that we cannot know what this really implies."

THE LINE THAT CAN'T BE CROSSED

The medical establishment is unlikely to develop any eagerness to assist in terminating pregnancies based on complex or cosmetic traits, especially those that are not even sure to happen. However, physicians are more likely to be responsive to wishes of parents who want to terminate a pregnancy involving a condition that will lead to a very early, painful death for an infant. For many of the other areas out in the middle ground, it is the job of the medical genetics community to educate the parents, to offer insights into possible consequences of different choices, to provide them with long-term in addition to short-term views of the situation they face, and to inform them about the best-case and worst-case scenarios. In the end, though, the decision falls to the parents, not to the medical genetics professionals.

Every case is as unique as the individuals caught up in the situation. The combination of the parents, their stages in life, their individual personal and medical histories, and the context in which they live, can result in very different decisions being made relative to exactly the same test result. Something people have to be able to take into account in making these decisions is that the person they are today, and the issues that drive their decisions today, will likely change over the course of a lifetime. Thus they may need assistance in seeing just how their current circumstances, as well as their possible future situations, color their decisions.

People have substantial differences in what they can tolerate, different limits beyond which they cannot go. Some hit their limit when faced with terminating a pregnancy. Others hit their limit when faced with watching their baby die a long, slow painful death. Some hit a limit when faced with raising a child whose needs they know they cannot meet in a society that too often fails to live up to the ideal of offering a loving alternative home for such a child. So many factors contribute, including financial status, presence or absence of a social safety net to help a family cope, mental illness, alcoholism, and the presence of other family members who are desperately ill. A diagnosis of Duchenne muscular dystrophy in a child might well mean one thing to a young, happy, healthy, financially secure couple going through their first pregnancy in the context of a very supportive extended family but mean something quite different to a couple struggling with unrelated heath problems in the parents, major financial problems, lack of support from family or friends, and another child in the family suffering from more advanced stages of the disease.

Even trisomy 18, which might seem like a simple case because it represents the severe end of the phenotype spectrum, can elicit a broad range of responses. In the story at the beginning of the chapter, the parents wished they had had the information they needed to save their baby from unnecessary pain. Other parents, when receiving the news of trisomy 18 as a result of prenatal testing, terminate the pregnancy because they feel that it would be tantamount to child abuse to put an infant through what they know is coming. However, we know another woman who bore her trisomy 18 daughter and went home to sit and rock with her until she passed away, a font of maternal love, calm, and acceptance that many others have no capacity to achieve.

Some see a decision to terminate a pregnancy as a selfish decision that does not consider what the child's perspective might be. Some, though, who see themselves wishing they could die quietly in their sleep instead of lingering on in lengthy pain, have that same perspective of their child's life. Several years ago, Julia met a young man who considered his parents to be guilty of criminal negligence for failing to terminate the pregnancy that produced him and, a whole host of medical problems, when they knew that any child of theirs would be at fifty percent risk of inheriting the trait that ran in the family. Yet his father must have had a very different view of the situation, suffering the consequences of a medical condition and deciding that the positive aspects of his life so outweighed the problems that he would be willing to pass both life and this trait along to his child.

For each of us, there is an ethical brick wall that would stop us dead in our tracks if we ran into it. Fortunately, most of us are never faced with our own brick wall, and thus we do not even know where our real limit would be if we were put to the test. For those who get the news that there is a problem, there are amazing differences in what that brick wall turns out to be for different people.

For some, the act of terminating a pregnancy would be the thing too terrible to contemplate, to take a life or to fail to at least give the child a chance. For others, to continue the pregnancy is the thing they would never be able to forgive themselves for, to bring a child into too brief a life, one filled with pain in which their needs are desperate but cannot be met. Either decision represents a heartbreaking way for the parents to say, "I love you."

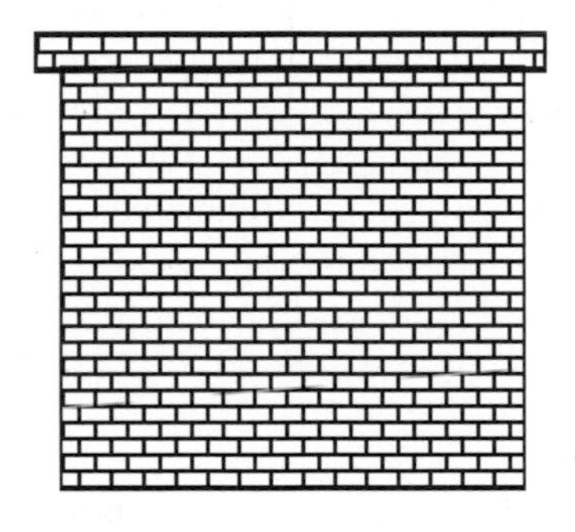

None of us will live with the consequences that fall upon someone else's family when they make reproductive decisions of this kind, invariably profound and painful in their far-reaching implications. So the role of modern medical genetics is to allow people to make decisions that are as fully informed as possible. However, the role of modern medical genetics is not to make the decisions for the family. Those who will live day in and day out, with the child or with the absence of the child, are the only ones who can know just where they will reach that boundary beyond which they cannot go.

American Sign Language sign for "I love you."

MAGIC BULLETS: THE POTENTIAL FOR GENE THERAPY

35

The scene on the screen in the conference room looked just like a home video, a movie showing a beautiful *briard dog* named Lancelot walking into a dimly lit room. The place seemed a bit crowded, with disarranged furniture scattered about. The audience in the conference room watched, spellbound, almost holding their breaths, as the dog made his way through the room, carefully avoiding objects as he swung his head around in an odd manner to scan the area ahead of him with his right eye. He daintily picked his way through the obstacle course, the film stopped, the lights came up, and a few quiet spontaneous cheers could be heard over the applause that broke out around the room. Several of the rational, objective researchers in the room had lumps in their throats as they listened to the conclusion to the presentation. Gene therapy treatment of Lancelot's right eye when he was four months old had effectively cured a canine model of *Leber congenital amaurosis*, a severe form of early childhood blindness that is incurable and may be diagnosed in humans in the first year of life. Those attending the talk had just witnessed a medical miracle: a "blind" dog that could walk through a crowded, unfamiliar room and successfully avoid contact with objects. Lancelot could see with his treated right eye! Lancelot and some of his relatives develop vision problems because of a defect in a gene called RPE65. Since both copies of the gene are defective, the obvious approach to gene therapy was to put a good copy of the RPE65 gene into the cells of Lancelot's eye. The strategy proved valid when the three blind puppies who were treated turned out to be cured, and they stayed cured! The movie starring Lancelot has played to audiences of scientists from around the world, and Lancelot has even visited Capitol Hill to attend a congressional briefing on gene therapy. To the scientists in the conference room, the concept of being able to use this approach to cure blind children was emotionally compelling in addition to being scientifically attractive. The general approach looked as if it might be usable for some other recessive forms of inherited retinal degenerations. However, many gene therapy projects have not yet arrived at such dramatic successes. Why can't all of the other diseases in need of gene therapy simply be treated in the same way as the briard dogs were treated? Not all diseases can be treated this way because there are a broad array of technical and strategic issues to be sorted out that differ from one disease to the next and from one gene to the next. In this chapter, we want to show you a bit about how gene therapy works and what some of the issues are that keep gene therapy researchers in their labs burning the midnight oil in search of answers.

After great expense of time and resources on the part of many really, really smart people, we finally know the sequence of the human genome (and many other genomes, as well). The genes have been found (well, many of them, anyway). We are starting to find out what some of the gene products do. Biochemical pathways are coming together that provide us broad conceptual insights into a variety of pathogenic processes. Those of us who consider this a beginning, not an end, now face the critical question: What do we do with all of this knowledge? How do we convert all of these advances into help for people who are not adequately helped by the current state of medical knowledge?

We have seen some ways in which insights into new genes and new biochemical pathways have led to dietary management in disorders such as PKU or standard pharmaceuticals derived from information gained from genes for some kinds of cancers. However, the hope that comes from successful gene hunts points in the direction of gene therapy, the therapeutic use of the discovered genes themselves, and not just the knowledge gained from finding those genes. Earlier, we talked about the use of hydroxyurea to turn on expression of fetal hemoglobin in individuals with sickle cell anemia. In the future, we expect that gene therapy will include both introduction of genes into cells in the body and use of inducing agents to alter the expression of genes that are already there.

WHICH GENE SHALL WE USE?

Gene Replacement Therapy

When the cystic fibrosis gene was discovered, the concept of *gene therapy* seemed pretty straightforward: put a copy of the CFTR gene back into individuals who have no functional copy of the CFTR gene. As soon as we say that, though, a lot of questions arise and we realize that there are actually many issues to be resolved in designing a gene therapy treatment. The first question that arises is: Which gene are we going to administer to the patient? It might seem as if the answer is obvious: put back a good copy of the gene that is defective. It may be that simple in the case of single-gene recessive disorders in which the disease results from loss of the function of the gene product. In fact, that is exactly what happened when the therapeutic version of the *RPE65* gene was put into Lancelot's eye: a simple replacement of something missing (Figure 35.1). There are several kinds of recessive retinal degeneration caused by defects in both copies of a single gene that could likely respond to almost exactly the same therapeutic protocol, with almost the only change being the choice of which gene to put into the eye. Another obvious situation for gene replacement therapy is that of individuals who are lacking one of the key blood clotting factors. Like the eye, disorders of the blood provide a more delimited treatment problem because it is possible to treat blood cells without having to treat the whole body. Although gene therapy for blood clotting disorders might not seem like the highest priority given the existence of treatment through administration of the protein, the need to improve treatment of blood clotting disorders is driven by a variety of clinical factors including the continuing potential for contamination of proteins isolated from blood products.

Gene Suppression Therapy

In the case of dominant diseases (remember the concept of the monkey wrench), there is already one good copy of the gene present in the cell and

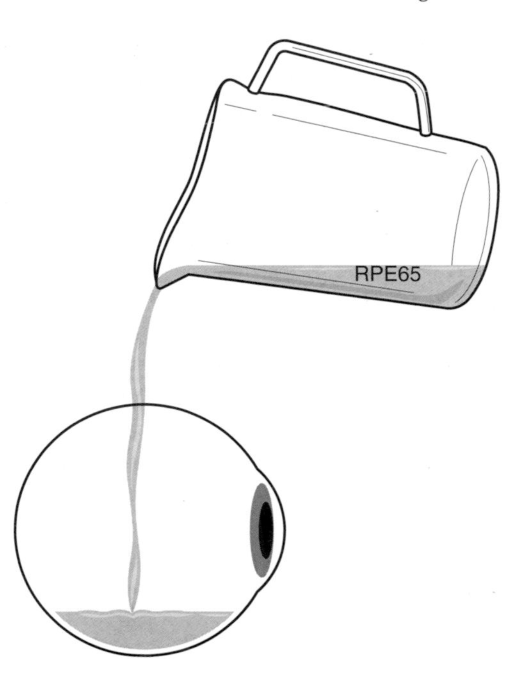

FIGURE 35.1 Gene replacement therapy adds back a functional copy of a gene in cases in which the disease results because defects in both copies of the gene result in loss of the cell's ability to carry on the functions normally handled by the product of that gene. In the case of the briard dog Lancelot, many good copies of the RPE65 gene were added into his eye in the vicinity of the retinal pigment epithelium cells that lacked the RPE65 protein activity that normally takes place there.

putting in more good copies of that gene may not help the situation. However, the situation can be helped by therapeutic approaches aimed at getting rid of the unwanted monkey wrench or the by-products of its misbehavior. So if the problem involves a toxic by-product, the use of gene therapy techniques to reduce the amount of a specific RNA can lead to reducing the amount of gene product being made. Scientists have successfully used an enzyme called a *ribozyme* (an RNA-based enzyme) to eliminate RNA from one gene while leaving other genes intact. In this case, the experiments were performed on cells growing in culture but these experiments showed that this approach can suppress the expression of an aberrant form of collagen that leads to osteogenesis imperfecta. In other approaches, small interfering RNA technology can reduce the amount of transcript coming from the offending gene by putting in a small RNA whose sequence is complementary to the sequence of the mRNA produced by the disease gene allele. Because of the sequence complementarity, the *small interfering RNA* can bind to the mutant transcript and get the cell to destroy the RNA coming from the disease gene (Figure 35.2). In some cases, it is conceivable that the small RNA can be designed so that the transcript from the disease allele will be destroyed at a higher rate than is the transcript from the normal allele, allowing for the possibility of reducing the amount of a monkey wrench while still allowing for some normal protein to carry out the normal function. Other strategies work at the level of the gene product, by adding in a gene whose product will chemically activate or inactivate the problem gene product.

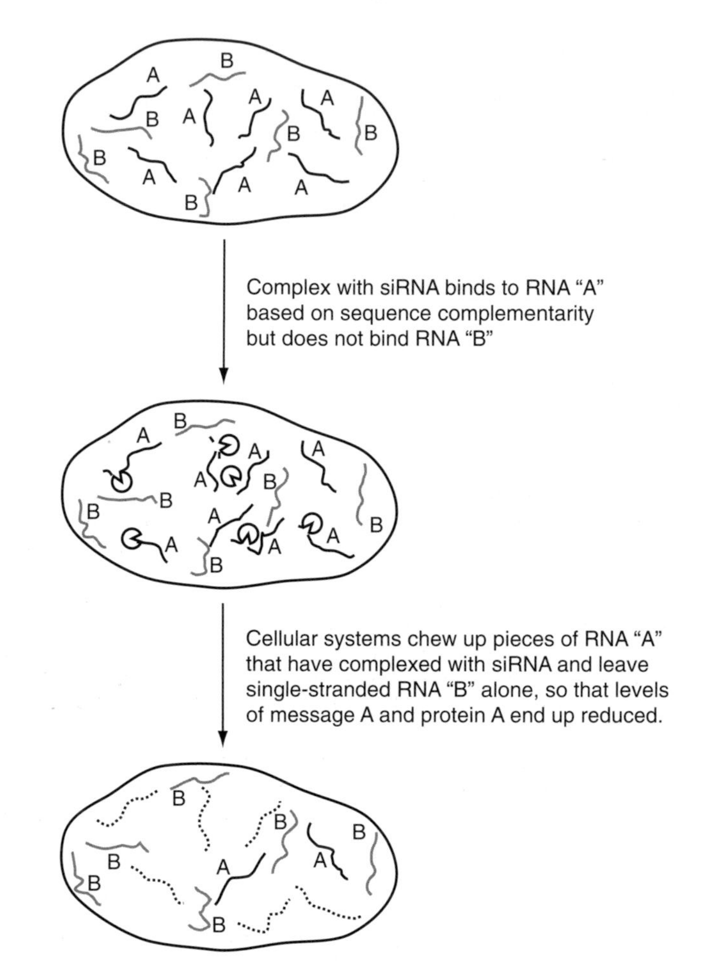

FIGURE 35.2 Gene suppression therapy. If the problem can best be solved by reducing the amount of a gene product (or its activity levels) a variety of technical approaches can be used. Small interfering RNAs can trick the cell into digesting and getting rid of RNA to which it binds and thus reduce the amount of the gene product in the cell. In some cases, it may be possible to design the small interfereing RNA so that it will selectively reduce the disease allele while leaving some of the normal allele present to produce the normal gene product. Someday it may be possible to fine-tune the use of promoter regions to control the level of expression of specific alleles of a gene.

End-Run Gene Therapy

In some cases, we may be trying to compensate for a problem that is too genetically complex to tackle at the point of the disease gene itself; in other cases, the trait may not even be genetic in its origins. In such cases, we may need to simply bypass the whole issue of which gene (or what else) is causing the disease, or even how many genes are involved, and target some other aspect of the disease pathology (Figure 35.3). Sometimes what is needed is to add a different gene that can supply a function that improves the body's ability to put up with the damage being caused, or that provides a mechanism to assist the body in recovering from damage that has been caused. An exciting example of this kind of "end-run" gene therapy that completely bypasses the

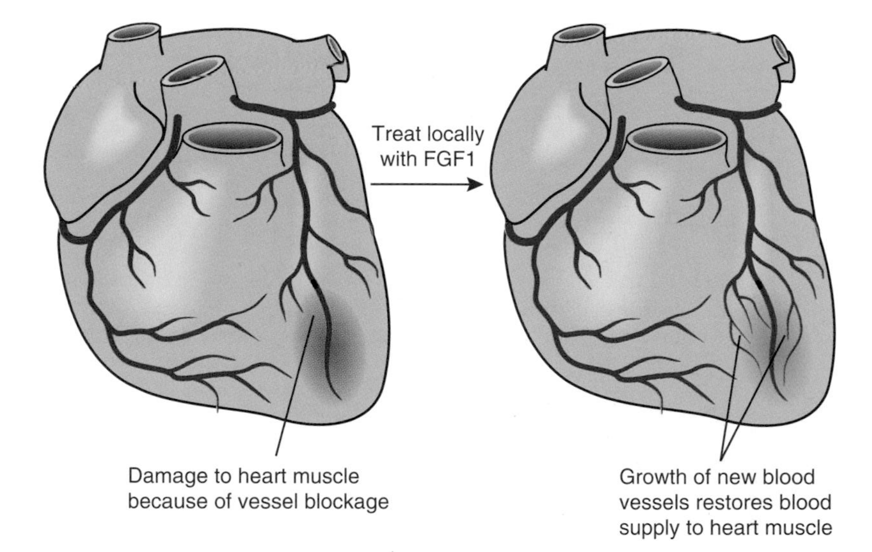

FIGURE 35.3 End-run gene therapy. In some cases, gene therapy can be used to treat a disease without going after the primary causes of the disease. In cardiovascular disease, gene therapy projects have encountered some interesting successes. One study used gene therapy to provide an APOE gene that produces an APOE protein that helps reduce "bad" cholesterol, resulting in disappearance of plaque attached to blood vessel walls. This artist's conception shows how uses of growth factor FGF1 can cause new blood vessel growth in a local area of the heart to restore blood supply to a region previously supplied by a blocked vessel. Successes of this kind have been seen in animal models, and some early human studies in gene therapy of cardiovascular diseases are ongoing.

original cause of the damage can be found in cardiovascular research. Researchers have shown that a growth factor called FGF1 can be used to stimulate local growth of new blood vessels to supply heart muscle in cases in which blockage is reducing the blood supply to the heart. In patients with blood vessel blockage, the combination of genetic and environmental factors contributing to blockage and damage to the heart muscle is likely complex and different for different individuals. Yet a single treatment approach that goes after the secondary problem of getting a blood supply to the heart could completely ignore the difference in underlying causes among the patients and successfully restore oxygenation of heart muscle.

Supplemental Gene Therapy

In some cases, tissues in the body simply need to be making more of something they already make. The item to be supplemented is not missing and the gene is not mutated. One of the situations in which this approach is being used is to get cells to make the proteins necessary for the formation of new bone material (Figure 35.4). In these cases, the patient does not have a defect in bone formation but rather has an injury of some kind that is more than his own body can heal. Gene therapy treatment of skin cells with *bone morphogenic protein* before placement of the cells into a region of bone erosion in periodontal disease can lead to formation of new bone in the region. Another approach places the gene therapy agents into a gel placed at the

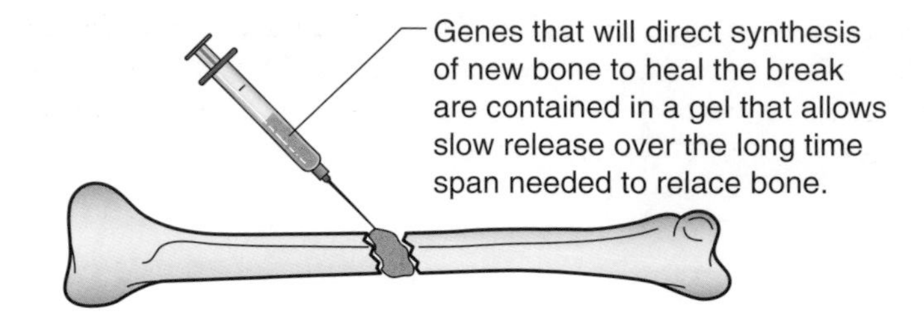

FIGURE 35.4 Gene supplementation therapy. An example of this strategy is the use of gene therapy agents that can induce cells in the bone to manufacture new bone. This is especially important in cases of severe fractures and fractures that do not heal well. By embedding the gene therapy agents in a gel at the site of the break, it is possible to have slow release of the DNA and gradual expression of the relevant genes over the extended time period needed for bone healing.

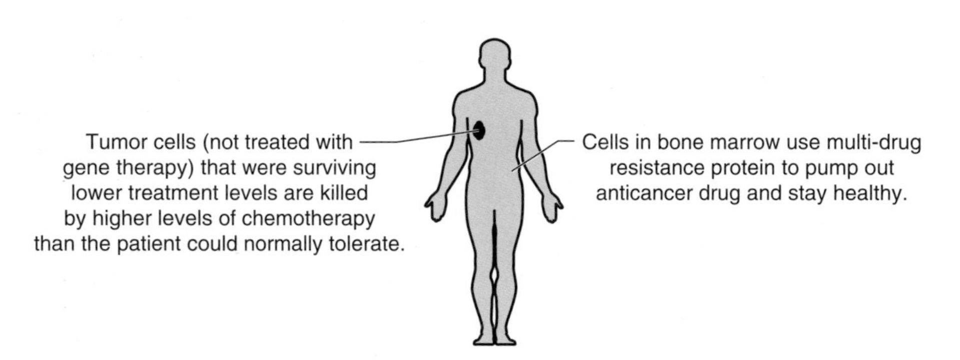

FIGURE 35.5 Supplemental gene therapy. Another use of supplemental gene therapy is to boost the ability of the patient to survive higher levels of chemotherapeutic agents being used to attack the tumor cells. A similar strategy might reduce the hazard of living or working in a contaminated environment.

point of a break in a bone, with gradual release over time resulting in sustained expression of the genes being used in the treatment.

Cleansing Therapy

During our lives, we suffer a variety of exposures that can be directly harmful or can increase our risk of things such as cancer. As we learn more about the normal mechanisms used by the body to eliminate toxic substances, more about biochemical pathways that can convert toxic substances into safe (or safer) substances, and more about ways to get compounds pumped out of cells or excreted from the body, we gain the potential to use gene therapy to protect us from exposure or to clean up our internal environments once we are exposed (Figure 35.5). An intriguing concept in cancer therapy is to put a gene into bone marrow cells that increases their resistance to the effects of anticancer drugs. This is important because some of the worst side effects in cancer treatment result because of key cells such as those in the bone marrow being damaged along with the cancer cells. If the bone marrow cells can resist the chemotherapeutic agents by pumping them out of the cell, the tumor cells could be attacked with higher doses of the drug than the patient would naturally be able to tolerate. Approaches of this kind have also been discussed as a preventive measure in cases in which occupational exposures to undesirable chemicals can be anticipated.

Magic Bullet Gene Therapy

In some cases, especially with cancer, what we really want is to be able to destroy specific cells while leaving the surrounding cells intact. An especially ingenious idea was developed by researchers who want to use their magic bullets to destroy malignant brain tumor cells while leaving the surrounding brain cells untouched. Brain cells are not usually thought of as growing or dividing, so a virus that infects only actively dividing cells can be used to deliver the gene therapy agent, which will only be taken up by the actively dividing tumor cells. Administration of an antiviral drug called *gancyclovir* will expose many of the brain cells to the gancylcovir, but it will specifically kill only those cells that have taken up the virus, so the tumor cells will die but surrounding tissues will remain intact (Figure 35.6). This concept, that cell death will occur only where two separate events coincide, resembles a process in current use in cancer treatment. In this process, low-level radiation administered from multiple different directions spares the surrounding tissues while killing only those cells present at the point where multiple radiation beams come together at the same place to result in a dose high enough to kill the cells.

None of these categories of gene therapy are categories that people in the field use when they talk about gene therapy. If you encounter a scientist at a cocktail party and start to talk to her about end-run gene therapy, you are likely to get a blank look because, as with our terms "absent essentials"

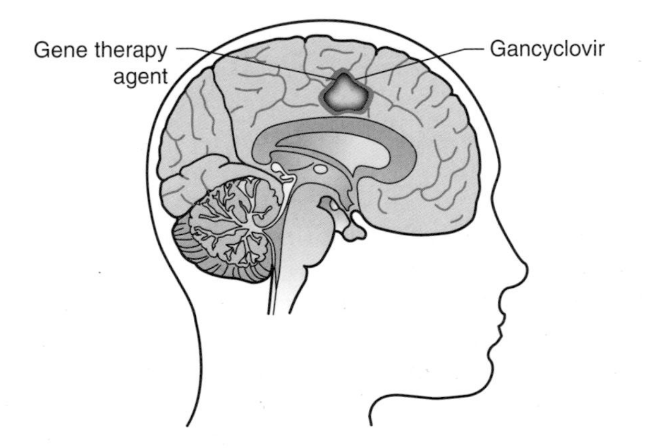

FIGURE 35.6 Magic bullet therapy. Many different strategies are being developed for being able to target therapy in such a way that only the tumor cells die while the normal cells remain healthy. One strategy is to use two different therapeutic agents that are each benign alone and kill cells only where both agents are present. Use of a gene therapy virus that can only infect dividing cells will tag tumor cells while sparing surrounding nondividing cells. A secondary treatment kills only tagged cells. This strategy would not work in many tissues of the body.

and "monkey wrenches," these are terms we use because we find them useful to encompass concepts that tell us about different gene therapy strategies. They are not terms that the other scientists use when they talk to each other about gene therapy.

HOW DO WE GET THE GENE INTO THE CELL?

One of the main gene therapy strategies involves the use of *viral vectors* that have been engineered to remove their disease-causing properties and to give them the ability to carry human genes along with their own DNA. There are several advantages to the use of such viruses. They greatly enhance the ability to get DNA into the cells, and they allow for mass production, which becomes important for quality control purposes. If you can make one gene therapy construct, prepare huge amounts of it, and do extensive testing on it, you can still have enough at the end to know that the batch you are using to treat people with is the batch that passed all of the safety testing. Another advantage is that sometimes it is possible to direct gene therapy into some cell types and away from other cell types, depending on which type of virus the vector is derived from. The biggest disadvantage is the tendency of the body to mount an immune reaction. In some cases, if what is desired is the destruction of a particular cell type, the use of vectors that invite an immune reaction may actually enhance the therapy, but it would not be useful in situations that call for repeated treatments because the virus would not be able to get to the target cell once the body develops an immune response to the virus.

There are other ways to get genes into cells that don't use viral vectors. In some cases, DNA copies of the gene can be packaged into *liposomes*, lipid packets that surround the DNA and help carry it into the cell. In other cases, direct injection of DNA can be carried out but would only get the DNA into a very limited set of cells. Some promising approaches call for removing cells from the body, carrying out the delivery of DNA in a cell culture dish and return of the cells to the body once the gene has been safely introduced.

Nuclear transfer technologies overlap with gene therapy approaches when we contemplate the following: If a nucleus is transferred from one of the patient's cells into a stem cell and the resulting cell is treated to repair the patient's genetic defect, the genetically altered nuclear-transfer cell can be brought through a set of differentiation steps to become the type of cell needed to cure the patient's problem. This complex set of events is not yet feasible for most problems we want to cure, but it is among the foreseeable approaches that we expect to be usable in our lifetime, if its development is allowed. Such approaches could potentially replace brain cells that have died in individuals with Huntington disease or Parkinson disease. Such approaches could potentially replace islet cells in the pancreas of a diabetic or even bone marrow in a patient with leukemia. Currently it looks as if nuclear transfer technology may end up blocked from development under the same laws that block cloning a human being. Yet this combination would overcome so many problems of therapeutic approaches currently being worked on, which each suffer from a technical flaw. Treatment of the cells outside of the body would avoid some of the problems with raising immune rejection of gene therapy agents.

Transfer of the patient's own nucleus into the cells will provide cells that are a perfect match for tissue transplantation antigens that are usually involved in transplant rejection when there is a mismatch. Growth of new tissue from stem cells will allow replacement of damaged tissue, a matter of great importance in quite a number of diseases in which death of cells takes the problem beyond the simple issue of fixing the genetic defect.

DO WE HAVE TO TREAT THE WHOLE BODY?

In many cases, we would actually prefer to avoid treating the whole body if we can, partly to help limit the immune reactions going on, partly because treating fewer cells means less risk of rare side effects, and partly because in some cases there will be cell types in the body that actually need to not be expressing the gene we are trying to get into one specific organ or cell type. Even for genes that are expressed throughout the body, disease resulting from a defect is often specific to a few organs or even one specific cell type. So we would prefer to limit the gene therapy agent very specifically to just the cells we want to treat.

One way to limit which cells end up with the gene therapy agent involves the selection of the type of gene therapy vector. Some vectors will treat only actively growing cells, whereas others will treat cells in any state of growth. Some vectors are derived from viruses that already have some specificity in terms of which cells in the human body they prefer. Thus, if we were wanting to treat the eye, we would want to ask whether we could build a vector from a virus known to infect the eye. If we wanted to treat cystic fibrosis, we would want to build our vector from a virus that infects lung cells. Now, in most cases, we do not have the luxury of starting with viruses that show absolute specificity for just the cell we want to target, but we can again do a least a bit of limiting where our treatment goes, depending on the vector we select.

In theory, we can also make our gene therapy construct contain not only the gene we want to express but also a promoter region next to it that will control where the gene will be expressed. So far, in studies of transgenic animals, all too often a promoter region placed artificially into a cell does not grant a pattern of gene expression identical to the natural pattern usually directed by that promoter. The promoter will give very specific expression in just one cell type when present in its natural location on the chromosome but the transgenic version of the promoter will not give expression in all cells of that type, and it may also give some expression in other cells when present as part of an external construct added to the cells. This may be happening in part because the *endogenous promoter* (the one that was there in the first place) is affected by other regional things, such as the structure of the chromosome in the local region, or other sequences present at some distance from the promoter itself. Thus, although use of a promoter specific to a rod cell may allow us to get something expressed in some of the rod cells, we do not yet have a way to get completeness and specificity in targeting simply through use of the promoter.

One strategy for treating only the cells you want to treat is to remove the target cells from the body, treat them in culture, and then return them to the body once they are fixed. This can be done with blood cells if you just need to end up with some treated cells and do not need to fix every single cell. However, if you need to treat every cell in the liver, this approach will not work.

Another strategy for limiting delivery is to deliver into a localized region. The eye and brain are expected to be good targets for gene therapy because of the ability to treat without exposing the rest of the body. In treating the liver, some efforts to limit delivery involve injecting into vessels that feed directly into the liver, but this results in much gene therapy agent ending up in other parts of the body. In treating bone, the clever use of a gel to hold the gene therapy agent in a localized position seems to work. In addition, one of the biggest current problems with gene therapy in humans is the tendency of the body to develop an immune reaction to the gene therapy agent or even the treated cells themselves. The eye and the brain are different from the rest of the body in that the normal immune surveillance of most of the body does not extend to the eye and brain. Thus some kinds of immune reactions that eliminate the gene therapy agent or kill the treated cells elsewhere in the body can potentially be avoided for eye and brain. On the other hand, it is a well-known phenomenon that the eye can end up being attacked by the immune system if it attracts too much attention from the immune system, so testing of gene therapy approaches to the eye and the brain have to be explored very carefully.

WHAT ARE THE BIGGEST PROBLEMS WITH GENE THERAPY?

One of the distressing early findings in gene therapy efforts was that, in many cases, positive results from treatment ended up being transient. In cases in which a repeat effort at treatment was tried, often the result the next time was much reduced or even nonexistent. This turns out to be the result of the action of the immune system that normally protects us from infection by bacteria and viruses. The immune system is very good at rapidly mounting a defense against such an infection. In many cases, the mechanism for getting the gene into the cell is an altered virus that can carry the gene into the cell. The use of cloning technology has allowed researchers to create gene therapy vectors that are derived from viruses that normally infect human cells but that have had the genes removed that make the virus able to cause disease. In place of the removed genes, the researchers place the gene that is due to be introduced into the cells. However, the protein coat that protects the viral DNA as it moves through the bloodstream and into the cell is the same protein coat that normally stimulates your *immune system* to attack the virus and keep you from becoming ill.

Researchers have worked to change those viral coat proteins to make them less visible to the immune system, but the ability of the body to eliminate viruses is rather amazing. The first time the gene therapy agent is administered, the viral particles avoid being eliminated but stimulate the beginnings of an immune response. If expression of the introduced gene drops off over the course of six months, the body is effectively well immunized against that virus by the time another attempt at treatment is made. The next time the same gene therapy construct is injected, the ability of the immune system to remove the virus may be so effective that none of the constructs will ever reach the cells that need to be treated.

The immune system may also recognize treated cells as foreign, which would result in the body trying to destroy the treated cells. This has turned out to be a problem in the treatment of Duchenne muscular dystrophy. During the 1990s, as frantic parents were asking how they could get their children into gene therapy trials, saying that they would be willing to do even very risky things rather than just sit and watch their children die, it was not possible to move ahead with trying gene therapy on the children because of concerns that the treatment not only would not cure them but actually would even make them worse if the immune system were to attack the treated cells. Researchers have been working at changing the gene therapy vectors to make them less likely to induce an immune response. There are some vectors that have fewer problems with creating an immune reaction, but they cannot carry the very large genes needed to treat traits such as Duchenne muscular dystrophy.

A very different issue is the problem of what happens to the human genome when a new gene is put in from the outside in such a way that the transgene integrates into chromosomal DNA. If the gene integrates into a region of *junk DNA* between genes, it might have little effect other than curing the cell's metabolic defect. However, if the transgene integrates into a gene in the chromosome and disrupts it, the consequences will depend on which gene gets disrupted. If it disrupts one copy of a gene encoding an enzyme involved in metabolism, it may have little effect or perhaps at the worst it will kill that one single cell.

However, if the gene therapy agent is delivered in vivo into a large number of existing cells in a human organ, each cell becomes a separate integration event. So even if a large number of cells receive transgenes that integrate safely between genes, it would take only one transgene integration into certain kinds of cancer genes to cause a problem. In the long run, optimal design of gene therapy will need to gain the ability to control where the transgene integrates, or at least to prevent certain kinds of integration events.

The first successful human gene therapy occurred in the treatment of adenosine deaminase (ADA) deficiency. Clearly, the need to protect young children from dying of immune deficiency offers a strong argument in favor of picking this as a treatment target. In addition, the fact that the treatment could be done to blood cells and that only some of the cells would need to be successfully treated makes ADA a potentially easier target for treatment than some other disorders. On the other hand, even though the image of the last days of Davy the bubble boy offer a compelling argument for treatment to those of us who lived with his story in the news as he grew up, treatments that help improve survival for these kids raise questions about the relative risks of the treatment vs. the risks of the disease if the ADA gene is not provided. Clearly, the balance of risks resulted in the development of the first successful human gene therapy, and families in several different countries have elected to enter their children into these treatment programs.

Unfortunately, after the initial rush of excitement at the news that the treatments were succeeding, two of the children developed leukemia-related disorders caused by the gene therapy treatment. All over the world, ADA gene therapy programs were stopped as an investigation began aimed at trying to find out why the treatment that fixed the ADA had then caused a new disorder. The answer is that the gene therapy vector being used does something that no one had expected: when it puts a copy of the ADA gene into the child's genome, it tends to do so by introducing it into the beginning sections of other genes rather than sticking it into the "garbage DNA" between the genes. In two cases, the ADA gene interjected itself into a gene that can cause leukemia-like illnesses. Understanding how this happens will help improve the safety of many gene therapy protocols being developed even as it helps improve ADA therapy. Clearly, one aspect of the problem is being caused by a gene that is part of the viral vector being used to insert the ADA gene. Thus it may be possible to avert this difficulty by simply altering the vector or using a different vector. With the new knowledge gained, it is hoped that scientists will be able to move ahead with a revised protocol that will be able to reduce such risks to the children being treated. Meanwhile, the two children are being treated for the illness caused by the gene therapy agent, and development of ADA gene therapy will continue.

SO WHO DO WE TREAT?

So who do we treat? The answer is a rather pragmatic one. We treat those with the most desperate need who are also "lucky" enough to have a trait involving a combination of genes and strategies that makes gene therapy development feasible. Some of the most desperate cases may not have gene therapy development going on because something about the needed therapy is not yet feasible. In some cases, traits that seem less terrible in their consequences may have ongoing gene therapy development because they seem as if they would be much easier situations to treat. By working on these more feasible cases, advances in gene therapy take place that end up being applicable to treatment of many other traits and potentially hasten the day when it will become feasible to treat something for which therapies are not currently workable.

Since many cancer genes are now known, since many other genes have been identified that could play a role in magic bullet strategies for treating cancer, and since cancer clearly rates a very high priority for the development of new treatments, we see that many of the gene therapy clinical trials listed at NIH involve cancer. Other traits for which various stages of gene therapy investigations are ongoing include things such as Alzheimer disease, heart disease, and diabetic neuropathy. A small number of simple, single-gene disorders, such as adenosine deaminase deficiency, can also be seen on the list. If we look at NIH funding for gene therapy research, we see a broader picture of what is going on. Many gene therapy research projects are spending time on further development of viral vectors and other aspects of delivery systems because that seems to be where many gene therapy efforts are stumbling. Many other projects have reached the stage of trying things out in an animal model system, but a variety of safety and efficacy questions usually have to be answered before the work can advance to the first phase of clinical trial testing in humans. We see the scientific world walking a fine line between the dangers of trying out untested treatments that could potentially harm or even kill vs. the clear danger of death for many patients if no further interventions are tried. However, additional factors determine which traits will advance to the point of testing treatments on human beings. We see cancer looming so large on the gene therapy trial list not only because of the desperate need for novel treatments, which is clearly there, but also because the kinds of strategies available to use in trying to kill cancer cells leave researchers dealing with very different problems from the folks who are trying to keep cells from dying.

One of the other key issues deals with the selection of specific individuals to participate in gene therapy trials. In some cases, the inclusion and exclusion criteria for a study may include only some stages of a disease. This can result in people who are excluded and don't understand why they can't join the study. In some cases, they may be excluded because the therapy that is currently feasible is not expected to work on their stage of the disease. In other cases, their stage of the disease may be considered to be at much higher risk of potential hazards of the study. Once again, the determining factors are often quite pragmatic. In the treatment of Huntington disease, one might imagine that the most advanced cases might offer the most compelling arguments for treatment, as well as the greatest opportunity to demonstrate gains from the therapy. However, when we look at the disease pathology we see that cells in the brain are dying, and quite frankly if the basis of the treatment is to put a neuroprotective gene into brain cells to help keep them alive, it is simply not going to work in cases where those cells are no longer there to be protected. Other strategies aimed at getting cells to grow and regenerate might work well in that same case, but that is irrelevant if the gene therapy trial you are wanting to join requires that you still have cells that you no longer have. So often simple issues of what can and cannot be made to work will override the seemingly dominant issue of who most needs the treatment.

In the first round of a clinical trial, when a small number of individuals are tested to determine whether the treatment is safe and perhaps to pin down the appropriate dosage, there are questions about who is most appropriate to treat. To many of us, it seems obvious that those with the most to lose without treatment and the most to gain from treatment would logically be the ones to take the risks in these early tests of safety. In a gene therapy study aimed at treating OTC deficiency, a bioethicist ruled that the most appropriate participants would not be infants at high risk of dying of OTC deficiency. Some might think it appropriate that those with the most to gain (or lose) would be the ones to take the largest risks. Instead it was decided that the pressures that the child's desperate health status place on the parents to put the child into the study, combined with the inability of the child to decide for himself if he is willing to be a study subject, seemed to make it ethically unacceptable to include these children in the first gene therapy tests. To some on the outside of the study, this seems surprising. Anyone participating in such studies is under great pressure to participate because of their health status, and anyone watching from the outside would wonder at how this supposed ethical dilemma is balanced against the ethical dilemma of expending a potentially lifesaving treatment on some unaffected individual who cannot benefit instead of offering it to an incredibly ill child who could potentially be saved if the therapy turned out to work.

Clearly, the complex situation in which a patient dying of cancer agrees to a treatment becomes incredibly more complex when the decision is being made by parents if the child cannot decide for himself. However, on some levels the issue is the same and the reasons in favor of participating are the same. A whole field of bioethics has grown to include very active consideration of very complicated situations such as these, and each new trait and treatment protocol seems to raise new questions about how to walk the fine line between treatment risk and disease risk, between informed consent and undue pressure to participate, or between death from nonintervention and the risk of death if there are unforeseen consequences of the intervention.

TRANSIENT VS. PERMANENT FIXES

Another issue that looms large is that of introduction of genes into the germ line. Currently, all approaches under development are aimed at a "somatic" fix for the problem, treating cells of the body while leaving the germ line untouched. In fact, one of the safety concerns that has to be dealt with in developing a gene therapy protocol is how to keep the treatment from reaching the germ line. Why not treat the germ line? That is, if we can successfully correct a genetic defect in a child's lungs or brain or liver, why not include that "fix" in the child's eggs or sperm so that they will not pass the problem along to their descendants?

We will offer this one example. If we put a copy of a gene permanently into a germ cell, we do so by putting a copy of it into one of the chromosomes. This is done by actually inserting its sequence of genetic letters into the long string of As, Cs, Gs, and Ts of one of the chromosomes. At this stage, we have little control over where the gene will integrate into the chromosome, and we have no expectation that the gene will cooperatively go sit down at the right spot to replace the endogenous copy of the gene. So if the gene we are working with is located on chromosome 1, and the new copy of the gene integrates into chromosome 4, we have a situation that works for that patient who was lacking any functional copies of the gene and now has a functional copy of the gene on chromosome 4. However, as soon as this person has children, there is no guarantee that the child who gets his defective copy on chromosome 1 will also get the copy of chromosome 4 that has the good copy (since the good copy did not integrate into both copies of chromosome 4). So he will pass along his defective copy of the gene, which may not even be a problem for his child if this is a recessive disease and he did not marry a carrier. At whatever point one of his descendants does marry a carrier, the good copy from grandpa's gene therapy–treated chromosome 4 may be nowhere in sight. In addition, as genes on different chromosomes assort independently we could find one of his great grandchildren suffering health effects from receiving a copy of the gene on chromosome 4 when she already has two other good copies of the gene. Thus the current technology that fails to target the location for the construct to integrate might not only fail to help affected members of the next generation, it might actively do harm to someone else who did not receive the disease gene but now instead has too many copies of the normal gene.

To go along with the fact that this treated copy of chromosome 4 might not segregate along with the disease allele in helpful ways, there is also the risk that introduction of the gene could disrupt one copy of a tumor suppressor gene or activate a proto-oncogene. This might not cause any problem to the originally treated patient, if one egg had one copy of this gene knocked out. However, if her children received the treated copy of chromosome 4, with the gene interjected into a tumor suppressor gene, they would only have one good copy of the cancer gene and would be subject to the elevated cancer risk that we see in inherited forms of cancer. Why? Because all that would be needed to turn a cell into a cancer cell would be one hit on that cancer gene, not the two hits that it would take for most people, and every cell in the body would be subject to the possibility of a single hit converting the cell into a cancer cell.

So does all of this mean that treatment targeted at the germ line should never take place? No, it just means that we do not yet know enough to do it safely. In fact, in the construction of transgenic animals, it is already possible to "knock in" a revised copy of a gene in place of the existing copy of the gene. If we can learn to control technologies of this kind so that gene therapy genes can be "knocked in" to the correct place on the chromosome, this will not only help prevent problems with the new gene segregating separately from the defect in subsequent generations, but it will also help keep the gene from integrating at some unwanted spot, such as into a cancer gene.

So we are not trying to raise some gloom-and-doom issue of gene therapy and cancer, nor to say that treating the germ line should never happen. What we are saying is that there are some aspects of gene therapy research that can move safely forward right now, and there are other aspects of gene therapy research that need more development. One of the things we find most heartening is the responsiveness of the scientific community to problems that have been encountered and the extent to which steps are being taken to try to limit ongoing steps to those that can be safely done, without shutting things down to such a slow pace that therapy for those in great need is needlessly delayed.

For now, we would settle for working towards the ability to treat each individual if that would truly fix their problems, while not tampering with their germ lines. Much work will be needed to make this possible, as well as many very creative ideas. Some of the smartest people in the world are working on the development of these technologies. Geneticists go after the right genes to use. Biochemists characterize the gene products and sort out the pathways. Molecular biologists design constructs that bring together human and viral DNA. Cell culture workers and animal model researchers test out preliminary ideas to pioneer new approaches and identify where improvements are needed. Virologists work to develop the vector systems for delivery of the genes. Immunologists study immune responses against the vectors. Biostatisticians evaluate the outcomes to help us tell whether something has actually worked, and help tell us how many subjects are needed in a study to be able to get a meaningful answer. Doctors work to improve systems for delivery of treatments and for monitoring the health status of treated individuals.

The other key to the whole process of developing gene therapy is the patients themselves, an often-unmentioned group that seem to us to be the real heroes in this story. In Chapter 36, we will tell you about one of these heroes and his dream that babies who die from OTC deficiency can be saved by gene therapy.

Section 10

FEARS, FAITH, AND FANTASIES

We began this book by discussing Mendel's laws and their biological bases. We have then systematically shown you case after case in which those laws were bent or broken, such as imprinting, linkage, sex linkage, nondisjunction, mutation, X-inactivation, quantitative traits, and complex traits. Even as the scientific field of genetics moves forward, we learn from studying and cherishing the exceptions that teach us what the real rules are. The last century has been an astonishing journey to those who have watched the infant field of genetics grow and move forward to the point at which we can contemplate effectively tinkering with our own heredity. The future offers us astonishing and wonderful possibilities if we manage to avoid the pitfalls waiting along with the wonders. As we contemplate those possibilities, let us take you into our last two chapters, which tackle the lessons to be learned from the actions of a modern medical hero and a discussion of our fears, faith, and fantasies about the near future of genetics.

HEROES AMONG US

36

"What's the worst that can happen to me? I die, and it's for the babies."

—*Jesse Gelsinger**

When Jesse Gelsinger was seventeen, he had a goal that was amazingly different from that of his high school classmates in Tucson, Arizona (Figure 36.1). Across North America, seniors in the spring of 1999 were struggling with many of the same choices that face high school students every year. They were worrying about whether they would graduate. They were talking about their chances of getting into college. They were applying for jobs, planning weddings, and deciding whether to enlist in the service. They planned major changes in their lives while struggling to see how those changes might change them. While the others waited for the freedoms they expected to come along with the term "adult," Jesse was waiting to turn eighteen because that was the magical age that would let him become a human subject in a gene therapy research project.

FIGURE 36.1 Jesse Gelsinger was an idealist who set out to do something that he thought would make a difference in the world. His dream of helping make OTC gene therapy a reality was a dream that extended beyond his own self-interest into the realm of the heroic. His sacrifice puts a name and face on the actions of more than 4000 heroes who have taken similar steps to make gene therapy a reality. (Courtesy of Jon Wolf Photography).

* The photograph of Jessie Gelsinger in front of the Rocky statue courtesy of Mickie Gelsinger.

How did this young man come to such an extraordinary, selfless view at a time when many his age are focused on themselves and the complex transitions going on in their lives? Some of the answer comes from Jesse's own medical history. Jesse suffered from a mild form of the same recessive disease, *ornithine transcarbamylase (OTC)* deficiency, that kills the severely affected babies he wanted to help. Jesse could identify with the danger to these infants, even though he had never met them. Jesse knew from firsthand experience that there is no cure.

JESSE'S TALE

Jesse's story begins about sixteen years before he died. At the age of two, Jesse Gelsinger was first diagnosed with a disease called OTC deficiency. Like Marlaina, whose story began Chapter 2, Jesse was unable to make enough of the OTC protein, which is part of the urea cycle that is used to remove excess nitrogen that enters our bodies when we consume proteins. If the urea cycle doesn't work, the nitrogen from the proteins accumulates in the form of ammonia that builds up in the blood and brain. High levels of ammonia can cause damage, especially to cells in the nervous system, and permanent brain damage can result. Since ammonia production is especially high in a body that is metabolizing meat and other proteins that have been consumed, treatment includes both a low-protein diet and medications that help keep ammonia levels from getting too high.

OTC is a severe disease resulting from a genetic defect. The one baby out of 25,000 who is born with OTC deficiency can usually be expected to go into a coma within days of birth. Even with medical help, many OTC children suffer permanent brain damage, and it is not uncommon for them to die young. Many die before they are one month old, and almost half die before the age of five years. The ones who die the earliest often have the most severe cases of the disease.

However, in some individuals like Jesse, the disease is less severe. Only some of their cells carry the genetic defect that causes this disease, so they are able to make some of the enzyme but not enough for their health to be completely normal. Most of the time, Jesse's low-protein diet, supplemented by thirty-two pills per day, kept him healthy enough to live a reasonably normal life. Although this may sound simple—take some pills, don't eat meat—OTC deficiency is a serious and potentially life-threatening illness. Pills and diet don't always keep things under control. When Jesse was seventeen, his ammonia levels got to be too high and he ended up in the hospital in a coma. He recovered and went back to school, once more the normal high school kid waiting to turn eighteen, waiting to graduate, waiting to figure out what he was going to do with his life.

So what was Jesse looking for when he decided to participate in a gene therapy trial? You might imagine that he was looking for a chance to rid himself of pills and maybe get to eat a steak sometimes. Consider this: the first stage of the gene therapy trial, the stage in which he participated, was not offering to cure him, only to engage his help in figuring out what therapeutic doses might be safe and useful. So perhaps he harbored some dreams of gene therapy that could apply to his own situation, but Jesse and other participants in the OTC gene therapy trial appear to have gone into it with rather more generous motives and less expectation of some immediate reward for themselves.

Looking back at him now, he seems a rather idealistic young man. According to his father, Jesse wanted to be a hero, to make a difference in this world through what he did. His OTC deficiency was well controlled, but he knew that babies were being born who would die and had died for lack of the OTC enzyme. He saw participation in the OTC gene therapy trial as a way that he could help, a way that he could make a difference in the world. According to one of his friends, Jesse said, "What's the worst that can happen to me? I die, and it's for the babies."

Jesse first heard about the OTC gene therapy trial when he was seventeen but was told that he had to wait until he turned eighteen to participate. The day he turned eighteen, he and his family left on a trip to Pennsylvania to check it out. He was told about the study and had to go through some tests before being accepted as a participant to ensure that he qualified for the study. Before he was started on the actual protocol, he also went through a process called informed consent. This type of informed consent process, which is required anytime a human being participates in research, was meant to tell Jesse about the possible risks and the potential benefits of the gene therapy protocol. Informed consent is supposed to provide enough information that the study participant can make a fully informed decision after balancing the potential gains against the potential risks.

On September 13, 1999, Jesse was sedated and a dose of the gene therapy agent was administered. By that evening, he was sick to his stomach. By the next morning, he was disoriented and his eyes were yellow with jaundice. By the afternoon of September 14, he had slipped into a coma with ammonia levels more than ten times normal. He was put on dialysis and at first seemed to be improving. By September 16, his kidneys had quit functioning and his other organs were failing. As his lungs went into respiratory distress, efforts were made to increase the amount of oxygen being transferred to his blood. By the next morning, tests showed that his brain was dead. A minister and relatives arrived for a brief service at his bedside before the life support machinery was shut off. He was pronounced dead at 2:30 in the afternoon on September 17, 1999, four days after he entered the hospital to begin his participation in the research protocol.

THE IMMEDIATE AFTERMATH

You can imagine the shock of his relatives who expected Jesse to spend a few days in Pennsylvania and then return to Arizona to get on with his life. You can imagine the shock of his father, who had read the consent forms along with Jesse and thought that his son was not heading into something all that risky. However, it is a fact that human beings can differ significantly in their reaction to any treatment, even known, well-tested treatments that have been through government approval processes and are used to treat large numbers of patients all over the world. Thus a November article in the *New York Times* indicates that Paul Gelsinger did not originally blame the doctors who had administered the gene therapy treatment or cared for his son during his final four days. In fact, the doctors expressed great distress over what had happened. The doctor that had administered the fatal dose flew to Arizona to join Jesse's friends and relatives as they scattered his ashes, poignantly carried in Jesse's own medicine bottles, from a rocky outcropping overlooking a deep gorge.

The initial announcement of Jesse's death made for headlines that shocked the public. The reaction within the scientific community can only be described as one of horror.

Everywhere we looked there were the same assurances that no one had done anything wrong. Then, as the normal process of investigating any such death proceeded, the facts that emerged sent genetics researchers around the country scurrying to confer with their colleagues with murmurs of, "No, that can't be true!" and, "Surely they didn't!" Gradually, increasing numbers of articles and editorials suggested that Jesse's death had been avoidable. They suggested that Jesse had not been properly informed of the risks of the procedure he underwent. They indicated that the approved version of the informed consent form had been changed to remove mention of animals and human subjects who had previously experienced problems from similar treatments. There was talk that Jesse's ammonia levels, just before the treatment trial began, were above the levels that were allowed for participants in this protocol.

You cannot imagine the distress this caused in the scientific community, where everyone seemed to be trying to wish the clock rolled back so this event could be undone. You cannot imagine the confusion this caused among scientists who had looked to the doctors involved as being the best of the best, the ones that others learned from and looked to for new ideas and breakthroughs. The head of this gene therapy trial was known as one of the top gene therapy researchers in the world, one of the smartest pioneers pushing into new territory into which others followed.

WHAT WENT WRONG?

As the investigation into Jesse's death continued, research institutions around the country started investigations into their own local research programs, asking not only about gene therapy trials going on but also about the rest of the research involving human subjects. Some other research programs, and in some cases whole institutions, had to stop their research and answer to charges that mistakes had been made, especially with regard to notifying proper authorities about cases in which "adverse events" had occurred. Although no other cases like Jesse's turned up during these investigations, heightened awareness among researchers led to changes in procedures aimed at keeping anything like this from happening again. While it became clear that the researchers at the University of Pennsylvania were not the only ones who failed of perfection in their actions and foresight, it also became clear that many people had participated in gene therapy trials safely and effectively. This offers real hope for what can be accomplished in the future.

There have been more than 4000 gene therapy trial participants from more than 300 studies. At the writing of this book, Jesse is the only one who has died as a result of a gene therapy protocol, although his death was not the only adverse event that has happened in the history of gene therapy. He was the youngest of the eighteen patients who were treated in this OTC gene therapy trial. Amidst the thousands of gene therapy trial participants who survived and went home, amidst the seventeen other individuals who were treated as part of this OTC gene therapy trial and went home, the question remains to this day: Why was Jesse the one individual who died? Although some insights into this problem have been gained, much remains to be learned about the way Jesse's body reacted to the gene therapy treatment.

Reports in the press, discussions in the scientific literature, and testimony before committees have pointed at many different possible problems with what happened. Some of the issues that have been raised deal with procedures—information reporting procedures, informed consent procedures, processes for avoiding conflict of interest. Some issues that have been raised have been ethical—who really should be participating in phase one clinical trials of gene therapy? Other issues have been medical and biological issues regarding the disease that results from OTC deficiency and the ways in which the human body responds to the introduction of gene therapy vectors.

However, we are left with this: even if the research team really did made every mistake of which they have been accused, would the prediction have been that Jesse would die? No. Realistically, who would have done the procedure if they had thought that Jesse was actually at any serious risk of dying?

Even if they made mistakes, how can we learn from them? If they were mistakes in judgment, we are still left with profound questions on a basic, biological, medical level: what killed Jesse Gelsinger, and why did he die when others who received the same treatment did not? We know that he died from multiple organ failure. We know that this happened in response to the infusion of the gene therapy vector. It would appear that there might have been some involvement of the idiosyncratic process by which a body with an OTC defect responds to infection in a way that leads to ammonia buildup. However, we are not going to be able to offer any single simple answer to this. He died of massive multiple organ failure, but how many underlying genetic factors combined with the gene therapy agent, the existence of elevated ammonia levels in his blood, and any of the errors that were made?

REACTIONS TO SUCH A LOSS

How has Jesse's family reacted, beyond their grief and their efforts to understand what happened? Not surprisingly, once they realized that what happened to Jesse might not have been some uncontrollable biological luck of the draw but at least partly the result of mistakes in judgment or mistakes in the system of fail-safes, a lawsuit was filed against the doctors and the university. Since the suit was settled out of court, we cannot comment on the outcome of it, other than to indicate that Jesse's father plans to apply money from the settlement towards improving oversight of gene therapy research. Interestingly, this represents one of the most effective ways to further Jesse's dream of making gene therapy for OTC into a reality. The more safely and effectively the research is conducted, the more likely it is to continue and to eventually succeed in Jesse's goal of providing a cure, or at least an effective treatment. None of this answers the big question: When will work on OTC gene therapy be resumed, and who will take on the project?

What will happen to the other children who currently struggle to survive with OTC? Do they and their families have any options? The National Urea Cycle Disorders Foundation website indicates that there is no current successful treatment. About half of the affected children will continue dying, and many others will suffer irreparable harm. After all that has happened, many still regard gene therapy as their greatest hope and they, like many others, see Jesse as a hero.

Jesse is not the only hero in this story. Seventeen other people took part in the OTC gene therapy trial, and they are also heroes. Like Jesse, they participated with the hope of making a difference in the world and with the specific hope of helping others who will be injured or die for lack of this one protein. The goal they were headed towards has not changed. The importance of making OTC gene therapy a reality has not gone away. As is true for so many disorders for which there are no cures available, the absence of other better answers continues to drive the need to solve the problems with gene therapy. Genetics offers hope for those who have little other direction to turn for such hope.

CAN WE PREVENT A SECOND TRAGEDY IN THIS TALE?

In the aftermath of this tragedy, we see that there was a second casualty, the death of the OTC gene therapy program that Jesse had believed in. The gene therapy institute in Pennsylvania has stopped working with human subjects and returned to work on animal model systems. There are some who feel relieved that they are no longer working with human subjects. Gene therapy programs are still working to develop cures to many other diseases, even some disorders involving other steps in the urea cycle, but for now the next stage in development of OTC gene therapy is not taking place. What will become of Jesse's dream of advancing OTC gene therapy to the point of being clinically useful?

Many people believe in the importance of gene therapy. They have demonstrated this belief by participating in gene therapy trials aimed at advancing the technology to the point of being able to cure or control a large number of different disorders. Many heroic individuals set out to make a difference for those who can't be helped by current medical practices. As is so often the case when heroism abounds, Jesse has emerged as the hero who symbolizes many other heroes because he was the one who died during the research protocol.

These days, it is surprising how often just mentioning Jesse's first name results in recognition of who we are talking about. It is reported that Jesse said he wanted to be a hero. He wanted to make a difference. And he did. He is a hero, a recognized hero among many, many unrecognized heroes, and his actions have made a difference.

We are telling Jesse's story because Jesse is a hero of modern times. He offers us lessons about ourselves and about the genetic foundations on which we are based. He came along in a time of peace and plenty and lengthening life span. He came along at a time when many of us take for granted the ability of modern medicine to do things that would have looked like miracles a century ago. He came along at a time when we expect a trip to the doctor to fix most things that can go wrong. He came along at a time when some people still suffer from conditions for which the American Medical Association and the Royal College of Surgeons offer no answers. He came along at a time when the cutting edge of medical research offered a chance to help those who cannot currently be cured. He saw the possibility of turning this generation's set of unreachable miracles into everyday reality, and he wanted to help.

We are telling Jesse's story because Jesse's dream points towards the greatest things that harvesting the human genome will offer us. His dream points in the direction of hope for those who currently lack hope. It would be so easy to turn this story into one of despair and anger. These emotions have haunted so many who know Jesse's tale, even those of us who never knew him in person; but to let Jesse's story lead us to cry gloom and doom would be to dismiss Jesse's dream, and the direction he was heading as he waited to turn eighteen. To let Jesse's death cause the death of the treatment program he dreamed of helping to create would result in a second great tragedy.

Jesse and those he wanted to help are the humanity in the Human Genome Project. Jesse's dream is the dream of all of us who spend our lives sifting through the genome in an effort to unlock its secrets, a dream that the astonishing wonders arising from the Human Genome Project and research going on all over the world can save lives and bring hope. However, Jesse's dream is lost unless we can learn the right lessons from what happened to Jesse, not that OTC gene therapy should be stopped but rather that there is a great need to improve approaches to OTC gene therapy, and gene therapy in general, as well as the processes of safeguarding research participants. If that can be done, eventually we will arrive at that seemingly magical moment when babies born with a terminal illness can be treated and sent home to grow up along with the other children who were born healthy.

Will you make me some magic with your own two hands?
Could you build an emerald city with these grains of sand?
—Jim Steinman

FEARS, FAITH, AND FANTASIES

37

"Once, human beings were as children, needing simple tales and naïve visions of pure truth. But in recent generations the Great Creator has been letting us pick up his tools and unroll blueprints, like apprentices preparing to work on our own. For some reason He's permitted us to learn the fundamental rules of nature and start tinkering with His craft. That's a fact as potent as any revelation. Oh, it's a heady thing, this apprenticeship and the powers that go with it. Perhaps in the long run, it will turn out to be a good thing. But that doesn't make us all-knowing."

—*David Brin*[1]

As we come to the end of this book, we offer you three different perspectives on human genetics, with a special concern for some of the scientific and ethical pitfalls that are ready to waylay some of the best intentions. Our backdrop for this discussion will be three hypothetical tales that effectively put the same fictitious young man into a time machine and transport him to three different eras—the beginning of the twentieth century, the beginning of the twenty-first century, and the beginning of the twenty-second century. While we have limited patience for some of what we see in the past and limited ability to project into the future, we hope that this discussion will let you follow the ethical issues and technical capabilities along the road from the past to the future. We stand at a crossroads where public policy decisions in the next few decades will decide the fates of many people who urgently need the help that genetic technologies could provide them, and it will be critical that our society as a whole find ways to proceed that avoid the major errors of the past while optimizing our ability to do the most good wherever possible.

The title of this chapter derives from our very mixed reactions to what we see along this road of genetic progress. The term "fears" reflects our concerns that, even as our society proceeds with the best of intentions, the course along the future road could still be influenced by attitudes similar to those that caused grievous errors and injustices in the no-so-distant past. The term "faith" reflects our belief that most of those involved in trying to sort out what can be done and what should be done have genuine interests in keeping the road running in an ethical direction, and that the current trends in scientific culture demand high standards of scientific practice that should help guard against some of the errors of the past. The term "fantasies" applies to the many truly wonderful possibilities that loom in such a near future that we expect some of them to become real within the next few decades. This combination—fears, faith, and fantasies—carries us forward, mindful of the historical mistakes to be avoided, earnest in wishes to find ways to make genetic technologies work for good, and excited at the possibilities unfolding as some of the fantasies become realities.

[1] From *Kiln People* by David Brin. 2002, Tor Books (New York).

FEARS

"Those who cannot remember the past are condemned to repeat it."
—*George Santayana*

It is 1931, and Allen's life is in an upheaval that he can barely understand. He had spent much of his early childhood at home, receiving little education because the local school could not cope with someone the teachers had declared an imbecile and his uneducated parents had no idea how to help teach someone who could not talk and rarely seemed to understand anything complex that was discussed. With little knowledge of the world outside of his parents' small farm, Allen has now found himself in trouble with the law. He was arrested for assault after he took a swing at the leader of a group of older boys who were taunting him (as they often did). Now, he finds himself standing in fearful confusion before a judge, and he cannot understand that this judge is making pronouncements of profound and far-reaching implications for the rest of the course of Allen's life. The judge feels a need to protect society from this subnormal and obviously violent individual. Allen offers no response as the judge informs him that he is to be placed in an institution (for his own good) and sterilized (for the good of humanity). When he is led away to face this terrible dual fate, he finally realizes that something is not right, but it is too late. His efforts to squirm free of the bailiffs simply confirm to the onlookers that the judge was right. In 1991, looking back on this tale from a perspective he only achieved many years after the event, Allen can only shake his head in disgust and residual anger and a very intelligent understanding of what had happened so long ago. Allen is not an "imbecile." Allen is deaf. As a mixed blessing, Allen's deafness was diagnosed by a doctor that he met in the hospital where he was forcibly sterilized. Unfortunately, Allen spent far too long in an institution before his parents figured out how to get him released and into an education program for the deaf. With a grace born of long practice and passionate conviction, his hands dance through the explanation that his form of deafness is recessive and so there was little expectation that he would have passed it along to his children. He pauses before adding that if he had fathered any children, he would have preferred that they were deaf because then they could have grown up in the deaf community and not been a part of the culture that could do the terrible things to people that had been done to him.

When Scott was about sixteen, he came across a section on marriage laws in an almanac and found that about seven states still had laws proscribing epileptics from marrying. This left a profound impression on Scott because he has epilepsy. These laws were fossils of the *American eugenics movement.* They were based on a now-discredited idea that epilepsy is associated with insanity and imbecility and that epileptics are dangerous. These rather odd views are themselves probably remnants of medieval beliefs suggesting that seizures were an exposition of demonic possession. The science was bad, but the laws were made anyway, something that should serve as a caution to all lawmakers who proceed to make laws on subjects they do not really understand.

We all stand on the threshold of a revolution that will change the lives of our species forever. We have learned how to assess at least some of the information contained in our genes, and we will continue getting better at this particular trick, much better. We are also rapidly developing the skills necessary to modify our genomes, surely in our somatic cells, possibly in our germ lines as well. There are truly many potential benefits of this technology, but there are also some major potential pitfalls. As an example, we want to digress into a bit of a history lesson. Specifically, we want to talk about a subject called the *science of eugenics* (Box 37.1). *Eugenics* is a term for the selective breeding of the human population for purposes of "improving the quality" of the human race.

BOX 37.1 THE AMERICAN EUGENICS MOVEMENT

A great deal of historical information exists to offer details of the eugenics movement's scientific investigations (and their flaws), the legal cases (and their impact), and the sociological arguments (and their underlying prejudices). Eventually, the theories and advisements of this movement made their way to Nazi Germany, where they contributed to a eugenics movement of even greater and more horrifying scale than what happened in the United States. A group at the Cold Spring Harbor Laboratories in New York has established an archive project to assemble information on the history of the American eugenics movement and its inherent flaws.[2] For some people in the United States, it can be quite an eye-opening experience to realize that a movement based on such a foul combination of error, fraud, and prejudice could have been allowed to develop so much power over the lives of the people of this country. For some people, the eye-opener is that this did not happen in some remote past or distant land but rather right here in our own sociological backyard.

Because all of us will watch in the near future as decisions are made about a variety of potential uses of information arising from genomic science, we want to talk about government-sponsored eugenics programs in the not-too-distant past, programs that made laws that deprived people of their ability to be parents. We're going to talk about governments that made laws about sterilization, incarceration, and even about people's right to be alive. The eugenics programs we are talking about happened here in the United States! It wasn't monsters who carried these programs out. It was done by people considered to be good, people who were seen as pillars of their community. They supposedly did these things in the name of God and the public good, and that makes it very scary.

In 1903 an organization called the American Breeders Association was formed. The association set out to bring Mendelian ideas to the United States. Much of what they did dealt with horses and other animals, but they also began to follow up some rather theoretical work on human breeding begun by a man named Galton in England. Shortly after the formation of the American Breeders Association, the American eugenics movement began. This movement evolved under the guidance of a federally funded agency located at Cold Spring Harbor, New York, called the eugenics record office. It was run by a Harvard professor named Charles Davenport. Davenport was considered to be one of the great liberal minds of his day.

The agency was interested in collecting data about the human population. They trained people to go out into the rest of the country and find pedigrees so they could gather evidence about how certain human traits or "diseases" were transmitted. The government was spending a lot of money for record keepers for finding appropriate families.

One particular bit of data they collected that stands out was two pedigrees that showed that "seafaringness" is an X-linked trait. But consider this: how

[2] We recommend a visit to www.eugenicsarchive.org/eugenics/.

could it not appear to be X linked at a time when virtually all sailors were men? There were pedigrees for idiocy, silliness, nomadism (the love of wandering), vagrancy, criminality, and more. Remember, people were paid to go out and create these pedigrees. They went to places such as prisons and mental institutions. There was purportedly a great deal of fraud and serious error involved.

Fueled by such information, the American eugenics movement quickly built up real steam. Eugenics booths and education programs were set up in county fairs and schools all over the country. Some brochures for the movement urged people to "wipe out idiocy, insanity, imbecility, epilepsy, and create a race of human thoroughbreds such as the world has never seen." These views on heredity seemed to fit with general common sense. People knew that some of these traits or behaviors did tend to run in families and that certain traits tended to occur in some families moreso than in others. In this sense, the American eugenics office was providing the so-called "evidence" to buttress well-established prejudices.

Unfortunately, much of these data were used as justification for new laws. Several states passed laws prohibiting people with certain traits from marrying. It seemed reasonable to many that one way to make a better society was to simply prevent marriages that were predicted to produce certain categories of "defective" progeny; state governments passed laws that idiots, criminals, and epileptics couldn't get married. In fact, related laws started being applied to other groups, including girls who had gotten pregnant out of wedlock. Eugenics-based marriage laws quickly became the norm in our growing country.

Buttressed by the eugenics movement, and fueled by prevailing racial prejudices, thirty-four states also passed laws making marriage illegal between people of different races, the so-called *antimiscegenation laws.* People worried and talked openly about the so-called dangers of "racial degeneracy." These things were not just happening in Nazi Germany; they were happening here.

Soon the laws would go beyond regulating marriage. In 1907, Indiana passed the first law requiring involuntary sterilization. It mandated that people with certain traits, including epilepsy, be sterilized; soon other states would follow suit with similar laws. By the 1930s, more than thirty states had passed laws of mandatory sterilization for an incredibly large number of traits. Between the 1920s and the 1940s, it is estimated that 30,000 to 35,000 people were sterilized involuntarily. This number is very likely to be a very gross *underestimate* because not all cases were reported. The people in the eugenics movement were deadly serious and were backing up their politics with the surgeon's scalpel.

Things got even worse around the 1920s and 1930s. Life got hard for people, and prosperity's infinite view was changing. Immigration was on the rise from all parts of the world. People here worried that some of these new arrivals were genetically inferior and that these genetically inferior people were bringing undesirable genetic traits into this country. Much of the testimony that helped the passage of the Immigration Restriction Act of 1934 was centered on arguments that high fractions of immigrants from certain countries were "feebleminded." Indeed, a progenitor of the IQ test was administered to newly arriving immigrants and suggested an enormous frequency of feeblemindedness among people from certain countries. People felt that such individuals should be denied entrance into this country because they were genetically inferior. However, the curious thing about these IQ tests was that they were administered in *English* to people who had just gotten to the United States and spoke not a word in English. Still, these data served as the basis for one of the most restrictive immigration acts in history, which stayed on the books until 1960.

As painfully crazy as all of this must seem to you, it is important to realize that these laws had wide backing throughout American society, even at the highest levels. A landmark case on involuntary sterilization went to the United States Supreme Court in 1924. The case was decided by none other than Oliver Wendell Holmes, known then as an independent vital force for social reform. Holmes was known as a kind, and intelligent man in many ways, but let us quote from the decision in which Holmes and his court upheld the rights of states to sterilize supposedly genetically inferior individuals against their will:

"We have seen more than once that the public welfare may call upon the best citizens for their lives. It would be strange then, if it could not call upon those of us who already sap the strength of the state for these lesser sacrifices, often not to be felt as such by those concerned, in order to prevent our being swamped with incompetence. It is better for all the world if instead of waiting to execute the degenerate offspring for crime or to let them starve from imbecility, societies can prevent the genetically unfit to continuing their own kind. The principles that sustain the compulsory vaccination are broad enough to cover the cutting of fallopian tubes. If we are willing to ask the best of us that they lay down their lives in the defense of their country, in the defense of liberty, and in the defense of freedom, why can not we ask from the weakest of us to voluntarily deprive themselves of the right to reproduce children?"

Again, these laws were passed and supported not by obvious monsters, but by people considered by those around them to be very good people. They did what they did in the name of right. That's what scares us—that these good people, acting for what they considered the good of others, with the full support of the church and of the state, were able to do so much evil to so many with so few voices being raised. It is easy to recognize and prevent evil when it comes with a gruesome countenance, a threatening manner, or overt ill will, but it slips by all too easily when concealed beneath civilized manners, a quiet demeanor, and gracious speech.

In the end, the wave of immigration changed this society. People finally began to realize, after twenty or thirty years, that immigrants from various parts of the world are more or less the same everywhere and that everyone had an enormous amount to contribute here. Common experience belied the messages of the eugenics movement.

More importantly, real genetics was blooming as a science. People were getting an idea of what genetics could and could not do. Good scientists were trying to do serious human genetics. They were discovering that nothing was as simple as the eugenics people said, and they were also discovering that the eugenics movement pedigree data could not be replicated. So by the end of World War II, most of the activities of the eugenics movement went away. What is left are reports in the history books, some fossilized laws about marriage in a handful of states, and far too many people who can remember what was done to them against their will.

FAITH

"Although the nightmare should be over, now some of the terrors are still intact."
—Jim Steinman

It is 2006. Alan is six months old, and his recent diagnosis of deafness has led to some very heated debates among his relatives, none of whom are deaf. Their debates about how to raise him are finally resolved with the decision to hire tutors and teach him to be bilingual, communicating via sign language and also through lip-reading and speaking. Alan's mother has started learning sign language and has found a nearby deaf family whose children include a deaf toddler that she hopes might become a playmate for her son as he grows up. Since the genetic defect causing Alan's deafness has been identified, relatives want to engage in the debate about whether there should be genetic testing during the next pregnancy, but they are gently edged out of a decision that Alan's parents feel falls to them alone. One grandfather has been angry ever since Alan was born. His calls and visits focus on his demands that the family enroll their child in an upcoming gene therapy trial that he has heard about. Alan's mother insists that such treatments still carry great enough risks that participation should only result from Alan deciding to participate some day when he is old enough to decide for himself. The grandfather tries begging, bribing, and even threats as he makes it clear that he himself would gladly take such risks, preferring death to a marginalized existence in society. Alan's mother counters that her son's existence is in no way marginal, and that he has every bit as much potential to live a good life and become a productive member of society as the rest of the babies in their town. This whole scene actually epitomizes medical genetics at the beginning of the twenty-first century. We can diagnose things that we cannot yet fix. There are a wealth of different ways we can help even if we cannot undo whatever event caused the problem in the first place. If we compare this story to Allen's tale a century ago, we see major advances on ethical and technical levels. Even if we can't do everything we would like to, we can treat Alan with respect and offer him options that will let him fully develop his abilities. We see the trend, not always practiced, towards thinking that the individual should be the final source of decisions about his own welfare. However, we also see that not everyone has dispensed with the less generous views of the past.

The eugenics movement grew out of a combination of bad science and prejudices that led people to buy into and act on that bad science because it was telling them what they wanted to hear. What protects us against more of the same?

One of the best current protections we have against the production of bad science is the peer review system, by which any piece of science that gets published in a reputable journal is subjected to detailed (and sometimes, to the authors, painful) scrutiny by others in the field. If major flaws are found, the paper is likely to be rejected. These reviewers are charged with finding any flaws they can in a piece of science, whether in the data, the analysis, or the logic being used in model building. Science is also pushed to higher standards by practices calling for replication of published work. Many journals require that authors agree to provide their cell lines or other materials to others who want to try to reproduce their work, so anyone who publishes knows that someone else could come along shortly and publish a paper indicating that their result cannot be replicated. In addition, funding agencies require a rigorous review of proposed work, including a detailed analysis of the feasibility of the work and the validity of the models and approaches. The result of all of this is that almost no one carries on research in a vaccum, and everyone's work is subject to a great deal of scrutiny by others.

The *National Institutes of Health (NIH)* is the largest funding agency for biomedical science in the United States. The NIH plays a major role in maintaining the quality of science by demanding that the scientists they fund meet the highest standards, and then backing their demands with a large financial stick. If someone is suspected of fraudulent scientific practices or of science that does not meet NIH standards, NIH conducts an investigation that can result in barring a scientist from receiving NIH funding, sometimes temporarily and sometimes for life. In some cases, investigations have lasted for years and involved extensive investigations complete with forensic analysis of lab notebooks. The result is a great deal of pressure on scientists to be honest and to live up to the standards set for them. Nothing will get a scientist's attention faster than threatening to take away the research funding that is as essential to the research as air is to the researcher himself.

One of the other factors that makes us more optimistic when we compare the practice of science at the beginning of the twentieth century and the practice of science at the beginning of the twenty-first century is the current scientific culture. Students passing through the higher education system in biomedical sciences receive signals from all levels of the people surrounding them that scientific rigor and honesty are simply expected. In rare cases in which someone does something wrong, the ripples expand into shock waves throughout the scientific community. A few years ago a graduate student was suspected of falsifying data, and his mentor immediately took steps to address the scientific community at large, as well as to contact NIH and bring them into an investigation of what had happened. The student was found to have committed scientific fraud and some very important publications were withdrawn, with much publicity, from some outstanding journals. The flurry of activity that took place, and the attitudes of horror expressed in lab after lab after lab around the country, offered an example of the kind of reinforcement for the norm of honesty in the field. Anyone growing up scientifically in this kind of atmosphere is growing up in a highly moral climate. This does not mean that we think no one will ever commit fraud again. It does not mean that we think that no one will make mistakes. However, it does mean that the scientists of the future are being trained in an atmosphere that fosters integrity, and to us that seems a hopeful thing for the science of the future.

One of the hazards that we face in the near future, as our body of knowledge grows and changes, is the danger of building laws on things that are not actually true. This could happen as it did before, when the science behind eugenics was simply bad (Box 37.2). In fact, it could happen now not because of fraud, but rather because we have failed to perceive just how complex some problems are. Even with the best efforts of good scientists, if the situation is terribly complex, there will be limits to our ability to really understand the underlying contributions to many complex traits, including behavioral and psychiatric traits. We are not arguing that the genetic components are not there. Rather, we are arguing that we must be very careful not to build policies based on oversimplified views. We must be very careful that any policies that develop are based on very good science and an adequately sophisticated understanding of the complex factors involved. It is so easy to catch a first glimpse of understanding of a situation that desperately needs help, and to then leap to erroneous conclusions that harm rather than help.

BOX 37.2 IS GENETIC DETERMINISM STILL AROUND?

Among many flaws in the eugenics movement was the concept of *genetic predetermination*. This concept is especially problematic when we look at complex traits, especially in the area of behavioral genetics. A few years ago, when Scott was a professor at a medical school, he heard a very famous physician give a lecture to a medical school class on human genetics. This fellow started his lecture by saying: "I want you to understand that genetics is not just an important course—it will be the most important course you are ever going to take, because everything you see is going to be genetic. Now, I see you don't believe that. You probably think that, at some point, you are going to be in an emergency room dealing with a gunshot wound, plugged by a police officer in the middle of a bad drug deal in the middle of the South Bronx. You think that this is not genetics, but you are wrong because I will argue to you that it was that person's genes that led him to be dealing drugs in the first place." In his book *On Human Nature*, Edward Wilson says, "The question is no longer whether human social behavior is genetically determined. It is to what extent. The accumulated evidence for a large hereditary component is more detailed and compelling than most persons, including even geneticists, realize. I will go further, it is decisive." In evaluating whether these statements are oversimplifying something terribly complex, we would refer you back to the chapter on MAOA (Chapter 32). A simple initial view of the data indeed offers the view of a gene defect that correlates with a behavior, but the fact that something contributes to a trait does not mean that it is sufficient to be the sole cause of the trait. Here we encounter one of the pitfalls to be avoided if we are to avoid wandering back into the kinds of scientific errors that plagued the eugenics movement. That pitfall is oversimplification of a complex problem. Now, it can happen because someone is taking too simplistic an approach to a complex problem, but the worrisome thing is that it can also happen because someone has not figured out the right questions to ask. Fortunately, when we look at work going on in behavioral genetics, we see smart, gifted researchers who do not seem to be oversimplifying the problems they are working on.

So in terms of the science, we would caution that we not build policy on bad science, and that we not build policies about things until we know enough to build intelligent policies that will be constructive for society and the individuals involved. We would caution that policy makers be certain they really understand the things they target with their laws. We would also caution that we not set policies in stone today that are fixes for temporary problems that are in the process of solving themselves through the natural progression in our scientific knowledge about the problems.

The other factor that contributed to the eugenics movement, along with bad science, was quite a display of questionable ethics. So one of the most heartening things we see coming out of the Human Genome Project has been the development of the program in *Ethical, Legal and Social Implications* of genome science technologies, affectionately known as *ELSI*. The *National Human Genome Research Institute* spends millions of dollars each year on ELSI research, in addition to a variety of educational programs aimed at trying to help improve the ability of the public to understand and contribute to the discussion of key ELSI issues.

ELSI projects touch on a variety of issues. We offer here just a sampling of some of these projects. A researcher in Maryland is tackling philosophical and policy issues relative to the integrity of the human genome and efforts to alter it. A group in Texas is investigating the ethical, legal, and social aspects of preimplantation diagnosis. A researcher in Kentucky is investigating the legal and social impact of identity testing on families. A project at the University of Michigan, called Engaging Minority Communities in Genetic Policy Making, works with fifteen different minority community organizations and a series of focus groups to identify genetic issues of special concern and importance to African-American and Hispanic populations. A study in Massachusetts aims to identify issues of importance to genetic dialogues between the scientific community and evangelical Christians. A group at the Hastings Center in New York will bring together groups including bioethicists and behavioral geneticists to develop resources with which to educate the public about behavioral genetics and to explore the associated ethical, legal, social, and scientific issues. A researcher in Oregon is studying how culture and social class affect communication about genetic information in breast cancer families. Other projects tackle legal and policy issues relative to insurance and use of genetic information, issues involved in genetic manipulation of the germ line, and efforts to enhance ELSI education at universities. The recurring theme we see in the NIH descriptions of these projects is the interactive nature of the projects. Clearly, major efforts are being made to engage representatives from a variety of different communities within and outside of science, and to bring feedback from the community back into the scientific processes. At the same time, major efforts are being made to improve the level of genetic education and awareness in the population. What we see in all of this is not a remote, paternalistic scientific society clinging to the right to decide things for everyone else; rather, we see a scientific community making major efforts to bridge gaps and bring the rest of the community into the decision-making processes.

One of the other areas to look at with concern is the area of public policy. The people who write the laws may be earnest in their efforts to write helpful laws, but do they know enough about some of the complex issues? Sometimes yes. When we see nuclear transfer technology being tossed out along with cloning human beings for apparently semantic reasons, we end up concerned that some of the dialogue and education processes in which the scientific community has invested may not be succeeding in putting across all of the important information and issues.

On the other hand, we see progress. The Genetic Information Nondiscrimination Act passed the American Senate in 2003. If it also passes the House of Representatives and gets signed into law it will prevent insurers and employers from discriminating based on genetic information. This means that an insurance company cannot deny you insurance or charge you more because you have a particular genotype, and that an employer cannot fire you or pay you less because you have a particular genotype.

The coming technologies offer tremendous potential for good or for ill, depending on whether or not they are used wisely and with respect for the rights of individuals to determine their own fates. It will fall to your generation and the next to figure out how to control this technology. The decisions cannot be made by the scientists alone, but they also cannot be made without the scientists. The good news is that people, scientists and community members alike, are talking very seriously about this problem. On many topics the feelings are strong and people hold views at substantial extremes from each other. However, we view the whole tapestry of this discussion, including the extremists on both ends, as very healthy. The more we as a society discuss the ethical implications of this new genetics, the better off we shall be.

FANTASIES

Any sufficiently advanced technology is indistinguishable from magic.
—*Arthur C. Clarke*[3]

It's the year 2102, and the activities in the delivery room are routine as the newborn infant Alain puts forth his first lusty cry of protest at the cold, bright environment that has just replaced the dark warmth that is all he has ever known. His hastily read genome sequence is implanted into a chip under his skin, and stem cells from his cord blood are transferred down the hall to the tissue engineering lab. There the cells will be remanufactured, through a combination of gene therapy and developmental induction, to become developmental precursor cells that can differentiate within his ears to replace the damaged cochlear hair cells responsible for a form of hereditary deafness that was detected in his genome sequence. Metabolic testing shows that he is not subject to any of the most common metabolic disorders that would potentially call for altering the infant's diet. By the time that Alain's happy parents are ready to take their new treasure home, they have in hand a list of the two common over-the-counter medications that won't work correctly in Alain's body and the eighteen allergens that they should eliminate from Alain's home environment to help avoid asthma. The new family heads happily home, knowing that modern genetic medicine has provided Alain with the sequence information that will let his doctor optimize his health care at every step along the way, and is busy repairing his hearing defect before it can ever have a chance to affect his interactions with the world. In the 1930s, Allen's deafness met with brutality and disrespect. In 2006, Alan's deafness faced limited options in a supportive environment. At the beginning of the twenty-second century, Alain may not ever even realize that he was born deaf and may never encounter the perspective that some might offer: that curing his deafness has deprived him of the opportunity to participate in a different, better culture. Is this a fantasy? Will future medicine offer the ability to whisk away the previously unsolvable problem's with little more trouble than we now expend on a headache or strep throat? Actually, we suspect that the most fantastic things in Alain's future are things we cannot talk about because they have not been dreamed up yet. And we also suspect that the day will come when traits such as epilepsy, "imbecility," and deafness that would have been sterilized and institutionalized 100 years ago or that would have been struggled with by the schools and hospitals of today, will indeed be dealt with so efficiently that they will become a dispensable point of curiosity, commented on much as we now will remark with wonder that a baby was born with a full head of hair or some baby teeth already in place. We have to wonder what loss of insight will accompany this freedom from adversity, and we also have to wonder how many will elect to decline to have such differences washed away in a wave of engineered stem cells. We also have to wonder at how substantial will be the disparaties in access to these wonders.

[3] From *Profiles of the Future, an Inquiry Into the Limits of the Possible* by Arthur C. Clark, 1984.

When Scott and Julia started graduate school, obtaining even a small piece of a gene was right out there on the cutting edge of genetic science. Now people talk about a thousand-dollar genome, a customized rendition of the order of As, Cs, Gs, and Ts in your genome to be worn in an implanted microchip for less than the cost of a down payment on most new cars. Some of the things we foresee in our wildest dreams will likely not come to pass, in some cases because of some unforeseen technical wall that will stop progress in a direction we think things should go, but more likely because some other new capabilities will arise that will take us in a direction that we cannot imagine. We live in an era of computers that can talk to us and send information over wireless network systems, an unimaginable form of magic to those who drew the cave paintings or those who used a stylus to press cuneiform letters into tablets made of clay from the Tigris and Euphrates rivers. We live in an era when people in need of organ transplants die for lack of organ donations. We laugh at the absurdity of a scene in a Star Trek movie in which Dr. McCoy pronounces surgery barbaric and gives a woman a new kidney by having her take a pill. Somewhere between the cave paintings and Dr. McCoy's kidney pill lie the realities that will bring about medical miracles we can only vaguely anticipate.

Some of our dreams for the outcomes of genome science and stem cell technology run far afield. Will we reach the day when genetic defects will simply be repaired at birth before they go on to cause cancer or heart disease or Alzheimer disease much later in life? Will we someday be able to give humans copies of the genes that let goldfish regenerate tissues that can't be repaired in humans? Will we someday find parents signing up to modify their children to have the ability to detect and follow the earth's magnetic field the way birds can? Can studies of animals taken into space point towards the genetic modifications we would need to truly adapt a human being to long-term existence in free fall? Could we turn on and off the right combination of genes to grow gills in addition to lungs? Will we be able to make human beings who can live without sleep? Currently, some of these ideas are the stuff of science fiction, topics whose social implications are tackled by authors such as Nancy Kress in *Beggars in Spain* or Lois McMaster Bujold in *Falling Free*.

Others of our dreams appear on the near horizon of our scientific view. Some of these ideas are actually being worked on currently by research groups actively trying to find out which genes in fish are responsible for regeneration of tissues that don't regenerate in humans. We already see companies working on a variety of gene-based strategies for enhancing the effectiveness of chemotherapeutic agents or protecting sensitive noncancerous cells from higher doses of the cancer drugs. Similarly, we see possible near-future breakthroughs in gene-based treatments for heart disease, neurodegenerative disorders, cystic fibrosis, and more.

However, that term, "breakthrough," looms as a large unknown. Research proceeds as a series of baby steps punctuated by occasional leaps. It is usually not possible to predict when the next leap will occur, but it also is usually possible to predict that it will occur. As more leaps move us forward into new technologies and ideas we cannot even guess at yet, being educated about genetics is one of the best ways to ensure that you will be in a position to understand the implications of those breakthroughs. If you want a say in preventing the mistakes of the past, you must engage in dialogues that will take place as society struggles to integrate major changes in ways that are beneficial and that do not create new problems to replace the problems just solved. Only through understanding the issues will we avoid the pitfalls of the past so that the best and brightest promises of today will carry us to the treasure pot at the end of the genomic rainbow. Surely there are wonders waiting there for our children and grandchildren if we can negotiate all of the ethical, legal, and social landmines, keep everyone engaged in the dialogue, and not succumb to unreasoning fears.

And so ends our book. In fact, we have only brushed the surface of this deep, complex topic. We hope that some of what we have told you has helped you to understand some things about yourself and your family. We hope that you have come away with questions that will lead you to further explore some of the topics we touched on. We think of the chapters as letters from us to you. If you get the chance, write to us in care of the publisher or send an e-mail message to Scott at rsh@stowers-institute.org or Julia at richj@umich.edu to let us know what you think.

We hold these truths to be self-evident, that all men are created equal, that they are endowed by their Creator with certain unalienable Rights, that among these are Life, Liberty, and the pursuit of Happiness.

—*Thomas Jefferson, The Declaration of Independence*

Credits

Page 3 — Man Fighting Dragon by Ann Boyajian. Copyright © 1999–2004 Getty Images. Used by permission. All rights reserved.

Page 3 — Lyrics from *Man of La Mancha.* Used by permission. Copyright © 1965. Lyrics by Joe Darion. Music by Mitch Leigh. Publishers Helens Music Company and Andrew Scott Music Company. All rights reserved.

Page 3 — Photo of Brenda Brodeur Knowles. Used by permission. Copyright James A. Knowles © 2004. All rights reserved.

Page 4 — War on Cancer quote appeared in the New York Times on December 9, 1969.

Page 7 — Quote from *"Repent, Harlequin!" Said the Ticktockman* by Harlan Ellison. Copyright © 1965 by Harlan Ellison. Renewed, © 1993 by The Kilimanjaro Corporation. Reprinted by arrangement with, and permission of, the Author and the Author's agent, Richard Curtis Associates, Inc., New York. All rights reserved. Harlan Ellison is a registered trademark of The Kilimanjaro Corporation. From The Essential Ellison: A 50 year Retrospective (Morpheus International, 2001).

Page 7 — Choosing Nut Shell. Copyright © 2003 by Corbis. Used by permission. All rights reserved.

Page 7 — Photo of Marlaina Susi. Copyright © 1999 by Michael and Paula Susi. Used by permission. All rights reserved.

Page 8 — Cell drawing by used by permission of the artist Edward H. Trager. Copyright © Edward H. Trager, 2004. All rights reserved.

Page 15 — Pease in a Pod by Stephen F. Hayes. Copyright © 1999–2004 by Getty Images. Used by permission. All rights reserved.

Page 15 — Quote fragment "In the beginning God . . ." as found in Genesis 1:1 in the Bible and the Torah.

Page 22 — The Mendel web site at http://www.mendelweb.org presents the information about the works of Mendel including the original German version of his paper *Versuche über Pflanzen-Hybriden* which was published in 1866 in *Verhandlungen des naturforschenden Vereins,* which was the proceedings volume of the natural history society of Brünn. The English translation of this work, *Experiments in Plant Hyrbidization,* has been published in a number of locations including William Bateson's *Mendel's Principles of Heredity: A Defence* published by Cambridge University Press in 1902.

Page 25 — Family Tree by Jacek Stachowski. Copyright © 2003 Corbis. Used by permission. All rights reserved.

Page 27 — Redefining Beauty image used by permission of the photographer, Rick Guidotti. Rick Guidotti for Positive Exposure at http://www.positiveexposure.org. This image and others by Mr. Guidotti can be found in an article on albinism in *Life* magazine, June 1998. All rights reserved.

Page 37 — Science and Agriculture image by Leon Zernitsky. Copyright © 2003 Corbis. Used by permission. All rights reserved.

Page 47 — Child Playing with DNA Model. Copyright © 2003 Corbis. Used by permission. All rights reserved.

Page 47 — Quote from Ursula K. LeGuin comes from her book *The Left Hand of Darkness*, 1969, 2003 re-issue (Ace Books).

Page 48 — Nucleotide images. Used by permission from D. M. Reed. Copyright © D. M. Reed, 2004. All rights reserved.

Page 50 — Nucleotide pairs images. Used by permission from D. M. Reed. Copyright © D. M. Reed 2004. All rights reserved.

Page 52 — Quote from J. D. Watson and F. H. C. Crick comes from "Molecular Structure of Nucleic Acids" as it appeared in *Nature, 171*. (London: Macmillan Publishers Ltd, 1953).

Page 53 — Images of two DNA helices. Used by permission from D. M. Reed. Copyright © 2004. All rights reserved.

Page 54 — View down through the center of a DNA helix. Used by permission from D. M. Reed. Copyright © 2004. All rights reserved.

Page 55 — Scientific Advancement by Dennis Harms. Copyright © 2003 Corbis. Used by permission. All rights reserved.

Page 55 — Quote "What hath God wrought?" is the message sent by Samuel Morse on May 24, 1844 to signal the opening of the first telegraph line between Washington, D. C. and Baltimore, MD. A variety of resources present the Morse code, including http://www.scphillips.com/morse.

Page 56 — Illustration of the OTC protein from Shi *et al.* Human ornithine transcarbamylase: crystallographic insights into substrate recognition and conformational changes. *Biochemical Journal* 2001;354:501–509. Used by permission of the author and Portland Press. Copyright © 2001. All rights reserved.

Page 62 — Codon usage diagram from page 294 in the *book Chemical Bioloy. Selected Papers of H. Gobind Khorana (with introductions)* by H. Gobind Khorana in the *World Scientific Series in 20th Century Biology Vol. 5* (World Scientific Publishing Company). Use by permission of the author.

Page 65 — Future Man by Herrmann Starke.

Page 71 — Test Tube Rainbow by Art Valero.

Page 71 — Fly head photos from *The Interactive Fly*, Copyright © 1995, 1996 by Thomas Brody. Web address for the Interactive Fly: http://sdb.bio.purdue.edu/fly/aimain/1aahome.htm. Photos by Anthony Mahowald and Rudi Turner. Used by permission. All rights reserved.

Page 73 — Photo of the University of Michigan Life Sciences Orchestra courtesy of the University of Michigan Health System Gifts of Art Program. Used by permission of Elaine Sims, Director of the University of Michigan Health System's Gifts of Arts Program. Copyright © 2001 University of Michigan. Photo by Lan Chang.

Page 77 — Rhodopsin promoter diagram courtesy of Kenneth PAGE Mitton, PhD, Oakland University Eye Research Institute, Rochester, MI. mitton@oakland.edu. Copyright © 2003 Kenneth PAGE Mitton. Used by permission. All rights reserved.

Page 87 — Thermograph of Chromosomes by Howard Sochurek.

Page 87 — Replica of van Leeuwenhoek's microscope courtesy of Alan Shinn. From web address http://www.mindspring.com/~alshinn/. Copyright © 1996 Alan Shinn. Used by permission. All rights reserved.

Page 89 — Intact metaphase cell image courtesy of the Clinical Cytogenetics Laboratory at the University of Michigan, Ann Arbor, MI, Diane Roulston, PhD, Director. Copyright © 2003 Diane Roulston. Used by permission. All rights reserved.

Page 90 — Female karyotype photo courtesy of the Clinical Cytogenetics Laboratory, University of Michigan, Ann Arbor, MI, Diane Roulston, PhD, Director. Copyright © 2003 Diane Roulston. Used by permission. All rights reserved.

Page 92 — Male karyotype photo courtesy of the Clinical Cytogenetics Laboratory, University of Michigan, Ann Arbor, MI, Diane Roulston, PhD, Director. Copyright © 2003 Diane Roulston. Used by permission. All rights reserved.

Page 93 — Multi-color FISH image courtesy of Octavian Henegariu, Yale University. Copyright © 2000 Octavian Henegariu. Used by permission. All rights reserved.

Page 93 — Single-color FISH image courtesy of Thomas Glover, University of Michigan. Copyright © 2004 Thomas Glover. Used by permission. All rights reserved.

Page 94 — Chromosome painting photo courtesy of Octavian Henegariu, Yale University. Copyright © 2000 Octavian Henegariu. Used by permission. All rights reserved.

Page 95 — DNA by Matthias Kulka. Copyright © 2003 Corbis. Used by permission. All rights reserved.

Page 106 — Collage of stages in mitosis courtesy of William C. Earnshaw, University of Edinburgh. Copyright © 2004 Willam C. Earnshaw. Used by permission. All rights reserved.

Page 107 — Sperm Swimming Towards Egg. Copyright © 2003 Corbis. Used by permission. All rights reserved.

Page 117 — Single-color FISH image courtesy of Thomas Glover, University of Michigan. Copyright © 2004 Thomas Glover. Used by permission. All rights reserved.

Page 117 — Diagram of chiasmata derived from Petronczki *et al.* Un Menage a Quatre: The Molecular Biology of Chromosome Segregation in Meiosis. *Cell* 2003;112:423–440. Photo showing chiasmata courtesy of Jasna Puizina. Used by permission of *Cell* and Jasna Puizina. All rights reserved.

Page 127 — Girl with Butterflies from the Palma Collection. Copyright © 1999–2004 Getty Images. Used by permission. All rights reserved.

Page 139 — Test-tube Baby by Tom and Dee Ann McCarthy. Copyright © 2003 Corbis. Used by permission. All rights reserved.

Page 140 — Definitions of *in vivo* and *in vitro* are derived from the definitions present in *Webster's New Collegiate Dictionary, Third College Edition* (Simon & Schuster, 1988).

Page 149 — Autoradiograph of DNA sequence courtesy of Kathleen Scott, University of Michigan. Copyright 1998 Kathleen Scott. All rights reserved.

Page 149 — Diagram of heterozygous sequence courtesy of Frank Rozsa, University of Michigan. From Rozsa *et al.* GLC1A mutations point to regions of potential functional importance on the TIGR/MYOC protein. Molecular Vision 1998;4:20. Used by permission of Frank Rozsa and Molecular Vision. All rights reserved.

Page 150 — DNA sequence trace courtesy of Rosa Ayala-Lugo. Copyright © 2002 Rosa Ayala-Lugo. Used by permission. All rights reserved.

Page 151 — Mutation sequence trace courtesy of Rosa Ayala-Lugo. Copyright © 2002 Rosa Ayala-Lugo. Used by permission. All rights reserved.

Page 153 — Heredity by Bernhard Bonhomme. Copyright © 2003 Corbis. Used by permission. All rights reserved.

Page 154 — Mirror drawing used by permission of the artist Sophia Tapio. Copyright © 2000 Sophia Tapio. All rights reserved.

Page 158 — MYOC protein-folding model use by permission of Frank Rozsa. Copyright © 1998 Frank Rozsa. All rights reserved.

Page 154 — The quote "Mirror, mirror on the wall, who is fairest of us all?" can be found from a variety of sources containing translated versions of *Grimm's Fairy Tales.* One 1898 version of the tale *Snow White* can be found on pages 9 through 20 of *Grimm's Fairy Tales* as translated by L. L. Weedon and illustrated by Ada Dennis, E. Stuart Hardy and others. (London: Ernest Nister, 1898), available online at
http://222.scils.Rutgers.edu/~kvander/snowwhitetext.html.

Page 163 — Chromosome 13 duplication karyotype photo courtesy of the Clinical Cytogenetics Laboratory, University of Michigan, Ann Arbor, MI, Diane Roulston, PhD, Director. Copyright © 2003 Diane Roulston. Used by permission. All rights reserved.

Page 169 — Milkman Delivering Milk by Paul Manz. Copyright © 2003 Corbis. Used by permission. All rights reserved.

Page 177 — People Looking at Double Helix. Copyright © 2003 Corbis. Used by permission. All rights reserved.

Page 178 — Table 18.1 derived from Table 4 in *Mendelian in Inheritance in Man. Catalogues of Autosomal Dominant, Autosomal Recessive and X-linked Phenotypes. Ninth Edition* by Victor A. McKusick. Copyright © 1990 Johns Hopkins University Press (Baltimore).

Page 179 — Titin figure from Bang *et al.* The complete gene sequence of titin, expression of an unusual approximately 700-kDa titin isoform and its interaction with obscurin identify a novel Z-line to I-band linking system. *Circulation Research* 2001; 89:1065–1072. Reprinted with permission from Lippincott Williams & Wilkins. Copyright © 2001. All rights reserved.

Page 187 — Picket Fence and Lake, Parque Nacional Torres del Paine in Chile. Copyright © 2003 Corbis. Used by permission. All rights reserved.

Page 189 — Photo of Nancy Sabin Wexler used by permission of the Hereditary Disease Foundation. Photo by Peter Ginter.

Page 199 — From The Tiresias Complex: Huntington's Disease as a Paradigm of Testing for Late-onset Disorders, *FASEB Journal 6*, 2820–2825, by N. S. Wexler. Copyright © 1992 by Federation of American Societies for Experimental BI. Reproduced with permission of Federation of American Societies for Experimental Biology in the format Other Book via Copyright Clearance Center. All rights reserved.

Page 205 — Male and Female Computer Key. Copyright © 2003 Corbis. Used by permission. All rights reserved.

Page 215 — DNA Tree by Adams and Adams. Copyright © 2003 Corbis. Used by permission. All rights reserved.

Pages 226–227 — Quote on sexual orientation comes from Simon LeVay and Dean H. Hamer, from Evidence for a Biological Influence in Male Homosexuality. *Scientific American*, May 1994, pp. 44–49. New York: Scientific American, Inc.

Page 229 — Pedigree with X-linked pattern of inheritance of male homosexuality adapted from figure 3 in "A Linkage Between DNA Markers on the X Chromosome and Male Sexual Orientation" by D. H. Hamer *et al.* as appeared in *Science* 1993;261:321–327.

Page 230 — Kinsey score figure adapted from figure 1 in "A Linkage Between DNA Markers on the X Chromosome and Male Sexual Orientation" by D. H. Hamer *et al.* as appeared in *Science* 1993;261:321–327.

Page 234 — Photo of Parisian fashion model courtesy of Alice Domurat Dreger. From *Hermaphrodites and the Medical Invention of Sex* by Alice Domurat Dreger, 1998. Used by permission. Harvard University Press (Cambridge, MA). Previously appeared in *Diseases of Women* by Lawson Tait (New York: William Wood & Company, 1879), p. 22

Page 237 — Man on Abacus by Rob Colvin. Copyright © 2003 Corbis. Used by permission. All rights reserved.

Page 241 — Trisomy 21 karyotype photo courtesy of the Clinical Cytogenetics Laboratory, University of Michigan, Ann Arbor, MI, Diane Roulston, PhD, Director. Copyright © 2003 Diane Roulston. Used by permission. All rights reserved.

Page 255 — Mother Kangaroo and Baby, Australia. Copyright © 2003 Corbis. Used by permission. All rights reserved.

Page 265 — Double Helix with People as Genes by Bek Shakirov. Copyright © 2003 Corbis. Used by permission. All rights reserved.

Page 279 — Photo of Neil Armstrong courtesy of NASA.

Page 285 — Photo of Francis Collins and Craig Venter used by permission of the Nature Publishing Group and Getty Images. Photo by Alex Wong. Copyright © 2001. All rights reserved.

Page 287 — Collage on Cloning from Digital Vision. Copyright 1999–2004 Getty Images. Used by Permission. All rights reserved.

Page 287 — Quote from the New York Times, February 23, 1997.

Page 287 — Photo of Dolly courtesy of the Roslin Institute, Edinburgh, Scotland. Copyright © 1998 Roslin Institute. All rights reserved.

Page 295 — Quote from Scott Gilbert used by permission. From *Developmental Biology* by Scott Gilbert (Sinauer Associates, 1993). All rights reserved.

Page 298 — Table 26.7 adapted from B. Vogelstein *et al. Science* 2002;295:1237. Copyright American Association for the Advancement of Science. Copyright © 2002. All rights reserved.

Page 299 — DNA Strands from Digital Vision. Copyright 1999–2004 Getty Images. Used by Permission. All rights reserved.

Page 299 — Two quotes from Francis S. Collins use by permission. From Contemplating the End of the Beginning. *Genome Research* 2001;11:641–643 All rights reserved.

Page 303 — Data in table 27.1 derived from information at the National Center for Biotechnology Information (NCBI) at the National Library of Medicine (NLM) and the National Institutes of Health (NIH). Data accessed in July of 2003.

Page 305 — Data in table 27.2 derived from two different databases. The euGenes database can be found at http://bio.indiana.edu:8089/man. The NCBI database can be found at http://www.ncbi.nlm.nih.gov/genome/seq. Data accessed in July of 2003.

Page 308 — Haplotype figure derived from a figure by Goncalo Abecasis, University of Michigan. Copyright © 2002 Goncalo Abecasis. Used by permission.

Page 311 — Researchers Working Together by Jude Maceren. Copyright © 2003 Corbis. Used by permission. All rights reserved.

Page 317 — Multipoint linkage graph courtesy of Edward H. Trager. Copyright © 2003 Edward H. Trager. Used by permission. All rights reserved.

Page 323 — Quoted with permission from Francis S. Collins, Director, National Human Genome Research Institute. From Contemplating the End of the Beginning. *Genome Research* 2001;11:641–643 All rights reserved.

Page 327 — DNA Helix with Eyeball by John Martin. Copyright © 2003 Corbis. Used by permission. All rights reserved.

Page 330 — Rhodopsin mutation image courtesy of Markus Preising and the Retina International Newletter at http://www.retina-international.com/sci-news/rhomut.htm. Used by permission. Copyright © 2000 Retina International and Markus Preising. All rights reserved.

Page 331 — Rhodopsin three-dimensional images courtesy of D. M. Reed. Copyright © 2003 D. M. Reed. Use by permission. All rights reserved.

Page 337 — Colorful DNA by Denis Scott. Copyright © 2003 Corbis. Used by permission. All rights reserved.

Page 338 — Ford auto assembly line photo appeared in the *Detroit News* on June 9, 2003. Copyright © 2003 Detroit News. Used by permission. All rights reserved.

Page 340 — Retinoid cycle redrawn based upon a drawing that appeared in a brochure advertising the Seventh Annual Vision Research Conference that took place May 2–3, 2003 in Ft. Lauderdale, FL sponsored by the Association for Research in Vision and Ophthalmology and Elsevier. A related drawing can be found in McBee *et al.* Confronting Complexity: The Interlink of Phototransduction and Retinoid Metabolism in the Vertebrate Retina. *Progress in Retinal and Eye Research* 2001;20:469–529.

Page 345 — Diversity by Kari Lehr, Birch Design Studios. Copyright © 2004 Fotosearch. Used by permission. All rights reserved.

Page 346 — Height graph adapted from Harrison *et al. Human Biology* 2nd *Edition* (Oxford University Press, Oxford, UK, 1977).

Page 353 — Multi-colored X-ray of skull with cogs and wheels in brain from Digital Vision. Copyright © 1999–2004 Getty Images. Used by permission. All rights reserved.

Page 354 — MAOA pedigree adapted from Brunner *et al.* X-linked Borderline Mental Retardation with Prominent Behavioral Disturbance: Phenotype, Genetic Localization, and Evidence for Disturbed Monoamine Metabolism. *American Journal of Human Genetics* 1993;52:1032–1039.

Page 356 — Data presented in Table 32.1 derive from Caspi *et al.* Role of Genotype in the Cycle of Violence in Maltreated Children. *Science* 2002; 297:851–854.

Page 359 — Woman and Health by Vadim Vahrameev. Copyright © 2003 Corbis. Used by permission. All rights reserved.

Page 370 — Diagram of BCR-ABL fusion genes adapted from Page Meltzer *et al.* Chromosome Alterations in Human Solid Tumors in *The Metabolic and Molecular Bases of Inherited Disease, Eighth Edition*, edited by C. R. Scriver, A. L. Beaudet, W. S. Sly, and D. Valle, (McGraw Hill, 2001), p. 558.

Page 374 — Microarray image from MacDonald *et al.* Expression Profiling of Medulloblastoma: PDGFRA and RAS/MAPK Pathway as Therapeutic Targets for Metastatic Disease. *Nature Genetics* 2001;29:143–152 used by permission of the Nature Publishing Group and Tobey MacDonald. http://www.nature.com. Copyright © 2001. All rights reserved.

Page 375 — Chromosome painting of chromosomes in breast cancer cells courtesy of Joanne Davidson, from Davidson *et al.* Molecular Cytogenetic Analaysis of Breast Cancer Cell Lines. *British Journal of Cancer* 2000; 83:1309–1317. Used by permission of the Nature Publishing GrouPage. All rights reserved.

Page 379 — DNA Models by Lawrence Lawry. Copyright © 1999–2004 Getty Images. Used by permission. All rights reserved.

Page 385 — Figure 34.1 adapted from N. Wald and H. Cuckle. *American Journal of Medical Genetics* 1988;31:197–209 and D. J. H. Brock, C. H. Rodeck and M. A. Ferguson-Smith Prenatal Diagnosis and Screening (Churchill Livingstone, London, 1992).

Page 403 — Hand Working on DNA Chain by Mario Cossu. Copyright © 2003 Corbis. Used by permission. All rights reserved.

Page 421 — High school graduation portrait of Jesse Gelsinger used by permission of Jon Wolf Photography. Copyright © 1999. All rights reserved.

Page 421 — Photo of Jesse Gelsinger in front of the statue of Rocky courtesy of Mickie Gelsinger. Used by permission. Copyright © 1999 Mickie Gelsinger. All rights reserved.

Page 428 — Quote from *I'd do Anything for Love but I Won't Do That* (Jim Steinman). © 1993 Edward B. Marks Music Company. Used by permission. All rights reserved.

Page 430 — Quote from George Santayana comes from his book *The Life of Reason, Reason in Common Sense* (Scribner's, 1905).

Page 433 — Quote from Oliver Wendell Holmes comes from the Supreme Court Case Buck vs. Bell in 1924.

Page 434 — From *Objects in the Rearview Mirror May Appear Closer Than They Are* (Jim Steinman). © 1993 Edward B. Marks Music Company. Used by permission. All rights reserved.

Page 438 — From *Profiles of the Future: an Inquiry into the Limits of the Possible* by Arthur Charles Clarke, 1984. Reprinted by permission of the author and the author's agents, Scovil Chichak Galen Literary Agency, Inc. Copyright © 1984 Arthur Charles Clarke. All rights reserved.

Page 429 — Image of a Father Lifting a Baby. Copyright © 2003 Corbis. Used by permission. All rights reserved.

Page 429 — Quote from David Brin, from *Kiln People*, by David Brin (New York: HarperCollins, 2002).

Page 440 — Quote from Thomas Jefferson comes from the Declaration of Independence of the United States of America, in a document with the title In Congress, July 4, 1776, The Unanimous Declaration of the Thirteen United States of America.

Index

Note: Page numbers in italics refer to figures outside of the main discussion.